Studies in Big Data

Volume 81

Series Editor

Janusz Kacprzyk, Polish Academy of Sciences, Warsaw, Poland

The series "Studies in Big Data" (SBD) publishes new developments and advances in the various areas of Big Data- quickly and with a high quality. The intent is to cover the theory, research, development, and applications of Big Data, as embedded in the fields of engineering, computer science, physics, economics and life sciences. The books of the series refer to the analysis and understanding of large, complex, and/or distributed data sets generated from recent digital sources coming from sensors or other physical instruments as well as simulations, crowd sourcing, social networks or other internet transactions, such as emails or video click streams and other. The series contains monographs, lecture notes and edited volumes in Big Data spanning the areas of computational intelligence including neural networks, evolutionary computation, soft computing, fuzzy systems, as well as artificial intelligence, data mining, modern statistics and Operations research, as well as self-organizing systems. Of particular value to both the contributors and the readership are the short publication timeframe and the world-wide distribution, which enable both wide and rapid dissemination of research output.

Indexed by zbMATH.

All books published in the series are submitted for consideration in Web of Science.

More information about this series at http://www.springer.com/series/11970

Weng-Long Chang · Athanasios V. Vasilakos

Fundamentals of Quantum Programming in IBM's Quantum Computers

 Springer

Weng-Long Chang
Department of Computer Science
and Information Engineering
National Kaohsiung University
of Science and Technology
Kaohsiung, Taiwan, Republic of China

Athanasios V. Vasilakos
School of Electrical and Data Engineering
University of Technology Sydney
Sydney, Australia

ISSN 2197-6503   ISSN 2197-6511   (electronic)
Studies in Big Data
ISBN 978-3-030-63585-5   ISBN 978-3-030-63583-1   (eBook)
https://doi.org/10.1007/978-3-030-63583-1

This Springer imprint is published by the registered company Springer Nature Switzerland AG
The registered company address is: Gewerbestrasse 11, 6330 Cham, Switzerland

Preface

In 2019, IBM's quantum computers own to 53 quantum bits. Intel's quantum computers have 49 quantum bits. Google's quantum computers own to 53 quantum bits. This is to say that quantum computers are no longer theoretical devices. To traditional digital computers, it will be true for that *quantum supremacy* is the watershed moment where a quantum computer completes one computation that would be intractable on a classical supercomputer and is imminent. The authors of this book ensure that the best use for quantum computers is to that experts of each domain use them to complete their experimental works. With that in mind, this book actually is a hands-on programmer's guide to who are interested in writing quantum programs in quantum computers with open quantum assemble language. The purpose of the book is to teach a reader to how to write his quantum programs to complete his works on quantum computers. Here is an outline of the chapters:

Chapter 1 explains quantum bits, superposition, entanglement, the Hilbert space, the composer and open quantum assemble language. It is shown that the **NOT** gate, the Hadamard gate, the Z gate, the Y gate, the S gate, the S^+ gate, the T gate, the T^+ gate, the identity gate, the **Controlled-NOT** gate, the **U1**(λ) gate, the **U2**(ϕ, λ) gate, the **U3**(θ, ϕ, λ) gate are all unitary operators (unitary matrices). Examples to each quantum gate are provided, which can be easily run in the backend *simulator* or *ibmqx4* in **IBM**'s quantum computers.

Chapter 2 shows that the Toffoli gate (the **CCNOT** gate) of three quantum bits is a unitary operator (a unitary matrix). Examples for using **NOT** gates, **CNOT** gates and **CCNOT** gates to implement logic operations including **NOT, OR, AND, NOR, NAND, Exclusive-OR (XOR)** and **Exclusive-NOR (XNOR)** are provided, which can be easily run in the backend *simulator* or *ibmqx4* in **IBM**'s quantum computers.

Chapter 3 introduces definition of the search problem, the satisfiability problem in n Boolean variables and m clauses and the clique problem in a graph with n vertices and θ edges. It explains how to complete data dependence analysis for the two famous **NP-complete** problems and shows core concepts of quantum search algorithm. Two examples that solve an instance of the two famous **NP-complete** problems are provided, which can be easily run in the backend *simulator* or *ibmqx4* in **IBM**'s quantum computers.

Chapter 4 illustrates core concepts of quantum Fourier transform and inverse quantum Fourier transform. It gives the reason of why quantum Fourier transform and inverse quantum Fourier transform are able to give exponential speed-up for fast Fourier transform. Two examples that compute the period and the frequency of two given oracular functions are provided, which can be easily run in the backend *simulator* or *ibmqx4* in **IBM**'s quantum computers.

Chapter 5 describes core concepts of Shor's order-finding algorithm. One example that completes the prime factorization to 15 is provided, which can be easily run in the backend *simulator* or *ibmqx4* in **IBM**'s quantum computers.

Chapter 6 explains core concepts of phase estimation and quantum counting. One example that computes eigenvalue of a $(2^2 \times 2^2)$ unitary matrix U with a $(2^2 \times 1)$ eigenvector $|u>$ is provided. Another example that computes the number of solution (s) in the independent-set problem in a graph with two vertices and one edge is also provided. The two examples can be easily run in the backend *simulator* or *ibmqx4* in **IBM**'s quantum computers.

This book contains extensive exercises at the end of each chapter. Solutions of exercises from Chaps. 1 through 6 can easily be completed by readers if they fully understand the contents of each chapter. Each reader can obtain solutions of all exercises when he (she) buys this book because the publisher copies answers of each exercise to a CD.

Power Point presentations have been developed for this book as an invaluable tool for learning. Each reader can obtain Power Point presentations because the publisher copies Power Point presentations to the same CD.

Finally, the authors would like to thank Renata who help us to do solution of exercise and to make Power Point presentations. Although the authors are responsible for all errors and omissions, this book benefited immensely from the invaluable feedback of a number of technical reviewers, consisting of Ju-Chin Chen, Kawuu Wei-Ching Li, Chih-Chiang Wang, Wen-Yu Chung, Chun-Yuan Hsiao, Mang Feng and Renata Wong.

Kaohsiung, Taiwan, Republic of China   Weng-Long Chang
Sydney, Australia   Athanasios V. Vasilakos
April 2020

Contents

Chapter 1
Introduction to Quantum Bits and Quantum Gates on IBM's Quantum Computer

Today from the viewpoint of computing characteristic, "Computer Science" in fact consists of traditional digital computers (Turing 1937; von Neumann 1956), bio-molecular computers (Adleman 1994) and quantum computers (Deutsch 1985). Today we can build traditional digital computers from integrated circuits that include thousands of millions of individual transistors. We call all of these traditional digital computers as *classical*. To traditional digital computers, quantum supremacy is the watershed moment where a quantum computer completes one computation that would be intractable on a classical supercomputer and is imminent (Aaronson and Chen 2017; Coles et al. 2018).

Because **IBM**'s quantum computers have become available as a cloud service to the public, the need of training a cohort of quantum programmers who have been developing classic computer programs for most of their career has arisen. With quantum assembly language on **IBM**'s quantum computers (Cross et al 2017; **IBM Q** 2016), we plan to study **quantum algorithms** that consists of some beautiful ideas that everyone interested in computation should know. Our goal is to explain quantum algorithms with vectors and matrices in linear algebra that is accessible to almost everyone. In this introductory chapter, we describe quantum bits and quantum gates operating quantum bits, and we explain how to use quantum assembly language on **IBM**'s quantum computers to implement them to solve any given a problem.

1.1 Quantum Bits

A *classical bit* has a state that is either zero (0) or one (1). A *quantum bit* (or a *qubit* for short) also has a state. Two possible states for a quantum bit are the states $|0\rangle$ and $|1\rangle$. Notation like '$|\ \rangle$' is called the *Dirac notation*. The states $|0\rangle$ and $|1\rangle$

© The Author(s), under exclusive license to Springer Nature Switzerland AG 2021
W.-L. Chang and A. V. Vasilakos, *Fundamentals of Quantum Programming in IBM's Quantum Computers*, Studies in Big Data 81,
https://doi.org/10.1007/978-3-030-63583-1_1

for a quantum bit correspond to the states 0 and 1 for a classical bit and are known as '*computational basis state vectors*' of thetwo-dimensional Hilbert space. The computational basis state vector to the state $|0\rangle$ of a quantum bit is represented as a (2×1) column vector $\begin{pmatrix} 1 \\ 0 \end{pmatrix}$ and the computational basis state vector to the state $|1\rangle$ of a quantum bit is also represented as a (2×1) column vector $\begin{pmatrix} 0 \\ 1 \end{pmatrix}$. They form an orthonormal basis for the two-dimensional Hilbert space.

The main difference between bits and quantum bits is that a quantum bit can be in a state other than $|0\rangle$ or $|1\rangle$. A quantum bit has two 'computational basis state vectors' $|0\rangle$ and $|1\rangle$ of the two-dimensional Hilbert space. Its arbitrary state $|\Phi\rangle$ is nothing else than a linearly weighted combination of the following computational basis state vectors, often called *superposition*:

$$|\Phi> \ = l_0|0> +l_1|1> \ = l_0\begin{pmatrix} 1 \\ 0 \end{pmatrix} + l_1\begin{pmatrix} 0 \\ 1 \end{pmatrix} = \begin{pmatrix} l_0 \\ l_1 \end{pmatrix}. \qquad (1.1)$$

The weighted factors l_0 and l_1 that are complex numbers are the so-called *probability amplitudes*. Thus they must satisfy $| l_0 |^2 + | l_1 |^2 = 1$. Put another way, the state of a quantum bit is a *unit* vector in the two-dimensional Hilbert space.

All the time, classical computers do examination of a bit to decide whether it is in the state 0 or 1 when they retrieve the content of the memory. However, quantum computers cannot check a quantum bit to decide its quantum state, that is, the values of l_0 and l_1. Instead, when after measuring a quantum bit from quantum computers, we obtain the result 0 with the probability $|l_0|^2$ or the result 1 with the probability $|l_1|^2$. This is to say that reading quantum bits is to *measure*, and the readout is in classical bits.

1.1.1 Multiple Quantum Bits

In a system of two classical bit, there are four possible states 00, 01, 10, and 11. Similarly, a system of two quantum bits has four states $|00\rangle$, $|01\rangle$, $|10\rangle$ and $|11\rangle$ that correspond to the classical four states 00, 01, 10, and 11 and are known as "*computational basis state vectors*" of the four-dimensional Hilbert space. The computational basis state vector to the state $|00\rangle$ of two quantum bits is represented as a (4×1) column vector $\begin{pmatrix} 1 \\ 0 \\ 0 \\ 0 \end{pmatrix}$. The computational basis state vector to the state $|01\rangle$ of two quantum bits is represented as a (4×1) column vector $\begin{pmatrix} 0 \\ 1 \\ 0 \\ 0 \end{pmatrix}$. The computational basis state

vector to the state $|10\rangle$ of two quantum bits is represented as a (4×1) column vector $\begin{pmatrix} 0 \\ 0 \\ 1 \\ 0 \end{pmatrix}$ and the computational basis state vector to the state $|11\rangle$ of two quantum bits

is represented as a (4×1) column vector $\begin{pmatrix} 0 \\ 0 \\ 0 \\ 1 \end{pmatrix}$. They form an orthonormal basis

for the four-dimensional Hilbert space. The arbitrary state of two quantum bits is nothing else than a linearly weighted combination of the following computational basis state vectors, often called *superposition*:

$$|\Phi > = l_0|00 > +l_1|01 > +l_2|10 > +l_3|11 > = l_0 \begin{pmatrix} 1 \\ 0 \\ 0 \\ 0 \end{pmatrix}$$

$$+ l_1 \begin{pmatrix} 0 \\ 1 \\ 0 \\ 0 \end{pmatrix} + l_2 \begin{pmatrix} 0 \\ 0 \\ 1 \\ 0 \end{pmatrix} + l_3 \begin{pmatrix} 0 \\ 0 \\ 0 \\ 1 \end{pmatrix} = \begin{pmatrix} l_0 \\ l_1 \\ l_2 \\ l_3 \end{pmatrix}. \tag{1.2}$$

The weighted factors l_0, l_1, l_2 and l_3 that are complex numbers are the so-called *probability amplitudes*. Therefore they must satisfy $|l_0|^2 + |l_1|^2 + |l_2|^2 + |l_3|^2 = \sum_k \in \{0, 1\}^2 |l_k|^2 = 1$, where the notation "$\{0, 1\}^2$" means "the set of strings of length two with each letter being either zero or one". Put another way, the state of two quantum bits is a *unit* vector in the four-dimensional Hilbert space.

Similarly, all the time, classical computers do examination of two bits to judge whether it is in the state $00, 01, 10$ or 11 when they retrieve the content of the memory. However, quantum computers cannot do examination of two quantum bits to judge its quantum state, that is, the values of l_0, l_1, l_2 and l_3. Instead, when after measuring two quantum bits from quantum computers, we obtain the result 00 with the probability $|l_0|^2$, the result 01 with the probability $|l_1|^2$, the result 10 with the probability $|l_2|^2$ or the result 11 with the probability $|l_3|^2$. This is to say that reading quantum bits is to *measure*, and the readout is in classical bits.

More generally, in a system of n classical bit, there are 2^n possible states 0, 1, 2, ... and $(2^n - 1)$ that are the decimal representation of n classical bits. Similarly, a system of n quantum bits has 2^n states $|0\rangle, |1\rangle, |2\rangle, ...$ and $|2^n - 1\rangle$ that are the decimal representation of n quantum bits and correspond to the classical 2^n states 0,

1, 2, … and $(2^n - 1)$. The 2^n states are to '*computational basis state vectors*' of the 2^n-dimensional Hilbert space. The first computational basis state vector to the state $|0\rangle$ of n quantum bits is represented as a $(2^n \times 1)$ column vector $\begin{pmatrix} 1 \\ 0 \\ \vdots \\ 0 \end{pmatrix}$. The second

computational basis state vector to the state $|1\rangle$ of n quantum bits is represented as a $(2^n \times 1)$ column vector $\begin{pmatrix} 0 \\ 1 \\ \vdots \\ 0 \end{pmatrix}$ and so on with that the last computational basis state

vector to the state $|2^n - 1\rangle$ of n quantum bits is represented as a $(2^n \times 1)$ column vector $\begin{pmatrix} 0 \\ \vdots \\ 0 \\ 1 \end{pmatrix}$. They form an orthonormal basis for the 2^n-dimensional Hilbert space. The

arbitrary state of n quantum bits is nothing else than a linearly weighted combination of the following computational basis state vectors, often called *superposition*:

$$|\Phi> = \sum_{k=0}^{2^n-1} l_k |k\rangle \qquad (1.3)$$

Each weighted factor l_k for $0 \leq k \leq (2^n - 1)$ that is a complex number is the so-called *probability amplitudes*. Hence they must satisfy $\sum_{k=0}^{2^n-1} |l_k|^2 = \sum_{k \in \{0,1\}^n} |l_k|^2 = 1$, where the notation "$\{0, 1\}^n$" means "the set of strings of length n with each letter being either zero or one". Put another way, the state of n quantum bit is a *unit* vector in the 2^n-dimensional Hilbert space.

Similarly, all the time, classical computers do examination of n bits to determine whether it is in the state 0, 1, 2, … and $(2^n - 1)$ when they retrieve the content of the memory. However, quantum computers cannot do examination of n quantum bits to determine its quantum state, that is, the value of each l_k for $0 \leq k \leq (2^n - 1)$. Instead, when after measuring n quantum bits from quantum computers, we obtain the result 0 with the probability $|l_0|^2$, the result 1 with the probability $|l_1|^2$, the result 2 with the probability $|l_2|^2$ or the last result $(2^n - 1)$ with the probability $|l_{2^n-1}|^2$. This is to say that reading n quantum bits is to *measure*, and the readout is in classical bits.

1.1.2 Declaration and Measurement of Multiple Quantum Bits

QASM is the abbreviation of quantum assembly language. Open QASM is a simple text language that illustrates generic quantum circuits. In Open QASM, the syntax of the human-readable form has elements of **C** and assembly languages. For an Open QASM program, the *first* (*non-comment*) line must be "**OPENQASM M.m;**" that indicates a *major* version **M** and *minor* version **m**. Because in the cloud on **IBM**'s quantum computers it supports version 2.0, we describe version 2.0 and use version 2.0 to write a quantum program. The version *keyword* cannot occur multiple times in a file. Using semicolons separates statements and ignore whitespace. Comments begin with a pair of forward slashes and end with a new line. The statement "**include** "**filename**";" continues parsing **filename** as if the contents of the file were pasted at the location of the **include** statement. Specifying the path is relative to the current working directory.

In Open QASM (version 2.0) the only storage types are classical and quantum registers that are, respectively, one-dimensional arrays of bits and quantum bits. The statement "**qreg name[size];**" declares an array of quantum bits (quantum register) with the given **name** and **size** that is the number of quantum bits to this quantum register. Identifiers, such as **name**, must start with a *lowercase* letter and can contain alphanumeric characters and underscores. The label (variable) **name[k]** refers to the kth quantum bit of this register for $0 \leq k \leq$ (**size**—1). The initial state of each quantum bit of this register is set to $|0\rangle$. Similarly, the statement "**creg name[size];**" declares an array of bits (classical register) with the given **name** and **size** that is the number of bits to this classical register. The label (variable) **name[k]** refers to the kth bit of this register for $0 \leq k \leq$ (**size**—1). Each bit of this classical register is initialized to 0. The statement "**measure qubit|qreg $\rightarrow$ bit|creg;**" measures the quantum bit(s) and records the measurement outcome(s) by overwriting the classical bit(s). Both arguments must be register-type, or both must be bit-type. If both arguments are register-type and have the same size, the statement "**measure a $\rightarrow$ b;**" means use measure **a[k] $\rightarrow$ b[k]**; for each index k of registers **a** and **b**.

For **IBM Q** Experience, the graphical representation of the measurement gate appears in Fig. 1.1.

It takes a quantum bit in a superposition of states as input and spits either a 1 or 0. Moreover, the output is not random. There is a probability of a 1 or 0 as output which depends on the original state of the quantum bit. It records the measurement outcome(s) by overwriting the classical bit(s).

Fig. 1.1 The graphical representation of the measurement gate

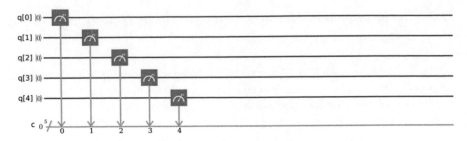

Fig. 1.2 The quantum circuit of declaring and measuring five quantum bits

In Listing 1.1, the program in the backend *ibmqx4* with five quantum bits in **IBM**'s.

Listing. 1.1 The program of declared and measured statements of five quantum bits.

```
1.  OPENQASM 2.0;
2.  include "qelib1.inc";
3.  qreg q[5];
4.  creg c[5];
5.  measure q[0] → c[0];
6.  measure q[1] → c[1];
7.  measure q[2] → c[2];
8.  measure q[3] → c[3];
9.  measure q[4] → c[4];
```

Quantum computer is the first example in which we describe how to declare quantum bits and to measure quantum bits. Figure 1.2 is the corresponding quantum circuit of the program in Listing 1.1. The statement "OPENQASM 2.0;" on line one of Listing 1.1 is to indicate that the program is written with version 2.0 of Open QASM. Next, the statement "include "qelib1.inc";" on line two of Listing 1.1 is to continue parsing the file "qelib1.inc" as if the contents of the file were pasted at the location of the include statement, where the file "qelib1.inc" is **Quantum Experience (QE) Standard Header** and the path is specified relative to the current working directory. The statement "qreg q[5];" on line three of Listing 1.1 is to declare that in the program there are five quantum bits. In the left top of Fig. 1.2, five quantum bits are subsequently q[0], q[1], q[2], q[3] and q[4]. The initial value of each quantum bit is set to $|0\rangle$. Next, the statement "creg c[5];" on line four of Listing 1.1 is to declare that in the program there are five classical bits. In the left bottom of Fig. 1.2, five classical bits are subsequently c[0], c[1], c[2], c[3] and c[4]. The initial value of each classical bit is set to 0. Classical bit c[4] is the most significant bit and classical bit c[0] is the least significant bit. The statement "measure q[0] → c[0];" on line five of Listing 1.1 is to measure the first quantum bit q[0] and to record the measurement

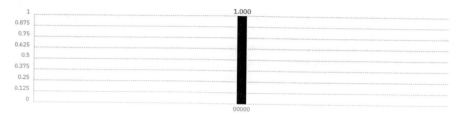

Fig. 1.3 After the measurement to the program in Listing 1.1 is completed, we obtain the answer 00000 with the probability 1.000

outcome by overwriting the first classical bit c[0]. Next, the statement "measure q[1] → c[1];" on line six of Listing 1.1 is to measure the second quantum bit q[1] and to record the measurement outcome by overwriting the second classical bit c[1]. The statement "measure q[2] → c[2];" on line seven of Listing 1.1 is to measure the third quantum bit q[2] and to record the measurement outcome by overwriting the third classical bit c[2]. Next, the statement "measure q[3] → c[3];" on line eight of Listing 1.1 is to measure the fourth quantum b-it q[3] and to record the measurement outcome by overwriting the fourth classical bit c[3]. The statement "measure q[4] → c[4];" on line nine of Listing 1.1 is to measure the fifth quantum bit q[4] and to record the measurement outcome by overwriting the fifth classical bit c[4]. In the backend *ibmqx4* with five quantum bits in **IBM**'s quantum computers, we use the command "simulate" to execute the program in Listing 1.1. The result appears in Fig. 1.3. From Fig. 1.3, we obtain the answer 00000 (c[4] = 0 = q[4] = $|0\rangle$, c[3] = 0 = q[3] = $|0\rangle$, c[2] = 0 = q[2] = $|0\rangle$, c[1] = 0 = q[1] = $|0\rangle$ and c[0] = 0 = q[0] = $|0\rangle$) with the probability one.

1.2 NOT Gate of Single Quantum Bit

Using an electrical circuit that includes wires and logic gates builds a classical computer. Similarly, using a quantum circuit that consists of wires and elementary quantum gates to complete and manipulate the quantum information builds a quantum computer. Using the language of *classical* computation can illustrate occurring of changing a classical state to another classical state. Analogous to the way, using the language of *quantum* computation can introduce occurring of changing a quantum state to another quantum state. In this section and in the later sections, we introduce quantum gates on **IBM**'s quantum computers, propose quantum circuits of many examples describing their application and describe how to use Open QASM to write programs for implementing those examples.

For a classical computer, its circuits contain *wires* and *logic gates*. We use the wires to carry information around the circuit and use the logic gates to complete manipulation of information that is to convert it from one state to another state. For example, we consider that one logic gate of classical single bit, **NOT** gate, whose operation is to convert the state 0 to another state 1 and the state 1 to another state 0. This is to say that the classical states 0 and 1 are interchanged.

Similarly, the quantum **NOT** gate takes the state $l_0 \, |0\rangle + l_1 \, |1\rangle$ to the corresponding state $l_0 \, |1\rangle + l_1 \, |0\rangle$, where the role of $|0\rangle$ and $|1\rangle$ have been interchanged. We assume that we denote a matrix **X** to represent the quantum **NOT** gate as follows:

$$\mathbf{X} = \begin{pmatrix} 0 & 1 \\ 1 & 0 \end{pmatrix}. \tag{1.4}$$

It is also assumed that $\mathbf{X}^+$ is the conjugate-transpose matrix of **X** and is equal to $(\mathbf{X}^*)^t = \begin{pmatrix} 0 & 1 \\ 1 & 0 \end{pmatrix}$, where the * indicates complex conjugation and the t points out the transpose operation. Because $\mathbf{X} \times (\mathbf{X}^*)^t = \begin{pmatrix} 0 & 1 \\ 1 & 0 \end{pmatrix} \times \begin{pmatrix} 0 & 1 \\ 1 & 0 \end{pmatrix} = (\mathbf{X}^*)^t \times \mathbf{X} = \begin{pmatrix} 0 & 1 \\ 1 & 0 \end{pmatrix} \times \begin{pmatrix} 0 & 1 \\ 1 & 0 \end{pmatrix} = \begin{pmatrix} 1 & 0 \\ 0 & 1 \end{pmatrix}$, **X** is a unitary matrix or a unitary operator. If the quantum state $l_0 \, |0\rangle + l_1 \, |1\rangle$ is written in a vector notation as

$$\begin{pmatrix} l_0 \\ l_1 \end{pmatrix}, \tag{1.5}$$

with the top entry is the amplitude for $|0\rangle$ and the bottom entry is the amplitude for $|1\rangle$, then the corresponding output from the quantum **NOT** gate is

$$\begin{pmatrix} 0 & 1 \\ 1 & 0 \end{pmatrix} \times \begin{pmatrix} l_0 \\ l_1 \end{pmatrix} = \begin{pmatrix} l_1 \\ l_0 \end{pmatrix} = l_1 \, |0 > + l_0 | 1\rangle. \tag{1.6}$$

Notice that the action of the quantum **NOT** gate is to that the state $|0\rangle$ is replaced by the state corresponding to the *first* column of the matrix **X** and the state $|1\rangle$ is also replaced by the state corresponding to the *second* column of the matrix **X**. Because $\mathbf{X}^2 = \mathbf{X} \times \mathbf{X} = \begin{pmatrix} 0 & 1 \\ 1 & 0 \end{pmatrix} \times \begin{pmatrix} 0 & 1 \\ 1 & 0 \end{pmatrix} = \begin{pmatrix} 1 & 0 \\ 0 & 1 \end{pmatrix}$, applying **X** twice to a state does nothing to it. For **IBM Q** Experience, the graphical representation of the quantum **NOT** gate appears in Fig. 1.4.

Fig. 1.4 The graphical representation of the quantum **NOT** gate

1.2.1 *Programming with NOT Gate of Single Quantum Bit*

In Listing 1.2, the program in the backend *ibmqx4* with five quantum bits in **IBM**'s quantum computer is the second example in which we introduce how to program with **NOT** gate operating one quantum bit. Figure 1.5 is the corresponding quantum circuit of the program in Listing 1.2. The statement "OPENQASM 2.0;" on line one of Listing 1.2 is to indicate that the program is written with version 2.0 of Open QASM. Then, the statement "include "qelib1.inc";" on line two of Listing 1.2 is to continue parsing the file "qelib1.inc" as if the contents of the file were pasted at the location of the include statement, where the file "qelib1.inc" is **Quantum Experience (QE) Standard Header** and the path is specified relative to the current working directory. The statement "qreg q[5];" on line three of Listing 1.2 is to declare that in the program there are five quantum bits. In the left top of Fig. 1.5, five quantum bits are subsequently q[0], q[1], q[2], q[3] and q[4]. The initial state of each quantum bit is set to $|0\rangle$. Next, the statement "creg c[5];" on line four of Listing 1.2 is to declare that there are five classical bits in the program. In the left bottom of Fig. 1.5, five classical bits are subsequently c[0], c[1], c[2], c[3] and c[4]. The initial value of each classical bit is set to 0. Classical bit c[4] is the most significant bit and classical bit c[0] is the least significant bit.

Listing. 1.2 The program to the use of five **NOT** gates operating five quantum bits.

```
1.  OPENQASM 2.0;
2.  include "qelib1.inc";
3.  qreg q[5];
4.  creg c[5];
5.  x q[0];
6.  x q[1];
7.  x q[2];
8.  x q[3];
9.  x q[4];
```

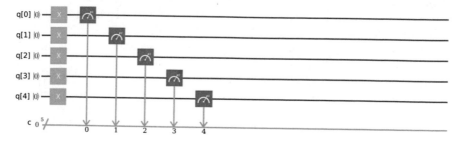

Fig. 1.5 The quantum circuit to five **NOT** gates operating five quantum bits

```
10.   measure q[0] → c[0];
11.   measure q[1] → c[1];
12.   measure q[2] → c[2];
13.   measure q[3] → c[3];
14.   measure q[4] → c[4];
```

The statement "x q[0];" on line five of Listing 1.2 actually implements $\begin{pmatrix} 0 & 1 \\ 1 & 0 \end{pmatrix} \times$
$\begin{pmatrix} 1 \\ 0 \end{pmatrix} = \begin{pmatrix} 0 \\ 1 \end{pmatrix}$. This indicates that the statement "x q[0];" on line five of Listing 1.2
is to use **NOT** gate to convert q[0] from one state $|0\rangle$ to another state $|1\rangle$, where
"x" is to represent **NOT** gate. Next, the statement "x q[1];" on line six of Listing
1.2 actually completes $\begin{pmatrix} 0 & 1 \\ 1 & 0 \end{pmatrix} \times \begin{pmatrix} 1 \\ 0 \end{pmatrix} = \begin{pmatrix} 0 \\ 1 \end{pmatrix}$. This is to say that the statement "x
q[1];" on line six of Listing 1.2 is to make use of **NOT** gate to convert q[1] from one
state $|0\rangle$ to another state $|1\rangle$. Next, the statement "x q[2];" on line seven of Listing
1.2 actually implements $\begin{pmatrix} 0 & 1 \\ 1 & 0 \end{pmatrix} \times \begin{pmatrix} 1 \\ 0 \end{pmatrix} = \begin{pmatrix} 0 \\ 1 \end{pmatrix}$. This implies that the statement "x
q[2];" on line seven of Listing 1.2 is to apply **NOT** gate to convert q[2] from one
state $|0\rangle$ to another state $|1\rangle$. Next, the statement "x q[3];" on line eight of Listing
1.2 actually completes $\begin{pmatrix} 0 & 1 \\ 1 & 0 \end{pmatrix} \times \begin{pmatrix} 1 \\ 0 \end{pmatrix} = \begin{pmatrix} 0 \\ 1 \end{pmatrix}$. This indicates that the statement
"x q[3];" on line eight of Listing 1.2 is to use **NOT** gate to convert q[3] from one
state $|0\rangle$ to another state $|1\rangle$. Next, the statement "x q[4];" on line nine of Listing
1.2 actually performs $\begin{pmatrix} 0 & 1 \\ 1 & 0 \end{pmatrix} \times \begin{pmatrix} 1 \\ 0 \end{pmatrix} = \begin{pmatrix} 0 \\ 1 \end{pmatrix}$. This is to say that the statement "x
q[4];" on line nine of Listing 1.2 is to make use of **NOT** gate to convert q[4] from
one state $|0\rangle$ to another state $|1\rangle$. After the five statements above are completed, the
state $|0\rangle$ of each quantum bit is converted as the state $|1\rangle$.

Next, the statement "measure q[0] → c[0];" on line ten of Listing 1.2 is to
measure the first quantum bit q[0] and to record the measurement outcome by
overwriting the first classical bit c[0]. The statement "measure q[1] → c[1];"
on line eleven of Listing 1.2 is to measure the second quantum bit q[1] and to
record the measurement outcome by overwriting the second classical bit c[1].
Next, the statement "measure q[2] → c[2];" on line 12 of Listing 1.2 is to
measure the third quantum bit q[2] and to record the measurement outcome by
overwriting the third classical bit c[2]. The statement "measure q[3] → c[3];"
on line 13 of Listing 1.2 is to measure the fourth quantum bit q[3] and to record
the measurement outcome by overwriting the fourth classical bit c[3]. Next, the
statement "measure q[4] → c[4];" on line 14 of Listing 1.2 is to measure the
fifth quantum bit q[4] and to record the measurement outcome by overwriting the
fifth classical bit c[4]. In the backend *ibmqx4* with five quantum bits in **IBM**'s
quantum computers, we use the command "simulate" to execute the program in

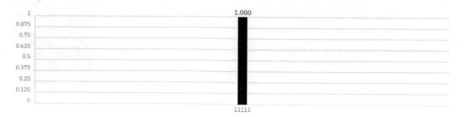

Fig. 1.6 After the measurement to the program in Listing 1.2 is completed, we obtain the answer 11111 with the probability 1.000

Listing 1.2. The result appears in Fig. 1.6. From Fig. 1.6, we obtain the answer 11111   ($c[4] = 1 = q[4] = |1\rangle$, $c[3] = 1 = q[3] = |1\rangle$, $c[2] = 1 = q[2] = |1\rangle$, $c[1] = 1 = q[1] = |1\rangle$, and $c[0] = 1 = q[0] = |1\rangle$) with the probability one.

1.3   The Hadamard Gate of Single Quantum Bit

The Hadamard gate of single quantum bit is

$$H = \frac{1}{\sqrt{2}}\begin{pmatrix} 1 & 1 \\ 1 & -1 \end{pmatrix} = \begin{pmatrix} \frac{1}{\sqrt{2}} & \frac{1}{\sqrt{2}} \\ \frac{1}{\sqrt{2}} & -\frac{1}{\sqrt{2}} \end{pmatrix}. \tag{1.7}$$

It is supposed that H^+ is the conjugate-transpose matrix of H and is equal to $(H^*)^t = \begin{pmatrix} \frac{1}{\sqrt{2}} & \frac{1}{\sqrt{2}} \\ \frac{1}{\sqrt{2}} & -\frac{1}{\sqrt{2}} \end{pmatrix}$, where the * indicates complex conjugation and the t indicates the transpose operation. Since $H \times (H^*)^t = \begin{pmatrix} \frac{1}{\sqrt{2}} & \frac{1}{\sqrt{2}} \\ \frac{1}{\sqrt{2}} & -\frac{1}{\sqrt{2}} \end{pmatrix} \times \begin{pmatrix} \frac{1}{\sqrt{2}} & \frac{1}{\sqrt{2}} \\ \frac{1}{\sqrt{2}} & -\frac{1}{\sqrt{2}} \end{pmatrix} = (H^*)^t \times$

$H = \begin{pmatrix} \frac{1}{\sqrt{2}} & \frac{1}{\sqrt{2}} \\ \frac{1}{\sqrt{2}} & -\frac{1}{\sqrt{2}} \end{pmatrix} \times \begin{pmatrix} \frac{1}{\sqrt{2}} & \frac{1}{\sqrt{2}} \\ \frac{1}{\sqrt{2}} & -\frac{1}{\sqrt{2}} \end{pmatrix} = \begin{pmatrix} 1 & 0 \\ 0 & 1 \end{pmatrix}$, H is a unitary matrix or a unitary operator. This is to say that the Hadamard gate H is one of quantum gates with single quantum bit. If the quantum state $l_0 |0\rangle + l_1 |1\rangle$ is written in a vector notation as

$$\begin{pmatrix} l_0 \\ l_1 \end{pmatrix}, \tag{1.8}$$

with the top entry is the amplitude for $|0\rangle$ and the bottom entry is the amplitude for $|1\rangle$, then the corresponding output from the Hadamard gate H is

$$\frac{1}{\sqrt{2}} \times \begin{pmatrix} 1 & 1 \\ 1 & -1 \end{pmatrix} \times \begin{pmatrix} l_0 \\ l_1 \end{pmatrix} = \begin{pmatrix} \frac{l_0+l_1}{\sqrt{2}} \\ \frac{l_0-l_1}{\sqrt{2}} \end{pmatrix} = \frac{l_0 + l_1}{\sqrt{2}} \ |0\rangle + \frac{l_0 - l_1}{\sqrt{2}} \ |1\rangle. \tag{1.9}$$

Fig. 1.7 The graphical
representation of the
Hadamard gate H

If in (1.8) the value of l_0 is equal to one and the value of l_1 is equal to zero, then the Hadamard gate H turns a $|0\rangle$ into $\frac{1}{\sqrt{2}}(|0\rangle + |1\rangle)$ (the first column of H), which is 'halfway' between $|0\rangle$ and $|1\rangle$. Similarly, if in (1.8) the value of l_0 is equal to zero and the value of l_1 is equal to one, then the Hadamard gate H turns a $|1\rangle$ into $\frac{1}{\sqrt{2}}(|0\rangle - |1\rangle)$ (the second column of H), which also is 'halfway' between $|0\rangle$ and

$|1\rangle$. Because $H^2 = H \times H = \begin{pmatrix} \frac{1}{\sqrt{2}} & \frac{1}{\sqrt{2}} \\ \frac{1}{\sqrt{2}} & -\frac{1}{\sqrt{2}} \end{pmatrix} \times \begin{pmatrix} \frac{1}{\sqrt{2}} & \frac{1}{\sqrt{2}} \\ \frac{1}{\sqrt{2}} & -\frac{1}{\sqrt{2}} \end{pmatrix} = \begin{pmatrix} 1 & 0 \\ 0 & 1 \end{pmatrix}$, using H twice

to a state does nothing to it. For **IBM Q** Experience, the graphical representation of the Hadamard gate H appears in Fig. 1.7.

1.3.1 Programmings with the Hadamard Gate of Single Quantum Bit

In Listing 1.3, the program in the backend *ibmqx4* with five quantum bits in **IBM**'s quantum computer is the third example in which we describe how to program with the Hadamard gate operating one quantum bit. Figure 1.8 is the corresponding quantum circuit of the program in Listing 1.3. The statement "OPENQASM 2.0;" on line one of Listing 1.3 is to point out that the program is written with version 2.0 of Open QASM.

Listing. 1.3 The program to the use of the Hadamard gate operating a quantum bit.

```
1.  OPENQASM 2.0;
2.  include "qelib1.inc";
3.  qreg q[5];
4.  creg c[5];
5.  h q[0];
6.  measure q[0] → c[0];
```

Then, the statement "include "qelib1.inc";" on line two of Listing 1.3 is to continue parsing the file "qelib1.inc" as if the contents of the file were pasted at the location of the include statement, where the file "qelib1.inc" is **Quantum Experience (QE) Standard Header** and the path is specified relative to the current working directory.

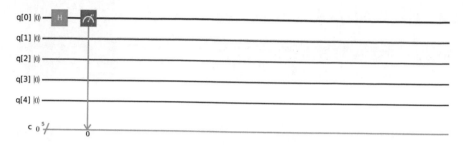

Fig. 1.8 The quantum circuit to the Hadamard gate operating a quantum bit

The statement "qreg q[5];" on line three of Listing 1.3 is to declare that in the program there are five quantum bits. In the left top of Fig. 1.8, five quantum bits are subsequently q[0], q[1], q[2], q[3] and q[4]. The initial state of each quantum bit is set to $|0\rangle$.

Next, the statement "creg c[5];" on line four of Listing 1.3 is to declare that there are five classical bits in the program. In the left bottom of Fig. 1.8, five classical bits are subsequently c[0], c[1], c[2], c[3] and c[4]. The initial value of each classical bit is set to 0. Classical bit c[4] is the most significant bit and classical bit c[0] is the least significant bit. The statement "h q[0];" on line five of Listing 1.3 actually

completes $\begin{pmatrix} \frac{1}{\sqrt{2}} & \frac{1}{\sqrt{2}} \\ \frac{1}{\sqrt{2}} & -\frac{1}{\sqrt{2}} \end{pmatrix} \times \begin{pmatrix} 1 \\ 0 \end{pmatrix} = \begin{pmatrix} \frac{1}{\sqrt{2}} \\ \frac{1}{\sqrt{2}} \end{pmatrix} = \frac{1}{\sqrt{2}}\begin{pmatrix} 1 \\ 1 \end{pmatrix} = \frac{1}{\sqrt{2}}\left(\begin{pmatrix} 1 \\ 0 \end{pmatrix} + \begin{pmatrix} 0 \\ 1 \end{pmatrix} \right) = $

$\frac{1}{\sqrt{2}}(|0\rangle + |1\rangle)$. This is to say that the statement "h q[0];" on line five of Listing 1.3 is to use the Hadamard gate to convert q[0] from one state $|0\rangle$ to another state $\frac{1}{\sqrt{2}}(|0\rangle + |1\rangle)$ (its superposition), where "h" is to represent the Hadamard gate.

Next, the statement "measure q[0] $\rightarrow$ c[0];" on line six of Listing 1.3 is to measure the first quantum bit q[0] and to record the measurement outcome by over-writing the first classical bit c[0]. In the backend *ibmqx4* with five quantum bits in **IBM**'s quantum computers, we apply the command "simulate" to execute the program in Listing 1.3. The result appears in Fig. 1.9. From Fig. 1.9, we obtain the answer 00001 (c[4] = 0, c[3] = 0, c[2] = 0, c[1] = 0 and c[0] = 1 = q[0] = $|1\rangle$) with the probability 0.520. Or we obtain the answer 00000 with the probability 0.480 (c[4] = 0, c[3] = 0, c[2] = 0, c[1] = 0 and c[0] = 0 = q[0] = $|0\rangle$).

Fig. 1.9 After the measurement to the program in Listing 1.3 is completed, we obtain the answer 00001 with the probability 0.520 or the answer 00000 with the probability 0.480

1.4   The Z Gate of Single Quantum Bit

The Z gate of single quantum bit is

$$Z = \begin{pmatrix} 1 & 0 \\ 0 & -1 \end{pmatrix} = \begin{pmatrix} 1 & 0 \\ 0 & e^{\sqrt{-1} \times \pi} \end{pmatrix}, \tag{1.10}$$

where $e^{\sqrt{-1} \times \pi}$ is equal to $\cos(\pi) + \sqrt{-1} \times \sin(\pi) = -1$. It is assumed that Z^+ is the conjugate-transpose matrix of Z and is equal to $(Z^*)^t = \begin{pmatrix} 1 & 0 \\ 0 & -1 \end{pmatrix}$, where the $*$ indicates complex conjugation and the t is the transpose operation. Because

$$Z \times (Z^*)^t = \begin{pmatrix} 1 & 0 \\ 0 & -1 \end{pmatrix} \times \begin{pmatrix} 1 & 0 \\ 0 & -1 \end{pmatrix} = (Z^*)^t \times Z = \begin{pmatrix} 1 & 0 \\ 0 & -1 \end{pmatrix} \times \begin{pmatrix} 1 & 0 \\ 0 & -1 \end{pmatrix} = \begin{pmatrix} 1 & 0 \\ 0 & 1 \end{pmatrix},$$

Z is a unitary matrix or a unitary operator. This implies that the Z gate is one of quantum gates with single quantum bit. If the quantum state $l_0 \, |0\rangle + l_1 \, |1\rangle$ is written in a vector notation as

$$\begin{pmatrix} l_0 \\ l_1 \end{pmatrix}, \tag{1.11}$$

with the top entry is the amplitude for $|0\rangle$ and the bottom entry is the amplitude for $|1\rangle$, then the corresponding output from the Z gate is

$$\begin{pmatrix} 1 & 0 \\ 0 & -1 \end{pmatrix} \times \begin{pmatrix} l_0 \\ l_1 \end{pmatrix} = \begin{pmatrix} l_0 \\ -l_1 \end{pmatrix} = l_0 \begin{pmatrix} 1 \\ 0 \end{pmatrix} + (-l_1) \begin{pmatrix} 0 \\ 1 \end{pmatrix} = l_0 |0\rangle + (-l_1)|1\rangle. \tag{1.12}$$

This indicates that the Z gate leaves $|0\rangle$ unchanged, and flips the sign of $|1\rangle$ to give $-|1\rangle$. Since $Z^2 = Z \times Z = \begin{pmatrix} 1 & 0 \\ 0 & -1 \end{pmatrix} \times \begin{pmatrix} 1 & 0 \\ 0 & -1 \end{pmatrix} = \begin{pmatrix} 1 & 0 \\ 0 & 1 \end{pmatrix}$, applying Z twice to a state does nothing to it. For **IBM Q** Experience, the graphical representation of the Z gate appears in Fig. 1.10.

Fig. 1.10 The graphical representation of the Z gate

1.4.1 Programming with the Z Gate of Single Quantum Bit

In Listing 1.4, the program in the backend *ibmqx4* with five quantum bits in **IBM**'s quantum computer is the fourth example in which we illustrate how to program with the Z gate that leaves $|0\rangle$ unchanged and flips the sign of $|1\rangle$ to give $-|1\rangle$. Figure 1.11 is the corresponding quantum circuit of the program in Listing 1.4. The statement "OPENQASM 2.0;" on line one of Listing 1.4 is to indicate that the program is written with version 2.0 of Open QASM. Next, the statement "include "qelib1.inc";" on line two of Listing 1.4 is to continue parsing the file "qelib1.inc" as if the contents of the file were pasted at the location of the include statement, where the file "qelib1.inc" is **Quantum Experience (QE) Standard Header** and the path is specified relative to the current working directory. The statement "qreg q[5];" on line three of Listing 1.4 is to declare that in the program there are five quantum bits. In the left top of Fig. 1.11, five quantum bits are subsequently q[0], q[1], q[2], q[3] and q[4]. The initial state of each quantum bit is set to $|0\rangle$. Next, the statement "creg c[5];" on line four of Listing 1.4 is to declare that there are five classical bits in the program. In the left bottom of Fig. 1.11, five classical bits are subsequently c[0], c[1], c[2], c[3] and c[4]. The initial value of each classical bit is set to 0. Classical bit c[4] is the most significant bit and classical bit c[0] is the least significant bit.

Listing. 1.4 The program to the use of the Z gate.

```
1.  OPENQASM 2.0;
2.  include "qelib1.inc";
3.  qreg q[5];
4.  creg c[5];
5.  h q[0];
6.  z q[0];
7.  measure q[0] → c[0];
```

Fig. 1.11 The corresponding quantum circuit of the program in Listing 1.4

Fig. 1.12 After the measurement to the program in Listing 1.4 is completed, we obtain the answer 00001 with the probability 0.490 or the answer 00000 with the probability 0.510

The statement "h q[0];" on line five of Listing 1.4 actually implements

$$\begin{pmatrix} \frac{1}{\sqrt{2}} & \frac{1}{\sqrt{2}} \\ \frac{1}{\sqrt{2}} & -\frac{1}{\sqrt{2}} \end{pmatrix} \times \begin{pmatrix} 1 \\ 0 \end{pmatrix} = \begin{pmatrix} \frac{1}{\sqrt{2}} \\ \frac{1}{\sqrt{2}} \end{pmatrix} = \frac{1}{\sqrt{2}}\begin{pmatrix} 1 \\ 1 \end{pmatrix} = \frac{1}{\sqrt{2}}\left(\begin{pmatrix} 1 \\ 0 \end{pmatrix} + \begin{pmatrix} 0 \\ 1 \end{pmatrix} \right) =$$

$\frac{1}{\sqrt{2}}(|0\rangle + |1\rangle)$. This indicates that the statement "h q[0];" on line five of Listing 1.4 is to use the Hadamard gate to convert q[0] from one state $|0\rangle$ to another state $\frac{1}{\sqrt{2}}(|0\rangle + |1\rangle)$ (its superposition), where "h" is to represent the Hadamard gate. Next, the statement "z q[0];" on line six of Listing 1.4 actually completes

$$\begin{pmatrix} 1 & 0 \\ 0 & -1 \end{pmatrix} \times \begin{pmatrix} \frac{1}{\sqrt{2}} \\ \frac{1}{\sqrt{2}} \end{pmatrix} = \begin{pmatrix} \frac{1}{\sqrt{2}} \\ -\frac{1}{\sqrt{2}} \end{pmatrix} = \frac{1}{\sqrt{2}}\begin{pmatrix} 1 \\ -1 \end{pmatrix} = \frac{1}{\sqrt{2}}\left(\begin{pmatrix} 1 \\ 0 \end{pmatrix} + \begin{pmatrix} 0 \\ -1 \end{pmatrix} \right) =$$

$\left(\frac{1}{\sqrt{2}}|0\rangle \right) + \left(-\frac{1}{\sqrt{2}} \right)(|1\rangle) = \frac{1}{\sqrt{2}}(|0\rangle - |1\rangle)$. This is to say that the statement "z q[0];" on line six of Listing 1.4 is to apply the Z gate to convert q[0] from one state $\frac{1}{\sqrt{2}}(|0\rangle + |1\rangle)$ to another state $\frac{1}{\sqrt{2}}(|0\rangle - |1\rangle)$.

Next, the statement "measure q[0] $\rightarrow$ c[0];" on line seven of Listing 1.4 is to measure the first quantum bit q[0] and to record the measurement outcome by overwriting the first classical bit c[0]. In the backend *ibmqx4* with five quantum bits in **IBM**'s quantum computers, we use the command "simulate" to execute the program in Listing 1.4. The result appears in Fig. 1.12. From Fig. 1.12, we obtain the answer 00001 ($c[4] = 0$, $c[3] = 0$, $c[2] = 0$, $c[1] = 0$ and $c[0] = 1 = q[0] = |1\rangle$) with the probability 0.490. Or we obtain the answer 00000 with the probability 0.510 ($c[4] = 0$, $c[3] = 0$, $c[2] = 0$, $c[1] = 0$ and $c[0] = 0 = q[0] = |0\rangle$).

1.5 The Y Gate of Single Quantum Bit

The Y gate of single quantum bit is

$$Y = \begin{pmatrix} 0 & -\sqrt{-1} \\ \sqrt{-1} & 0 \end{pmatrix} = \begin{pmatrix} 0 & -i \\ i & 0 \end{pmatrix} = \begin{pmatrix} 0 & e^{-\sqrt{-1}\times\frac{\pi}{2}} \\ e^{\sqrt{-1}\times\frac{\pi}{2}} & 0 \end{pmatrix}, \qquad (1.13)$$

where $i = \sqrt{-1}$ is known as the imaginary unit and $e^{-\sqrt{-1}\times\frac{\pi}{2}} = \cos\left(\frac{\pi}{2}\right) + \sqrt{-1}\times\sin\left(\frac{\pi}{2}\right) = \sqrt{-1}$ and $e^{-\sqrt{-1}\times\frac{\pi}{2}} = \cos\left(-\frac{\pi}{2}\right) + \sqrt{-1}\times\sin\left(-\frac{\pi}{2}\right) = -\sqrt{-1}$. It is supposed that Y^+ is the conjugate-transpose matrix of Y and is equal to

Fig. 1.13 The graphical representation of the Y gate

$(Y^*)^t = \begin{pmatrix} 0 & -\sqrt{-1} \\ \sqrt{-1} & 0 \end{pmatrix} = \begin{pmatrix} 0 & -i \\ i & 0 \end{pmatrix}$, where the * indicates complex conjuga-

tion and the t is the transpose operation. Since $Y \times (Y^*)^t = \begin{pmatrix} 0 & -\sqrt{-1} \\ \sqrt{-1} & 0 \end{pmatrix} \times$

$\begin{pmatrix} 0 & -\sqrt{-1} \\ \sqrt{-1} & 0 \end{pmatrix} = (Y^*)^t \times Y = \begin{pmatrix} 0 & -\sqrt{-1} \\ \sqrt{-1} & 0 \end{pmatrix} \times \begin{pmatrix} 0 & -\sqrt{-1} \\ \sqrt{-1} & 0 \end{pmatrix} = \begin{pmatrix} 1 & 0 \\ 0 & 1 \end{pmatrix}$,

Y is a unitary matrix or a unitary operator. This is to say that the Y gate is one of quantum gates with single quantum bit. If the quantum state $l_0 |0\rangle + l_1 |1\rangle$ is written in a vector notation as

$$\begin{pmatrix} l_0 \\ l_1 \end{pmatrix}, \tag{1.14}$$

with the top entry is the amplitude for $|0\rangle$ and the bottom entry is the amplitude for $|1\rangle$, then the corresponding output from the Y gate is

$$\begin{pmatrix} 0 & -\sqrt{-1} \\ \sqrt{-1} & 0 \end{pmatrix} \times \begin{pmatrix} l_0 \\ l_1 \end{pmatrix} = \begin{pmatrix} -\sqrt{-1}l_1 \\ \sqrt{-1}l_0 \end{pmatrix} = \left(-\sqrt{-1}l_1\right)\begin{pmatrix} 1 \\ 0 \end{pmatrix} + \left(\sqrt{-1}l_0\right)\begin{pmatrix} 0 \\ 1 \end{pmatrix}$$
$$= \left(-\sqrt{-1}l_1\right)|0\rangle + \left(\sqrt{-1}l_0\right)|1\rangle. \tag{1.15}$$

This indicates that the Y gate converts single quantum bit from one state $l_0 |0\rangle + l_1 |1\rangle$ to another state $\left(-\sqrt{-1}l_1\right)|0\rangle + \left(\sqrt{-1}l_0\right)|1\rangle$. Since $Y^2 = Y \times Y$

$= \begin{pmatrix} 0 & -\sqrt{-1} \\ \sqrt{-1} & 0 \end{pmatrix} \times \begin{pmatrix} 0 & -\sqrt{-1} \\ \sqrt{-1} & 0 \end{pmatrix} = \begin{pmatrix} 1 & 0 \\ 0 & 1 \end{pmatrix}$, using Y twice to a state does

nothing to it. For **IBM Q** Experience, the graphical representation of the Y gate appears in Fig. 1.13.

1.5.1 Programming with the Y Gate of Single Quantum Bit

In Listing 1.5, the program in the backend *ibmqx4* with five quantum bits in **IBM**'s quantum computer is the fifth example in which we introduce how to program with the Y gate that converts single quantum bit from one state $\frac{1}{\sqrt{2}} |0\rangle + \frac{1}{\sqrt{2}} |1\rangle$ to another state $\left(-\sqrt{-1}\frac{1}{\sqrt{2}}\right)|0\rangle + \left(\sqrt{-1}\frac{1}{\sqrt{2}}\right)|1\rangle$. Figure 1.14 is the corresponding quantum circuit of the program in Listing 1.5. The statement "OPENQASM 2.0;" on line one

Fig. 1.14 The corresponding quantum circuit of the program in Listing 1.5

of Listing 1.5 is to point out that the program is written with version 2.0 of Open QASM. Next, the statement "include "qelib1.inc";" on line two of Listing 1.5 is to continue parsing the file "qelib1.inc" as if the contents of the file were pasted at the location of the include statement, where the file "qelib1.inc" is **Quantum Experience (QE) Standard Header** and the path is specified relative to the current working directory. The statement "qreg q[5];" on line three of Listing 1.5 is to declare that in the program there are five quantum bits. In the left top of Fig. 1.14, five quantum bits are subsequently q[0], q[1], q[2], q[3] and q[4]. The initial state of each quantum bit is set to $|0\rangle$. Next, the statement "creg c[5];" on line four of Listing 1.5 is to declare that there are five classical bits in the program. In the left bottom of Fig. 1.14, five classical bits are subsequently c[0], c[1], c[2], c[3] and c[4]. The initial value of each classical bit is set to zero (0). Classical bit c[4] is the most significant bit and classical bit c[0] is the least significant bit.

Listing. 1.5 The program to the use of the Y gate.

```
1.  OPENQASM 2.0;
2.  include "qelib1.inc";
3.  qreg q[5];
4.  creg c[5];
5.  h q[0];
6.  y q[0];
7.  measure q[0] → c[0];
```

The statement "h q[0];" on line five of Listing 1.5 actually completes

$$\begin{pmatrix} \frac{1}{\sqrt{2}} & \frac{1}{\sqrt{2}} \\ \frac{1}{\sqrt{2}} & -\frac{1}{\sqrt{2}} \end{pmatrix} \times \begin{pmatrix} 1 \\ 0 \end{pmatrix} = \begin{pmatrix} \frac{1}{\sqrt{2}} \\ \frac{1}{\sqrt{2}} \end{pmatrix} = \frac{1}{\sqrt{2}}\begin{pmatrix} 1 \\ 1 \end{pmatrix} = \frac{1}{\sqrt{2}}\left(\begin{pmatrix} 1 \\ 0 \end{pmatrix} + \begin{pmatrix} 0 \\ 1 \end{pmatrix}\right) =$$

$\frac{1}{\sqrt{2}}(|0\rangle + |1\rangle)$. This is to say that the statement "h q[0];" on line five of Listing 1.5 is to apply the Hadamard gate to convert q[0] from one state $|0\rangle$ to another state $\frac{1}{\sqrt{2}}(|0\rangle + |1\rangle)$ (its superposition), where "h" is to represent the Hadamard

Fig. 1.15 After the measurement to the program in Listing 1.5 is completed, we obtain the answer 00001 with the probability 0.520 or the answer 00000 with the probability 0.480

gate. Next, the statement "y q[0];" on line six of Listing 1.5 actually imple-

ments $\begin{pmatrix} 0 & -\sqrt{-1} \\ \sqrt{-1} & 0 \end{pmatrix} \times \begin{pmatrix} \frac{1}{\sqrt{2}} \\ \frac{1}{\sqrt{2}} \end{pmatrix} = \begin{pmatrix} -\sqrt{-1} \times \frac{1}{\sqrt{2}} \\ \sqrt{-1} \times \frac{1}{\sqrt{2}} \end{pmatrix} = \frac{1}{\sqrt{2}} \begin{pmatrix} -\sqrt{-1} \\ \sqrt{-1} \end{pmatrix} =$

$\frac{1}{\sqrt{2}} \left(\begin{pmatrix} -\sqrt{-1} \\ 0 \end{pmatrix} + \begin{pmatrix} 0 \\ \sqrt{-1} \end{pmatrix} \right)$

$= \left(-\sqrt{-1}\frac{1}{\sqrt{2}} \right) \begin{pmatrix} 1 \\ 0 \end{pmatrix} + \left(\sqrt{-1}\frac{1}{\sqrt{2}} \right) \begin{pmatrix} 0 \\ 1 \end{pmatrix} = \left(-\sqrt{-1}\frac{1}{\sqrt{2}} \right) |0\rangle + \left(-\sqrt{-1}\frac{1}{\sqrt{2}} \right) |1\rangle$

$= \frac{1}{\sqrt{2}}(-\sqrt{-1}|0\rangle + \sqrt{-1}|1\rangle)$. This indicates that the statement "y q[0];" on line six of Listing 1.5 is to apply the Y gate to convert q[0] from one state $\frac{1}{\sqrt{2}}(|0\rangle + |1\rangle)$ to another state $\frac{1}{\sqrt{2}}(-\sqrt{-1}|0\rangle + \sqrt{-1}|1\rangle)$.

Next, the statement "measure q[0] → c[0];" on line seven of Listing 1.5 is to measure the first quantum bit q[0] and to record the measurement outcome by overwriting the first classical bit c[0]. In the backend *ibmqx4* with five quantum bits in **IBM**'s quantum computers, we apply the command "simulate" to execute the program in Listing 1.5. The result appears in Fig. 1.15. From Fig. 1.15, we obtain the answer 00001 (c[4] = 0, c[3] = 0, c[2] = 0, c[1] = 0 and c[0] = 1 = q[0] = |1⟩) with the probability 0.520. Or we obtain the answer 00000 with the probability 0.480 (c[4] = 0, c[3] = 0, c[2] = 0, c[1] = 0 and c[0] = 0 = q[0] = |0⟩).

1.6 The S Gate of Single Quantum Bit

The S gate of single quantum bit that is the square root of the Z gate is

$$S = \begin{pmatrix} 1 & 0 \\ 0 & \sqrt{-1} \end{pmatrix} = \begin{pmatrix} 1 & 0 \\ 0 & i \end{pmatrix} = \begin{pmatrix} 1 & 0 \\ 0 & e^{\sqrt{-1} \times \frac{\pi}{2}} \end{pmatrix} \tag{1.16}$$

where $i = \sqrt{-1}$ is known as the imaginary unit and $e^{\sqrt{-1} \times \frac{\pi}{2}} = \cos\left(\frac{\pi}{2}\right) + \sqrt{-1} \times \sin\left(\frac{\pi}{2}\right) = \sqrt{-1}$. It is assumed that S^+ is the conjugate-transpose matrix of S and is

Fig. 1.16 The graphical
representation of the S gate

equal to $(S^*)^t = \begin{pmatrix} 1 & 0 \\ 0 & -\sqrt{-1} \end{pmatrix} = \begin{pmatrix} 1 & 0 \\ 0 & -i \end{pmatrix}$, where the * points out complex conjuga-

tion and the t indicates the transpose operation. Because $S \times (S^*)^t = \begin{pmatrix} 1 & 0 \\ 0 & \sqrt{-1} \end{pmatrix} \times$

$\begin{pmatrix} 1 & 0 \\ 0 & -\sqrt{-1} \end{pmatrix} = (S^*)^t \times S = \begin{pmatrix} 1 & 0 \\ 0 & -\sqrt{-1} \end{pmatrix} \times \begin{pmatrix} 1 & 0 \\ 0 & \sqrt{-1} \end{pmatrix} = \begin{pmatrix} 1 & 0 \\ 0 & 1 \end{pmatrix}$, S is a unitary

matrix or a unitary operator. This implies that the S gate is one of quantum gates with single quantum bit. If the quantum state $l_0 |0\rangle + l_1 |1\rangle$ is written in a vector notation as

$$\begin{pmatrix} l_0 \\ l_1 \end{pmatrix}, \tag{1.17}$$

with the top entry is the amplitude for $|0\rangle$ and the bottom entry is the amplitude for $|1\rangle$, then the corresponding output from the S gate is

$$\begin{pmatrix} 1 & 0 \\ 0 & \sqrt{-1} \end{pmatrix} \times \begin{pmatrix} l_0 \\ l_1 \end{pmatrix} = \begin{pmatrix} l_0 \\ \sqrt{-1}l_1 \end{pmatrix} = (l_0)\begin{pmatrix} 1 \\ 0 \end{pmatrix} + \left(\sqrt{-1}l_1\right)\begin{pmatrix} 0 \\ 1 \end{pmatrix}$$
$$= (l_0)|0> + \left(\sqrt{-1}l_1\right)|1> = (l_0)|0> + \left(e^{\sqrt{-1}\times\frac{\pi}{2}} \times l_1\right)|1\rangle. \tag{1.18}$$

This indicates that the S gate converts single quantum bit from one state $l_0 |0\rangle + l_1 |1\rangle$ to another state $(l_0) |0\rangle + \left(\sqrt{-1}\,l_1\right) |1\rangle = (l_0) |0\rangle + \left(e^{\sqrt{-1}\times\frac{\pi}{2}} \times l_1\right) |1\rangle$. This is also to say that the S gate leaves $|0\rangle$ unchanged and modifies the phase of $|1\rangle$ to give $\left(\sqrt{-1}\right) |1\rangle = \left(e^{\sqrt{-1}\times\frac{\pi}{2}} |1\rangle\right)$. The probability of measuring a $|0\rangle$ or $|1\rangle$ is unchanged after applying the S gate, however it modifies the phase of the quantum state. Because $S^2 = S \times S = \begin{pmatrix} 1 & 0 \\ 0 & \sqrt{-1} \end{pmatrix} \times \begin{pmatrix} 1 & 0 \\ 0 & \sqrt{-1} \end{pmatrix} = \begin{pmatrix} 1 & 0 \\ 0 & -1 \end{pmatrix}$, applying S twice to a state is equivalent to do the Z gate to it. For **IBM Q** Experience, the graphical representation of the S gate appears in Fig. 1.16.

1.6.1 Programming with the S Gate of Single Quantum Bit

In Listing 1.6, the program in the backend *ibmqx4* with five quantum bits in **IBM's** quantum computer is the sixth example in which we describe how to program with

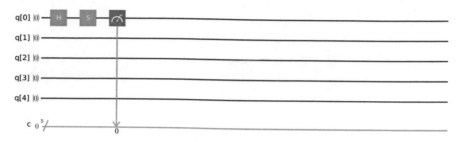

Fig. 1.17 The corresponding quantum circuit of the program in Listing 1.6

the S gate that converts single quantum bit from one state $\frac{1}{\sqrt{2}}(|0\rangle + |1\rangle)$ to another state $\left(\frac{1}{\sqrt{2}}\right)\left(|0\rangle + \left(\sqrt{-1}\right)|1\rangle\right)$. Figure 1.17 is the corresponding quantum circuit of the program in Listing 1.6. The statement "OPENQASM 2.0;" on line one of Listing 1.6 is to indicate that the program is written with version 2.0 of Open QASM. Next, the statement "include "qelib1.inc";" on line two of Listing 1.6 is to continue parsing the file "qelib1.inc" as if the contents of the file were pasted at the location of the include statement, where the file "qelib1.inc" is **Quantum Experience (QE) Standard Header** and the path is specified relative to the current working directory. The statement "qreg q[5];" on line three of Listing 1.6 is to declare that in the program there are five quantum bits. In the left top of Fig. 1.17, five quantum bits are subsequently q[0], q[1], q[2], q[3] and q[4]. The initial state of each quantum bit is set to $|0\rangle$. Next, the statement "creg c[5];" on line four of Listing 1.6 is to declare that there are five classical bits in the program. In the left bottom of Fig. 1.17, five classical bits are subsequently c[0], c[1], c[2], c[3] and c[4]. The initial value of each classical bit is set to 0. Classical bit c[4] is the most significant bit and classical bit c[0] is the least significant bit.

Listing. 1.6 The program to the use of the S gate.

```
1.  OPENQASM 2.0;
2.  include "qelib1.inc";
3.  qreg q[5];
4.  creg c[5];
5.  h q[0];
6.  s q[0];
7.  measure q[0] → c[0];
```

The statement "h q[0];" on line five of Listing 1.6 actually implements

$$\begin{pmatrix} \frac{1}{\sqrt{2}} & \frac{1}{\sqrt{2}} \\ \frac{1}{\sqrt{2}} & -\frac{1}{\sqrt{2}} \end{pmatrix} \times \begin{pmatrix} 1 \\ 0 \end{pmatrix} = \begin{pmatrix} \frac{1}{\sqrt{2}} \\ \frac{1}{\sqrt{2}} \end{pmatrix} = \frac{1}{\sqrt{2}}\begin{pmatrix} 1 \\ 1 \end{pmatrix} = \frac{1}{\sqrt{2}}\left(\begin{pmatrix} 1 \\ 0 \end{pmatrix} + \begin{pmatrix} 0 \\ 1 \end{pmatrix}\right) =$$

$\frac{1}{\sqrt{2}}(|0\rangle + |1\rangle)$. This indicates that the statement "h q[0];" on line five of

Fig. 1.18 After the measurement to the program in Listing 1.6 is completed, we obtain the answer 00001 with the probability 0.550 or the answer 00000 with the probability 0.450

Listing 1.6 is to use the Hadamard gate to convert q[0] from one state $|0\rangle$ to another state $\frac{1}{\sqrt{2}}(|0\rangle + |1\rangle)$ (its superposition), where "h" is to represent the Hadamard gate. Next, the statement "s q[0];" on line six of Listing 1.6 actually completes $\begin{pmatrix} 1 & 0 \\ 0 & \sqrt{-1} \end{pmatrix} \times \begin{pmatrix} \frac{1}{\sqrt{2}} \\ \frac{1}{\sqrt{2}} \end{pmatrix} = \begin{pmatrix} \frac{1}{\sqrt{2}} \\ \sqrt{-1} \times \frac{1}{\sqrt{2}} \end{pmatrix} = \frac{1}{\sqrt{2}} \begin{pmatrix} 1 \\ \sqrt{-1} \end{pmatrix} =$

$\frac{1}{\sqrt{2}} \left(\begin{pmatrix} 1 \\ 0 \end{pmatrix} + \begin{pmatrix} 0 \\ \sqrt{-1} \end{pmatrix} \right) = \begin{pmatrix} \frac{1}{\sqrt{2}} \end{pmatrix} \left(\begin{pmatrix} 1 \\ 0 \end{pmatrix} + (\sqrt{-1}) \begin{pmatrix} 0 \\ 1 \end{pmatrix} \right) = \frac{1}{\sqrt{2}}(|0\rangle + \sqrt{-1}|1\rangle).$

This is to say that the statement "s q[0];" on line six of Listing 1.6 is to use the S gate to convert q[0] from one state $\frac{1}{\sqrt{2}} (|0\rangle + |1\rangle)$ to another state $\frac{1}{\sqrt{2}}(|0\rangle + \sqrt{-1}\ |1\rangle)$.

Next, the statement "measure q[0] $\rightarrow$ c[0];" on line seven of Listing 1.6 is to measure the first quantum bit q[0] and to record the measurement outcome by overwriting the first classical bit c[0]. In the backend *ibmqx4* with five quantum bits in **IBM**'s quantum computers, we use the command "simulate" to execute the program in Listing 1.6. The result appears in Fig. 1.18. From Fig. 1.18, we obtain the answer 00001 (c[4] = 0, c[3] = 0, c[2] = 0, c[1] = 0 and c[0] = 1 = q[0] = $|1\rangle$) with the probability 0.550. Or we obtain the answer 00000 with the probability 0.450 (c[4] = 0, c[3] = 0, c[2] = 0, c[1] = 0 and c[0] = 0 = q[0] = $|0\rangle$).

1.7   The S^+ Gate of Single Quantum Bit

The S^+ gate of single quantum bit that is the conjugate-transpose matrix of the S gate is

$$S^+ = \begin{pmatrix} 1 & 0 \\ 0 & -\sqrt{-1} \end{pmatrix} = \begin{pmatrix} 1 & 0 \\ 0 & -i \end{pmatrix} = \begin{pmatrix} 1 & 0 \\ 0 & -e^{\sqrt{-1} \times \frac{\pi}{2}} \end{pmatrix} = \begin{pmatrix} 1 & 0 \\ 0 & e^{-\sqrt{-1} \times \frac{\pi}{2}} \end{pmatrix}, \quad (1.19)$$

where $i = \sqrt{-1}$ is known as the imaginary unit and $e^{-\sqrt{-1} \times \frac{\pi}{2}} = \cos\left(-\frac{\pi}{2}\right) + \sqrt{-1} \times \sin\left(-\frac{\pi}{2}\right) = -\sqrt{-1}$. It is supposed that $(S^+)^+$ is the conjugate-transpose matrix of S^+ and is equal to $((S^+)^*)^t = \begin{pmatrix} 1 & 0 \\ 0 & \sqrt{-1} \end{pmatrix} = \begin{pmatrix} 1 & 0 \\ 0 & i \end{pmatrix} = \begin{pmatrix} 1 & 0 \\ 0 & e^{\sqrt{-1} \times \frac{\pi}{2}} \end{pmatrix},$ where the * indicates complex conjugation and the t is the transpose operation.

Fig. 1.19 The graphical
representation of the S^+ gate

Since $S^+ \times ((S^+)^*)^t = \begin{pmatrix} 1 & 0 \\ 0 & -\sqrt{-1} \end{pmatrix} \times \begin{pmatrix} 1 & 0 \\ 0 & \sqrt{-1} \end{pmatrix} = ((S^+)^*)^t \times S^+ = \begin{pmatrix} 1 & 0 \\ 0 & \sqrt{-1} \end{pmatrix}$

$\times \begin{pmatrix} 1 & 0 \\ 0 & -\sqrt{-1} \end{pmatrix} = \begin{pmatrix} 1 & 0 \\ 0 & 1 \end{pmatrix}$, S^+ is a unitary matrix or a unitary operator. This is to
say that the S^+ gate is one of quantum gates with single quantum bit. If the quantum
state $l_0 |0\rangle + l_1 |1\rangle$ is written in a vector notation as

$$\begin{pmatrix} l_0 \\ l_1 \end{pmatrix}, \tag{1.20}$$

with the top entry is the amplitude for $|0\rangle$ and the bottom entry is the amplitude
for $|1\rangle$, then the corresponding output from the S^+ gate is

$$\begin{pmatrix} 1 & 0 \\ 0 & -\sqrt{-1} \end{pmatrix} \times \begin{pmatrix} l_0 \\ l_1 \end{pmatrix} = \begin{pmatrix} l_0 \\ -\sqrt{-1}l_1 \end{pmatrix} = (l_0)\begin{pmatrix} 1 \\ 0 \end{pmatrix} + (-\sqrt{-1}l_1)\begin{pmatrix} 0 \\ 1 \end{pmatrix}$$

$$= (l_0)|1\rangle + \left(-\sqrt{-1}l_1\right)|1\rangle = (l_0)|1\rangle + \left(e^{-\sqrt{-1}\times\frac{\pi}{2}} \times l_1\right)|1\rangle. \tag{1.21}$$

This implies that the S^+ gate converts single quantum bit from one state $l_0 |0\rangle + l_1 |1\rangle$
to another state $(l_0) |0\rangle + \left(-\sqrt{-1}\,l_1\right)|1\rangle = (l_0) |0\rangle + \left(e^{-\sqrt{-1}\times\frac{\pi}{2}} \times l_1\right)|1\rangle$. This also
indicates that the S^+ gate leaves $|0\rangle$ unchanged and modifies the phase of $|1\rangle$ to give
$(-\sqrt{-1})|1\rangle = \left(e^{-\sqrt{-1}\times\frac{\pi}{2}} |1\rangle\right)$. The probability of measuring a $|0\rangle$ or $|1\rangle$ is unchanged
after applying the S^+ gate, however it modifies the phase of the quantum state. Because
$(S^+)^2 = S^+ \times S^+ = \begin{pmatrix} 1 & 0 \\ 0 & -\sqrt{-1} \end{pmatrix} \times \begin{pmatrix} 1 & 0 \\ 0 & -\sqrt{-1} \end{pmatrix} = \begin{pmatrix} 1 & 0 \\ 0 & -1 \end{pmatrix}$, applying S^+ twice
to a state is equivalent to do the Z gate to it. For **IBM Q** Experience, the graphical
representation of the S^+ gate appears in Fig. 1.19.

1.7.1 Programming with the S^+ Gate of Single Quantum Bit

In Listing 1.7, the program in the backend *ibmqx4* with five quantum bits in **IBM**'s
quantum computer is the seventh example in which we illustrate how to program with
the S^+ gate that converts single quantum bit from one state $\frac{1}{\sqrt{2}}(|0\rangle + |1\rangle)$ to another
state $\left(\frac{1}{\sqrt{2}}\right)(|0\rangle + (-\sqrt{-1})|1\rangle)$. Figure 1.20 is the corresponding quantum circuit of

Fig. 1.20 The corresponding quantum circuit of the program in Listing 1.7

the program in Listing 1.7. The statement "OPENQASM 2.0;" on line one of Listing 1.7 is to point out that the program is written with version 2.0 of Open QASM. Next, the statement "include "qelib1.inc";" on line two of Listing 1.7 is to continue parsing the file "qelib1.inc" as if the contents of the file were pasted at the location of the include statement, where the file "qelib1.inc" is **Quantum Experience (QE) Standard Header** and the path is specified relative to the current working directory. The statement "qreg q[5];" on line three of Listing 1.7 is to declare that in the program there are five quantum bits. In the left top of Fig. 1.20, five quantum bits are subsequently q[0], q[1], q[2], q[3] and q[4]. The initial state of each quantum bit is set to $|0\rangle$. Next, the statement "creg c[5];" on line four of Listing 1.7 is to declare that there are five classical bits in the program. In the left bottom of Fig. 1.20, five classical bits are subsequently c[0], c[1], c[2], c[3] and c[4]. The initial value of each classical bit is set to 0. Classical bit c[4] is the most significant bit and classical bit c[0] is the least significant bit.

Listing. 1.7 The program to the use of the S^+ gate.

```
1.  OPENQASM 2.0;
2.  include "qelib1.inc";
3.  qreg q[5];
4.  creg c[5];
5.  h q[0];
6.  sdg q[0];
7.  measure q[0] → c[0];
```

The statement "h q[0];" on line five of Listing 1.7 actually completes
$$\begin{pmatrix} \frac{1}{\sqrt{2}} & \frac{1}{\sqrt{2}} \\ \frac{1}{\sqrt{2}} & -\frac{1}{\sqrt{2}} \end{pmatrix} \times \begin{pmatrix} 1 \\ 0 \end{pmatrix} = \begin{pmatrix} \frac{1}{\sqrt{2}} \\ \frac{1}{\sqrt{2}} \end{pmatrix} = \frac{1}{\sqrt{2}} \begin{pmatrix} 1 \\ 1 \end{pmatrix} = \frac{1}{\sqrt{2}} \left(\begin{pmatrix} 1 \\ 0 \end{pmatrix} + \begin{pmatrix} 0 \\ 1 \end{pmatrix} \right) =$$
$\frac{1}{\sqrt{2}}(|0\rangle + |1\rangle)$. This is to say that the statement "h q[0];" on line five of Listing 1.7 is to apply the Hadamard gate to convert q[0] from one state $|0\rangle$ to another state $\frac{1}{\sqrt{2}}(|0\rangle + |1\rangle)$ (its superposition), where "h" is to represent the Hadamard gate. Next,

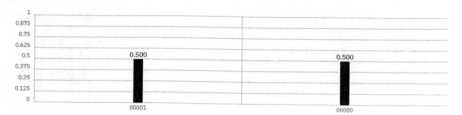

Fig. 1.21 After the measurement to the program in Listing 1.7 is completed, we obtain the answer 00001 with the probability 0.500 or the answer 00000 with the probability 0.500

the statement "sdg q[0];" on line six of Listing 1.7 actually performs $\begin{pmatrix} 1 & 0 \\ 0 & -\sqrt{-1} \end{pmatrix} \times$

$$\begin{pmatrix} \frac{1}{\sqrt{2}} \\ \frac{1}{\sqrt{2}} \end{pmatrix} = \begin{pmatrix} \frac{1}{\sqrt{2}} \\ -\sqrt{-1} \times \frac{1}{\sqrt{2}} \end{pmatrix} = \frac{1}{\sqrt{2}} \begin{pmatrix} 1 \\ -\sqrt{-1} \end{pmatrix} = \frac{1}{\sqrt{2}} \left(\begin{pmatrix} 1 \\ 0 \end{pmatrix} + \begin{pmatrix} 0 \\ -\sqrt{-1} \end{pmatrix} \right) =$$

$\frac{1}{\sqrt{2}} \left(\begin{pmatrix} 1 \\ 0 \end{pmatrix} + (-\sqrt{-1}) \begin{pmatrix} 0 \\ 1 \end{pmatrix} \right) = \left(\frac{1}{\sqrt{2}} \right) \left(|0\rangle + (-\sqrt{-1}) |1\rangle \right)$. This indicates that the statement "sdg q[0];" on line six of Listing 1.7 is to apply the S^+ gate to convert q[0] from one state $\frac{1}{\sqrt{2}}(|0\rangle + |1\rangle)$ to another state $\frac{1}{\sqrt{2}}(|0\rangle + (-\sqrt{-1})|1\rangle)$.

Next, the statement "measure q[0] → c[0];" on line seven of Listing 1.7 is to measure the first quantum bit q[0] and to record the measurement outcome by over-writing the first classical bit c[0]. In the backend *ibmqx4* with five quantum bits in **IBM**'s quantum computers, we apply the command "simulate" to execute the program in Listing 1.7. The result appears in Fig. 1.21. From Fig. 1.21, we obtain the answer 00001 (c[4] = 0, c[3] = 0, c[2] = 0, c[1] = 0 and c[0] = 1 = q[0] = 1)) with the probability 0.500. Or we obtain the answer 00000 with the probability 0.500 (c[4] = 0, c[3] = 0, c[2] = 0, c[1] = 0 and c[0] = 0 = q[0] = 0)).

1.8 The T Gate of Single Quantum Bit

The T gate of single quantum bit that is the square root of the S gate is

$$T = \begin{pmatrix} 1 & 0 \\ 0 & e^{\sqrt{-1} \times \frac{\pi}{4}} \end{pmatrix} = \begin{pmatrix} 1 & 0 \\ 0 & \frac{1+\sqrt{-1}}{\sqrt{2}} \end{pmatrix} = \begin{pmatrix} 1 & 0 \\ 0 & \frac{1+i}{\sqrt{2}} \end{pmatrix}, \tag{1.22}$$

where $i = \sqrt{-1}$ is known as the imaginary unit and $e^{\sqrt{-1} \times \frac{\pi}{4}} = \cos\left(\frac{\pi}{4}\right) + \sqrt{-1} \times \sin\left(\frac{\pi}{4}\right) = \frac{1+\sqrt{-1}}{2} = \frac{1+i}{2}$. It is assumed that T^+ is the conjugate-transpose matrix of T and is equal to $(T^*)^t = \begin{pmatrix} 1 & 0 \\ 0 & e^{-\sqrt{-1} \times \frac{\pi}{4}} \end{pmatrix} = \begin{pmatrix} 1 & 0 \\ 0 & \frac{1-\sqrt{-1}}{\sqrt{2}} \end{pmatrix} = \begin{pmatrix} 1 & 0 \\ 0 & \frac{1-i}{\sqrt{2}} \end{pmatrix}$, where the * indicates complex conjugation and the t is the transpose operation. Since $T \times (T^*)^t$

Fig. 1.22 The graphical
representation of the T gate

$$= \begin{pmatrix} 1 & 0 \\ 0 & e^{\sqrt{-1} \times \frac{\pi}{4}} \end{pmatrix} \times \begin{pmatrix} 1 & 0 \\ 0 & e^{-\sqrt{-1} \times \frac{\pi}{4}} \end{pmatrix} = (T^*)^t \times T = \begin{pmatrix} 1 & 0 \\ 0 & e^{-\sqrt{-1} \times \frac{\pi}{4}} \end{pmatrix} \times \begin{pmatrix} 1 & 0 \\ 0 & e^{\sqrt{-1} \times \frac{\pi}{4}} \end{pmatrix}$$

$$= \begin{pmatrix} 1 & 0 \\ 0 & 1 \end{pmatrix},$$ T is a unitary matrix or a unitary operator. This indicates that the T gate
is one of quantum gates with single quantum bit. If the quantum state $l_0 |0\rangle + l_1 |1\rangle$ is
written in a vector notation as

$$\begin{pmatrix} l_0 \\ l_1 \end{pmatrix}, \tag{1.23}$$

with the top entry is the amplitude for $|0\rangle$ and the bottom entry is the amplitude
for $|1\rangle$, then the corresponding output from the T gate is

$$\begin{pmatrix} 1 & 0 \\ 0 & e^{\sqrt{-1} \times \frac{\pi}{4}} \end{pmatrix} \times \begin{pmatrix} l_0 \\ l_1 \end{pmatrix} = \begin{pmatrix} l_0 \\ e^{\sqrt{-1} \times \frac{\pi}{4}} \times l_1 \end{pmatrix}$$

$$= (l_0) \begin{pmatrix} 1 \\ 0 \end{pmatrix} + \left(e^{\sqrt{-1} \times \frac{\pi}{4}} \times l_1 \right) \begin{pmatrix} 0 \\ 1 \end{pmatrix} = (l_0) |0\rangle + \left(e^{\sqrt{-1} \times \frac{\pi}{4}} \times l_1 \right) |1\rangle. \tag{1.24}$$

This is to say that the T gate converts single quantum bit from one state $l_0 |0\rangle +$
$l_1 |1\rangle$ to another state $(l_0) |0\rangle + \left(e^{\sqrt{-1} \times \frac{\pi}{4}} \times l_1 \right) |1\rangle$. This also is to say that the T
gate leaves $|0\rangle$ unchanged and modifies the phase of $|1\rangle$ to give $\left(e^{\sqrt{-1} \times \frac{\pi}{4}} \right) |1\rangle$. The
probability of measuring a $|0\rangle$ or $|1\rangle$ is unchanged after using the T gate, however it
modifies the phase of the quantum state. Because $(T)^2 = T \times T = \begin{pmatrix} 1 & 0 \\ 0 & e^{\sqrt{-1} \times \frac{\pi}{4}} \end{pmatrix} \times$

$\begin{pmatrix} 1 & 0 \\ 0 & e^{\sqrt{-1} \times \frac{\pi}{4}} \end{pmatrix} = \begin{pmatrix} 1 & 0 \\ 0 & e^{\sqrt{-1} \times \frac{\pi}{2}} \end{pmatrix} = \begin{pmatrix} 1 & 0 \\ 0 & \sqrt{-1} \end{pmatrix}$, using T twice to a state is equivalent
to do the S gate to it. For **IBM Q** Experience, the graphical representation of the T
gate appears in Fig. 1.22.

1.8.1 *Programming with the T Gate of Single Quantum Bit*

In Listing 1.8, the program in the backend *ibmqx4* with five quantum bits in **IBM**'s
quantum computer is the eighth example in which we describe how to program with
the T gate that converts single quantum bit from one state $\frac{1}{\sqrt{2}}(|0\rangle + |1\rangle)$ to another

Fig. 1.23 The corresponding quantum circuit of the program in Listing 1.8

state $\left(\frac{1}{\sqrt{2}}\right)\left(|0\rangle + \left(e^{\sqrt{-1} \times \frac{\pi}{4}}\right)|1\rangle\right)$. Figure 1.23 is the corresponding quantum circuit of the program in Listing 1.8. The statement "OPENQASM 2.0;" on line one of Listing 1.8 is to indicate that the program is written with version 2.0 of Open QASM. Next, the statement "include "qelib1.inc";" on line two of Listing 1.8 is to continue parsing the file "qelib1.inc" as if the contents of the file were pasted at the location of the include statement, where the file "qelib1.inc" is **Quantum Experience (QE) Standard Header** and the path is specified relative to the current working directory. The statement "qreg q[5];" on line three of Listing 1.8 is to declare that in the program there are five quantum bits. In the left top of Fig. 1.23, five quantum bits are subsequently q[0], q[1], q[2], q[3] and q[4]. The initial state of each quantum bit is set to $|0\rangle$. Next, the statement "creg c[5];" on line four of Listing 1.8 is to declare that there are five classical bits in the program. In the left bottom of Fig. 1.23, five classical bits are subsequently c[0], c[1], c[2], c[3] and c[4]. The initial value of each classical bit is set to 0. Classical bit c[4] is the most significant bit and classical bit c[0] is the least significant bit.

Listing. 1.8 The program to the use of the T gate.

```
1.  OPENQASM 2.0;
2.  include "qelib1.inc";
3.  qreg q[5];
4.  creg c[5];
5.  h q[0];
6.  t q[0];
7.  measure q[0] → c[0];
```

The statement "h q[0];" on line five of Listing 1.8 actually implements
$$\begin{pmatrix} \frac{1}{\sqrt{2}} & \frac{1}{\sqrt{2}} \\ \frac{1}{\sqrt{2}} & -\frac{1}{\sqrt{2}} \end{pmatrix} \times \begin{pmatrix} 1 \\ 0 \end{pmatrix} = \begin{pmatrix} \frac{1}{\sqrt{2}} \\ \frac{1}{\sqrt{2}} \end{pmatrix} = \frac{1}{\sqrt{2}} \begin{pmatrix} 1 \\ 1 \end{pmatrix} = \frac{1}{\sqrt{2}} \left(\begin{pmatrix} 1 \\ 0 \end{pmatrix} + \begin{pmatrix} 0 \\ 1 \end{pmatrix} \right) = \frac{1}{\sqrt{2}} (|0\rangle + |1\rangle).$$
This implies that the statement "h q[0];" on line five of Listing 1.8 is to use the Hadamard gate to convert q[0] from one state $|0\rangle$ to another state $\frac{1}{\sqrt{2}}(|0\rangle + |1\rangle)$ (its

Fig. 1.24 After the measurement to the program in Listing 1.8 is completed, we obtain the answer 00001 with the probability 0.480 or the answer 00000 with the probability 0.520

superposition), where "h" is to represent the Hadamard gate. Next, the statement "t q[0];" on line six of Listing 1.8 actually completes $\begin{pmatrix} 1 & 0 \\ 0 & e^{\sqrt{-1} \times \frac{\pi}{4}} \end{pmatrix} \times \begin{pmatrix} \frac{1}{\sqrt{2}} \\ \frac{1}{\sqrt{2}} \end{pmatrix} =$

$\begin{pmatrix} \frac{1}{\sqrt{2}} \\ e^{\sqrt{-1} \times \frac{\pi}{4}} \times \frac{1}{\sqrt{2}} \end{pmatrix} = \frac{1}{\sqrt{2}} \begin{pmatrix} 1 \\ e^{\sqrt{-1} \times \frac{\pi}{4}} \end{pmatrix} = \frac{1}{\sqrt{2}} (\begin{pmatrix} 1 \\ 0 \end{pmatrix} + \begin{pmatrix} 0 \\ e^{\sqrt{-1} \times \frac{\pi}{4}} \end{pmatrix}) = (\frac{1}{\sqrt{2}}) (\begin{pmatrix} 1 \\ 0 \end{pmatrix} +$

$(e^{\sqrt{-1} \times \frac{\pi}{4}}) \begin{pmatrix} 0 \\ 1 \end{pmatrix}) = \frac{1}{\sqrt{2}} (|0\rangle + (e^{\sqrt{-1} \times \frac{\pi}{4}}) |1\rangle)$. This is to say that the statement "t q[0];" on line six of Listing 1.8 is to use the T gate to convert q[0] from one state $\frac{1}{\sqrt{2}}$ $(|0\rangle + |1\rangle)$ to another state $\frac{1}{\sqrt{2}} (|0\rangle + (e^{\sqrt{-1} \times \frac{\pi}{4}})|1\rangle)$.

Next, the statement "measure q[0] $\rightarrow$ c[0];" on line seven of Listing 1.8 is to measure the first quantum bit q[0] and to record the measurement outcome by over-writing the first classical bit c[0]. In the backend *ibmqx4* with five quantum bits in **IBM**'s quantum computers, we use the command "simulate" to execute the program in Listing 1.8. The result appears in Fig. 1.24. From Fig. 1.24, we obtain the answer 00001 (c[4] = 0, c[3] = 0, c[2] = 0, c[1] = 0 and c[0] = 1 = q[0] = 1$\rangle$) with the probability 0.480. Or we obtain the answer 00000 with the probability 0.520 (c[4] = 0, c[3] = 0, c[2] = 0, c[1] = 0 and c[0] = 0 = q[0] = 0$\rangle$).

1.9 The T⁺ Gate of Single Quantum Bit

The T^+ gate of single quantum bit that is the conjugate-transpose matrix of the T gate is

$$T^+ = \begin{pmatrix} 1 & 0 \\ 0 & e^{-\sqrt{-1} \times \frac{\pi}{4}} \end{pmatrix} = \begin{pmatrix} 1 & 0 \\ 0 & \frac{1-\sqrt{-1}}{\sqrt{2}} \end{pmatrix} = \begin{pmatrix} 1 & 0 \\ 0 & \frac{1-i}{\sqrt{2}} \end{pmatrix}, \tag{1.25}$$

where $i = \sqrt{-1}$ is known as the imaginary unit and $e^{-\sqrt{-1} \times \frac{\pi}{4}} = \cos(-\frac{\pi}{4}) + \sqrt{-1}$ $\times \sin(-\frac{\pi}{4}) = \frac{1-\sqrt{-1}}{\sqrt{2}} = \frac{1-i}{\sqrt{2}}$. It is supposed that $(T^+)^+$ is the conjugate-transpose matrix of T^+ and is equal to $((T^+)^*)^t = \begin{pmatrix} 1 & 0 \\ 0 & e^{\sqrt{-1} \times \frac{\pi}{4}} \end{pmatrix} = \begin{pmatrix} 1 & 0 \\ 0 & \frac{1+\sqrt{-1}}{\sqrt{2}} \end{pmatrix} = \begin{pmatrix} 1 & 0 \\ 0 & \frac{1+i}{\sqrt{2}} \end{pmatrix}$,

where the $*$ is the complex conjugation and the t is the transpose operation. Since

$$T^+ \times ((T^+)^*)^t = \begin{pmatrix} 1 & 0 \\ 0 & e^{-\sqrt{-1} \times \frac{\pi}{4}} \end{pmatrix} \times \begin{pmatrix} 1 & 0 \\ 0 & e^{\sqrt{-1} \times \frac{\pi}{4}} \end{pmatrix} = ((T^+)^*)^t \times T^+ = \begin{pmatrix} 1 & 0 \\ 0 & e^{\sqrt{-1} \times \frac{\pi}{4}} \end{pmatrix}$$

$$\times \begin{pmatrix} 1 & 0 \\ 0 & e^{-\sqrt{-1} \times \frac{\pi}{4}} \end{pmatrix} = \begin{pmatrix} 1 & 0 \\ 0 & 1 \end{pmatrix}, \; T^+ \text{ is a unitary matrix or a unitary operator. This is to}$$

say that the T^+ gate is one of quantum gates with single quantum bit. If the quantum state $l_0 |0\rangle + l_1 |1\rangle$ is written in a vector notation as

$$\begin{pmatrix} l_0 \\ l_1 \end{pmatrix}, \tag{1.26}$$

with the top entry is the amplitude for $|0\rangle$ and the bottom entry is the amplitude for $|1\rangle$, then the corresponding output from the T^+ gate is

$$\begin{pmatrix} 1 & 0 \\ 0 & e^{-\sqrt{-1} \times \frac{\pi}{4}} \end{pmatrix} \times \begin{pmatrix} l_0 \\ l_1 \end{pmatrix} = \begin{pmatrix} l_0 \\ e^{-\sqrt{-1} \times \frac{\pi}{4}} \times l_1 \end{pmatrix}$$

$$= (l_0) \begin{pmatrix} 1 \\ 0 \end{pmatrix} + \left(e^{-\sqrt{-1} \times \frac{\pi}{4}} \times l_1 \right) \begin{pmatrix} 0 \\ 1 \end{pmatrix} = (l_0) |0\rangle + \left(e^{-\sqrt{-1} \times \frac{\pi}{4}} \times l_1 \right) |1\rangle. \tag{1.27}$$

This indicates that the T^+ gate converts single quantum bit from one state $l_0 |0\rangle + l_1 |1\rangle$ to another state $(l_0) |0\rangle + \left(e^{-\sqrt{-1} \times \frac{\pi}{4}} \times l_1 \right) |1\rangle$. This also is to say that the T^+ gate leaves $|0\rangle$ unchanged and modifies the phase of $|1\rangle$ to give $\left(e^{-\sqrt{-1} \times \frac{\pi}{4}} \right) |1\rangle$. The probability of measuring a $|0\rangle$ or $|1\rangle$ is unchanged after applying the T^+ gate, however it modifies the phase of the quantum state. Because $(T^+)^2 = T^+ \times T^+ =$

$$\begin{pmatrix} 1 & 0 \\ 0 & e^{-\sqrt{-1} \times \frac{\pi}{4}} \end{pmatrix} \times \begin{pmatrix} 1 & 0 \\ 0 & e^{-\sqrt{-1} \times \frac{\pi}{4}} \end{pmatrix} = \begin{pmatrix} 1 & 0 \\ 0 & e^{-\sqrt{-1} \times \frac{\pi}{2}} \end{pmatrix} = \begin{pmatrix} 1 & 0 \\ 0 & -\sqrt{-1} \end{pmatrix}, \text{ applying } T^+$$

twice to a state is equivalent to do the S^+ gate to it. For **IBM Q** Experience, the graphical representation of the T^+ gate appears in Fig. 1.25.

Fig. 1.25 The graphical representation of the T^+ gate

1.9.1 Programming with the T^+ Gate of Single Quantum Bit

In Listing 1.9, the program in the backend *ibmqx4* with five quantum bits in **IBM**'s quantum computer is the ninth example in which we introduce how to program with the T^+ gate that converts single quantum bit from one state $\frac{1}{\sqrt{2}}\left(|0\rangle + |1\rangle\right)$ to another state $\left(\frac{1}{\sqrt{2}}\right)\left(|0\rangle + \left(e^{-\sqrt{-1}\times\frac{\pi}{4}}\right)|1\rangle\right)$. Figure 1.26 is the corresponding quantum circuit of the program in Listing 1.9. The statement "OPENQASM 2.0;" on line one of Listing 1.9 is to point out that the program is written with version 2.0 of Open QASM. Next, the statement "include "qelib1.inc";" on line two of Listing 1.9 is to continue parsing the file "qelib1.inc" as if the contents of the file were pasted at the location of the include statement, where the file "qelib1.inc" is **Quantum Experience (QE) Standard Header** and the path is specified relative to the current working directory. The statement "qreg q[5];" on line three of Listing 1.9 is to declare that in the program thereare five quantum bits. In the left top of Fig. 1.26, five quantum bits are subsequently q[0], q[1], q[2], q[3] and q[4]. The initial state of each quantum bit is set to $|0\rangle$. Next, the statement "creg c[5];" on line four of Listing 1.9 is to declare that there are five classical bits in the program. In the left bottom of Fig. 1.26, five classical bits are subsequently c[0], c[1], c[2], c[3] and c[4]. The initial value of each classical bit is set to 0. Classical bit c[4] is the most significant bit and classical bit c[0] is the least significant bit.

Listing. 1.9 The program to the use of the T^+ gate.

```
1.  OPENQASM 2.0;
2.  include "qelib1.inc";
3.  qreg q[5];
4.  creg c[5];
5.  h q[0];
6.  tdg q[0];
7.  measure q[0] → c[0];
```

Fig. 1.26 The corresponding quantum circuit of the program in Listing 1.9

Fig. 1.27 After the measurement to the program in Listing 1.9 is completed, we obtain the answer 00001 with the probability 0.550 or the answer 00000 with the probability 0.450

The statement "h q[0];" on line five of Listing 1.9 actually completes $\begin{pmatrix} \frac{1}{\sqrt{2}} & \frac{1}{\sqrt{2}} \\ \frac{1}{\sqrt{2}} & -\frac{1}{\sqrt{2}} \end{pmatrix}$

$\times \begin{pmatrix} 1 \\ 0 \end{pmatrix} = \begin{pmatrix} \frac{1}{\sqrt{2}} \\ \frac{1}{\sqrt{2}} \end{pmatrix} = \frac{1}{\sqrt{2}} \begin{pmatrix} 1 \\ 1 \end{pmatrix} = \frac{1}{\sqrt{2}} (\begin{pmatrix} 1 \\ 0 \end{pmatrix} + \begin{pmatrix} 0 \\ 1 \end{pmatrix}) = \frac{1}{\sqrt{2}} (|0\rangle + |1\rangle)$. This is to say that the statement "h q[0];" on line five of Listing 1.9 is to use the Hadamard gate to convert q[0] from one state $|0\rangle$ to another state $\frac{1}{\sqrt{2}} (|0\rangle + |1\rangle)$ (its superposition), where "h" is to represent the Hadamard gate. Next, the statement "tdg q[0];" on line six of Listing 1.9 actually implements $\begin{pmatrix} 1 & 0 \\ 0 & e^{-\sqrt{-1} \times \frac{\pi}{4}} \end{pmatrix} \times \begin{pmatrix} \frac{1}{\sqrt{2}} \\ \frac{1}{\sqrt{2}} \end{pmatrix} = $

$\begin{pmatrix} \frac{1}{\sqrt{2}} \\ e^{-\sqrt{-1} \times \frac{\pi}{4}} \times \frac{1}{\sqrt{2}} \end{pmatrix} = \frac{1}{\sqrt{2}} \begin{pmatrix} 1 \\ e^{-\sqrt{-1} \times \frac{\pi}{4}} \end{pmatrix} = \frac{1}{\sqrt{2}} (\begin{pmatrix} 1 \\ 0 \end{pmatrix} + \begin{pmatrix} 0 \\ e^{-\sqrt{-1} \times \frac{\pi}{4}} \end{pmatrix}) = \frac{1}{\sqrt{2}}$

$(\begin{pmatrix} 1 \\ 0 \end{pmatrix} + (e^{-\sqrt{-1} \times \frac{\pi}{4}}) \begin{pmatrix} 0 \\ 1 \end{pmatrix}) = \frac{1}{\sqrt{2}} (|0\rangle + (e^{-\sqrt{-1} \times \frac{\pi}{4}})|1\rangle)$. This indicates that the statement "tdg q[0];" on line six of Listing 1.9 is to use the T^+ gate to convert q[0] from one state $\frac{1}{\sqrt{2}} (|0\rangle + |1\rangle)$ to another state $\frac{1}{\sqrt{2}} (|0\rangle + (e^{-\sqrt{-1} \times \frac{\pi}{4}})|1\rangle)$.

Next, the statement "measure q[0] → c[0];" on line seven of Listing 1.9 is to measure the first quantum bit q[0] and to record the measurement outcome by overwriting the first classical bit c[0]. In the backend *ibmqx4* with five quantum bits in **IBM**'s quantum computers, we apply the command "simulate" to execute the program in Listing 1.9. The result appears in Fig. 1.27. From Fig. 1.27, we obtain the answer 00001 (c[4] = 0, c[3] = 0, c[2] = 0, c[1] = 0 and c[0] = 1 = q[0] = $|1\rangle$) with the probability 0.550. Or we obtain the answer 00000 with the probability 0.450 (c[4] = 0, c[3] = 0, c[2] = 0, c[1] = 0 and c[0] = 0 = q[0] = $|0\rangle$).

1.10 The Identity Gate of Single Quantum Bit

The *identity* gate *id* of single quantum bit is

$$id = \begin{pmatrix} 1 & 0 \\ 0 & 1 \end{pmatrix}. \tag{1.28}$$

It is assumed that id^+ is the conjugate-transpose matrix of id and is equal to $((id)^*)^t = \begin{pmatrix} 1 & 0 \\ 0 & 1 \end{pmatrix}$, where the $*$ is the complex conjugation and the t is the transpose operation. Because $id \times ((id)^*)^t = \begin{pmatrix} 1 & 0 \\ 0 & 1 \end{pmatrix} \times \begin{pmatrix} 1 & 0 \\ 0 & 1 \end{pmatrix} = ((id)^*)^t \times id = \begin{pmatrix} 1 & 0 \\ 0 & 1 \end{pmatrix} \times \begin{pmatrix} 1 & 0 \\ 0 & 1 \end{pmatrix} = \begin{pmatrix} 1 & 0 \\ 0 & 1 \end{pmatrix}$, id is a unitary matrix or a unitary operator. This indicates that the identity gate id is one of quantum gates with single quantum bit. If the quantum state $l_0 |0\rangle + l_1 |1\rangle$ is written in a vector notation as

$$\begin{pmatrix} l_0 \\ l_1 \end{pmatrix}, \tag{1.29}$$

with the top entry is the amplitude for $|0\rangle$ and the bottom entry is the amplitude for $|1\rangle$, then the corresponding output from the identity gate id is

$$\begin{pmatrix} 1 & 0 \\ 0 & 1 \end{pmatrix} \times \begin{pmatrix} l_0 \\ l_1 \end{pmatrix} = \begin{pmatrix} l_0 \\ l_1 \end{pmatrix} = (l_0)\begin{pmatrix} 1 \\ 0 \end{pmatrix} + (l_1)\begin{pmatrix} 0 \\ 1 \end{pmatrix} = (l_0)|0> +(l_1)|1>. \tag{1.30}$$

This is to say that the identity gate id converts single quantum bit from one state $l_0 |0\rangle + l_1 |1\rangle$ to another state $(l_0) |0\rangle + (l_1) |1\rangle$. This also is to say that the identity gate id does not change $|0\rangle$ and $|1\rangle$ and only performs an idle operation on single quantum bit for a time equal to one unit of time. The probability of measuring a $|0\rangle$ or $|1\rangle$ is unchanged after using the identity gate id. Since $(id)^2 = id \times id = \begin{pmatrix} 1 & 0 \\ 0 & 1 \end{pmatrix} \times \begin{pmatrix} 1 & 0 \\ 0 & 1 \end{pmatrix} = \begin{pmatrix} 1 & 0 \\ 0 & 1 \end{pmatrix}$, applying the identity gate id twice to a state is equivalent to do nothing to it. For **IBM Q** Experience, the graphical representation of the identity gate id appears in Fig. 1.28.

Fig. 1.28 The graphical representation of the identity gate id

Fig. 1.29 The corresponding quantum circuit of the program in Listing 1.10

1.10.1   Programming with the Identity Gate of Single Quantum Bit

In Listing 1.10, the program in the backend *ibmqx4* with five quantum bits in **IBM**'s quantum computer is the tenth example in which we introduce how to program with the identity gate *id* that converts single quantum bit q[0] from one state $\frac{1}{\sqrt{2}}(|0\rangle + |1\rangle)$ to another state $\frac{1}{\sqrt{2}}(|0\rangle + |1\rangle)$. Actually, the identity gate *id* only completes an idle operation on q[0] for a time equal to one unit of time. Figure 1.29 is the corresponding quantum circuit of the program in Listing 1.10. The statement "OPENQASM 2.0;" on line one of Listing 1.10 is to indicate that the program is written with version 2.0 of Open QASM. Next, the statement "include "qelib1.inc";" on line two of Listing 1.10 is to continue parsing the file "qelib1.inc" as if the contents of the file were pasted at the location of the include statement, where the file "qelib1.inc" is **Quantum Experience (QE) Standard Header** and the path is specified relative to the current working directory. The statement "qreg q[5];" on line three of Listing 1.10 is to declare that in the program there are five quantum bits. In the left top of Fig. 1.29, five quantum bits

Listing. 1.10 The program to the use of the identity gate.

```
1.  OPENQASM 2.0;
2.  include "qelib1.inc";
3.  qreg q[5];
4.  creg c[5];
5.  h q[0];
6.  id q[0];
7.  measure q[0] → c[0];
```

are subsequently q[0], q[1], q[2], q[3] and q[4]. The initial state of each quantum bit is set to $|0\rangle$. Next, the statement "creg c[5];" on line four of Listing 1.10 is to declare that there are five classical bits in the program. In the left bottom of Fig. 1.29,

Fig. 1.30 After the measurement to the program in Listing 1.10 is completed, we obtain the answer 00001 with the probability 0.460 or the answer 00000 with the probability 0.540

five classical bits are subsequently c[0], c[1], c[2], c[3] and c[4]. The initial value of each classical bit is set to 0. Classical bit c[4] is the most significant bit and classical bit c[0] is the least significant bit.

The statement "h q[0];" on line five of Listing 1.10 actually performs $\begin{pmatrix} \frac{1}{\sqrt{2}} & \frac{1}{\sqrt{2}} \\ \frac{1}{\sqrt{2}} & -\frac{1}{\sqrt{2}} \end{pmatrix}$

$\times \begin{pmatrix} 1 \\ 0 \end{pmatrix} = \begin{pmatrix} \frac{1}{\sqrt{2}} \\ \frac{1}{\sqrt{2}} \end{pmatrix} = \frac{1}{\sqrt{2}} \begin{pmatrix} 1 \\ 1 \end{pmatrix} = \frac{1}{\sqrt{2}} \left(\begin{pmatrix} 1 \\ 0 \end{pmatrix} + \begin{pmatrix} 0 \\ 1 \end{pmatrix} \right) = \frac{1}{\sqrt{2}} (|0\rangle + |1\rangle).$ This implies

that the statement "h q[0];" on line five of Listing 1.10 is to apply the Hadamard gate to convert q[0] from one state $|0\rangle$ to another state $\frac{1}{\sqrt{2}} (|0\rangle + |1\rangle)$ (its superposition), where "h" is to represent the Hadamard gate. Next, the statement "id q[0];" on line six of Listing 1.10 actually completes $\begin{pmatrix} 1 & 0 \\ 0 & 1 \end{pmatrix} \times \begin{pmatrix} \frac{1}{\sqrt{2}} \\ \frac{1}{\sqrt{2}} \end{pmatrix} = \begin{pmatrix} \frac{1}{\sqrt{2}} \\ \frac{1}{\sqrt{2}} \end{pmatrix} = \frac{1}{\sqrt{2}} \begin{pmatrix} 1 \\ 1 \end{pmatrix} = \frac{1}{\sqrt{2}}$

$(\begin{pmatrix} 1 \\ 0 \end{pmatrix} + \begin{pmatrix} 0 \\ 1 \end{pmatrix}) = \frac{1}{\sqrt{2}} (|0\rangle + |1\rangle).$ This is to say that the statement "id q[0];" on line

six of Listing 1.10 is to use the identity gate *id* to convert q[0] from one state $\frac{1}{\sqrt{2}}$ $(|0\rangle + |1\rangle)$ to another state $\frac{1}{\sqrt{2}} (|0\rangle + |1\rangle)$. This also is to say that the identity gate *id* only completes an idle operation on q[0] for a time equal to one unit of time.

Next, the statement "measure q[0] $\rightarrow$ c[0];" on line seven of Listing 1.10 is to measure the first quantum bit q[0] and to record the measurement outcome by overwriting the first classical bit c[0]. In the backend *ibmqx4* with five quantum bits in **IBM**'s quantum computers, we apply the command "simulate" to execute the program in Listing 1.10. The result appears in Fig. 1.30. From Fig. 1.30, we obtain the answer 00001 (c[4] = 0, c[3] = 0, c[2] = 0, c[1] = 0 and c[0] = 1 = q[0] = $|1\rangle$) with the probability 0.460. Or we obtain the answer 00000 with the probability 0.540 (c[4] = 0, c[3] = 0, c[2] = 0, c[1] = 0 and c[0] = 0 = q[0] = $|0\rangle$).

1.11   The Controlled-NOT Gate of Two Quantum Bits

The *controlled-NOT* or *CNOT* gate of two quantum bits is

$$U_{CN} = \begin{pmatrix} 1\,0\,0\,0 \\ 0\,1\,0\,0 \\ 0\,0\,0\,1 \\ 0\,0\,1\,0 \end{pmatrix}. \tag{1.31}$$

It is supposed that $U_{CN}{}^{+}$ is the conjugate-transpose matrix of U_{CN} and is equal to $((U_{CN})^{*})^{t} = \begin{pmatrix} 1\,0\,\,0\,0 \\ 0\,1\,\,0\,0 \\ 0\,0\,\,0\,1 \\ 0\,0\,\,1\,0 \end{pmatrix}$, where the $*$ is the complex conjugation and the t is

the transpose operation. Since $U_{CN} \times ((U_{CN})^{*})^{t} = \begin{pmatrix} 1\,0\,\,0\,0 \\ 0\,1\,\,0\,0 \\ 0\,0\,\,0\,1 \\ 0\,0\,\,1\,0 \end{pmatrix} \times \begin{pmatrix} 1\,0\,\,0\,0 \\ 0\,1\,\,0\,0 \\ 0\,0\,\,0\,1 \\ 0\,0\,\,1\,0 \end{pmatrix} =$

$((U_{CN})^{*})^{t} \times U_{CN} = \begin{pmatrix} 1\,0\,\,0\,0 \\ 0\,1\,\,0\,0 \\ 0\,0\,\,0\,1 \\ 0\,0\,\,1\,0 \end{pmatrix} \times \begin{pmatrix} 1\,0\,\,0\,0 \\ 0\,1\,\,0\,0 \\ 0\,0\,\,0\,1 \\ 0\,0\,\,1\,0 \end{pmatrix} = \begin{pmatrix} 1\,0\,\,0\,0 \\ 0\,1\,\,0\,0 \\ 0\,0\,\,1\,0 \\ 0\,0\,\,0\,1 \end{pmatrix}$, U_{CN} is a unitary

matrix or a unitary operator. This is to say that the **controlled-NOT** or **CNOT** gate U_{CN} is one of quantum gates with two quantum bits. If the quantum state $l_{0}\,|00\rangle + l_{1}\,|01\rangle + l_{2}\,|10\rangle + l_{3}\,|11\rangle$ is written in a vector notation as

$$\begin{pmatrix} l_{0} \\ l_{1} \\ l_{2} \\ l_{3} \end{pmatrix}, \tag{1.32}$$

with the first entry l_{0} is the amplitude for $|00\rangle$, the second entry l_{1} is the amplitude for $|01\rangle$, the third entry l_{2} is the amplitude for $|10\rangle$ and the fourth entry l_{3} is the amplitude for $|11\rangle$, then the corresponding output from the **CNOT** gate U_{CN} is

$$\begin{pmatrix} 1\,0\,\,0\,0 \\ 0\,1\,\,0\,0 \\ 0\,0\,\,0\,1 \\ 0\,0\,\,1\,0 \end{pmatrix} \times \begin{pmatrix} l_{0} \\ l_{1} \\ l_{2} \\ l_{3} \end{pmatrix} = \begin{pmatrix} l_{0} \\ l_{1} \\ l_{3} \\ l_{2} \end{pmatrix} = l_{0}\begin{pmatrix} 1 \\ 0 \\ 0 \\ 0 \end{pmatrix} + l_{1}\begin{pmatrix} 0 \\ 1 \\ 0 \\ 0 \end{pmatrix} + l_{3}\begin{pmatrix} 0 \\ 0 \\ 1 \\ 0 \end{pmatrix} + l_{2}\begin{pmatrix} 0 \\ 0 \\ 0 \\ 1 \end{pmatrix}$$

$$= l_{0}|00> + l_{1}|01> + l_{3}|10> + l_{2}|11>. \tag{1.33}$$

This indicates that the **CNOT** gate U_{CN} converts two quantum bits from one state $l_{0}\,|00\rangle + l_{1}\,|01\rangle + l_{2}\,|10\rangle + l_{3}\,|11\rangle$ to another state $l_{0}\,|00\rangle + l_{1}\,|01\rangle + l_{3}\,|10\rangle + l_{2}\,|11\rangle$. This is to say that in the **CNOT** gate U_{CN} if the control quantum bit (the *first* quantum bit) is set to 0, then the target quantum bit (the *second* quantum bit) is left alone. If the control quantum bit (the *first* quantum bit) is set to 1, then the target quantum

Fig. 1.31 The graphical
representation of the **CNOT**
gate U_{CN}

bit (the *second* quantum bit) is flipped. The probability of measuring a $|00\rangle$ or $|01\rangle$
is unchanged, the probability of measuring a $|10\rangle$ is $|l_3|^2$ and the probability of
measuring a $|11\rangle$ is $|l_2|^2$ after applying the **CNOT** gate U_{CN}. Because $(U_{CN})^2 = U_{CN}$

$$\times\ U_{CN} = \begin{pmatrix} 1\ 0\ 0\ 0 \\ 0\ 1\ 0\ 0 \\ 0\ 0\ 0\ 1 \\ 0\ 0\ 1\ 0 \end{pmatrix} \times \begin{pmatrix} 1\ 0\ 0\ 0 \\ 0\ 1\ 0\ 0 \\ 0\ 0\ 0\ 1 \\ 0\ 0\ 1\ 0 \end{pmatrix} = \begin{pmatrix} 1\ 0\ 0\ 0 \\ 0\ 1\ 0\ 0 \\ 0\ 0\ 1\ 0 \\ 0\ 0\ 0\ 1 \end{pmatrix}, \text{applying the \textbf{CNOT} gate}$$

U_{CN} twice to a state is equivalent to do nothing to it. For **IBM Q** Experience, the
graphical representation of the **CNOT** gate U_{CN} appears in Fig. 1.31.

In the graphical representation of the **CNOT** gate U_{CN}, the top wire carries the
controlled quantum bit and the bottom wire carries the target quantum bit.

1.11.1 Connectivity of the Controlled-NOT Gate in IBMQX4

Those authors that wrote textbooks write quantum algorithms with a fully connected
hardware in which one can apply a quantum gate of two quantum bits to any pair of
two quantum bits. In practice, *ibmqx4* that is a real quantum computer may not have
full connectivity. In the *ibmqx4* with five quantum bits, there are six connections. Six
connections of a **CNOT** gate appears in Fig. 1.32. The *first* **CNOT** gate in Fig. 1.32 has
the controlled quantum bit q[1] and the target quantum bit q[0] and the corresponding
instruction in version 2.0 of Open QASM is "cx q[1],q[0];", where cx is to represent
the **CNOT** gate. The *second* **CNOT** gate in Fig. 1.32 has the controlled.

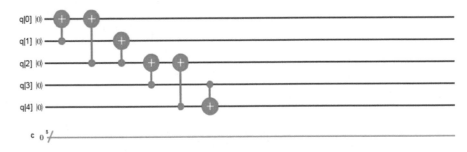

Fig. 1.32 A **CNOT** gate has six connections in *ibmqx4* on **IBM** quantum computers

quantum bit q[2] and the target quantum bit q[0] and the corresponding instruction in version 2.0 of Open QASM is "cx q[2],q[0];", where cx is to represent the **CNOT** gate.

The *third* **CNOT** gate in Fig. 1.32 has the controlled quantum bit q[2] and the target quantum bit q[1] and the corresponding instruction in version 2.0 of Open QASM is "cx q[2],q[1];", where cx is to represent the **CNOT** gate. The *fourth* **CNOT** gate in Fig. 1.32 has the controlled quantum bit q[3] and the target quantum bit q[2] and the corresponding instruction in version 2.0 of Open QASM is "cx q[3],q[2];", where cx is to represent the **CNOT** gate. The *fifth* **CNOT** gate in Fig. 1.32 has the controlled quantum bit q[4] and the target quantum bit q[2] and the corresponding instruction in version 2.0 of Open QASM is "cx q[4],q[2];", where cx is to represent the **CNOT** gate. The *sixth* **CNOT** gate in Fig. 1.32 has the controlled quantum bit q[3] and the target quantum bit q[4] and the corresponding instruction in version 2.0 of Open QASM is "cx q[3],q[4];", where cx is to represent the **CNOT** gate.

In contrast, a fully connected hardware with five quantum bits would allow a **CNOT** gate to apply to twenty different pairs of any two-quantum bits. This indicates that there are fourteen "missing connections". Fortunately, there are different ways to yield connections by means of using clever gate sequences. For example, a **CNOT** gate that has the controlled quantum bit q[j] and the target quantum bit q[k] for $0 \leq j$ and $k \leq 4$ can be reversed by means of applying Hadamard gates on each quantum bit both before and after the **CNOT** gate. This is to say that the new instruction (the new connection) "cx q[k], q[j]" is implemented by means of applying the five instructions that are subsequently "h q[j];", "h q[k]", "cx q[j], q[k]", "h q[j]" and "h q[k]" for $0 \leq j$ and $k \leq 4$. Similarly, there exists a gate sequence to make a **CNOT** gate with the controlled bit q[j] and the target bit q[l] if one has connections between the controlled bit q[j] and the target bit q[k], and the controlled bit q[k] and the target bit q[l] for $0 \leq j, k$ and $l \leq 4$. This indicates that the new instruction (the new connection) "cx q[j], q[l]" is implemented by means of applying the four instructions that are subsequently "cx q[k], q[l]", "cx q[j], q[k]", "cx q[k], q[l]" and "cx q[j], q[k]" for $0 \leq j, k$ and $l \leq 4$.

1.11.2 Implementing a Copy Machine of One Bit with the CNOT Gate

Offered its data input is initialized permanently with $|0\rangle$. Then, the **CNOT** gate emits a copy of the controlled input on each output. Therefore, the **CNOT** gate actually is a copy machine of one bit. In Listing 1.11, the program in the backend *ibmqx4* with five quantum bits in **IBM**'s quantum computer is the eleventh example in which we describe how to program with the **CNOT** gate that converts the controlled bit q[3] and the target bit q[4] from one state $\left(\frac{1}{\sqrt{2}}(|0\rangle + |1\rangle) \right) (|0\rangle) = \left(\frac{1}{\sqrt{2}}(|00\rangle + |10\rangle) \right)$ to another state $\left(\frac{1}{\sqrt{2}} \right)(|00\rangle + |11\rangle)$. Actually, the **CNOT** gate emits a copy of the

Fig. 1.33 The corresponding quantum circuit of the program in Listing 1.11

controlled input on each output. Figure 1.33 is the corresponding quantum circuit of the program in Listing 1.11. The statement "OPENQASM 2.0;" on line one of Listing 1.11 is to point out that the program is written with version 2.0 of Open QASM. Then, the statement "include "qelib1.inc";" on line two of Listing 1.11 is to continue parsing the file "qelib1.inc" as if the contents of the file were pasted at the location of the include statement, where the file "qelib1.inc" is **Quantum Experience (QE) Standard Header** and the path is specified relative to the current working directory.

The statement "qreg q[5];" on line three of Listing 1.11 is to declare that in the program there are five quantum bits. In the left top of Fig. 1.33, five quantum bits are subsequently q[0], q[1], q[2], q[3] and q[4]. The initial value of each quantum bit is set to $|0\rangle$. Next, the statement "creg c[5];" on line four of Listing 1.11 is to declare that there are five classical bits in the program. In the left bottom of Fig. 1.33, five classical bits are subsequently c[0], c[1], c[2], c[3] and c[4]. The initial value of each classical bit is set to 0. Classical bit c[4] is the most significant bit and classical bit c[0] is the least significant bit.

Listing. 1.11 Implementing a copy machine of one bit with the **CNOT** gate.

```
1.   OPENQASM 2.0;
2.   include "qelib1.inc";
3.   qreg q[5];
4.   creg c[5];
5.   h q[3];
6.   cx q[3],q[4];
7.   measure q[3] → c[3];
8.   measure q[4] → c[4];
```

The statement "h q[3];" on line five of Listing 1.11 actually completes

$$\begin{pmatrix} \frac{1}{\sqrt{2}} & \frac{1}{\sqrt{2}} \\ \frac{1}{\sqrt{2}} & -\frac{1}{\sqrt{2}} \end{pmatrix} \times \begin{pmatrix} 1 \\ 0 \end{pmatrix} = \begin{pmatrix} \frac{1}{\sqrt{2}} \\ \frac{1}{\sqrt{2}} \end{pmatrix} = \frac{1}{\sqrt{2}} \begin{pmatrix} 1 \\ 1 \end{pmatrix} = \frac{1}{\sqrt{2}} \left(\begin{pmatrix} 1 \\ 0 \end{pmatrix} + \begin{pmatrix} 0 \\ 1 \end{pmatrix} \right) = \frac{1}{\sqrt{2}} \left(|0\rangle + |1\rangle \right).$$

This is to say that the statement "h q[3];" on line five of Listing 1.11 is to use the

Hadamard gate to convert q[3] from one state $|0\rangle$ to another state $\frac{1}{\sqrt{2}}(|0\rangle + |1\rangle)$ (its superposition), where "h" is to represent the Hadamard gate. Next, the statement "cx q[3],q[4];" on line six of Listing 1.11 actually completes $\begin{pmatrix} 1 & 0 & 0 & 0 \\ 0 & 1 & 0 & 0 \\ 0 & 0 & 0 & 1 \\ 0 & 0 & 1 & 0 \end{pmatrix} \times \begin{pmatrix} \frac{1}{\sqrt{2}} \\ 0 \\ \frac{1}{\sqrt{2}} \\ 0 \end{pmatrix} =$

$\begin{pmatrix} \frac{1}{\sqrt{2}} \\ 0 \\ 0 \\ \frac{1}{\sqrt{2}} \end{pmatrix} = \frac{1}{\sqrt{2}} \begin{pmatrix} 1 \\ 0 \\ 0 \\ 1 \end{pmatrix} = \frac{1}{\sqrt{2}} \left(\begin{pmatrix} 1 \\ 0 \\ 0 \\ 0 \end{pmatrix} + \begin{pmatrix} 0 \\ 0 \\ 0 \\ 1 \end{pmatrix} \right) = \frac{1}{\sqrt{2}}(|00\rangle + |11\rangle)$. This indicates

that the statement "cx q[3],q[4];" on line six of Listing 1.11 is to apply the **CNOT** gate to emit a copy of the controlled input q[3] on each output.

Next, the statement "measure q[3] → c[3];" on line seven of Listing 1.11 is to measure the fourth quantum bit q[3] and to record the measurement outcome by overwriting the fourth classical bit c[3]. The statement "measure q[4] → c[4];" on line eight of Listing 1.11 is to measure the fifth quantum bit q[4] and to record the measurement outcome by overwriting the fifth classical bit c[4].In the backend *ibmqx4* with five quantum bits in **IBM**'s quantum computers, we use the command "simulate" to execute the program in Listing 1.11. The result appears in Fig. 1.34. From Fig. 1.34,

we obtain the answer 00000 (c[4] = 0 = q[4] = $|0\rangle$, c[3] = 0 = q[3] = $|0\rangle$, c[2] = 0, c[1] = 0 and c[0] = 0) with the probability 0.540. Or we obtain the answer 11000 with the probability 0.460 (c[4] = 1 = q[4] = $|1\rangle$, c[3] = 1 = q[3] = $|1\rangle$, c[2] = 0, c[1] = 0 and c[0] = 0). If the answer is 00000, then this imply that the **CNOT** gate copy the value 0 of the controlled input q[3] to the target bit q[4]. If the answer is 11000, then this indicates that the **CNOT** gate copy the value 1 of the controlled input q[3] to the target bit q[4].

Fig. 1.34 After the measurement to the program in Listing 1.11 is completed, we obtain the answer 00000 with the probability 0.540 or the answer 11000 with the probability 0.460

1.12   The U1(λ) Gate of Single Quantum Bit with One Parameter

The $U1(\lambda)$ gate that is the first physical gate of the Quantum Experience and is a phase gate of single quantum bit of one parameter with zero duration is

$$U1(\lambda) = U1(\text{lambda}) = \begin{pmatrix} 1 & 0 \\ 0 & e^{\sqrt{-1}\times\lambda} \end{pmatrix}, \tag{1.34}$$

where λ (lambda) is a real value. It is assumed that $(U1(\lambda))^+$ $(U1(\text{lambda}))^+$ is the conjugate-transpose matrix of $U1(\lambda)$ $(U1(\text{lambda}))$ and is equal to $(((U1(\lambda)))^*)^t$
$= \begin{pmatrix} 1 & 0 \\ 0 & e^{-\sqrt{-1}\times\lambda} \end{pmatrix}$, where the $*$ is the complex conjugation and the t is the transpose

operation. Because $U1(\lambda) \times (U1(\lambda))^+ = U1(\lambda) \times (((U1(\lambda)))^*)^t = \begin{pmatrix} 1 & 0 \\ 0 & e^{\sqrt{-1}\times\lambda} \end{pmatrix} \times$

$\begin{pmatrix} 1 & 0 \\ 0 & e^{-\sqrt{-1}\times\lambda} \end{pmatrix} = (U1(\lambda))^+ \times U1(\lambda) = (((U1(\lambda)))^*)^t \times U1(\lambda) = \begin{pmatrix} 1 & 0 \\ 0 & e^{-\sqrt{-1}\times\lambda} \end{pmatrix}$

$\times \begin{pmatrix} 1 & 0 \\ 0 & e^{\sqrt{-1}\times\lambda} \end{pmatrix} = \begin{pmatrix} 1 & 0 \\ 0 & 1 \end{pmatrix}$, $U1(\lambda)$ $(U1(\text{lambda}))$ is a unitary matrix or a unitary

operator. This implies that the phase gate $U1(\lambda)$ $(U1(\text{lambda}))$ is one of quantum gates that is a phase gate of single quantum bit of one parameter with zero duration. If the quantum state $l_0 \, |0\rangle + l_1 \, |1\rangle$ is written in a vector notation as

$$\begin{pmatrix} l_0 \\ l_1 \end{pmatrix}, \tag{1.35}$$

with the first entry l_0 is the amplitude for $|0\rangle$ and the second entry l_1 is the amplitude for $|1\rangle$, then the corresponding output from the phase gate $U1(\lambda)$ $(U1(\text{lambda}))$ is

$$\begin{pmatrix} 1 & 0 \\ 0 & e^{\sqrt{-1}\times\lambda} \end{pmatrix} \times \begin{pmatrix} l_0 \\ l_1 \end{pmatrix} = \begin{pmatrix} l_0 \\ e^{\sqrt{-1}\times\lambda} \times l_1 \end{pmatrix} = l_0 \begin{pmatrix} 1 \\ 0 \end{pmatrix} + \left(e^{\sqrt{-1}\times\lambda} \times l_1\right)\begin{pmatrix} 0 \\ 1 \end{pmatrix}$$

$$= l_0 |0\rangle + \left(e^{\sqrt{-1}\times\lambda} \times l_1\right)|1\rangle . \tag{1.36}$$

This is to say that the phase gate $U1(\lambda)$ $(U1(\text{lambda}))$ converts one quantum bit from one state $l_0 \, |0\rangle + l_1 \, |1\rangle$ to another state $l_0 \, |0\rangle + \left(e^{\sqrt{-1}\times\lambda} \times l_1\right) |1\rangle$. This indicates that the phase gate $U1(\lambda)$ $(U1(\text{lambda}))$ leaves $|0\rangle$ unchanged and modifies the phase of $|1\rangle$ to give $(e^{\sqrt{-1}\times\lambda}) |1\rangle$. The probability of measuring a $|0\rangle$ or $|1\rangle$ is unchanged after using the phase gate $U1(\lambda)$ $(U1(\text{lambda}))$, however it modifies the

phase of the quantum state. Because $(U1(\lambda))^2 = U1(\lambda) \times U1(\lambda) = \begin{pmatrix} 1 & 0 \\ 0 & e^{\sqrt{-1}\times\lambda} \end{pmatrix} \times$

Fig. 1.35 The graphical representation of the phase gate $U1(\lambda)$ ($U1$(lambda)).

$$\begin{pmatrix} 1 & 0 \\ 0 & e^{\sqrt{-1}\times\lambda} \end{pmatrix} = \begin{pmatrix} 1 & 0 \\ 0 & e^{\sqrt{-1}\times2\times\lambda} \end{pmatrix},$$ using the phase gate $U1(\lambda)$ ($U1$(lambda)) twice to a state is equivalent to do that leaves $|0\rangle$ unchanged and modifies the phase of $|1\rangle$ to give $(e^{\sqrt{-1}\times2\times\lambda})$ $|1\rangle$ to it. For **IBM Q** Experience, the graphical representation of the phase gate $U1(\lambda)$ ($U1$(lambda)) appears in Fig. 1.35.

1.12.1 Programming with the U1(λ) Gate with One Parameter

In Listing 1.12, in the backend *ibmqx4* with five quantum bits in **IBM**'s quantum computer, the program is the *twelfth* example in which we introduce how to program with the phase gate $U1(2 * pi)$ that converts single quantum bit q[0] from one state $(\frac{1}{\sqrt{2}})$ ($|0\rangle + |1\rangle$) to another state $(\frac{1}{\sqrt{2}})$ ($|0\rangle + |1\rangle$). Because the input value of the first parameter lambda for the phase gate $U1$(lambda) is $(2 * pi)$ and $U1(2 * pi)$ is equal to $\begin{pmatrix} 1 & 0 \\ 0 & e^{\sqrt{-1}\times2\times\pi} \end{pmatrix} = \begin{pmatrix} 1 & 0 \\ 0 & 1 \end{pmatrix}$, the phase gate $U1(2 * pi)$ actually implements one identity gate. Figure 1.36 is the corresponding quantum circuit of the program in Listing 1.12.

The statement "OPENQASM 2.0;" on line one of Listing 1.12 is to indicate that the program is written with version 2.0 of Open QASM. Next, the statement "include "qelib1.inc";" on line two of Listing 1.12 is to continue parsing the file "qelib1.inc" as if the contents of the file were pasted at the location of the include statement, where the file "qelib1.inc" is **Quantum Experience (QE) Standard Header** and the path is specified relative to the current working directory. The statement "qreg q[5];" on line three of Listing 1.12 is to declare that in the program there are five

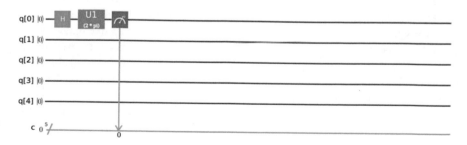

Fig. 1.36 The corresponding quantum circuit of the program in Listing 1.12

quantum bits. In the left top of Fig. 1.36, five quantum bits are subsequently q[0], q[1], q[2], q[3] and q[4]. The initial value of each quantum bit is set to $|0\rangle$. Then, the statement "creg c[5];" on line four of Listing 1.12 is to declare that there are five classical bits in the program. In the left bottom of Fig. 1.36, five classical bits are subsequently c[0], c[1], c[2], c[3] and c[4]. The initial value of each classical bit is set to 0. Classical bit c[4] is the most significant bit and classical bit c[0] is the least significant bit.

Listing. 1.12 Program of using the phase gate $U1(2 * pi)$ with one parameter.

```
1.  OPENQASM 2.0;
2.  include "qelib1.inc";
3.  qreg q[5];
4.  creg c[5];
5.  h q[0];
6.  u1(2 * pi) q[0];
7.  measure q[0] → c[0];
```

The statement "h q[0];" on line five of Listing 1.12 actually implements

$$\begin{pmatrix} \frac{1}{\sqrt{2}} & \frac{1}{\sqrt{2}} \\ \frac{1}{\sqrt{2}} & -\frac{1}{\sqrt{2}} \end{pmatrix} \times \begin{pmatrix} 1 \\ 0 \end{pmatrix} = \begin{pmatrix} \frac{1}{\sqrt{2}} \\ \frac{1}{\sqrt{2}} \end{pmatrix} = \frac{1}{\sqrt{2}} \begin{pmatrix} 1 \\ 1 \end{pmatrix} = \frac{1}{\sqrt{2}} (\begin{pmatrix} 1 \\ 0 \end{pmatrix} + \begin{pmatrix} 0 \\ 1 \end{pmatrix}) = \frac{1}{\sqrt{2}} (|0\rangle +$$

$|1\rangle$). This is to say that the statement "h q[0];" on line five of Listing 1.12 is to apply the Hadamard gate to convert q[0] from one state $|0\rangle$ to another state $\frac{1}{\sqrt{2}}(|0\rangle + |1\rangle)$ (its superposition), where "h" is to represent the Hadamard gate. Next, the statement "u1(2 * pi) q[0];" on line six of Listing 1.12 actually completes $\begin{pmatrix} 1 & 0 \\ 0 & e^{\sqrt{-1} \times 2 \times \pi} \end{pmatrix} \times$

$$\begin{pmatrix} \frac{1}{\sqrt{2}} \\ \frac{1}{\sqrt{2}} \end{pmatrix} = \begin{pmatrix} 1 & 0 \\ 0 & 1 \end{pmatrix} \times \begin{pmatrix} \frac{1}{\sqrt{2}} \\ \frac{1}{\sqrt{2}} \end{pmatrix} = \begin{pmatrix} \frac{1}{\sqrt{2}} \\ \frac{1}{\sqrt{2}} \end{pmatrix} = \frac{1}{\sqrt{2}} (\begin{pmatrix} 1 \\ 0 \end{pmatrix} + \begin{pmatrix} 0 \\ 1 \end{pmatrix}) = \frac{1}{\sqrt{2}} (|0\rangle + |1\rangle).$$

This is to say that the statement "u1(2 * pi) q[0];" on line six of Listing 1.12 is to implement one identity gate to q[0].

Next, the statement "measure q[0] → c[0];" on line seven of Listing 1.12 is to measure the first quantum bit q[0] and to record the measurement outcome by overwriting the first classical bit c[0]. In the backend *ibmqx4* with five quantum bits in **IBM**'s quantum computers, we apply the command "simulate" to execute the program in Listing 1.12. The result appears in Fig. 1.37. From Fig. 1.37, we obtain the answer 00000 (c[4] = 0, c[3] = 0, c[2] = 0, c[1] = 0 and c[0] = 0 = q[0] = $|0\rangle$) with the probability 0.530. Or we obtain the answer 00001 (c[4] = 0, c[3] = 0, c[2] = 0, c[1] = 0 and c[0] = 1 = q[0] = $|1\rangle$) with the probability 0.470.

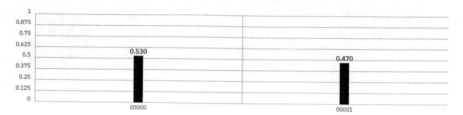

Fig. 1.37 After the measurement to the program in Listing 1.12 is completed, we obtain the answer 00000 with the probability 0.530 or the answer 00001 with the probability 0.470

1.13 The U2(ϕ, λ) Gate of Single Quantum Bit with Two Parameters

The phase gate $U2(\phi, \lambda)$ ($U2$(phi, lambda)) that is the second physical gate of the Quantum Experience and is a phase gate of single quantum bit of two parameters with duration one unit of gate time is

$$U2(\phi, \lambda) = U2(\text{phi, lambda}) = \begin{pmatrix} \frac{1}{\sqrt{2}} & \frac{-e^{\sqrt{-1}\times\lambda}}{\sqrt{2}} \\ \frac{e^{\sqrt{-1}\times\phi}}{\sqrt{2}} & \frac{e^{\sqrt{-1}\times(\lambda+\phi)}}{\sqrt{2}} \end{pmatrix}, \qquad (1.37)$$

where ϕ and λ (phi and lambda) are both real numbers. It is assumed that $(U2(\phi, \lambda))^+$ $((U2(\text{phi, lambda}))^+)$ is the conjugate-transpose matrix of $U2(\phi, \lambda)$ ($U2$(phi, lambda)) and is equal to $((U2(\phi, \lambda))^*)^t = \begin{pmatrix} \frac{1}{\sqrt{2}} & \frac{e^{-\sqrt{-1}\times\phi}}{\sqrt{2}} \\ \frac{-e^{-\sqrt{-1}\times\lambda}}{\sqrt{2}} & \frac{e^{-\sqrt{-1}\times(\lambda+\phi)}}{\sqrt{2}} \end{pmatrix}$, where the $*$ is the complex conjugation and the t is the transpose operation. Because $U2(\phi, \lambda) \times (U2(\phi,$

$$\lambda))^+ = U2(\phi, \lambda) \times ((U2(\phi, \lambda))^*)^t = \begin{pmatrix} \frac{1}{\sqrt{2}} & \frac{-e^{\sqrt{-1}\times\lambda}}{\sqrt{2}} \\ \frac{e^{\sqrt{-1}\times\phi}}{\sqrt{2}} & \frac{e^{\sqrt{-1}\times(\lambda+\phi)}}{\sqrt{2}} \end{pmatrix} \times \begin{pmatrix} \frac{1}{\sqrt{2}} & \frac{e^{-\sqrt{-1}\times\phi}}{\sqrt{2}} \\ \frac{-e^{-\sqrt{-1}\times\lambda}}{\sqrt{2}} & \frac{e^{-\sqrt{-1}\times(\lambda+\phi)}}{\sqrt{2}} \end{pmatrix}$$

$$= (U2(\phi, \lambda))^+ \times U2(\phi, \lambda) = ((U2(\phi, \lambda))^*)^t \times U2(\phi, \lambda) = \begin{pmatrix} \frac{1}{\sqrt{2}} & \frac{e^{-\sqrt{-1}\times\phi}}{\sqrt{2}} \\ \frac{-e^{-\sqrt{-1}\times\lambda}}{\sqrt{2}} & \frac{e^{-\sqrt{-1}\times(\lambda+\phi)}}{\sqrt{2}} \end{pmatrix}$$

$$\times \begin{pmatrix} \frac{1}{\sqrt{2}} & \frac{-e^{\sqrt{-1}\times\lambda}}{\sqrt{2}} \\ \frac{e^{\sqrt{-1}\times\phi}}{\sqrt{2}} & \frac{e^{\sqrt{-1}\times(\lambda+\phi)}}{\sqrt{2}} \end{pmatrix} = \begin{pmatrix} 1 & 0 \\ 0 & 1 \end{pmatrix}, U2(\phi, \lambda)$$ ($U2$(phi, lambda)) is a unitary matrix or

a unitary operator. This implies that the phase gate $U2(\phi, \lambda)$ ($U2$(phi, lambda)) is one of quantum gates that is a phase gate of single quantum bit of two parameters with duration one unit of time. If the quantum state $l_0 |0\rangle + |1\rangle$n is written in a vector notation as

$$\begin{pmatrix} l_0 \\ l_1 \end{pmatrix}, \qquad (1.38)$$

Fig. 1.38 The graphical representation of the phase gate $U2(\phi, \lambda)$

with the first entry l_0 is the amplitude for $|0\rangle$ nand the second entry l_1 is the amplitude for $|1\rangle$, then the corresponding output from the phase gate $U2(\phi, \lambda)$ ($U2(phi, lambda)$) is

$$\begin{pmatrix} \frac{1}{\sqrt{2}} & \frac{-e^{\sqrt{-1}\times\lambda}}{\sqrt{2}} \\ \frac{e^{\sqrt{-1}\times\phi}}{\sqrt{2}} & \frac{e^{\sqrt{-1}\times(\lambda+\phi)}}{\sqrt{2}} \end{pmatrix} \times \begin{pmatrix} l_0 \\ l_1 \end{pmatrix} = \begin{pmatrix} \frac{l_0 - e^{\sqrt{-1}\times\lambda}\times l_1}{\sqrt{2}} \\ \frac{e^{\sqrt{-1}\times\phi}\times l_0 + e^{\sqrt{-1}\times(\lambda+\phi)}\times l_1}{\sqrt{2}} \end{pmatrix}$$

$$= \frac{l_0 - e^{\sqrt{-1}\times\lambda}\times l_1}{\sqrt{2}}|0\rangle + \frac{e^{\sqrt{-1}\times\phi}\times l_0 + e^{\sqrt{-1}\times(\lambda+\phi)}\times l_1}{\sqrt{2}}|1\rangle. \qquad (1.39)$$

This is to say that the phase gate $U2(\phi, \lambda)$ ($U2(phi, lambda)$) converts single quantum bit from one state $l_0|0\rangle + l_1|1\rangle$ to another state $\frac{l_0 - e^{\sqrt{-1}\times\lambda}\times l_1}{\sqrt{2}}|0\rangle + \frac{e^{\sqrt{-1}\times\phi}\times l_0 + e^{\sqrt{-1}\times(\lambda+\phi)}\times l_1}{\sqrt{2}}|1\rangle$. Since $(U2(\phi, \lambda))^2 =$

$$U2(\phi, \lambda) \times U2(\phi, \lambda) = \begin{pmatrix} \frac{1}{\sqrt{2}} & \frac{-e^{\sqrt{-1}\times\lambda}}{\sqrt{2}} \\ \frac{e^{\sqrt{-1}\times\phi}}{\sqrt{2}} & \frac{e^{\sqrt{-1}\times(\lambda+\phi)}}{\sqrt{2}} \end{pmatrix} \times \begin{pmatrix} \frac{1}{\sqrt{2}} & \frac{-e^{\sqrt{-1}\times\lambda}}{\sqrt{2}} \\ \frac{e^{\sqrt{-1}\times\phi}}{\sqrt{2}} & \frac{e^{\sqrt{-1}\times(\lambda+\phi)}}{\sqrt{2}} \end{pmatrix} =$$

$$\begin{pmatrix} \frac{1-e^{\sqrt{-1}\times(\lambda+\phi)}}{2} & \frac{-e^{\sqrt{-1}\times\lambda}\times\left(1+e^{\sqrt{-1}\times(\lambda+\phi)}\right)}{2} \\ \frac{e^{\sqrt{-1}\times\phi}\times\left(1+e^{\sqrt{-1}\times(\lambda+\phi)}\right)}{2} & \frac{\left(1-e^{\sqrt{-1}\times(\lambda+\phi)}\right)\times\left(-e^{\sqrt{-1}\times(\lambda+\phi)}\right)}{2} \end{pmatrix}$$, using the phase gate $U2(\phi,$

$\lambda)$ ($U2(phi, lambda)$) twice to a state is equivalent to modify the amplitude to it. For **IBM Q** Experience, the graphical representation of the phase gate $U2(\phi, \lambda)$ ($U2(phi, lambda)$) appears in Fig. 1.38.

1.13.1   Programming with the $U2(\phi, \lambda)$ Gate with Two Parameters

For the phase gate $U2(phi, lambda)$ the input values of the first parameter phi and the second parameter lambda are respectively (0 * pi) and (1 * pi), so $U2(0*pi, 1*pi)$ is

equal to $\begin{pmatrix} \frac{1}{\sqrt{2}} & \frac{-e^{\sqrt{-1}\times 1\times\pi}}{\sqrt{2}} \\ \frac{e^{\sqrt{-1}\times 0\times\pi}}{\sqrt{2}} & \frac{e^{\sqrt{-1}\times(1\times\pi+0\times\pi)}}{\sqrt{2}} \end{pmatrix} = \begin{pmatrix} \frac{1}{\sqrt{2}} & \frac{1}{\sqrt{2}} \\ \frac{1}{\sqrt{2}} & -\frac{1}{\sqrt{2}} \end{pmatrix}$. Therefore, the phase gate $U2(0*pi,$

$1*pi)$ actually implements one Hadamard gate. In Listing 1.13, in the backend *ibmqx4* with five quantum bits in **IBM**'s quantum computer, the program is the *thirteenth* example in which we illustrate how to program with the phase gate $U2(0*pi, 1*pi)$ that converts single quantum bit q[0] from one state $\left(\frac{1}{\sqrt{2}}\right)(|0\rangle + |1\rangle)$ to another state

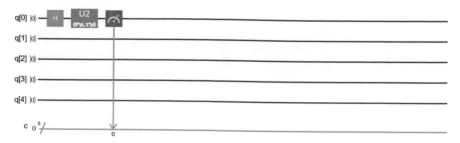

Fig. 1.39 The corresponding quantum circuit of the program in Listing 1.13

($|0\rangle$). Figure 1.39 is the corresponding quantum circuit of the program in Listing 1.13.

The statement "OPENQASM 2.0;" on line one of Listing 1.13 is to point out that the program is written with version 2.0 of Open QASM. Then, the statement "include "qelib1.inc";" on line two of Listing 1.13 is to continue parsing the file "qelib1.inc" as if the contents of the file were pasted at the location of the include statement, where the file "qelib1.inc" is **Quantum Experience (QE) Standard Header** and the path is specified relative to the current working directory. The statement "qreg q[5];" on line three of Listing 1.13 is to declare that in the program there are five quantum bits. In the left top of Fig. 1.39, five quantum bits are subsequently q[0], q[1], q[2], q[3] and q[4]. The initial value of each quantum bit is set to $|0\rangle$. Next, the statement "creg c[5];" on line four of Listing 1.13 is to declare that there are five classical bits in the program. In the left bottom of Fig. 1.39, five classical bits are subsequently c[0], c[1], c[2], c[3] and c[4]. The initial value of each classical bit is set to 0. Classical bit c[4] is the most significant bit and classical bit c[0] is the least significant bit.

Listing. 1.13 nProgram of using the phase gate $U2(0*pi, 1*pi)$ with two parameters.

```
1.  OPENQASM 2.0;
2.  include "qelib1.inc";
3.  qreg q[5];
4.  creg c[5];
5.  h q[0];
6.  u2(0*pi,1*pi) q[0];
7.  measure q[0] → c[0];
```

The statement "h q[0];" on line five of Listing 1.13 actually completes
$$\begin{pmatrix} \frac{1}{\sqrt{2}} & \frac{1}{\sqrt{2}} \\ \frac{1}{\sqrt{2}} & -\frac{1}{\sqrt{2}} \end{pmatrix} \times \begin{pmatrix} 1 \\ 0 \end{pmatrix} = \begin{pmatrix} \frac{1}{\sqrt{2}} \\ \frac{1}{\sqrt{2}} \end{pmatrix} = \frac{1}{\sqrt{2}} \begin{pmatrix} 1 \\ 1 \end{pmatrix} = \frac{1}{\sqrt{2}} \left(\begin{pmatrix} 1 \\ 0 \end{pmatrix} + \begin{pmatrix} 0 \\ 1 \end{pmatrix} \right) = \frac{1}{\sqrt{2}} (|0\rangle + |1\rangle).$$
This is to say that the statement "h q[0];" on line five of Listing 1.13 is to

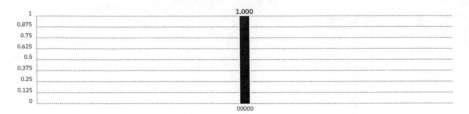

Fig. 1.40 After the measurement to the program in Listing 1.13 is completed, we obtain the answer 00000 with the probability 1.000

use the Hadamard gate to convert q[0] from one state $|0\rangle$ to another state $\frac{1}{\sqrt{2}}$ $(|0\rangle + |1\rangle)$ (its superposition), where "h" is to represent the Hadamard gate. Next, the statement "u2(0*pi,1*pi) q[0];" on line six of Listing 1.13 actually implements $\begin{pmatrix} \frac{1}{\sqrt{2}} & \frac{-e^{\sqrt{-1}\times 1\times \pi}}{\sqrt{2}} \\ \frac{e^{\sqrt{-1}\times 0\times \pi}}{\sqrt{2}} & \frac{e^{\sqrt{-1}\times(1\times\pi+0\times\pi)}}{\sqrt{2}} \end{pmatrix} \times \begin{pmatrix} \frac{1}{\sqrt{2}} \\ \frac{1}{\sqrt{2}} \end{pmatrix} = \begin{pmatrix} \frac{1}{\sqrt{2}} & \frac{1}{\sqrt{2}} \\ \frac{1}{\sqrt{2}} & -\frac{1}{\sqrt{2}} \end{pmatrix} \times \begin{pmatrix} \frac{1}{\sqrt{2}} \\ \frac{1}{\sqrt{2}} \end{pmatrix} = \begin{pmatrix} 1 \\ 0 \end{pmatrix} = |0\rangle$. This indicates that the statement "u2(0*pi,1*pi) q[0];" on line six of Listing 1.13 is to complete one Hadamard gate to q[0]. Therefore, applying the Hadamard gate twice from line five and line six of Listing 1.13 to q[0] does nothing to it.

Next, the statement "measure q[0] → c[0];" on line seven of Listing 1.13 is to measure the first quantum bit q[0] and to record the measurement outcome by overwriting the first classical bit c[0]. In the backend *ibmqx4* with five quantum bits in **IBM**'s quantum computers, we use the command "simulate" to execute the program in Listing 1.13. The result appears in Fig. 1.40. From Fig. 1.40, we obtain the answer 00000 (c[4] = 0, c[3] = 0, c[2] = 0, c[1] = 0 and c[0] = 0 = q[0] = $|0\rangle$) with the probability 1.000.

1.14   The U3(θ, φ, λ) Gate of Single Quantum Bit of Three Parameters

The phase gate $U3(\theta, \phi, \lambda)$ ($U3$(theta, phi, lambda)) that is the third physical gate of the Quantum Experience and is a phase gate of single quantum bit of three parameters with duration two units of gate time is

$$U3(\theta, \phi, \lambda) = \begin{pmatrix} \cos\left(\frac{\theta}{2}\right) & -e^{\sqrt{-1}\times\lambda} \times \sin\left(\frac{\theta}{2}\right) \\ e^{\sqrt{-1}\times\phi} \times \sin\left(\frac{\theta}{2}\right) & e^{\sqrt{-1}\times(\lambda+\phi)} \times \cos\left(\frac{\theta}{2}\right) \end{pmatrix}, \quad (1.40)$$

where θ, φ and λ (theta, phi and lambda) are all real numbers. It is supposed that $(U3(\theta, \phi, \lambda))^+$ $((U3$(theta, phi, lambda))$^+)$ is the conjugate-transpose matrix of $U3(\theta, \phi, \lambda)$ ($U3$(theta, phi, lambda)) and is equal to

$$((U3(\theta, \phi, \lambda))^*)^t = \begin{pmatrix} \cos\left(\frac{\theta}{2}\right) & e^{-\sqrt{-1}\times\phi} \times \sin\left(\frac{\theta}{2}\right) \\ -e^{-\sqrt{-1}\times\lambda} \times \sin\left(\frac{\theta}{2}\right) & e^{-\sqrt{-1}\times(\lambda+\phi)} \times \cos\left(\frac{\theta}{2}\right) \end{pmatrix},$$ where the

* is the complex conjugation and the t is the transpose operation. Because

$$U3(\theta, \phi, \lambda) \times ((U3(\theta, \phi, \lambda))^*)^t = \begin{pmatrix} \cos\left(\frac{\theta}{2}\right) & -e^{\sqrt{-1}\times\lambda} \times \sin\left(\frac{\theta}{2}\right) \\ e^{\sqrt{-1}\times\phi} \times \sin\left(\frac{\theta}{2}\right) & e^{\sqrt{-1}\times(\lambda+\phi)} \times \cos\left(\frac{\theta}{2}\right) \end{pmatrix}$$

$$\times \begin{pmatrix} \cos\left(\frac{\theta}{2}\right) & e^{-\sqrt{-1}\times\phi} \times \sin\left(\frac{\theta}{2}\right) \\ -e^{-\sqrt{-1}\times\lambda} \times \sin\left(\frac{\theta}{2}\right) & e^{-\sqrt{-1}\times(\lambda+\phi)} \times \cos\left(\frac{\theta}{2}\right) \end{pmatrix} = ((U3(\theta, \phi, \lambda))^*)^t$$

$$\times \quad U3(\theta, \phi, \lambda) = \begin{pmatrix} \cos\left(\frac{\theta}{2}\right) & e^{-\sqrt{-1}\times\phi} \times \sin\left(\frac{\theta}{2}\right) \\ -e^{-\sqrt{-1}\times\lambda} \times \sin\left(\frac{\theta}{2}\right) & e^{-\sqrt{-1}\times(\lambda+\phi)} \times \cos\left(\frac{\theta}{2}\right) \end{pmatrix} \times$$

$$\begin{pmatrix} \cos\left(\frac{\theta}{2}\right) & -e^{\sqrt{-1}\times\lambda} \times \sin\left(\frac{\theta}{2}\right) \\ e^{\sqrt{-1}\times\phi} \times \sin\left(\frac{\theta}{2}\right) & e^{\sqrt{-1}\times(\lambda+\phi)} \times \cos\left(\frac{\theta}{2}\right) \end{pmatrix} = \begin{pmatrix} 1 & 0 \\ 0 & 1 \end{pmatrix},$$ $U3(\theta, \phi, \lambda)$ ($U3$(theta,

phi, lambda)) is a unitary matrix or a unitary operator. This implies that the phase gate $U3(\theta, \phi, \lambda)$ ($U3$(theta, phi, lambda)) is one of quantum gates that is a phase gate of single quantum bit of three parameters with duration two units of gate time. If the quantum state $l_0\,|0\rangle + l_1\,|1\rangle$ is written in a vector notation as

$$\begin{pmatrix} l_0 \\ l_1 \end{pmatrix}, \tag{1.41}$$

with the first entry l_0 is the amplitude for $|0\rangle$ and the second entry l_1 is the amplitude for $|1\rangle$, then the corresponding output from the phase gate $U3(\theta, \phi, \lambda)$ ($U3$(theta, phi, lambda)) is

$$\begin{pmatrix} \cos\left(\frac{\theta}{2}\right) & -e^{\sqrt{-1}\times\lambda} \times \sin\left(\frac{\theta}{2}\right) \\ e^{\sqrt{-1}\times\phi} \times \sin\left(\frac{\theta}{2}\right) & e^{\sqrt{-1}\times(\lambda+\phi)} \times \cos\left(\frac{\theta}{2}\right) \end{pmatrix} \times \begin{pmatrix} l_0 \\ l_1 \end{pmatrix}$$

$$= \begin{pmatrix} l_0 \times \cos\left(\frac{\theta}{2}\right) - l_1 \times e^{\sqrt{-1}\times\lambda} \times \sin\left(\frac{\theta}{2}\right) \\ l_0 \times e^{\sqrt{-1}\times\phi} \times \sin\left(\frac{\theta}{2}\right) + l_1 \times e^{\sqrt{-1}\times(\lambda+\phi)} \times \cos\left(\frac{\theta}{2}\right) \end{pmatrix}$$

$$= \left(l_0 \times \cos\left(\frac{\theta}{2}\right) - l_1 \times e^{\sqrt{-1}\times\lambda} \times \sin\left(\frac{\theta}{2}\right) \right) |0\rangle$$

$$+ \left(l_0 \times e^{\sqrt{-1}\times\phi} \times \sin\left(\frac{\theta}{2}\right) + l_1 \times e^{\sqrt{-1}\times(\lambda+\phi)} \times \cos\left(\frac{\theta}{2}\right) \right) |1\rangle. \tag{1.42}$$

This indicates that the phase gate $U3(\theta, \phi, \lambda)$ ($U3$(theta, phi, lambda)) converts single quantum bit from one state $l_0\,|0\rangle + l_1\,|1\rangle$ to another state $\left(l_0 \times \cos\left(\frac{\theta}{2}\right) - l_1 \times e^{\sqrt{-1}\times\lambda} \times \sin\left(\frac{\theta}{2}\right) \right)$ $|0\rangle$ + $(l_0 \times e^{\sqrt{-1}\times\phi} \times$ $\sin\left(\frac{\theta}{2}\right) + l_1 \times e^{\sqrt{-1}\times(\lambda+\phi)} \times \cos\left(\frac{\theta}{2}\right))$ $|1\rangle$. Because $(U3(\theta, \phi, \lambda))^2 =$

Fig. 1.41 The graphical representation of the phase gate $U3(\theta, \phi, \lambda)$

$$\begin{pmatrix} \cos^2\left(\dfrac{\theta}{2}\right) - e^{\sqrt{-1}\times(\lambda+\phi)} \times \sin^2\left(\dfrac{\theta}{2}\right) \\[8pt] -\cos\left(\dfrac{\theta}{2}\right) \times \sin\left(\dfrac{\theta}{2}\right) \times e^{\sqrt{-1}\times\lambda} \times \left(1 + e^{\sqrt{-1}\times(\lambda+\phi)}\right) \\[8pt] \cos\left(\dfrac{\theta}{2}\right) \times \sin\left(\dfrac{\theta}{2}\right) \times e^{\sqrt{-1}\times\phi} \times \left(1 + e^{\sqrt{-1}\times(\lambda+\phi)}\right) \\[8pt] \left(\sin^2\left(\dfrac{\theta}{2}\right) - \cos^2\left(\dfrac{\theta}{2}\right)\right) \times e^{\sqrt{-1}\times(\lambda+\phi)} \times \left(-e^{\sqrt{-1}\times(\lambda+\phi)}\right) \end{pmatrix},$$ applying the

phase gate $U3(\theta, \phi, \lambda)$ ($U3$(theta, phi, lambda)) twice to a state is equivalent to modify the amplitude to it. For **IBM Q** Experience, the graphical representation of the phase gate $U3(\theta, \phi, \lambda)$ ($U3$(theta, phi, lambda)) appears in Fig. 1.41.

1.14.1 Programming with the U3(θ, ϕ, λ) Gate with Three Parameters

For the phase gate $U3$(theta, phi, lambda), the input value of the first parameter theta is (0.5 * pi), the input value of the second parameter phi is (0 * pi) and the input value of the third parameter lambda is (1 * pi), so $U3$(0.5*pi, 0*pi, 1*pi) is equal

to $\begin{pmatrix} \cos\left(\frac{\pi}{4}\right) & -e^{\sqrt{-1}\times1\times\pi} \times \sin\left(\frac{\pi}{4}\right) \\ e^{\sqrt{-1}\times0\times\pi} \times \sin\left(\frac{\pi}{4}\right) & e^{\sqrt{-1}\times(1\times\pi+0\times\pi)} \times \cos\left(\frac{\pi}{4}\right) \end{pmatrix} = \begin{pmatrix} \frac{1}{\sqrt{2}} & \frac{-e^{\sqrt{-1}\times1\times\pi}}{\sqrt{2}} \\ \frac{e^{\sqrt{-1}\times0\times\pi}}{\sqrt{2}} & \frac{e^{\sqrt{-1}\times(1\times\pi+0\times\pi)}}{\sqrt{2}} \end{pmatrix} = $

$\begin{pmatrix} \frac{1}{\sqrt{2}} & \frac{1}{\sqrt{2}} \\ \frac{1}{\sqrt{2}} & -\frac{1}{\sqrt{2}} \end{pmatrix}$. Hence, the phase gate $U3$(0.5*pi, 0*pi, 1*pi) actually completes one

Hadamard gate. In Listing 1.14, in the backend *ibmqx4* with five quantum bits in **IBM**'s quantum computer, the program is the *fourteenth* example in which we describe how to program with the phase gate $U3$(0.5*pi, 0*pi, 1*pi) that converts single quantum bit q[0] from one state $\left(\frac{1}{\sqrt{2}}\right)$ ($|0\rangle + |1\rangle$) to another state ($|0\rangle$). Figure 1.42 is the corresponding quantum circuit of the program in Listing 1.14.

The statement "OPENQASM 2.0;" on line one of Listing 1.14 is to indicate that the program is written with version 2.0 of Open QASM. Next, the statement "include "qelib1.inc";" on line two of Listing 1.14 is to continue parsing the file "qelib1.inc" as if the contents of the file were pasted at the location of the include statement, where the file "qelib1.inc" is **Quantum Experience (QE) Standard Header** and the path is specified relative to the current working directory. The statement "qreg q[5];" on line three of Listing 1.14 is to declare that in the program there are five quantum bits. In the left top of Fig. 1.42, five quantum bits are subsequently q[0],

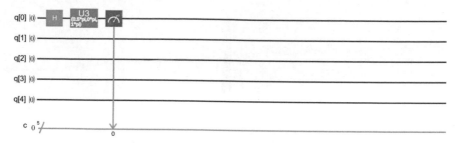

Fig. 1.42 The corresponding quantum circuit of the program in Listing 1.14

q[1], q[2], q[3] and q[4]. The initial value of each quantum bit is set to $|0\rangle$. Next, the statement "creg c[5];" on line four of Listing 1.14 is to declare that there are five classical bits in the program. In the left bottom of Fig. 1.42, five classical bits are subsequently c[0], c[1], c[2], c[3] and c[4]. The initial value of each classical bit is set to zero (0). Classical bit c[4] is the most significant bit and classical bit c[0] is the least significant bit.

Listing. 1.14 nProgram of using the phase gate $U3(0.5*pi, 0*pi, 1*pi)$ of three parameters.

```
1.  OPENQASM 2.0;
2.  include "qelib1.inc";
3.  qreg q[5];
4.  creg c[5];
5.  h q[0];
6.  u3(0.5*pi,0*pi,1*pi) q[0];
7.  measure q[0] → c[0];
```

The statement "h q[0];" on line five of Listing 1.14 actually performs $\begin{pmatrix} \frac{1}{\sqrt{2}} & \frac{1}{\sqrt{2}} \\ \frac{1}{\sqrt{2}} & -\frac{1}{\sqrt{2}} \end{pmatrix}$

$\times \begin{pmatrix} 1 \\ 0 \end{pmatrix} = \begin{pmatrix} \frac{1}{\sqrt{2}} \\ \frac{1}{\sqrt{2}} \end{pmatrix} = \frac{1}{\sqrt{2}} \begin{pmatrix} 1 \\ 1 \end{pmatrix} = \frac{1}{\sqrt{2}} \left(\begin{pmatrix} 1 \\ 0 \end{pmatrix} + \begin{pmatrix} 0 \\ 1 \end{pmatrix} \right) = \frac{1}{\sqrt{2}} (|0\rangle + |1\rangle)$. This

indicates that the statement "h q[0];" on line five of Listing 1.14 is to apply the Hadamard gate to convert q[0] from one state $|0\rangle$ to another state $\frac{1}{\sqrt{2}} (|0\rangle + |1\rangle)$ (its superposition), where "h" is to represent the Hadamard gate. Next, the statement "u3(0.5*pi,0*pi,1*pi) q[0];"

on   line   six   of   Listing   1.14   actually   completes

$\begin{pmatrix} \cos\left(\frac{\pi}{4}\right) & -e^{\sqrt{-1}\times 1\times\pi} \times \sin\left(\frac{\pi}{4}\right) \\ e^{\sqrt{-1}\times 0\times\pi} \times \sin\left(\frac{\pi}{4}\right) & e^{\sqrt{-1}\times(1\times\pi+0\times\pi)} \times \cos\left(\frac{\pi}{4}\right) \end{pmatrix}$ $\times$ $\begin{pmatrix} \frac{1}{\sqrt{2}} \\ \frac{1}{\sqrt{2}} \end{pmatrix}$ $=$

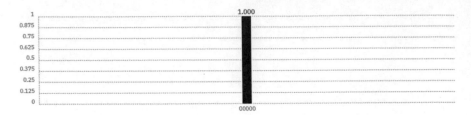

Fig. 1.43 After the measurement to the program in Listing 1.14 is completed, we obtain the answer 00000 with the probability 1.000

$$\begin{pmatrix} \frac{1}{\sqrt{2}} & \frac{-e^{\sqrt{-1}\times 1\times\pi}}{\sqrt{2}} \\ \frac{e^{\sqrt{-1}\times 0\times\pi}}{\sqrt{2}} & \frac{e^{\sqrt{-1}\times(1\times\pi+0\times\pi)}}{\sqrt{2}} \end{pmatrix} \times \begin{pmatrix} \frac{1}{\sqrt{2}} \\ \frac{1}{\sqrt{2}} \end{pmatrix} = \begin{pmatrix} \frac{1}{\sqrt{2}} & \frac{1}{\sqrt{2}} \\ \frac{1}{\sqrt{2}} & -\frac{1}{\sqrt{2}} \end{pmatrix} \times \begin{pmatrix} \frac{1}{\sqrt{2}} \\ \frac{1}{\sqrt{2}} \end{pmatrix} = \begin{pmatrix} 1 \\ 0 \end{pmatrix} = |0\rangle.$$

This is to say that the statement "u3(0.5*pi,0*pi,1*pi) q[0];" on line six of Listing 1.14 is to complete one Hadamard gate to q[0]. Hence, using the Hadamard gate twice from line five and line six of Listing 1.14 to q[0] does nothing to it.

Next, the statement "measure q[0] → c[0];" on line seven of Listing 1.14 is to measure the first quantum bit q[0] and to record the measurement outcome by overwriting the first classical bit c[0]. In the backend *ibmqx4* with five quantum bits in **IBM**'s quantum computers, we use the command "simulate" to execute the program in Listing 1.14. The result appears in Fig. 1.43. From Fig. 1.43, we obtain the answer 00000 ($c[4] = 0$, $c[3] = 0$, $c[2] = 0$, $c[1] = 0$ and $c[0] = 0 = q[0] = |0\rangle$) with the probability 1.000.

1.15   Summary

In this chapter, we introduced single quantum bit, multiple quantum bits and their superposition. We also described two statements of declaration and measurement for quantum bits and classical bits in Open QASM (version 2.0) in the backend *ibmqx4* with five quantum bits in **IBM**'s quantum computers. We illustrated all of the quantum gates with single quantum bit and the ***controlled-NOT*** or ***CNOT*** gate of two quantum bits in the backend *ibmqx4* with five quantum bits in **IBM**'s quantum computers. Simultaneously, we also in detail introduced connectivity of the ***controlled-NOT*** gate in the backend *ibmqx4* with five quantum bits in **IBM**'s quantum computers. We introduced how to program with each quantum gate of single quantum bit completing each different kind of application and how to execute each quantum program in the backend *ibmqx4* with five quantum bits in **IBM**'s quantum computers. We also described how to program with the ***controlled-NOT*** gate to implement a copy machine of one bit.

1.16   Bibliographical Notes

A famous article that gives a detailed technical definition for quantum supremacy is (Aaronson and Chen 2017). Popular textbooks (Nielsen and Chuang 2000; Imre and Balazs 2007; Lipton and Regan 2014) give an excellent introduction for quantum bits and quantum gates. A popular textbook (Silva 2018), a famous project (**IBM Q** 2016) and two famous articles (Cross et al 2017; Coles et al 2018) give many excellent examples to write quantum programs with quantum assembly language in Open QASM (version 2.0) in the backend *ibmqx4* with five quantum bits in **IBM**'s quantum computers.

1.17   Exercises

1.1   Please compute the values of three parameters θ, ϕ, and λ in the U3(θ, ϕ, λ) gate so that the U3(θ, ϕ, λ) gate with three parameters is equivalent to the *NOT* gate.

1.2   Please calculate the values of three parameters θ, ϕ, and λ in the U3(θ, ϕ, λ) gate such that the U3(θ, ϕ, λ) gate with three parameters is equivalent to the Hadamard gate.

1.3   Please figure out the values of three parameters θ, ϕ, and λ in the U3(θ, ϕ, λ) gate so that the U3(θ, ϕ, λ) gate with three parameters is equivalent to the *Z* gate.

1.4   Please determine the values of three parameters θ, ϕ, and λ in the U3(θ, ϕ, λ) gate such that the U3(θ, ϕ, λ) gate with three parameters is equivalent to the *Y* gate.

1.5   Please compute the values of three parameters θ, ϕ, and λ in the U3(θ, ϕ, λ) gate so that the U3(θ, ϕ, λ) gate with three parameters is equivalent to the *S* gate.

1.6   Please calculate the values of three parameters θ, ϕ, and λ in the U3(θ, ϕ, λ) gate such that the U3(θ, ϕ, λ) gate with three parameters is equivalent to the S^+ gate.

1.7   Please figure out the values of three parameters θ, ϕ, and λ in the U3(θ, ϕ, λ) gate so that the U3(θ, ϕ, λ) gate with three parameters is equivalent to the *T* gate.

1.8   Please determine the values of three parameters θ, ϕ, and λ in the U3(θ, ϕ, λ) gate such that the U3(θ, ϕ, λ) gate with three parameters is equivalent to the T^+ gate.

1.9   Please compute the values of three parameters θ, ϕ, and λ in the U3(θ, ϕ, λ) gate so that the U3(θ, ϕ, λ) gate with three parameters is equivalent to the *identity* gate.

1.10 Please calculate the values of three parameters θ, ϕ, and λ in the $U3(\theta, \phi, \lambda)$ gate such that the $U3(\theta, \phi, \lambda)$ gate with three parameters is equivalent to the $U1(\lambda)$ gate with one parameter.

1.11 Please figure out the values of three parameters θ, ϕ, and λ in the $U3(\theta, \phi, \lambda)$ gate so that the $U3(\theta, \phi, \lambda)$ gate with three parameters is equivalent to the $U2(\phi, \lambda)$ gate with two parameters.

References

Aaronson, S., Chen, L.: Complexity-theoretic foundations of quantum supremacy experiments. In: The 32nd Computational Complexity Conference, vol. 79, pp. 22:1–22:67 (2017)

Adleman, L.M.: Molecular computation of solutions to combinatorial problems. Science **226**, 10211024 (1994)

Coles, P.J., Eidenbenz, S., Pakin, S., Adedoyin, A., Ambrosiano, J., Anisimov, P., Casper, W., Chennupati, G., Coffrin, C., Djidjev, H., Gunter, D., Karra, S., Lemons, N., Lin, S., Lokhov, A., Malyzhenkov, A.,Mascarenas, D., Mniszewski, S., Nadiga, B., O'Malley, D., Oyen, D., Prasad, L., Roberts, R., Romero, P., Santhi, N., Sinitsyn, N., Swart, P., Vuffray, M., Wendelberger, J., Yoon, B., Zamora, R., Zhu, W.: Quantum algorithm implementations for beginners. https://arxiv.org/abs/1804.03719 (2018)

Cross, A.W., Bishop, L.S., Smolin, J. A., Gambetta, J.M.: Open quantum assembly language. https://arxiv.org/abs/1707.03429 (2017).

Deutsch, D.: Quantum theory, the Church-turing principle and the universal quantum computer. In: The Proceedings of Royal Society London A (1985), pp. 400497.

Imre, S., Balazs, F.: Quantum Computation and Communications: An Engineering Approach. Wiley, UK (2007). ISBN-10: 047086902X and ISBN-13: 978-0470869024, 2005.

Lipton, R.J., Regan, K.W.: Quantum Algorithms via Linear Algebra: a Primer. The MIT Press (2014). ISBN 978-0-262-02839-4.

Nielsen, M.A., Chuang, I.L.: Quantum Computation and Quantum Information. Cambridge University Press, New York, NY (2000). ISBN-10: 9781107002173 and ISBN-13: 978-1107002173.

Silva, V.: Practical Quantum Computing for Developers: Programming Quantum Rigs in the Cloud using Python, Quantum Assembly Language and IBM Q Experience. Apress, December 13, 2018, ISBN-10: 1484242173 and ISBN-13: 978-1484242179.

The IBM Quantum Experience. https://www.research.ibm.com/quantum/, accessed November 2016.

Turing, A.: On computable numbers, with an application to the entscheidungsproblem. In: The Proceedings of the London Mathematical Society, Ser. 2, vol. 42 (19367), pp. 230265; corrections, Ibid, vol. 43, pp. 544546 (1937)

von Neumann, J.: Probabilistic Logics and the Synthesis of Reliable Organisms from Unreliable Components, pp. 329378. Princeton University Press (1956)

Chapter 2
Boolean Algebra and Its Applications

Using a field of mathematics called *modern algebra* designs and maintains *classical* computers. Over a hundred years, algebraists have studied for mathematical systems and they call them as *Boolean algebras*. The name *Boolean algebra* honors a fascinating English mathematician, George Boole because in 1854 he published a *classic* book, *An Investigation of the Laws of Thought, on which Are Founded Mathematical Theories of Logic and Probabilities*. Boole that has stated objective (intention) was to complete a mathematical analysis of logic. Boole's investigation has enlightened the *calculus of propositions* and *algebra of sets*. In this book, we designate the algebra now used in the design and maintenance of quantum logical circuitry as *Boolean algebra*.

There are several advantages in having a mathematical technique for the illustration of the internal workings of a quantum algorithm (circuit) for solving each different kind of applications in **IBM**'s quantum computers. The first advantage is to that it is often far more convenient to calculate with algebraic expressions used to describe the internal workings of a quantum algorithm (circuit) than it is to apply schematic or even logical diagrams. The second advantage is to that an ordinary algebraic expression that describes the internal workings of a quantum algorithm (circuit) may be reduced or simplified. This enables that the designer of quantum algorithms (circuits) achieves economy of construction and reliability of quantum operation. Boolean algebra also provides an economical and straightforward way of designing quantum algorithms (circuits) for solving each different kind of applications. In all, a knowledge of Boolean algebra is indispensable in the computing field. In this chapter, we describe how to complete logic operations that appear in Fig. 2.1 and include **NOT, AND, NAND, OR, NOR, Exclusive-OR (XOR)** and **Exclusive-NOR (XNOR)** with quantum logic gates in the backend *ibmqx4* or a simulator in **IBM**'s quantum computers. We also illustrate how to complete several applications from Boolean algebra.

© The Author(s), under exclusive license to Springer Nature Switzerland AG 2021
W.-L. Chang and A. V. Vasilakos, *Fundamentals of Quantum Programming in IBM's Quantum Computers*, Studies in Big Data 81,
https://doi.org/10.1007/978-3-030-63583-1_2

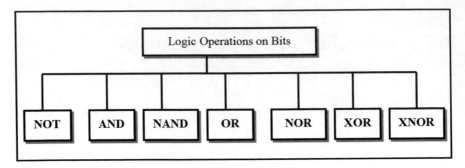

Fig. 2.1 Logic operations on bits

2.1 Illustration to NOT Operation

The **NOT** operation acquires a single input and yields one single output. It inverts the value of a bit into the one's complement of the bit. This is to say that the **NOT** operation for a bit provides the following result:

$$\textbf{NOT}\,1 = 0$$
$$\textbf{NOT}\,0 = 1 \tag{2.1}$$

The value of a Boolean variable (a bit) is only zero (0) or one (1). Therefore, **NOT** of a Boolean variable (a bit) q[0], written as $\overline{q[0]}$ is equal to one (1) if and only if q[0] is equal to zero (0). Similarly, $\overline{q[0]}$ is equal to zero (0) if and only if q[0] is equal to one (1). The rules in (2.1) for the **NOT** operation may also be expressed in the form of a truth table that appears in Table 2.1.

From (2.1) and Table 2.1, the **NOT** operation of a bit is to invert the value of the bit into its one's complement. The **NOT** operation of n bits is to provide the corresponding one's complement for each input in n inputs by means of implementing the **NOT** operation of a bit of n times. The following subsections will be used to illustrate how to design the quantum programs to complete the **NOT** operation of a bit and the **NOT** operation of two bits.

Table 2.1 The truth table for the **NOT** operation

Input	Output
q[0]	$\overline{q[0]}$
0	1
1	0

2.1.1 Quantum Program to the One's Complement of a Bit

Consider that two values for unsigned integer of one bit are, respectively, $0(0_{10})$ and $1(1_{10})$, where 0_{10} is the decimal representation of zero and 1_{10} is the decimal representation of one. We want to implement simultaneously the one's complement of those two values.

In Listing 2.1, the program in the backend *ibmqx4* with five quantum bits in **IBM**'s quantum computer is the *first* example of the *second* chapter in which we illustrate how to write a quantum program to invert $0(0_{10})$ and $1(1_{10})$ into their one's complement. Figure 2.2 is the corresponding quantum circuit of the program in Listing 2.1. The statement "OPENQASM 2.0;" on line one of Listing 2.1 is to indicate that the program is written with version 2.0 of Open QASM. Then, the statement "include "qelib1.inc";" on line two of Listing 2.1 is to continue parsing the file "qelib1.inc" as if the contents of the file were pasted at the location of the include statement, where the file "qelib1.inc" is **Quantum Experience (QE) Standard Header** and the path is specified relative to the current working directory.

Listing 2.1 The program of taking the one's complement to the input of a bit.

```
1.  OPENQASM 2.0;
2.  include "qelib1.inc";
3.  qreg q[5];
4.  creg c[5];
5.  h q[0];
6.  x q[0];
7.  measure q[0] → c[0];
```

Next, the statement "qreg q[5];" on line three of Listing 2.1 is to declare that in the program there are five quantum bits. In the left top of Fig. 2.2, five quantum bits are subsequently q[0], q[1], q[2], q[3] and q[4]. The initial value of each quantum bit is set to $|0\rangle$. We use a quantum bit q[0] to encode the input of a bit that is unsigned

Fig. 2.2 The corresponding quantum circuit of the program in Listing 2.1

integer of a bit. Next, the statement "creg c[5];" on line four of Listing 2.1 is to declare that there are five classical bits in the program. In the left bottom of Fig. 2.2, five classical bits are respectively c[0], c[1], c[2], c[3] and c[4]. The initial value of each classical bit is set to zero (0). Classical bit c[4] is the most significant bit and classical bit c[0] is the least significant bit.

Next, the statement "h q[0];" on line five of Listing 2.1 actually completes

$$\begin{pmatrix} \frac{1}{\sqrt{2}} & \frac{1}{\sqrt{2}} \\ \frac{1}{\sqrt{2}} & -\frac{1}{\sqrt{2}} \end{pmatrix} \times \begin{pmatrix} 1 \\ 0 \end{pmatrix} = \begin{pmatrix} \frac{1}{\sqrt{2}} \\ \frac{1}{\sqrt{2}} \end{pmatrix} = \frac{1}{\sqrt{2}} \begin{pmatrix} 1 \\ 1 \end{pmatrix} = \frac{1}{\sqrt{2}} \left(\begin{pmatrix} 1 \\ 0 \end{pmatrix} + \begin{pmatrix} 0 \\ 1 \end{pmatrix} \right) = \frac{1}{\sqrt{2}}(|0\rangle + |1\rangle).$$

This is to say that the statement "h q[0];" on line five of Listing 2.1 is to apply the Hadamard gate to convert q[0] from one state $|0\rangle$ to another state $\frac{1}{\sqrt{2}}(|0\rangle + |1\rangle)$ (its superposition). In its superposition, $|0\rangle$ with the amplitude $\frac{1}{\sqrt{2}}$ encodes the value zero (0) to the input of a bit and $|1\rangle$ with the amplitude $\frac{1}{\sqrt{2}}$ encodes the value 1 (one) to the input of a bit. Next, the statement "x q[0];" on line six of Listing 2.1 actually completes $\begin{pmatrix} 0 & 1 \\ 1 & 0 \end{pmatrix} \times \begin{pmatrix} \frac{1}{\sqrt{2}} \\ \frac{1}{\sqrt{2}} \end{pmatrix} = \begin{pmatrix} \frac{1}{\sqrt{2}} \\ \frac{1}{\sqrt{2}} \end{pmatrix} = \frac{1}{\sqrt{2}} \begin{pmatrix} 1 \\ 1 \end{pmatrix} = \frac{1}{\sqrt{2}} \left(\begin{pmatrix} 0 \\ 1 \end{pmatrix} + \begin{pmatrix} 1 \\ 0 \end{pmatrix} \right) = \frac{1}{\sqrt{2}}(|1\rangle + |0\rangle).$ This indicates that the statement "x q[0];" on line six of Listing 2.1 inverts $|0\rangle$ with the amplitude $\frac{1}{\sqrt{2}}$ (the input zero of a bit) into $|1\rangle$ with the amplitude $\frac{1}{\sqrt{2}}$ (its corresponding one's complement) and inverts $|1\rangle$ with the amplitude $\frac{1}{\sqrt{2}}$ (the input one of a bit) into $|0\rangle$ with the amplitude $\frac{1}{\sqrt{2}}$ (its corresponding one's complement). This also implies that two instructions (two **NOT** operations) of taking one's complement to the input of a bit are completed by means of using one quantum instruction "x q[0];".

Next, the statement "measure q[0] $\rightarrow$ c[0];" on line seven of Listing 2.1 is to measure the first quantum bit q[0] and to record the measurement outcome by overwriting the first classical bit c[0]. In the backend *ibmqx4* with five quantum bits in **IBM**'s quantum computers, we use the command "simulate" to execute the program in Listing 2.1. The measured result appears in Fig. 2.3. From Fig. 2.3, we obtain the answer 00001 (c[4] = 0, c[3] = 0, c[2] = 0, c[1] = 0 and c[0] = 1 = q[0] = |1\rangle) with the probability 0.530. This is to say that we obtain the one's complement (q[0] = |1\rangle) with the probability 0.530 to the input zero (0) of a bit. Or we obtain the answer 00000 with the probability 0.470 (c[4] = 0, c[3] = 0, c[2] = 0, c[1] = 0 and c[0] =

Fig. 2.3 After the measurement to the program in Listing 2.1 is completed, we obtain the answer 00001 with the probability 0.530 or the answer 00000 with the probability 0.470

$0 = q[0] = |0\rangle$). This is also to say that we obtain the one's complement ($q[0] = |0\rangle$) with the probability 0.470 to the input one (1) of a bit.

2.1.2   Quantum Program to the One's Complement of Two Bits

Consider that four values to unsigned integer of two bits are subsequently $00(0_{10})$, $01(1_{10})$, $10 (2_{10})$ and $11(3_{10})$, where 0_{10} is the decimal representation of zero, 1_{10} is the decimal representation of one, 2_{10} is the decimal representation of two and 3_{10} is the decimal representation of three. We want to implementsimultaneously the one's complement of those four values.

In Listing 2.2, the program in the backend *ibmqx4* with five quantum bits in **IBM**'s quantum computer is the *second* example of the *second* chapter in which we describe how to write a quantum program to take the one's complement of $00(0_{10})$, $01(1_{10})$, $10 (2_{10})$ and $11 (3_{10})$. Figure 2.4 is the corresponding quantum circuit of the program in Listing 2.2. The statement "OPENQASM 2.0;" on line one of Listing 2.2 is to point to that the program is written with version 2.0 of Open QASM. Next, the statement "include "qelib1.inc";" on line two of Listing 2.2 is to continue parsing the file "qelib1.inc" as if the contents of the file were pasted at the location of the include statement, where the file "qelib1.inc" is **Quantum Experience (QE) Standard Header** and the path is specified relative to the current working directory.

Listing 2.2 The program of taking the one's complement to the input of two bits.

```
1.   OPENQASM 2.0;
2.   include "qelib1.inc";
3.   qreg q[5];
4.   creg c[5];
5.   h q[0];
6.   h q[1];
```

Fig. 2.4 The corresponding quantum circuit of the program in Listing 2.2

```
 7.   x q[0];
 8.   x q[1];
 9.   measure q[0] → c[0];
10.   measure q[1] → c[1];
```

Then, the statement "qreg q[5];" on line three of Listing 2.2 is to declare that in the program there are five quantum bits. In the left top of Fig. 2.4, five quantum bits are subsequently q[0], q[1], q[2], q[3] and q[4]. The initial value of each quantum bit is set to $|0\rangle$. We use two quantum bits q[0] and q[1] to encode the input of two bits that are unsigned integer of two bits. Next, the statement "creg c[5];" on line four of Listing 2.2 is to declare that there are five classical bits in the program. In the left bottom of Fig. 2.4, five classical bits are respectively c[0], c[1], c[2], c[3] and c[4]. The initial value of each classical bit is set to zero (0). Classical bit c[4] is the most significant bit and classical bit c[0] is the least significant bit.

Then, the statement "h q[0];" on line five of Listing 2.2 actually completes

$$\begin{pmatrix} \frac{1}{\sqrt{2}} & \frac{1}{\sqrt{2}} \\ \frac{1}{\sqrt{2}} & -\frac{1}{\sqrt{2}} \end{pmatrix} \times \begin{pmatrix} 1 \\ 0 \end{pmatrix} = \begin{pmatrix} \frac{1}{\sqrt{2}} \\ \frac{1}{\sqrt{2}} \end{pmatrix} = \frac{1}{\sqrt{2}} \begin{pmatrix} 1 \\ 1 \end{pmatrix} = \frac{1}{\sqrt{2}} \left(\begin{pmatrix} 1 \\ 0 \end{pmatrix} + \begin{pmatrix} 0 \\ 1 \end{pmatrix} \right) = \frac{1}{\sqrt{2}} (|0\rangle + $$

$|1\rangle)$. This indicates that the statement "h q[0];" on line five of Listing 2.2 is to apply the Hadamard gate to convert q[0] from one state $|0\rangle$ to another state $\frac{1}{\sqrt{2}}(|0\rangle + |1\rangle)$ (its superposition). Next, the statement "h q[1];" on line six of Listing 2.2 actually

completes $\begin{pmatrix} \frac{1}{\sqrt{2}} & \frac{1}{\sqrt{2}} \\ \frac{1}{\sqrt{2}} & -\frac{1}{\sqrt{2}} \end{pmatrix} \times \begin{pmatrix} 1 \\ 0 \end{pmatrix} = \begin{pmatrix} \frac{1}{\sqrt{2}} \\ \frac{1}{\sqrt{2}} \end{pmatrix} = \frac{1}{\sqrt{2}} \begin{pmatrix} 1 \\ 1 \end{pmatrix} = \frac{1}{\sqrt{2}} \left(\begin{pmatrix} 1 \\ 0 \end{pmatrix} + \begin{pmatrix} 0 \\ 1 \end{pmatrix} \right) =$

$\frac{1}{\sqrt{2}}(|0\rangle + |1\rangle)$. This implies that the statement "h q[1];" on line six of Listing 2.2 is to use the Hadamard gate to convert q[1] from one state $|0\rangle$ to another state $\frac{1}{\sqrt{2}}(|0\rangle + |1\rangle)$ (its superposition). Hence, the superposition of the two quantum bits q[0] and q[1] is $(\frac{1}{\sqrt{2}}(|0\rangle + |1\rangle))$ $(\frac{1}{\sqrt{2}}(|0\rangle + |1\rangle)) = \frac{1}{2}(|0\rangle |0\rangle + |0\rangle |1\rangle + |1\rangle |0\rangle + |1\rangle |1\rangle) = \frac{1}{2}(|00\rangle + |01\rangle + |10\rangle + |11\rangle)$. In the superposition, state $|00\rangle$ with the amplitude $\frac{1}{2}$ encodes the value 00 (zero) to the input of two bits. State $|01\rangle$ with the amplitude $\frac{1}{2}$ encodes the value 1 (one) to the input of two bits. State $|10\rangle$ with the amplitude $\frac{1}{2}$ encodes the value 2 (two) to the input of two bits and $|11\rangle$ with the amplitude $\frac{1}{2}$ encodes the value 3 (three) to the input of two bits.

Next, the statement "x q[0];" on line seven of Listing 2.2 actually completes

$$\begin{pmatrix} 0 & 1 \\ 1 & 0 \end{pmatrix} \times \begin{pmatrix} \frac{1}{\sqrt{2}} \\ \frac{1}{\sqrt{2}} \end{pmatrix} = \begin{pmatrix} \frac{1}{\sqrt{2}} \\ \frac{1}{\sqrt{2}} \end{pmatrix} = \frac{1}{\sqrt{2}} \begin{pmatrix} 1 \\ 1 \end{pmatrix} = \frac{1}{\sqrt{2}} \left(\begin{pmatrix} 0 \\ 1 \end{pmatrix} + \begin{pmatrix} 1 \\ 0 \end{pmatrix} \right) = \frac{1}{\sqrt{2}} (|1\rangle + |0\rangle)$$

and the statement "x q[1];" on line eight of Listing 2.2 actually completes $\begin{pmatrix} 0 & 1 \\ 1 & 0 \end{pmatrix}$

$$\times \begin{pmatrix} \frac{1}{\sqrt{2}} \\ \frac{1}{\sqrt{2}} \end{pmatrix} = \begin{pmatrix} \frac{1}{\sqrt{2}} \\ \frac{1}{\sqrt{2}} \end{pmatrix} = \frac{1}{\sqrt{2}} \begin{pmatrix} 1 \\ 1 \end{pmatrix} = \frac{1}{\sqrt{2}} \left(\begin{pmatrix} 0 \\ 1 \end{pmatrix} + \begin{pmatrix} 1 \\ 0 \end{pmatrix} \right) = \frac{1}{\sqrt{2}} (|1\rangle + |0\rangle)$$. This is to

say that the two statements "x q[0];" and "x q[1];" on line seven and line eight of Listing 2.2 inverts $|00\rangle$ with the amplitude $\frac{1}{2}$ (the input zero of two bits) into $|11\rangle$ with the amplitude $\frac{1}{2}$ (its one's complement). They inverts $|01\rangle$ with the amplitude $\frac{1}{2}$ (the input one of two bits) into $|10\rangle$ with the amplitude $\frac{1}{2}$ (its one's complement), inverts $|10\rangle$ with the amplitude $\frac{1}{2}$ (the input two of two bits) into $|01\rangle$ with the amplitude $\frac{1}{2}$ (its one's complement) and inverts $|11\rangle$ with the amplitude $\frac{1}{2}$ (the input three of two bits) into $|00\rangle$ with the amplitude $\frac{1}{2}$ (its one's complement). This indicates that eight instructions (eight **NOT** operations) of taking one's complement to the input of two bits are completed by means of applying two quantum operations "x q[0];" and "x q[1];".

Next, the statement "measure q[0] $\rightarrow$ c[0];" on line nine of Listing 2.2 is to measure the first quantum bit q[0] and to record the measurement outcome by overwriting the first classical bit c[0]. The statement "measure q[1] $\rightarrow$ c[1];"on line ten of Listing 2.2 is to measure the second quantum bit q[1] and to record the measurement outcome by overwriting the second classical bit c[1]. In the backend *ibmqx4* with five quantum bits in **IBM**'s quantum computers, we apply the command "simulate" to execute the program in Listing 2.2. The measured result appears in Fig. 2.5. From Fig. 2.5, we obtain the answer 00001 (c[4] = 0, c[3] = 0, c[2] = 0, c[1] = 0 = q[1] = $|0\rangle$ and c[0] = 1 = q[0] = $|1\rangle$) with the probability 0.330. This is to say that we obtain the one's complement (q[1] = $|0\rangle$ and q[0] = $|1\rangle$) with the probability 0.330 to the input two (10) of two bits. Or we obtain the answer 00010 (c[4] = 0, c[3] = 0, c[2] = 0, c[1] = 1 = q[1] = $|1\rangle$ and c[0] = 0 = q[0] = $|0\rangle$) with the probability 0.260. This indicates that we obtain the one's complement (q[1] = $|1\rangle$ and q[0] = $|0\rangle$) with the probability 0.260 to the input one (01) of two bits. Or we obtain the answer 00011 (c[4] = 0, c[3] = 0, c[2] = 0, c[1] = 1 = q[1] = $|1\rangle$ and c[0] = 1 = q[0] = $|1\rangle$) with the probability 0.210. This implies that we obtain the one's complement (q[1] = $|1\rangle$ and q[0] = $|1\rangle$) with the probability 0.210 to the input zero (00) of two bits. Or we obtain the answer 00000 (c[4] = 0, c[3] = 0, c[2] = 0, c[1] = 0 = q[1] = $|0\rangle$ and c[0] = 0 = q[0] = $|0\rangle$) with the probability 0.200. This is to say that we obtain the one's complement (q[1] = $|0\rangle$ and q[0] = $|0\rangle$) with the probability 0.200 to the input three (11) of two bits.

Fig. 2.5 After the measurement to the program in Listing 2.2 is completed, we obtain the answer 00001 with the probability 0.330 or the answer 00010 with the probability 0.260 or the answer 00011 with the probability 0.210 or the answer 00000 with the probability 0.200

2.2   The Toffoli Gate of Three Quantum Bits

The Toffoli gate that is also known as the ***controlled-controlled-NOT*** or ***CCNOT*** gate of three quantum bits is

$$U_{CCN} = \begin{pmatrix} 1\,0\,0\,0\,0\,0\,0\,0 \\ 0\,1\,0\,0\,0\,0\,0\,0 \\ 0\,0\,1\,0\,0\,0\,0\,0 \\ 0\,0\,0\,1\,0\,0\,0\,0 \\ 0\,0\,0\,0\,1\,0\,0\,0 \\ 0\,0\,0\,0\,0\,1\,0\,0 \\ 0\,0\,0\,0\,0\,0\,0\,1 \\ 0\,0\,0\,0\,0\,0\,1\,0 \end{pmatrix}. \tag{2.2}$$

It is assumed that $U_{CCN}{}^{+}$ is the conjugate-transpose matrix of U_{CCN} and is equal to

$$((U_{CCN})^{*})^{t} = \begin{pmatrix} 1\,0\,0\,0\,0\,0\,0\,0 \\ 0\,1\,0\,0\,0\,0\,0\,0 \\ 0\,0\,1\,0\,0\,0\,0\,0 \\ 0\,0\,0\,1\,0\,0\,0\,0 \\ 0\,0\,0\,0\,1\,0\,0\,0 \\ 0\,0\,0\,0\,0\,1\,0\,0 \\ 0\,0\,0\,0\,0\,0\,0\,1 \\ 0\,0\,0\,0\,0\,0\,1\,0 \end{pmatrix}, \text{ where the } * \text{ is the complex conjugation and the}$$

t is the transpose operation. Because $U_{CCN} \times ((U_{CCN})^{*})^{t} =$

$$\begin{pmatrix} 1\,0\,0\,0\,0\,0\,0\,0 \\ 0\,1\,0\,0\,0\,0\,0\,0 \\ 0\,0\,1\,0\,0\,0\,0\,0 \\ 0\,0\,0\,1\,0\,0\,0\,0 \\ 0\,0\,0\,0\,1\,0\,0\,0 \\ 0\,0\,0\,0\,0\,1\,0\,0 \\ 0\,0\,0\,0\,0\,0\,0\,1 \\ 0\,0\,0\,0\,0\,0\,1\,0 \end{pmatrix}$$

$$\times \begin{pmatrix} 1\,0\,0\,0\,0\,0\,0\,0 \\ 0\,1\,0\,0\,0\,0\,0\,0 \\ 0\,0\,1\,0\,0\,0\,0\,0 \\ 0\,0\,0\,1\,0\,0\,0\,0 \\ 0\,0\,0\,0\,1\,0\,0\,0 \\ 0\,0\,0\,0\,0\,1\,0\,0 \\ 0\,0\,0\,0\,0\,0\,0\,1 \\ 0\,0\,0\,0\,0\,0\,1\,0 \end{pmatrix} = ((U_{CCN})^{*})^{t} \times U_{CCN} = \begin{pmatrix} 1\,0\,0\,0\,0\,0\,0\,0 \\ 0\,1\,0\,0\,0\,0\,0\,0 \\ 0\,0\,1\,0\,0\,0\,0\,0 \\ 0\,0\,0\,1\,0\,0\,0\,0 \\ 0\,0\,0\,0\,1\,0\,0\,0 \\ 0\,0\,0\,0\,0\,1\,0\,0 \\ 0\,0\,0\,0\,0\,0\,0\,1 \\ 0\,0\,0\,0\,0\,0\,1\,0 \end{pmatrix} \times$$

$$
\begin{pmatrix}
1 & 0 & 0 & 0 & 0 & 0 & 0 & 0 \\
0 & 1 & 0 & 0 & 0 & 0 & 0 & 0 \\
0 & 0 & 1 & 0 & 0 & 0 & 0 & 0 \\
0 & 0 & 0 & 1 & 0 & 0 & 0 & 0 \\
0 & 0 & 0 & 0 & 1 & 0 & 0 & 0 \\
0 & 0 & 0 & 0 & 0 & 1 & 0 & 0 \\
0 & 0 & 0 & 0 & 0 & 0 & 0 & 1 \\
0 & 0 & 0 & 0 & 0 & 0 & 1 & 0
\end{pmatrix}
=
\begin{pmatrix}
1 & 0 & 0 & 0 & 0 & 0 & 0 & 0 \\
0 & 1 & 0 & 0 & 0 & 0 & 0 & 0 \\
0 & 0 & 1 & 0 & 0 & 0 & 0 & 0 \\
0 & 0 & 0 & 1 & 0 & 0 & 0 & 0 \\
0 & 0 & 0 & 0 & 1 & 0 & 0 & 0 \\
0 & 0 & 0 & 0 & 0 & 1 & 0 & 0 \\
0 & 0 & 0 & 0 & 0 & 0 & 1 & 0 \\
0 & 0 & 0 & 0 & 0 & 0 & 0 & 1
\end{pmatrix}
$$, U_{CCN} is a unitary matrix or a unitary

operator. This indicates that the **controlled-controlled-NOT** or **CCNOT** gate (the Toffoli gate) U_{CCN} is one of quantum gates with three quantum bits. The quantum state $l_0|000\rangle + l_1|001\rangle + l_2|010\rangle + l_3|011\rangle + l_4|100\rangle + l_5|101\rangle + l_6|110\rangle + l_7|111\rangle$ is written in a vector notation as

$$
\begin{pmatrix}
l_0 \\
l_1 \\
l_2 \\
l_3 \\
l_4 \\
l_5 \\
l_6 \\
l_7
\end{pmatrix}
\tag{2.3}
$$

The first entry l_0 is the amplitude for $|000\rangle$. The second entry l_1 is the amplitude for $|001\rangle$. The third entry l_2 is the amplitude for $|010\rangle$. The fourth entry l_3 is the amplitude for $|011\rangle$. The fifth entry l_4 is the amplitude for $|100\rangle$. The sixth entry l_5 is the amplitude for $|101\rangle$. The seventh entry l_6 is the amplitude for $|110\rangle$ and the eighth entry l_7 is the amplitude for $|111\rangle$. The **CCNOT** gate U_{CCN} takes it as its input state vector and generates the output to be

$$
\begin{pmatrix}
1 & 0 & 0 & 0 & 0 & 0 & 0 & 0 \\
0 & 1 & 0 & 0 & 0 & 0 & 0 & 0 \\
0 & 0 & 1 & 0 & 0 & 0 & 0 & 0 \\
0 & 0 & 0 & 1 & 0 & 0 & 0 & 0 \\
0 & 0 & 0 & 0 & 1 & 0 & 0 & 0 \\
0 & 0 & 0 & 0 & 0 & 1 & 0 & 0 \\
0 & 0 & 0 & 0 & 0 & 0 & 0 & 1 \\
0 & 0 & 0 & 0 & 0 & 0 & 1 & 0
\end{pmatrix}
\times
\begin{pmatrix}
l_0 \\
l_1 \\
l_2 \\
l_3 \\
l_4 \\
l_5 \\
l_6 \\
l_7
\end{pmatrix}
=
\begin{pmatrix}
l_0 \\
l_1 \\
l_2 \\
l_3 \\
l_4 \\
l_5 \\
l_7 \\
l_6
\end{pmatrix}
= l_0
\begin{pmatrix}
1 \\
0 \\
0 \\
0 \\
0 \\
0 \\
0 \\
0
\end{pmatrix}
$$

$$+ l_1 \begin{pmatrix} 0 \\ 1 \\ 0 \\ 0 \\ 0 \\ 0 \\ 0 \\ 0 \end{pmatrix} + l_2 \begin{pmatrix} 0 \\ 0 \\ 1 \\ 0 \\ 0 \\ 0 \\ 0 \\ 0 \end{pmatrix} + l_3 \begin{pmatrix} 0 \\ 0 \\ 0 \\ 1 \\ 0 \\ 0 \\ 0 \\ 0 \end{pmatrix} + l_4 \begin{pmatrix} 0 \\ 0 \\ 0 \\ 0 \\ 1 \\ 0 \\ 0 \\ 0 \end{pmatrix}$$

$$+ l_5 \begin{pmatrix} 0 \\ 0 \\ 0 \\ 0 \\ 0 \\ 0 \\ 1 \\ 0 \\ 0 \end{pmatrix} + l_7 \begin{pmatrix} 0 \\ 0 \\ 0 \\ 0 \\ 0 \\ 0 \\ 1 \\ 0 \end{pmatrix} + l_6 \begin{pmatrix} 0 \\ 0 \\ 0 \\ 0 \\ 0 \\ 0 \\ 0 \\ 1 \end{pmatrix}$$

$$= l_0 |000\rangle + l_1 |001\rangle + l_2 |010\rangle + l_3 |011\rangle$$
$$+ l_4 |100\rangle + l_5 |101\rangle + l_7 |110\rangle + l_6 |111\rangle. \quad (2.4)$$

This is to say that the **CCNOT** gate U_{CCN} converts three quantum bits from one state $l_0 |000\rangle + l_1 |001\rangle + l_2 |010\rangle + l_3 |011\rangle + l_4 |100\rangle + l_5 |101\rangle + l_6 |110\rangle + l_7 |111\rangle$ to another state $l_0 |000\rangle + l_1 |001\rangle + l_2 |010\rangle + l_3 |011\rangle + l_4 |100\rangle + l_5 |101\rangle + l_7 |110\rangle + l_6 |111\rangle$. This implies that in the **CCNOT** gate U_{CCN} if the two control quantum bits (the *first* quantum bit and the second quantum bit) is set to 0, then the target quantum bit (the *third* quantum bit) is left alone. If the two control quantum bits (the *first* quantum bit and the *second* quantum bit) is both set to 1, then the **CCNOT** gate U_{CCN} flips the target quantum bit (the *third* quantum bit). The probability of measuring a $|0\rangle$ $|000\rangle, |1\rangle |001\rangle, |010\rangle, |011\rangle, |100\rangle$ or $|101\rangle$ is unchanged, the probability of measuring a $|0\rangle$ $|110\rangle$ is $|l_7|^2$ and the probability of measuring a $|0\rangle$ $|111\rangle$ is $|l_6|^2$ after using the

CCNOT gate U_{CCN}. Since $(U_{CCN})^2 = U_{CCN} \times U_{CCN} = \begin{pmatrix} 1 & 0 & 0 & 0 & 0 & 0 & 0 & 0 \\ 0 & 1 & 0 & 0 & 0 & 0 & 0 & 0 \\ 0 & 0 & 1 & 0 & 0 & 0 & 0 & 0 \\ 0 & 0 & 0 & 1 & 0 & 0 & 0 & 0 \\ 0 & 0 & 0 & 0 & 1 & 0 & 0 & 0 \\ 0 & 0 & 0 & 0 & 0 & 1 & 0 & 0 \\ 0 & 0 & 0 & 0 & 0 & 0 & 0 & 1 \\ 0 & 0 & 0 & 0 & 0 & 0 & 1 & 0 \end{pmatrix} \times$

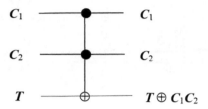

Fig. 2.6 The circuit representation of the **CCNOT** gate

$$\begin{pmatrix} 1\,0\,0\,0\,0\,0\,0\,0 \\ 0\,1\,0\,0\,0\,0\,0\,0 \\ 0\,0\,1\,0\,0\,0\,0\,0 \\ 0\,0\,0\,1\,0\,0\,0\,0 \\ 0\,0\,0\,0\,1\,0\,0\,0 \\ 0\,0\,0\,0\,0\,1\,0\,0 \\ 0\,0\,0\,0\,0\,0\,0\,1 \\ 0\,0\,0\,0\,0\,0\,1\,0 \end{pmatrix} = \begin{pmatrix} 1\,0\,0\,0\,0\,0\,0\,0 \\ 0\,1\,0\,0\,0\,0\,0\,0 \\ 0\,0\,1\,0\,0\,0\,0\,0 \\ 0\,0\,0\,1\,0\,0\,0\,0 \\ 0\,0\,0\,0\,1\,0\,0\,0 \\ 0\,0\,0\,0\,0\,1\,0\,0 \\ 0\,0\,0\,0\,0\,0\,1\,0 \\ 0\,0\,0\,0\,0\,0\,0\,1 \end{pmatrix}$$, using the **CCNOT** gate U_{CCN} twice to

a state is equivalent to do nothing to it. The graphical representation of the **CCNOT** gate U_{CCN} appears in Fig. 2.6. In Fig. 2.6, the left top two wires are *control bits* that are unchanged by the action of the **CCNOT** gate. The bottom wire is a *target bit* that the action of the **CCNOT** gate flips it if both control bits are set to one (1), and otherwise is left alone, where $\oplus$ is addition modulo two.

2.2.1 Implementing the Toffoli Gate of Three Quantum Bits

The Toffoli gate has three input bits and three output bits. Its truth table appears in Table 2.2. In **IBM Q** Experience, in the backend *simulator*, it now provides one quantum instruction (operation) of implementing the **CCNOT** gate (the Toffoli gate) with three quantum bits. But in the backend *ibmqx4*, it does not support the Toffoli gate. We decompose **CCNOT** gate into *six* **CNOT** gates and *nine* gates of one quantum bit that appear in Fig. 2.7. In Fig. 2.7, H is the Hadamard gate, $T = \begin{bmatrix} 1 & 0 \\ 0 & e^{\sqrt{-1} \times \frac{\pi}{4}} \end{bmatrix}$ and $T^+ = \begin{bmatrix} 1 & 0 \\ 0 & e^{-1 \times \sqrt{-1} \times \frac{\pi}{4}} \end{bmatrix}$. In **IBM Q** Experience, the available gates are that **CNOT** is the only gate with two quantum bits and the other gates act on single quantum bit and they are introduced in the previous chapter. In the backend *ibmqx4* with five quantum bits, there are only six pairs of **CNOT** gates. Connectivity of the **CNOT** gate in the backend *ibmqx4* with five quantum bits appears in Fig. 1.32 in Sect. 1.11.1.

Table 2.2   The truth table for the Toffoli gate with three input bits and three output bits

Input			Output		
C_1	C_2	T	C_1	C_2	T
0	0	0	0	0	0
0	1	0	0	1	0
1	0	0	1	0	0
1	1	0	1	1	1
0	0	1	0	0	1
0	1	1	0	1	1
1	0	1	1	0	1
1	1	1	1	1	0

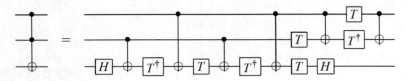

Fig. 2.7   Decomposing **CCNOT** gate into six **CNOT** gates and nine gates of one bit

In Listing 2.3, the program in the backend *ibmqx4* with five quantum bits in **IBM**'s quantum computer is the *third* example of the *second* chapter in which we illustrate how to write a quantum program to implement a Toffoli gate (a **CCNOT** gate) of three quantum bits. Figure 2.8 is the corresponding quantum circuit of the program in Listing 2.3. The statement "OPENQASM 2.0;" on line one of Listing 2.3 is to indicate that the program is written with version 2.0 of Open QASM. Next, the statement "include "qelib1.inc";" on line two of Listing 2.3 is to continue parsing the file "qelib1.inc" as if the contents of the file were pasted at the location of the include statement, where the file "qelib1.inc" is **Quantum Experience (QE) Standard Header** and the path is specified relative to the current working directory.

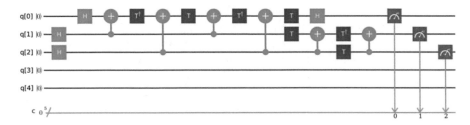

Fig. 2.8   The corresponding quantum circuit of the program in Listing 2.3

Listing 2.3 The program of implementing a **CCNOT** gate of three quantum bits.

```
1.   OPENQASM 2.0;
2.   include "qelib1.inc";
3.   qreg q[5];
4.   creg c[5];
5.   h q[1];
6.   h q[2];
7.   h q[0];
8.   cx q[1], q[0];
9.   tdg q[0];
10.  cx q[2], q[0];
11.  t q[0];
12.  cx q[1], q[0];
13.  tdg q[0];
14.  cx q[2], q[0];
15.  t q[0];
16.  t q[1];
17.  h q[0];
18.  cx q[2], q[1];
19.  tdg q[1];
20.  t q[2];
21.  cx q[2], q[1];
22.  measure q[0] → c[0];
23.  measure q[1] → c[1];
24.  measure q[2] → c[2];
```

Next, the statement "qreg q[5];" on line three of Listing 2.3 is to declare that in the program there are five quantum bits. In the left top of Fig. 2.8, five quantum bits are subsequently q[0], q[1], q[2], q[3] and q[4]. The initial value of each quantum bit is set to $|0\rangle$. We apply three quantum bits q[2], q[1] and q[0] to encode subsequently the first control bit, the second control bit and the target bit. For the convenience of our explanation, $q[k]^0$ for $0 \leq k \leq 4$ is to represent the value 0 of q[k] and $q[k]^1$ for $0 \leq k \leq 4$ is to represent the value 1 of q[k]. Similarly, for the convenience of our explanation, an initial state vector of implementing a Toffoli gate is as follows:

$$|\Phi_0\rangle = \left|q[2]^0\right\rangle\left|q[1]^0\right\rangle\left|q[0]^0\right\rangle = |0\rangle|0\rangle|0\rangle = |000\rangle.$$

Then, the statement "creg c[5];" on line four of Listing 2.3 is to declare that there are five classical bits in the program. In the left bottom of Fig. 2.8, five classical bits are respectively c[0], c[1], c[2], c[3] and c[4]. The initial value of each classical bit is set to zero (0). Classical bit c[4] is the most significant bit and classical bit c[0] is the least significant bit.

Next, the two statements "h q[1];" and "h q[2];"on line five and line six of Listing 2.3 implement two Hadamard gates of the *first* time slot of the quantum circuit in Fig. 2.8 and both actually complete $\begin{pmatrix} \frac{1}{\sqrt{2}} & \frac{1}{\sqrt{2}} \\ \frac{1}{\sqrt{2}} & -\frac{1}{\sqrt{2}} \end{pmatrix} \times \begin{pmatrix} 1 \\ 0 \end{pmatrix} = \begin{pmatrix} \frac{1}{\sqrt{2}} \\ \frac{1}{\sqrt{2}} \end{pmatrix} = \frac{1}{\sqrt{2}} \begin{pmatrix} 1 \\ 1 \end{pmatrix} = \frac{1}{\sqrt{2}} \left(\begin{pmatrix} 1 \\ 0 \end{pmatrix} + \begin{pmatrix} 0 \\ 1 \end{pmatrix} \right) = \frac{1}{\sqrt{2}}(|0\rangle + |1\rangle))$. This is to say that converting q[1] from one state $|0\rangle$ to another state $\frac{1}{\sqrt{2}}(|0\rangle + |1\rangle))$ (its superposition) and converting q[2] from one state $|0\rangle$ to another state $\frac{1}{\sqrt{2}}(|0\rangle + |1\rangle))$ (its superposition) are completed. Thus, the superposition of the two quantum bits q[2] and q[1] is $(\frac{1}{\sqrt{2}}(|0\rangle + |1\rangle)) (\frac{1}{\sqrt{2}}(|0\rangle + |1\rangle)) = \frac{1}{2}(|0\rangle |0\rangle + |0\rangle |1\rangle + |1\rangle |0\rangle + |1\rangle |1\rangle) = \frac{1}{2}(|00\rangle + |01\rangle + |10\rangle + |11\rangle)$. Because in the *first* time slot of the quantum circuit in Fig. 2.8 there is no quantum gate to act on the quantum bit q[0], its state $|0\rangle$ is not changed. Therefore, after using the two statements "h q[1];" and "h q[2];"on line five and line six of Listing 2.3 implement two Hadamard gates in the *first* time slot of the quantum circuit in Fig. 2.8, the following new state vector is

$$|\Phi_1\rangle = \frac{1}{2}(|0\rangle |0\rangle |0\rangle + |0\rangle |1\rangle |0\rangle + |1\rangle |0\rangle |0\rangle + |1\rangle |1\rangle |0\rangle)$$

$$= \frac{1}{2}(|000\rangle + |010\rangle + |100\rangle + |110\rangle).$$

The next 12 time slots in the quantum circuit of Fig. 2.8 implement a Toffoli gate. Next, the statement "h q[0];" on line seven of Listing 2.3 takes the new state vector $|\Phi_1\rangle = \frac{1}{2}(|000\rangle + |010\rangle + |100\rangle + |110\rangle)$ as its input and completes one Hadamard gate for q[0] in the *second* time slot. This is to say that the statement "h q[0];" converts q[0] from one state $|0\rangle$ to another state $\frac{1}{\sqrt{2}}(|0\rangle + |1\rangle))$ (its superposition). Because there is no other quantum gate in the *second* time slot to act on quantum bits q[2] and q[1], their states are not changed. Therefore, after using the statement "h q[0];" on line seven of Listing 2.3 completes one Hadamard gate for q[0] in the *second* time slot, the following new state vector is

$$|\Phi_2\rangle = \frac{1}{2} \left(|0\rangle |0\rangle \frac{1}{\sqrt{2}}(|0\rangle + |1\rangle) + |0\rangle |1\rangle \frac{1}{\sqrt{2}}(|0\rangle + |1\rangle) \right.$$

$$\left. + |1\rangle |0\rangle \frac{1}{\sqrt{2}}(|0\rangle + |1\rangle) + |1\rangle |1\rangle \frac{1}{\sqrt{2}}(|0\rangle + |1\rangle) \right)$$

$$= \frac{1}{2\sqrt{2}}(|000\rangle + |001\rangle + |010\rangle + |011\rangle$$

$$+ |100\rangle + |101\rangle + |110\rangle + |111\rangle).$$

Next, the statement "cx q[1], q[0];" on line eight of Listing 2.3 takes the new state vector $|\Phi_2\rangle = \frac{1}{2\sqrt{2}}(|000\rangle + |001\rangle + |010\rangle + |011\rangle + |100\rangle + |101\rangle + |110\rangle + |111\rangle)$ as its input and completes one **CNOT** gate for q[1] and q[0] in the *third* time slot. If the value of the control bit q[1] is equal to one, then the statement "cx q[1], q[0];" flips the value of the target bit q[0]. Otherwise, it is not changed. Because there is no other quantum gate in the *third* time slot to act on quantum bit q[2], its state is not changed. Hence, after using the statement "cx q[1], q[0];" on line eight of Listing 2.3 implements one **CNOT** gate for q[1] and q[0] in the *third* time slot, the following new state vector is

$$|\Phi_3\rangle = \frac{1}{2\sqrt{2}}(|000\rangle + |001\rangle + |011\rangle + |010\rangle$$
$$+ |100\rangle + |101\rangle + |111\rangle + |110\rangle).$$

Next, the statement "tdg q[0];" on line nine of Listing 2.3 takes the new state vector $|\Phi_3\rangle = \frac{1}{2\sqrt{2}}(|000\rangle + |001\rangle + |011\rangle + |010\rangle + |100\rangle + |101\rangle + |111\rangle + |110\rangle)$ as its input and completes one T^+ gate for q[0] in the *fourth* time slot. If the value of q[0] is equal to one, then the statement "tdg q[0];" changes its phase as $(e^{-1\times\sqrt{-1}\times\frac{\pi}{4}})$. Otherwise, its phase is not changed. There is no other quantum gate in the *fourth* time slot to act on quantum bits q[2] and q[1], so their states are not changed. Thus, after using the statement "tdg q[0];" on line nine of Listing 2.3 completes one T^+ gate for q[0] in the *fourth* time slot, the following new state vector is

$$|\Phi_4\rangle = \frac{1}{2\sqrt{2}}\Big(|000\rangle + e^{-1\times\sqrt{-1}\times\frac{\pi}{4}}|001\rangle + e^{-1\times\sqrt{-1}\times\frac{\pi}{4}}|011\rangle + |010\rangle$$
$$+ |100\rangle + e^{-1\times\sqrt{-1}\times\frac{\pi}{4}}|101\rangle + e^{-1\times\sqrt{-1}\times\frac{\pi}{4}}|111\rangle + |110\rangle\Big).$$

Next, the statement "cx q[2], q[0];" on line ten of Listing 2.3 takes the new state vector $|\Phi_4\rangle = \frac{1}{2\sqrt{2}}(|000\rangle + e^{-1\times\sqrt{-1}\times\frac{\pi}{4}}|001\rangle + e^{-1\times\sqrt{-1}\times\frac{\pi}{4}}|011\rangle + |010\rangle + |100\rangle + e^{-1\times\sqrt{-1}\times\frac{\pi}{4}}|101\rangle + e^{-1\times\sqrt{-1}\times\frac{\pi}{4}}|111\rangle + |110\rangle)$ as its input and completes one **CNOT** gate for q[2] and q[0] in the *fifth* time slot. If the value of the control bit q[2] is equal to one, then the statement "cx q[2], q[0];" flips the value of the target bit q[0]. Otherwise, it is not changed. Because there is no other quantum gate in the *fifth* time slot to act on quantum bit q[1], its state is not changed. Thus, after using the statement "cx q[2], q[0];" on line ten of Listing 2.3 completes one **CNOT** gate for q[2] and q[0] in the *fifth* time slot, the following new state vector is

$$|\Phi_5\rangle = \frac{1}{2\sqrt{2}}\Big(|000\rangle + e^{-1\times\sqrt{-1}\times\frac{\pi}{4}}|001\rangle + e^{-1\times\sqrt{-1}\times\frac{\pi}{4}}|011\rangle + |010\rangle + |101\rangle$$
$$+ e^{-1\times\sqrt{-1}\times\frac{\pi}{4}}|100\rangle + e^{-1\times\sqrt{-1}\times\frac{\pi}{4}}|110\rangle + |111\rangle\Big).$$

Next, the statement "t q[0];" on line eleven of Listing 2.3 takes the new state vector $|\Phi_5\rangle = \frac{1}{2\sqrt{2}}(|000\rangle + e^{-1\times\sqrt{-1}\times\frac{\pi}{4}}|001\rangle + e^{-1\times\sqrt{-1}\times\frac{\pi}{4}}|011\rangle + |010\rangle + |101\rangle + e^{-1\times\sqrt{-1}\times\frac{\pi}{4}}|100\rangle + e^{-1\times\sqrt{-1}\times\frac{\pi}{4}}|110\rangle + |111\rangle)$ as its input and completes one T gate for q[0] in the *sixth* time slot. If the value of q[0] is equal to one, then the statement "t q[0];" changes its phase as ($e^{1\times\sqrt{-1}\times\frac{\pi}{4}}$). Otherwise, its phase is not changed. Since there is no other quantum gate in the *sixth* time slot to act on quantum bits q[2] and q[1], their states are not changed. Hence, after using the statement "t q[0];" completes one *T* gate for q[0] in the *sixth* time slot on line eleven of Listing 2.3, the following new state vector is

$$|\Phi_6\rangle = \frac{1}{2\sqrt{2}}\Big(|000\rangle + \Big(e^{1\times\sqrt{-1}\times\frac{\pi}{4}} \times e^{-1\times\sqrt{-1}\times\frac{\pi}{4}}\Big)|001\rangle$$
$$+ \Big(e^{1\times\sqrt{-1}\times\frac{\pi}{4}} \times e^{-1\times\sqrt{-1}\times\frac{\pi}{4}}\Big)|011\rangle + |010\rangle + e^{1\times\sqrt{-1}\times\frac{\pi}{4}}|101\rangle$$
$$+ e^{-1\times\sqrt{-1}\times\frac{\pi}{4}}|100\rangle + e^{-1\times\sqrt{-1}\times\frac{\pi}{4}}|110\rangle + e^{1\times\sqrt{-1}\times\frac{\pi}{4}}|111\rangle$$
$$= \frac{1}{2\sqrt{2}}\Big(|000\rangle + |001\rangle + |011\rangle + |010\rangle + e^{1\times\sqrt{-1}\times\frac{\pi}{4}}|101\rangle$$
$$+ e^{-1\times\sqrt{-1}\times\frac{\pi}{4}}|100\rangle + e^{-1\times\sqrt{-1}\times\frac{\pi}{4}}|110\rangle + e^{1\times\sqrt{-1}\times\frac{\pi}{4}}|111\rangle\Big).$$

Next, the statement "cx q[1], q[0];" on line twelve of Listing 2.3 takes the new state vector $|\Phi_6\rangle = \frac{1}{2\sqrt{2}}(|000\rangle + |001\rangle + |011\rangle + |010\rangle + e^{1\times\sqrt{-1}\times\frac{\pi}{4}}|101\rangle + e^{-1\times\sqrt{-1}\times\frac{\pi}{4}}|100\rangle + e^{-1\times\sqrt{-1}\times\frac{\pi}{4}}|110\rangle + e^{1\times\sqrt{-1}\times\frac{\pi}{4}}|111\rangle)$ as its input and performs one **CNOT** gate for q[1] and q[0] in the *seventh* time slot. If the value of the control bit q[1] is equal to one, then the statement "cx q[1], q[0];" flips the value of the target bit q[0]. Otherwise, it is not changed. There is no other quantum gate in the *seventh* time slot to act on quantum bit q[2], so its state is not changed. Therefore, after using the statement "cx q[1], q[0];" on line twelve of Listing 2.3 implements one **CNOT** gate for q[1] and q[0] in the *seventh* time slot, the following new state vector is

$$|\Phi_7\rangle = \frac{1}{2\sqrt{2}}\Big(|000\rangle + |001\rangle + |010\rangle + |011\rangle + e^{1\times\sqrt{-1}\times\frac{\pi}{4}}|101\rangle$$
$$+ e^{-1\times\sqrt{-1}\times\frac{\pi}{4}}|100\rangle + e^{-1\times\sqrt{-1}\times\frac{\pi}{4}}|111\rangle + e^{1\times\sqrt{-1}\times\frac{\pi}{4}}|110\rangle\Big).$$

Next, the statement "tdg q[0];" on line thirteen of Listing 2.3 takes the new state vector $|\Phi_7\rangle = \frac{1}{2\sqrt{2}}(|000\rangle + |001\rangle + |010\rangle + |011\rangle + e^{1\times\sqrt{-1}\times\frac{\pi}{4}}|101\rangle + e^{-1\times\sqrt{-1}\times\frac{\pi}{4}}$ $|100\rangle + e^{-1\times\sqrt{-1}\times\frac{\pi}{4}}|111\rangle + e^{1\times\sqrt{-1}\times\frac{\pi}{4}}|110\rangle)$ as its input and finish one T^+ gate for q[0] in the *eighth* time slot. If the value of q[0] is equal to one, then the statement "tdg q[0];" changes its phase as ($e^{-1\times\sqrt{-1}\times\frac{\pi}{4}}$). Otherwise, its phase is not changed. Because there is no other quantum gate in the *eighth* time slot to act on quantum bits q[2] and q[1], their states are not changed. Hence, after using the statement "tdg q[0];" on line thirteen of Listing 2.3 completes one T^+ gate for q[0] in the *eighth* time slot, the following new state vector is

$$|\Phi_8\rangle = \frac{1}{2\sqrt{2}}\Big(|000\rangle + e^{-1\times\sqrt{-1}\times\frac{\pi}{4}}|001\rangle + |010\rangle + e^{-1\times\sqrt{-1}\times\frac{\pi}{4}}|011\rangle$$

$$+ \left(e^{-1 \times \sqrt{-1} \times \frac{\pi}{4}} \times e^{1 \times \sqrt{-1} \times \frac{\pi}{4}}\right)|101\rangle + e^{-1 \times \sqrt{-1} \times \frac{\pi}{4}}|100\rangle$$

$$+ \left(e^{-1 \times \sqrt{-1} \times \frac{\pi}{4}} \times e^{-1 \times \sqrt{-1} \times \frac{\pi}{4}}\right)|111\rangle + e^{1 \times \sqrt{-1} \times \frac{\pi}{4}}|110\rangle \Bigg).$$

Next, the statement "cx q[2], q[0];" on line fourteen of Listing 2.3 takes the new state vector $|\Phi_8\rangle = \frac{1}{2\sqrt{2}}(|000\rangle + e^{-1 \times \sqrt{-1} \times \frac{\pi}{4}}|001\rangle + |010\rangle + e^{-1 \times \sqrt{-1} \times \frac{\pi}{4}}|011\rangle + (e^{-1 \times \sqrt{-1} \times \frac{\pi}{4}} \times e^{1 \times \sqrt{-1} \times \frac{\pi}{4}})|101\rangle + e^{-1 \times \sqrt{-1} \times \frac{\pi}{4}}|100\rangle + (e^{-1 \times \sqrt{-1} \times \frac{\pi}{4}} \times e^{-1 \times \sqrt{-1} \times \frac{\pi}{4}})|111\rangle + e^{1 \times \sqrt{-1} \times \frac{\pi}{4}}|110\rangle)$ as its input and completes one **CNOT** gate for q[2] and q[0] in the *ninth* time slot. If the value of the control bit q[2] is equal to one, then the statement "cx q[2], q[0];" flips the value of the target bit q[0]. Otherwise, it is not changed. Since there is no other quantum gate in the *ninth* time slot to act on quantum bit q[1], its state is not changed. Thus, after using the statement "cx q[2], q[0];" on line fourteen of Listing 2.3 completes one **CNOT** gate for q[2] and q[0] in the *ninth* time slot, we obtain the following new state vector

$$|\Phi_9\rangle = \frac{1}{2\sqrt{2}}\Bigg(|000\rangle + e^{-1 \times \sqrt{-1} \times \frac{\pi}{4}}|001\rangle + |010\rangle + e^{-1 \times \sqrt{-1} \times \frac{\pi}{4}}|011\rangle$$

$$+ |100\rangle + e^{-1 \times \sqrt{-1} \times \frac{\pi}{4}}|101\rangle + \left(e^{-1 \times \sqrt{-1} \times \frac{\pi}{4}} \times e^{-1 \times \sqrt{-1} \times \frac{\pi}{4}}\right)|110\rangle$$

$$+ e^{1 \times \sqrt{-1} \times \frac{\pi}{4}}|111\rangle\Bigg).$$

Next, the statement "t q[0];" on line fifteen of Listing 2.3 takes the new state vector $|\Phi_9\rangle = \frac{1}{2\sqrt{2}}(|000\rangle + e^{-1 \times \sqrt{-1} \times \frac{\pi}{4}}|001\rangle + |010\rangle + e^{-1 \times \sqrt{-1} \times \frac{\pi}{4}}|011\rangle + |100\rangle + e^{-1 \times \sqrt{-1} \times \frac{\pi}{4}}|101\rangle + (e^{-1 \times \sqrt{-1} \times \frac{\pi}{4}} \times e^{-1 \times \sqrt{-1} \times \frac{\pi}{4}})|110\rangle + e^{1 \times \sqrt{-1} \times \frac{\pi}{4}}|111\rangle)$ as its input and performs one T gate for q[0] in the *tenth* time slot. If the value of q[0] is equal to one, then the statement "t q[0];" changes its phase as $(e^{1 \times \sqrt{-1} \times \frac{\pi}{4}})$. Otherwise, its phase is not changed. There is no other quantum gate in the *tenth* time slot to act on quantum bit q[2], so its state is not changed. Hence, after using the statement "t q[0];" on line fifteen of Listing 2.3 completes one T gate for q[0] in the *tenth* time slot, the following new state vector is

$$|\Phi_{10}\rangle = \frac{1}{2\sqrt{2}}\Bigg(|000\rangle + \left(e^{1 \times \sqrt{-1} \times \frac{\pi}{4}} \times e^{-1 \times \sqrt{-1} \times \frac{\pi}{4}}\right)|001\rangle + |010\rangle$$

$$+ \left(e^{1 \times \sqrt{-1} \times \frac{\pi}{4}} \times e^{-1 \times \sqrt{-1} \times \frac{\pi}{4}}\right)|011\rangle + |100\rangle$$

$$+ \left(e^{1 \times \sqrt{-1} \times \frac{\pi}{4}} \times e^{-1 \times \sqrt{-1} \times \frac{\pi}{4}}\right)|101\rangle$$

$$+ \left(e^{-1 \times \sqrt{-1} \times \frac{\pi}{4}} \times e^{-1 \times \sqrt{-1} \times \frac{\pi}{4}}\right)|110\rangle + \left(e^{1 \times \sqrt{-1} \times \frac{\pi}{4}} \times e^{1 \times \sqrt{-1} \times \frac{\pi}{4}}\right)|111\rangle\Bigg)$$

$$= \frac{1}{2\sqrt{2}}(|000\rangle + |001\rangle + |010\rangle + |011\rangle + |100\rangle + |101\rangle$$

$$+ \left(e^{-1 \times \sqrt{-1} \times \frac{\pi}{4}} \times e^{-1 \times \sqrt{-1} \times \frac{\pi}{4}}\right)|110\rangle + \left(e^{1 \times \sqrt{-1} \times \frac{\pi}{4}} \times e^{1 \times \sqrt{-1} \times \frac{\pi}{4}}\right)|111\rangle).$$

Next, the statement "t q[1];" on line sixteen of Listing 2.3 takes the new state vector $|\Phi_{10}\rangle = \frac{1}{2\sqrt{2}}(|000\rangle + |001\rangle + |010\rangle + |011\rangle + |100\rangle + |101\rangle + (e^{-1\times\sqrt{-1}\times\frac{\pi}{4}}$ $\times e^{-1\times\sqrt{-1}\times\frac{\pi}{4}})|110\rangle + (e^{1\times\sqrt{-1}\times\frac{\pi}{4}} \times e^{1\times\sqrt{-1}\times\frac{\pi}{4}})|111\rangle)$ as its input and completes one T gate for q[1] in the *tenth* time slot. If the value of q[1] is equal to one, then the statement "t q[1];" changes its phase as $(e^{1\times\sqrt{-1}\times\frac{\pi}{4}})$. Otherwise, its phase is not changed. Because there is no other quantum gate in the *tenth* time slot to act on quantum bit q[2], so its state is not changed. Therefore, after using the statement "t q[1];" on line sixteen of Listing 2.3 completes one T gate for q[1] in the *tenth* time slot, the following new state vector is

$$
\begin{aligned}
|\Phi_{11}\rangle = \frac{1}{2\sqrt{2}}\Big(&|000\rangle + |001\rangle + e^{1\times\sqrt{-1}\times\frac{\pi}{4}}|010\rangle \\
&+ e^{1\times\sqrt{-1}\times\frac{\pi}{4}}|011\rangle + |100\rangle + |101\rangle \\
&+ \Big(e^{1\times\sqrt{-1}\times\frac{\pi}{4}} \times e^{-1\times\sqrt{-1}\times\frac{\pi}{4}} \times e^{-1\times\sqrt{-1}\times\frac{\pi}{4}}\Big)|110\rangle \\
&+ \Big(e^{1\times\sqrt{-1}\times\frac{\pi}{4}} \times e^{1\times\sqrt{-1}\times\frac{\pi}{4}} \times e^{1\times\sqrt{-1}\times\frac{\pi}{4}}\Big)|111\rangle\Big).
\end{aligned}
$$

Next, the statement "h q[0];" on line seventeen of Listing 2.3 takes the new state vector $|\Phi_{11}\rangle = \frac{1}{2\sqrt{2}}(|000\rangle + |001\rangle + e^{1\times\sqrt{-1}\times\frac{\pi}{4}}|010\rangle + e^{1\times\sqrt{-1}\times\frac{\pi}{4}}|011\rangle + |100\rangle + |101\rangle + (e^{1\times\sqrt{-1}\times\frac{\pi}{4}} \times e^{-1\times\sqrt{-1}\times\frac{\pi}{4}} \times e^{-1\times\sqrt{-1}\times\frac{\pi}{4}})|110\rangle + (e^{1\times\sqrt{-1}\times\frac{\pi}{4}} \times e^{1\times\sqrt{-1}\times\frac{\pi}{4}})|111\rangle)$ as its input and completes one Hadamard gate for q[0] in the *eleventh* time slot. If the value of q[0] is equal to one, then its state is changed as $\frac{1}{\sqrt{2}}(|0\rangle - |1\rangle)$. Otherwise, its state is changed as $\frac{1}{\sqrt{2}}(|0\rangle + |1\rangle)$. Hence, after using the statement "h q[0];" on line seventeen of Listing 2.3 completes one Hadamard gate for q[0] in the *eleventh* time slot, the following new state vector is

$$
\begin{aligned}
|\Phi_{12}\rangle &= \frac{1}{2\sqrt{2}}\Big(\frac{2}{\sqrt{2}}|000\rangle + \Big(\frac{2}{\sqrt{2}} \times e^{1\times\sqrt{-1}\times\frac{\pi}{4}}\Big)|010\rangle \\
&\qquad + \frac{2}{\sqrt{2}}|100\rangle + \Big(e^{-1\times\sqrt{-1}\times\frac{\pi}{4}} \times \frac{2}{\sqrt{2}}\Big)|111\rangle\Big) \\
&= \frac{1}{2}\Big(|000\rangle + e^{1\times\sqrt{-1}\times\frac{\pi}{4}}|010\rangle + |100\rangle + e^{-1\times\sqrt{-1}\times\frac{\pi}{4}}|111\rangle\Big).
\end{aligned}
$$

Next, the statement "cx q[2], q[1];" on line eighteen of Listing 2.3 takes the new state vector $|\Phi_{12}\rangle = \frac{1}{2}(|000\rangle + e^{1\times\sqrt{-1}\times\frac{\pi}{4}}|010\rangle + |100\rangle + e^{-1\times\sqrt{-1}\times\frac{\pi}{4}}|111\rangle)$ as its input and performs one **CNOT** gate for q[2] and q[1] in the *eleventh* time slot. If the value of the control bit q[2] is equal to one, then the statement "cx q[2], q[1];" flips the value of the target bit q[1]. Otherwise, its value is not changed. Hence, after using the statement "cx q[2], q[1];" on line eighteen of Listing 2.3 completes one **CNOT** gate for q[2] and q[1] in the *eleventh* time slot, the following new state vector is

$$|\Phi_{13}\rangle = \frac{1}{2}\left(|000\rangle + e^{1\times\sqrt{-1}\times\frac{\pi}{4}}|010\rangle + |110\rangle + e^{-1\times\sqrt{-1}\times\frac{\pi}{4}}|101\rangle\right).$$

Next, the statement "tdg q[1];" on line nineteen of Listing 2.3 takes the new state vector $|\Phi_{13}\rangle = \frac{1}{2}(|000\rangle + e^{1\times\sqrt{-1}\times\frac{\pi}{4}}|010\rangle + |110\rangle + e^{-1\times\sqrt{-1}\times\frac{\pi}{4}}|101\rangle)$ as its input and completes one T^+ gate for q[1] in the *twelfth* time slot. If the value of q[1] is equal to one, then the statement "tdg q[1];" changes its phase as $(e^{-1\times\sqrt{-1}\times\frac{\pi}{4}})$. Otherwise, its phase is not changed. There is no other quantum gate in the *twelfth* time slot to act on quantum bit q[0], so its state is not changed. Hence, after using the statement "tdg q[1];" on line nineteen of Listing 2.3 completes one T^+ gate for q[1] in the *twelfth* time slot, the following new state vector is

$$|\Phi_{14}\rangle = \frac{1}{2}\left(|000\rangle + \left(e^{-1\times\sqrt{-1}\times\frac{\pi}{4}} \times e^{1\times\sqrt{-1}\times\frac{\pi}{4}}\right)|010\rangle\right.$$
$$\left. + e^{-1\times\sqrt{-1}\times\frac{\pi}{4}}|110\rangle + e^{-1\times\sqrt{-1}\times\frac{\pi}{4}}|101\rangle\right).$$

Next, the statement "t q[2];" on line twenty of Listing 2.3 takes the new state vector $|\Phi_{14}\rangle = \frac{1}{2}(|000\rangle + (e^{-1\times\sqrt{-1}\times\frac{\pi}{4}} \times e^{1\times\sqrt{-1}\times\frac{\pi}{4}})|010\rangle + e^{-1\times\sqrt{-1}\times\frac{\pi}{4}}|110\rangle + e^{-1\times\sqrt{-1}\times\frac{\pi}{4}}|101\rangle)$ as its input and completes one T gate for q[2] in the *twelfth* time slot. If the value of q[2] is equal to one, then the statement "t q[2];" changes its phase as $(e^{1\times\sqrt{-1}\times\frac{\pi}{4}})$. Otherwise, its phase is not changed. There is no other quantum gate in the *twelfth* time slot to act on quantum bit q[0], so its state is not changed. Hence, after using the statement "t q[2];" on line twenty of Listing 2.3 completes one T gate for q[2] in the *twelfth* time slot, the following new state vector is

$$|\Phi_{15}\rangle = \frac{1}{2}\left(|000\rangle + |010\rangle + \left(e^{1\times\sqrt{-1}\times\frac{\pi}{4}} \times e^{-1\times\sqrt{-1}\times\frac{\pi}{4}}\right)|110\rangle\right.$$
$$\left. + \left(e^{1\times\sqrt{-1}\times\frac{\pi}{4}} \times e^{-1\times\sqrt{-1}\times\frac{\pi}{4}}\right)|101\rangle\right)$$
$$= \frac{1}{2}(|000\rangle + |010\rangle + |110\rangle + |101\rangle).$$

Next, the statement "cx q[2], q[1];" on line twenty-one of Listing 2.3 takes the new state vector $|\Phi_{15}\rangle = \frac{1}{2}(|000\rangle + |010\rangle + |110\rangle + |101\rangle)$ as its input and performs one *CNOT* gate for q[2] and q[1] in the *thirteenth* time slot. If the value of the control bit q[2] is equal to one, then the statement "cx q[2], q[1];" flips the value of the target bit q[1]. Otherwise, its value is not changed. Because there is no other quantum gate in the *thirteenth* time slot to act on quantum bit q[0], its state is not changed. Thus, after one *CNOT* gate for q[2] and q[1] in the *thirteenth* time slot is completed by the statement "cx q[2], q[1];" on line twenty-one of Listing 2.3, the following new state vector is obtained:

$$|\Phi_{16}\rangle = \frac{1}{2}(|000\rangle + |010\rangle + |100\rangle + |111\rangle).$$

Fig. 2.9 After the measurement to the program in Listing 2.3 is completed, we obtain the answer 00100 with the probability 0.260 or the answer 00010 with the probability 0.250 or the answer 00111 with the probability 0.250 or the answer 00000 with the probability 0.240

Next, the statement "measure q[0] → c[0];" on line twenty-two of Listing 2.3 is to measure the first quantum bit q[0] and to record the measurement outcome by overwriting the first classical bit c[0]. The statement "measure q[1] → c[1];" on line twenty-three of Listing 2.3 is to measure the second quantum bit q[1] and to record the measurement outcome by overwriting the second classical bit c[1]. Next, the statement "measure q[2] → c[2];" on line twenty-four of Listing 2.3 is to measure the third quantum bit q[2] and to record the measurement outcome by overwriting the third classical bit c[2]. In the backend *ibmqx4* with five quantum bits in **IBM**'s quantum computers, we use the command "simulate" to execute the program in Listing 2.3. The measured result appears in Fig. 2.9.

From Fig. 2.9, we obtain the answer 00100 ($c[4] = 0$, $c[3] = 0$, $c[2] = 1 = q[2]$ = |1⟩, $c[1] = 0 = q[1] = $ |0⟩ and $c[0] = 0 = q[0] = $ |0⟩) with the probability 0.260. Because the value of the first control bit q[2] is equal to one and the value of the second control bit q[1] is equal to zero, the value of the target bit q[0] is not changed and is equal to zero with the probability 0.260. Or we obtain the answer 00010 ($c[4]$ = 0, $c[3] = 0$, $c[2] = 0 = q[2] = $ |0⟩, $c[1] = 1 = q[1] = $ |1⟩ and $c[0] = 0 = q[0] = $ |0⟩) with the probability 0.250. The value of the first control bit q[2] is equal to zero and the value of the second control bit q[1] is equal to one, so the value of the target bit q[0] is not changed and is equal to zero with the probability 0.250.

Or we obtain the answer 00111 ($c[4] = 0$, $c[3] = 0$, $c[2] = 1 = q[2] = $ |1⟩, $c[1] = $ 1 = q[1] = |1⟩ and $c[0] = 1 = q[0] = $ |1⟩) with the probability 0.250. The value of the first control bit q[2] is equal to one and the value of the second control bit q[1] is also equal to one. Therefore, the **CCNOT** gate that implements one **AND** operation flips the value of the target bit q[0] and the value of the target bit q[0] is equal to one with the probability 0.250. Or we obtain the answer 00000 ($c[4] = 0$, $c[3] = 0$, $c[2] = 0 = q[2] = $ |0⟩, $c[1] = 0 = q[1] = $ |0⟩ and $c[0] = 0 = q[0] = $ |0⟩) with the probability 0.240. Because the value of the first control bit q[2] is equal to zero and the value of the second control bit q[1] is equal to zero, the value of the target bit q[0] is not changed and is equal to zero with the probability 0.240.

2.3 Introduction to AND Operation

The **AND** operation of a bit obtains two inputs of a bit and produces one single output of a bit. If the value of the first input is one (1) and the value of the second input is one (1), then it generates a result (output) of one (1). Otherwise, the result is zero (0). A symbol "$\wedge$" is used to represent the **AND** operation. Therefore, the **AND** operation of a bit that has two inputs of a bit is the following four possible combinational results:

$$0 \wedge 0 = 0$$
$$0 \wedge 1 = 0$$
$$1 \wedge 0 = 0$$
$$1 \wedge 1 = 1 \qquad (2.5)$$

The value of a Boolean variable (a bit) is only zero (0) or one (1). Thus, **AND** of two Boolean variables (two inputs of a bit) q[2] and q[1], written as q[2] $\wedge$ q[1] is equal to one (1) if and only if q[2] and q[1] are both one (1). Similarly, q[2] $\wedge$ q[1] is equal to zero (0) if and only if either q[2] or q[1], or both, are zero (0). We use a truth table with logic operation to represent all possible combinations of inputs and the corresponding outputs. Hence, the rules in (2.5) for the **AND** operation of a bit that has two inputs of a bit and generates one output of a bit may be expressed in the form of a truth table that appears in Table 2.3.

2.3.1 Quantum Program of Implementing AND Operation

We use one **CCNOT** gate that has three quantum input bits and three quantum output bits to implement **AND** operation of a *classical* bit that has two inputs of a classical bit and generates one output of a classical bit. We use the two control bits C_1 and C_2 of the **CCNOT** gate to encode two inputs q[2] and q[1] of a classical bit in **AND** operation of a *classical* bit. We apply the target bit T of the **CCNOT** gate to store one output q[2] $\wedge$ q[1] of a classical bit in **AND** operation of a *classical* bit. Using the

Table 2.3 The truth table for the **AND** operation of a bit that has two inputs of a bit and generates one output of a bit

Input		Output
q[2]	q[1]	q[2] $\wedge$ q[1]
0	0	0
0	1	0
1	0	0
1	1	1

form of a truth table that appears in Table 2.4 expresses the rule of using one **CCNOT** gate to implement **AND** operation. Its graph representation appears in Fig. 2.10. The initial state of the target bit in the **CCNOT** gate in Fig. 2.10 is set to $|0\rangle$.

In Listing 2.4, the program in the backend *ibmqx4* with five quantum bits in **IBM**'s quantum computer is the *fourth* example of the *second* chapter in which we describe how to write a quantum program to implement **AND** operation of a classical bit by means of using one **CCNOT** gate of three quantum bits. Figure 2.11 is the corresponding quantum circuit of the program in Listing 2.4. The statement "OPEN-QASM 2.0;" on line one of Listing 2.4 is to point out that the program is written with version 2.0 of Open QASM. Next, the statement "include "qelib1.inc";" on line two of Listing 2.4 is to continue parsing the file "qelib1.inc" as if the contents of the file were pasted at the location of the include statement, where the file "qelib1.inc" is **Quantum Experience (QE) Standard Header** and the path is specified relative to the current working directory.

Table 2.4 The truth table of using one **CCNOT** gate to implement **AND** operation

Input			Output		
C_1	C_2	T	C_1	C_2	$T = q[2] \wedge q[1]$
0	0	0	0	0	0
0	1	0	0	1	0
1	0	0	1	0	0
1	1	0	1	1	1

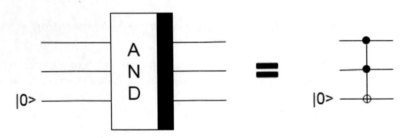

Fig. 2.10 The quantum circuit of implementing **AND** operation of a *classical* bit

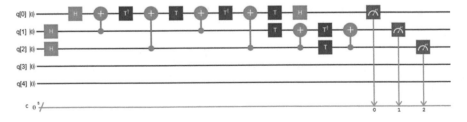

Fig. 2.11 The corresponding quantum circuit of the program in Listing 2.4

Listing 2.4 The program of using one *CCNOT* gate to implement **AND** operation.

```
1.   OPENQASM 2.0;
2.   include "qelib1.inc";
3.   qreg q[5];
4.   creg c[5];
5.   h q[1];
6.   h q[2];
7.   h q[0];
8.   cx q[1], q[0];
9.   tdg q[0];
10.  cx q[2], q[0];
11.  t q[0];
12.  cx q[1], q[0];
13.  tdg q[0];
14.  cx q[2], q[0];
15.  t q[0];
16.  t q[1];
17.  h q[0];
18.  cx q[2], q[1];
19.  tdg q[1];
20.  t q[2];
21.  cx q[2], q[1];
22.  measure q[0] → c[0];
23.  measure q[1] → c[1];
24.  measure q[2] → c[2];
```

Then, the statement "qreg q[5];" on line three of Listing 2.4 is to declare that in the program there are five quantum bits. In the left top of Fig. 2.11, five quantum bits are subsequently q[0], q[1], q[2], q[3] and q[4]. The initial value of each quantum bit is set to $|0\rangle$. We use three quantum bits q[2], q[1] and q[0] to encode respectively the first control bit, the second control bit and the target bit. This is to say that we apply quantum bits q[2] and q[1] to encode two inputs of a classical bit in **AND** operation of a *classical* bit and use quantum bit q[0] to store the result of **AND** operation of a *classical* bit. For the convenience of our explanation, $q[k]^0$ for $0 \leq k \leq 4$ is to represent the value of q[k] to be zero (0) and $q[k]^1$ for $0 \leq k \leq 4$ is to represent the value of q[k] to be one (1). Similarly, for the convenience of our explanation, an initial state vector of implementing **AND** operation of a classical bit is as follows:

$$|A_0\rangle = \left|q[2]^0\right\rangle\left|q[1]^0\right\rangle\left|q[0]^0\right\rangle = |0\rangle|0\rangle|0\rangle = |000\rangle.$$

Next, the statement "creg c[5];" on line four of Listing 2.4 is to declare that there are five classical bits in the program. In the left bottom of Fig. 2.11, five classical bits are subsequently c[0], c[1], c[2], c[3] and c[4]. The initial value of each classical bit is set to zero (0). Classical bit c[4] is the most significant bit and classical bit c[0] is the least significant bit.

Next, the two statements "h q[1];" and "h q[2];"on line five and line six of Listing 2.4 implement two Hadamard gates of the *first* time slot of the quantum circuit in Fig. 2.11. This implies that the statement "h q[1];" converts q[1] from one state $|0\rangle$ to another state $\frac{1}{\sqrt{2}}(|0\rangle + |1\rangle)$ (its superposition) and the statement "h q[2];" converts q[2] from one state $|0\rangle$ to another state $\frac{1}{\sqrt{2}}(|0\rangle + |1\rangle)$ (its superposition). In the *first* time slot of the quantum circuit in Fig. 2.11 there is no quantum gate to act on the quantum bit q[0], so its state $|0\rangle$ is not changed. Thus, after using the two statements "h q[1];" and "h q[2];"on line five and line six of Listing 2.4 implement two Hadamard gates in the *first* time slot of the quantum circuit in Fig. 2.11, the following new state vector is

$$|A_1\rangle = \frac{1}{2}(|0\rangle|0\rangle|0\rangle + |0\rangle|1\rangle|0\rangle + |1\rangle|0\rangle|0\rangle + |1\rangle|1\rangle|0\rangle)$$

$$= \frac{1}{2}(|000\rangle + |010\rangle + |100\rangle + |110\rangle).$$

In the new state vector $|A_1\rangle$, four combinational states of quantum bits q[2] and q[1] with that the amplitude of each combinational state is $\frac{1}{2}$ encode all of the possible inputs for **AND** operation of a classical bit. The initial state of quantum bit q[0] in four combinational states of quantum bits q[2] and q[1] is $|0\rangle$ and it stores the result for **AND** operation of a classical bit.

The next 12 time slots in the quantum circuit of Fig. 2.11 implement **AND** operation of a classical bit by means of implementing one *CCNOT* gate. There are fifteen statements from line seven through line twenty-one in Listing 2.4. The front eight statements are "h q[0];", "cx q[1], q[0];", "tdg q[0];", "cx q[2], q[0];", "t q[0];", "cx q[1], q[0];", "tdg q[0];" and "cx q[2], q[0];". The last seven statements are "t q[0];", "t q[1];", "h q[0];", "cx q[2], q[1];", "tdg q[1];", "t q[2];" and "cx q[2], q[1];". They implement each quantum gate from the *second* time slot through the *thirteenth* time slot in Fig. 2.11. They take the new state vector $|A_1\rangle = \frac{1}{2}(|000\rangle + |010\rangle + |100\rangle + |110\rangle)$ as the input in the *second* time slot and complete **AND** operation of a classical bit. This gives that the following new state vector is

$$|A_{16}\rangle = \frac{1}{2}(|000\rangle + |010\rangle + |100\rangle + |111\rangle).$$

Next, three measurements from the *fourteenth* time slot through the *sixteenth* time slot in Fig. 2.11 were implemented by the three statements "measure q[0] $\rightarrow$ c[0];", "measure q[1] $\rightarrow$ c[1];" and "measure q[2] $\rightarrow$ c[2];" on line twenty-two through line twenty-four of Listing 2.4. They are to measure the first quantum bit q[0], the second quantum bit q[1] and the third quantum bit q[2]. They are to record the measurement outcome by overwriting the first classical bit c[0], the second classical bit c[1] and the third classical bit c[2]. In the backend *ibmqx4* with five quantum bits in **IBM**'s

Fig. 2.12 After the measurement to the program in Listing 2.4 is completed, we obtain the answer 00010 with the probability 0.290 or the answer 00111 with the probability 0.280 or the answer 00100 with the probability 0.240 or the answer 00000 with the probability 0.190

quantum computers, we use the command "simulate" to execute the program in Listing 2.4. The measured result appears in Fig. 2.12.

From Fig. 2.12, we obtain the answer 00010 ($c[4] = 0$, $c[3] = 0$, $c[2] = 0 = q[2]$ $= |0\rangle$, $c[1] = 1 = q[1] = |1\rangle$ and $c[0] = 0 = q[0] = |0\rangle$) with the probability 0.290. Since the value of the first control bit $q[2]$ is equal to zero and the value of the second control bit $q[1]$ is equal to one, the value of the target bit $q[0]$ is not changed and is equal to zero with the probability 0.290. Or we obtain the answer 00111 ($c[4] =$ 0, $c[3] = 0$, $c[2] = 1 = q[2] = |1\rangle$, $c[1] = 1 = q[1] = |1\rangle$ and $c[0] = 1 = q[0] =$ $|1\rangle$) with the probability 0.280. The value of the first control bit $q[2]$ is equal to one and the value of the second control bit $q[1]$ is equal to one, so the **CCNOT** gate that implements one **AND** operation flips the value of the target bit $q[0]$ so that the value is equal to one with the probability 0.280.

Or we obtain the answer 00100 ($c[4] = 0$, $c[3] = 0$, $c[2] = 1 = q[2] = |1\rangle$), $c[1]$ $= 0 = q[1] = |0\rangle$ and $c[0] = 0 = q[0] = |0\rangle$) with the probability 0.240. Because the value of the first control bit $q[2]$ is equal to one and the value of the second control bit $q[1]$ is equal to zero, the value of the target bit $q[0]$ is not changed and is equal to zero with the probability 0.240. Or we obtain the answer 00000 ($c[4] = 0$, $c[3] =$ 0, $c[2] = 0 = q[2] = |0\rangle$), $c[1] = 0 = q[1] = |0\rangle$ and $c[0] = 0 = q[0] = |0\rangle$) with the probability 0.190. The value of the first control bit $q[2]$ is equal to zero and the value of the second control bit $q[1]$ is equal to zero, so the value of the target bit $q[0]$ is not changed and is equal to zero with the probability 0.190.

2.4 Introduction to NAND Operation

The **NAND** operation of a bit takes two inputs of a bit and generates one single output of a bit. If the value of the first input is one (1) and the value of the second input is one (1), then it yields a result (output) of zero (0). Otherwise, the result (output) is one (1). A symbol "$\overline{\wedge}$" is applied to represent the **NAND** operation. Hence, the **NAND** operation of a bit that has two inputs of a bit is the following four possible combinational results:

$$\overline{0 \wedge 0} = 1$$

$$\overline{0 \wedge 1} = 1$$

$$\overline{1 \wedge 0} = 1$$
$$\overline{1 \wedge 1} = 0 \tag{2.6}$$

The value of a Boolean variable (a bit) is only one (1) or zero (0). Hence, **NAND** of two Boolean variables (two inputs of a bit) q[2] and q[1], written as $\overline{q[2] \wedge q[1]}$ is equal to zero (0) if and only if q[2] and q[1] are both one (1). Similarly, $\overline{q[2] \wedge q[1]}$ is equal to one (1) if and only if either q[2] or q[1], or both, are zero (0). Usually using a truth table with logic operation is to represent all possible combinations of inputs and the corresponding outputs. Therefore, using the form of a truth table that appears in Table 2.5 may express the rules in (2.6) for the **NAND** operation of a bit that has two inputs of a bit and produces one single output of a bit.

2.4.1 Quantum Program of Implementing NAND Operation

We apply one **CCNOT** gate that has three quantum input bits and three quantum output bits to implement **NAND** operation of a *classical* bit that has two inputs of a classical bit and produces one output of a classical bit. We make use of the two control bits C_1 and C_2 of the **CCNOT** gate to encode two inputs q[2] and q[1] of a classical bit in **NAND** operation of a *classical* bit and use the target bit T of the **CCNOT** gate to store one output $\overline{q[2] \wedge q[1]}$ of a classical bit in **NAND** operation of a *classical* bit. Using the form of a truth table that appears in Table 2.6 may express the rule of applying one **CCNOT** gate to implement **NAND** operation. Its graph

Table 2.5 The truth table for the **NAND** operation of a bit that has two inputs of a bit and produces one output of a bit

Input		Output
q[2]	q[1]	$\overline{q[2] \wedge q[1]}$
0	0	1
0	1	1
1	0	1
1	1	0

Table 2.6 The truth table of applying one **CCNOT** gate to implement **NAND** operation

Input			Output		
C_1	C_2	T	C_1	C_2	$T = \overline{q[2] \wedge q[1]}$
0	0	1	0	0	1
0	1	1	0	1	1
1	0	1	1	0	1
1	1	1	1	1	0

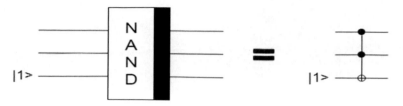

Fig. 2.13 The quantum circuit of implementing **NAND** operation of a *classical* bit

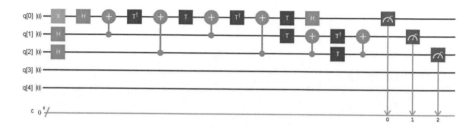

Fig. 2.14 The corresponding quantum circuit of the program in Listing 2.5

representation appears in Fig. 2.13. The initial state of the target bit in the **CCNOT** gate in Fig. 2.13 is set to $|1\rangle$.

In Listing 2.5, the program in the backend *ibmqx4* with five quantum bits in **IBM**'s quantum computer is the *fifth* example of the *second* chapter in which we illustrate how to write a quantum program to implement **NAND** operation of a classical bit by means of applying one **CCNOT** gate of three quantum bits. Figure 2.14 is the corresponding quantum circuit of the program in Listing 2.5. The statement "OPENQASM 2.0;" on line one of Listing 2.5 is to indicate that the program is written with version 2.0 of Open QASM. Then, the statement "include "qelib1.inc";" on line two of Listing 2.5 is to continue parsing the file "qelib1.inc" as if the contents of the file were pasted at the location of the include statement, where the file "qelib1.inc" is **Quantum Experience (QE) Standard Header** and the path is specified relative to the current working directory.

Listing 2.5 The program of applying one **CCNOT** gate to implement **NAND** operation.

```
1.  OPENQASM 2.0;
2.  include "qelib1.inc";
3.  qreg q[5];
4.  creg c[5];
5.  x q[0];
6.  h q[1];
```

```
 7.   h q[2];
 8.   h q[0];
 9.   cx q[1], q[0];
10.   tdg q[0];
11.   cx q[2], q[0];
12.   t q[0];
13.   cx q[1], q[0];
14.   tdg q[0];
15.   cx q[2], q[0];
16.   t q[0];
17.   t q[1];
18.   h q[0];
19.   cx q[2], q[1];
20.   tdg q[1];
21.   t q[2];
22.   cx q[2], q[1];
23.   measure q[0] → c[0];
24.   measure q[1] → c[1];
25.   measure q[2] → c[2];
```

Next, the statement "qreg q[5];" on line three of Listing 2.5 is to declare that in the program there are five quantum bits. In the left top of Fig. 2.14, five quantum bits are subsequently q[0], q[1], q[2], q[3] and q[4]. The initial value of each quantum bit is set to $|0\rangle$. We make use of three quantum bits q[2], q[1] and q[0] to encode respectively the first control bit, the second control bit and the target bit. This is to say that we use quantum bits q[2] and q[1] to encode two inputs of a classical bit in **NAND** operation of a *classical* bit and apply quantum bit q[0] to store the result of **NAND** operation of a *classical* bit. For the convenience of our explanation, $q[k]^0$ for $0 \le k \le 4$ is to represent the value of q[k] to be zero (0) and $q[k]^1$ for $0 \le k \le 4$ is to represent the value of q[k] to be one (1). Similarly, for the convenience of our explanation, an initial state vector of implementing **NAND** operation of a classical bit is as follows:

$$|B_0\rangle = |q[2]^0\rangle|q[1]^0\rangle|q[0]^0\rangle = |0\rangle|0\rangle|0\rangle = |000\rangle.$$

Next, the statement "creg c[5];" on line four of Listing 2.5 is to declare that there are five classical bits in the program. In the left bottom of Fig. 2.14, five classical bits are subsequently c[0], c[1], c[2], c[3] and c[4]. The initial value of each classical bit is set to zero (0). Classical bit c[4] is the most significant bit and classical bit c[0] is the least significant bit.

Next, the three statements "x q[0];", "h q[1];" and "h q[2];"on line five through line seven of Listing 2.5 implement one *NOT* gate and two Hadamard gates of the *first* time slot of the quantum circuit in Fig. 2.14. This is to say that the statement "x q[0];" converts q[0] from one state $|0\rangle$ to another state $|1\rangle$ (its negation). The statement "h q[1];" converts q[1] from one state $|0\rangle$ to another state $\frac{1}{\sqrt{2}}(|0\rangle + |1\rangle)$ (its superposition) and the statement "h q[2];" converts q[2] from one state $|0\rangle$ to another state $\frac{1}{\sqrt{2}}(|0\rangle + |1\rangle)$ (its superposition). Therefore, after one *NOT* gate and two Hadamard gates in the *first* time slot of the quantum circuit in Fig. 2.14 are implemented by means of applying the three statements "x q[0];", "h q[1];" and "h q[2];"on line five through line seven of Listing 2.5, the following new state vector is obtained:

$$|B_1\rangle = \frac{1}{2}(|0\rangle|0\rangle|1\rangle + |0\rangle|1\rangle|1\rangle + |1\rangle|0\rangle|1\rangle + |1\rangle|1\rangle|1\rangle)$$

$$= \frac{1}{2}(|001\rangle + |011\rangle + |101\rangle + |111\rangle).$$

In the new state vector $|B_1\rangle$, four combinational states of quantum bits q[2] and q[1] with that the amplitude of each combinational state is $\frac{1}{2}$ encode all of the possible inputs for **NAND** operation of a classical bit. The initial state of quantum bit q[0] in four combinational states of quantum bits q[2] and q[1] is $|1\rangle$ and it stores the result for **NAND** operation of a classical bit.

The next 12 time slots in the quantum circuit of Fig. 2.14 implement **NAND** operation of a classical bit by means of implementing one *CCNOT* gate. There are fifteen statements from line eight through line twenty-two in Listing 2.5. The front eight statements are "h q[0];", "cx q[1], q[0];", "tdg q[0];", "cx q[2], q[0];", "t q[0];", "cx q[1], q[0];", "tdg q[0];" and "cx q[2], q[0];". The rear seven statements are "t q[0];", "t q[1];", "h q[0];", "cx q[2], q[1];", "tdg q[1];", "t q[2];" and "cx q[2], q[1];". They implemented each quantum gate from the *second* time slot through the *thirteenth* time slot in Fig. 2.14. They take the new state vector $|B_1\rangle = \frac{1}{2}(|001\rangle + |011\rangle + |101\rangle + |111\rangle)$ as the input in the *second* time slot and complete **NAND** operation of a classical bit. This gives that the following new state vector is

$$|B_{16}\rangle = \frac{1}{2}(|001\rangle + |011\rangle + |101\rangle + |110\rangle).$$

Then, three measurements from the *fourteenth* time slot through the *sixteenth* time slot in Fig. 2.14 were implemented by the three statements "measure q[0] $\rightarrow$ c[0];", "measure q[1] $\rightarrow$ c[1];" and "measure q[2] $\rightarrow$ c[2];" on line twenty-three through line twenty-five of Listing 2.5. They are to measure the first quantum bit q[0], the second quantum bit q[1] and the third quantum bit q[2]. They record the measurement outcome by overwriting the first classical bit c[0], the second classical bit c[1] and the third classical bit c[2]. In the backend *ibmqx4* with five quantum bits in **IBM**'s quantum computers, we make use of the command "simulate" to execute the program in Listing 2.5. The measured result appears in Fig. 2.15. From Fig. 2.15, we obtain the answer 00011 (c[4] = 0, c[3] = 0, c[2] = 0 = q[2] = $|0\rangle$), c[1] = 1 = q[1] = $|1\rangle$ and c[0] = 1 = q[0] = $|1\rangle$) with the probability 0.340. Because the value of the first control bit q[2] is equal to zero (0) and the value of the second control bit q[1]

Fig. 2.15 After the measurement to the program in Listing 2.5 is completed, we obtain the answer 00011 with the probability 0.340 or the answer 00001 with the probability 0.300 or the answer 00101 with the probability 0.200 or the answer 00110 with the probability 0.160

is equal to one (1), the value of the target bit q[0] is not changed and is equal to one (1) with the probability 0.340.

Or we obtain the answer 00001 (c[4] = 0, c[3] = 0, c[2] = 0 = q[2] = |0⟩, c[1] = 0 = q[1] = |0⟩ and c[0] = 1 = q[0] = |1⟩) with the probability 0.300. Since the value of the first control bit q[2] is equal to zero (0) and the value of the second control bit q[1] is equal to zero (0), the value of the target bit q[0] is not changed and is equal to one (1) with the probability 0.300. Or we obtain the answer 00101 (c[4] = 0, c[3] = 0, c[2] = 1 = q[2] = |1⟩, c[1] = 0 = q[1] = |0⟩ and c[0] = 1 = q[0] = |1⟩) with the probability 0.200. The value of the first control bit q[2] is equal to one (1) and the value of the second control bit q[1] is equal to zero (0), so the value of the target bit q[0] is not changed and is equal to one (1) with the probability 0.200. Or we obtain the answer 00110 (c[4] = 0, c[3] = 0, c[2] = 1 = q[2] = |1⟩, c[1] = 1 = q[1] = |1⟩ and c[0] = 0 = q[0] = |0⟩) with the probability 0.160. The value of the first control bit q[2] is equal to one (1) and the value of the second control bit q[1] is also equal to one (1). Therefore, the **CCNOT** gate that implements one **NAND** operation flips the value of the target bit q[0] so that the value is equal to zero (0) with the probability 0.160.

2.5 Introduction to OR Operation

The **OR** operation of a bit acquires two inputs of a bit and yields one single output of a bit. If the value of the first input is one (1) or the value of the second input is also one (1) or their values are both one (1), then it produces a result (output) of 1 (one). Otherwise, the result (output) is zero (0). A symbol "$\vee$" is used to represent the **OR** operation. Thus, the **OR** operation of a bit that takes two inputs of a bit is the following four possible combinational results:

$$0 \vee 0 = 0$$
$$0 \vee 1 = 1$$
$$1 \vee 0 = 1$$
$$1 \vee 1 = 1 \qquad (2.7)$$

Table 2.7 The truth table for the **OR** operation of a bit that takes two inputs of a bit and generates one single output of a bit

Input		Output
q[2]	q[1]	q[2] ∨ q[1]
0	0	0
0	1	1
1	0	1
1	1	1

The value of a Boolean variable (a bit) is only one (1) or zero (0). Therefore, **OR** of two Boolean variables (two inputs of a bit) q[2] and q[1], written as q[2] ∨ q[1] is equal to one (1) if and only if the value of q[2] is one (1) or the value of q[1] is one (1) or their values are both one (1). Similarly, q[2] ∨ q[1] is equal to zero (0) if and only if the value of q[2] and the value of q[1] are both zero (0). Often using a truth table with logic operation is to represent all possible combinations of inputs and the corresponding outputs. Hence, the rules in (2.7) for the **OR** operation of a bit that obtains two inputs of a bit and generates one single output of a bit may also be expressed in the form of a truth table that appears in Table 2.7.

2.5.1 Quantum Program of Implementing OR Operation

We make use of one **CCNOT** gate that has three quantum input bits and three quantum output bits to implement **OR** operation of a *classical* bit that obtains two inputs of a classical bit and yields one output of a classical bit. We use the two control bits C_1 and C_2 of the **CCNOT** gate to encode two inputs q[2] and q[1] of a classical bit in **OR** operation of a *classical* bit. We apply the target bit T of the **CCNOT** gate to store one output q[2] ∨ q[1] = $\overline{\overline{q[2] \vee q[1]}}$ = $\overline{\overline{q[2]} \wedge \overline{q[1]}}$ of a classical bit in **OR** operation of a *classical* bit.

Using the form of a truth table that appears in Table 2.8 may express the rule of using one **CCNOT** gate to implement **OR** operation. Its graph representation appears in Fig. 2.16. In Fig. 2.16, the first control bit (the top first wire) C_1 and the second

Table 2.8 The truth table of applying one **CCNOT** gate to implement **OR** operation

Input			Output		
C_1	C_2	T	C_1	C_2	$T = q[2] \vee q[1] = \overline{\overline{q[2]} \wedge \overline{q[1]}}$
0	0	1	0	0	0
0	1	1	0	1	1
1	0	1	1	0	1
1	1	1	1	1	1

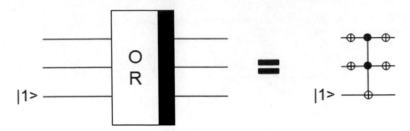

Fig. 2.16 The quantum circuit of implementing **OR** operation of a *classical* bit

control bit (the second wire) C_2 of the **CCNOT** gate respectively encode the first input q[2] and the second input q[1] of a classical bit in **OR** operation of a classical bit. In Fig. 2.16, the target bit (the bottom wire) T of the **CCNOT** gate is to store one output q[2] $\lor$ q[1] $= \overline{\overline{q[2]} \land \overline{q[1]}}$ in **OR** operation of a classical bit.

The initial state of the target bit T in the **CCNOT** gate in Fig. 2.16 is set to $|1\rangle$. Implementing **OR** operation of a *classical* bit takes two inputs q[2] and q[1] of a classical bit and produces one output $\overline{\overline{q[2]} \land \overline{q[1]}}$ of a classical bit. It is equivalent to implement **NAND** operation of a classical bit that takes two inputs $\overline{q[2]}$ and $\overline{q[1]}$ of a classical bit and yields one output $\overline{(\overline{q[2]} \land \overline{q[1]})}$. Therefore, in Fig. 2.16, we use two **NOT** gates to operate the two control bits C_1 and C_2 of the **CCNOT** gate that encode two inputs q[2] and q[1] of a classical bit and to generate their negations $\overline{q[2]}$ and $\overline{q[1]}$. Next, in Fig. 2.16, we apply one **CCNOT** gate to take their negations $\overline{q[2]}$ and $\overline{q[1]}$ as the input and to complete **NAND** operation of a classical bit. From Table 2.8, two inputs q[2] and q[1] of a classical bit in **OR** operation of a *classical* bit that is encoded by the two control bits C_1 and C_2 of the **CCNOT** gate in Fig. 2.16 are not changed. Next, we again make use of two **NOT** gates to operate the two control bits C_1 and C_2 of the **CCNOT** gate in Fig. 2.16 and to generate the result $\overline{\overline{q[2]}} =$ q[2] and $\overline{\overline{q[1]}} =$ q[1]. This is to say that using **NOT** gate twice to the first control bit C_1 and the second control bit C_2 of the **CCNOT** gate in Fig. 2.16 does nothing to them.

In Listing 2.6, the program in the backend *ibmqx4* with five quantum bits in **IBM**'s quantum computer is the *sixth* example of the *second* chapter. We introduce how to write a quantum program to implement **OR** operation of a classical bit by means of using one **CCNOT** gate of three quantum bits and four **NOT** gates of one quantum bit. Figure 2.17 is the corresponding quantum circuit of the program in Listing 2.6. The statement "OPENQASM 2.0;" on line one of Listing 2.6 is to point out that the program is written with version 2.0 of Open QASM. Next, the statement "include "qelib1.inc";" on line two of Listing 2.6 is to continue parsing the file "qelib1.inc" as if the contents of the file were pasted at the location of the include statement, where the file "qelib1.inc" is **Quantum Experience (QE) Standard Header** and the path is specified relative to the current working directory.

Listing 2.6 The program of using one **CCNOT** gate and four **NOT** gates to implement **OR** operation.

```
1.    OPENQASM 2.0;
2.    include "qelib1.inc";
3.    qreg q[5];
4.    creg c[5];
5.    x q[0];
6.    h q[1];
7.    h q[2];
8.    x q[1];
9.    x q[2];
10.   h q[0];
11.   cx q[1], q[0];
12.   tdg q[0];
13.   cx q[2], q[0];
14.   t q[0];
15.   cx q[1], q[0];
16.   tdg q[0];
17.   cx q[2], q[0];
18.   t q[0];
19.   t q[1];
20.   h q[0];
21.   cx q[2], q[1];
22.   tdg q[1];
23.   t q[2];
24.   cx q[2], q[1];
25.   x q[1];
26.   x q[2];
27.   measure q[0] → c[0];
28.   measure q[1] → c[1];
29.   measure q[2] → c[2];
```

Then, the statement "qreg q[5];" on line three of Listing 2.6 is to declare that in the program there are five quantum bits. In the left top of Fig. 2.17, five quantum bits are subsequently q[0], q[1], q[2], q[3] and q[4]. The initial value of each quantum

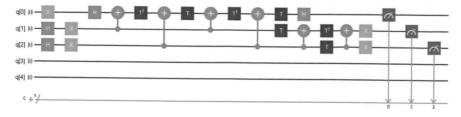

Fig. 2.17 The corresponding quantum circuit of the program in Listing 2.6

bit is set to $|0\rangle$. We use three quantum bits q[2], q[1] and q[0] to encode respectively the first control bit, the second control bit and the target bit. This indicates that we make use of quantum bits q[2] and q[1] to encode two inputs of a classical bit in **OR** operation of a *classical* bit and apply quantum bit q[0] to store the result of **OR** operation of a *classical* bit. For the convenience of our explanation, q[k]^0 for $0 \leq k \leq 4$ is to represent the value of q[k] to be zero (0) and q[k]^1 for $0 \leq k \leq 4$ is to represent the value of q[k] to be one (1). Similarly, for the convenience of our explanation, an initial state vector of implementing **OR** operation of a classical bit is as follows:

$$|C_0\rangle = \left|q[2]^0\right\rangle\left|q[1]^0\right\rangle\left|q[0]^0\right\rangle = |0\rangle|0\rangle|0\rangle = |000\rangle.$$

Next, the statement "creg c[5];" on line four of Listing 2.6 is to declare that there are five classical bits in the program. In the left bottom of Fig. 2.17, five classical bits are subsequently c[0], c[1], c[2], c[3] and c[4]. The initial value of each classical bit is set to zero (0). Classical bit c[4] is the most significant bit and classical bit c[0] is the least significant bit.

Next, the three statements "x q[0];", "h q[1];" and "h q[2];"on line five through line seven of Listing 2.6 implement one *NOT* gate and two Hadamard gates of the *first* time slot of the quantum circuit in Fig. 2.17. This implies that the statement "x q[0];" converts q[0] from one state $|0\rangle$ to another state $|1\rangle$ (its negation), the statement "h q[1];" converts q[1] from one state $|0\rangle$ to another state $\frac{1}{\sqrt{2}}(|0\rangle + |1\rangle)$ (its superposition) and the statement "h q[2];" converts q[2] from one state $|0\rangle$ to another state $\frac{1}{\sqrt{2}}(|0\rangle + |1\rangle)$ (its superposition). Hence, after one *NOT* gate and two Hadamard gates in the *first* time slot of the quantum circuit in Fig. 2.17 are implemented by means of using the three statements "x q[0];", "h q[1];" and "h q[2];"on line five through line seven of Listing 2.6, the following new state vector is obtained:

$$|C_1\rangle = \frac{1}{2}(|0\rangle|0\rangle|1\rangle + |0\rangle|1\rangle|1\rangle + |1\rangle|0\rangle|1\rangle + |1\rangle|1\rangle|1\rangle)$$

$$= \frac{1}{2}(|001\rangle + |011\rangle + |101\rangle + |111\rangle).$$

In the new state vector $|C_1\rangle$, four combinational states of quantum bits q[2] and q[1] with that the amplitude of each combinational state is $\frac{1}{2}$ encode all of the possible inputs for **OR** operation of a classical bit. The initial state of quantum bit q[0] in four combinational states of quantum bits q[2] and q[1] is $|1\rangle$ and it stores the result for **OR** operation of a classical bit.

Then, the two statements "x q[1];" and "x q[2];" on line eight through line nine of Listing 2.6 implement two *NOT* gates of the *second* time slot of the quantum circuit in Fig. 2.17. They take the new state vector $|C_1\rangle = \frac{1}{2}(|001\rangle + |011\rangle + |101\rangle + |111\rangle)$ as the input in the *second* time slot of Fig. 2.17. This is to say that in the new state vector $|C_1\rangle$ the state $(|0\rangle + |1\rangle)$ of q[2] is converted into the state $(|1\rangle + |0\rangle)$ and the state $(|0\rangle + |1\rangle)$ of q[1] is converted into the state $(|1\rangle + |0\rangle)$. Because there is no gate to act on q[0], its state is not changed. Therefore, after two *NOT* gates in the *second* time slot of the quantum circuit in Fig. 2.17 are implemented by means of applying the two statements "x q[1];" and "x q[2];" on line eight through line nine of Listing 2.6, the following new state vector is obtained:

$$|C_2\rangle = \frac{1}{2}(|1\rangle|1\rangle|1\rangle + |1\rangle|0\rangle|1\rangle + |0\rangle|1\rangle|1\rangle + |0\rangle|0\rangle|1\rangle)$$

$$= \frac{1}{2}(|111\rangle + |101\rangle + |011\rangle + |001\rangle).$$

The next 12 time slots in the quantum circuit of Fig. 2.17 implement **OR** operation ($q[2] \vee q[1] = \overline{\overline{q[2]} \wedge \overline{q[1]}}$) of a classical bit that is equivalent to implement **NAND** operation of a classical bit with two inputs $\overline{q[2]}$ and $\overline{q[1]}$ by means of implementing one **CCNOT** gate. There are fifteen statements from line ten through line twenty-four in Listing 2.6. The front eight statements are "h q[0];", "cx q[1], q[0];", "tdg q[0];", "cx q[2], q[0];", "t q[0];", "cx q[1], q[0];", "tdg q[0];" and "cx q[2], q[0];". The rear seven statements are "t q[0];", "t q[1];", "h q[0];", "cx q[2], q[1];", "tdg q[1];", "t q[2];" and "cx q[2], q[1];". They implemented each quantum gate from the *third* time slot through the *fourteenth* time slot in Fig. 2.17. They take the new state vector $|C_2\rangle = \frac{1}{2}(|111\rangle + |101\rangle + |011\rangle + |001\rangle)$ as the input in the *third* time slot and complete **OR** operation ($q[2] \vee q[1] = \overline{\overline{q[2]} \wedge \overline{q[1]}}$) of a classical bit. This gives that the following new state vector is

$$|C_{17}\rangle = \frac{1}{2}(|110\rangle + |101\rangle + |011\rangle + |001\rangle).$$

Next, the two statements "x q[1];" and "x q[2];" on line twenty-five through line twenty-six of Listing 2.6 implement two **NOT** gates of the *fifteenth* time slot of the quantum circuit in Fig. 2.17. They take the new state vector $|C_{17}\rangle = \frac{1}{2}(|110\rangle + |101\rangle + |011\rangle + |001\rangle)$ as the input in the *fifteenth* time slot of Fig. 2.17. This is to say that in the new state vector $|C_{17}\rangle$ the state ($|110\rangle$) is converted into the state ($|000\rangle$), the state ($|101\rangle$) is converted into the state ($|011\rangle$), the state ($|011\rangle$) is converted into the state ($|101\rangle$) and the state ($|001\rangle$) is converted into the state ($|111\rangle$). Because there is no gate to act on q[0], its state is not changed. Thus, after two **NOT** gates in the *fifteenth* time slot of the quantum circuit in Fig. 2.17 were implemented by means of applying the two statements "x q[1];" and "x q[2];" on line twenty-five through line twenty-six of Listing 2.6, the following new state vector is obtained:

$$|C_{18}\rangle = \frac{1}{2}(|000\rangle + |011\rangle + |101\rangle + |111\rangle).$$

Next, three measurements from the *sixteenth* time slot through the *eighteenth* time slot in Fig. 2.17 were implemented by the three statements "measure q[0] → c[0];", "measure q[1] → c[1];" and "measure q[2] → c[2];" on line twenty-seven through line twenty-nine of Listing 2.6. They are to measure the first quantum bit q[0], the second quantum bit q[1] and the third quantum bit q[2]. They record the measurement

Fig. 2.18 After the measurement to the program in Listing 2.6 is completed, we obtain the answer 00101 with the probability 0.290 or the answer 00011 with the probability 0.270 or the answer 00000 with the probability 0.250 or the answer 00111 with the probability 0.190

outcome by overwriting the first classical bit $c[0]$, the second classical bit $c[1]$ and the third classical bit $c[2]$. In the backend *ibmqx4* with five quantum bits in **IBM**'s quantum computers, we use the command "simulate" to execute the program in Listing 2.6. The measured result appears in Fig. 2.18. From Fig. 2.18, we obtain the answer 00101 ($c[4] = 0$, $c[3] = 0$, $c[2] = 1 = q[2] = |1\rangle$, $c[1] = 0 = q[1] = |0\rangle$ and $c[0] = 1 = q[0] = |1\rangle$) with the probability 0.290. In **OR** operation of a classical bit, the value of the first input (the first control bit) $q[2]$ is equal to one (1) and the value of the second input (the second control bit) $q[1]$ is equal to zero (0). Therefore, the value of the output (the target bit) $q[0]$ is equal to one (1) with the probability 0.290.

Or we obtain the answer 00011 ($c[4] = 0$, $c[3] = 0$, $c[2] = 0 = q[2] = |0\rangle$, $c[1] = 1 = q[1] = |1\rangle$ and $c[0] = 1 = q[0] = |1\rangle$) with the probability 0.270. In **OR** operation of a classical bit, the value of the first input (the first control bit) $q[2]$ is equal to zero (0) and the value of the second input (the second control bit) $q[1]$ is equal to one (1). Hence, the value of the output (the target bit) $q[0]$ is equal to one (1) with the probability 0.270. Or we obtain the answer 00000 ($c[4] = 0$, $c[3] = 0$, $c[2] = 0 = q[2] = |0\rangle$, $c[1] = 0 = q[1] = |0\rangle$ and $c[0] = 0 = q[0] = |0\rangle$) with the probability 0.250. In **OR** operation of a classical bit, the value of the first input (the first control bit) $q[2]$ is equal to zero (0) and the value of the second input (the second control bit) $q[1]$ is also equal to zero (0). Thus, the value of the output (the target bit) $q[0]$ is equal to zero (0) with the probability 0.250. Or we obtain the answer 00111 ($c[4] = 0$, $c[3] = 0$, $c[2] = 1 = q[2] = |1\rangle$, $c[1] = 1 = q[1] = |1\rangle$ and $c[0] = 1 = q[0] = |1\rangle$) with the probability 0.190. In **OR** operation of a classical bit, the value of the first input (the first control bit) $q[2]$ is equal to one (1) and the value of the second input (the second control bit) $q[1]$ is also equal to one (1). Therefore, the value of the output (the target bit) is equal to one (1) with the probability 0.190.

2.6   Introduction of NOR Operation

The **NOR** operation of a bit obtains two inputs of a bit and produces one single output of a bit. If the value of the first input is zero (0) and the value of the second input is zero (0), then it produces a result (output) of one (1). However, if either the value of the first input or the value of the second input, or both of them, are one (1), then it yields a result (output) of zero (0). A symbol "$\overline{\vee}$" is applied to represent the

Table 2.9 The truth table for the **NOR** operation of a bit that takes two inputs of a bit and generates one single output of a bit

Input		Output
q[2]	q[1]	$\overline{q[2] \vee q[1]}$
0	0	1
0	1	0
1	0	0
1	1	0

NOR operation. Therefore, the **NOR** operation of a bit that acquires two inputs of a bit is the following four possible combinational results:

$$\overline{0 \vee 0} = 1$$

$$\overline{0 \vee 1} = 0$$

$$\overline{1 \vee 0} = 0$$

$$\overline{1 \vee 1} = 0 \tag{2.8}$$

The value of a Boolean variable (a bit) is only zero (0) or one (1). Hence, **NOR** operation of two Boolean variables (two inputs of a bit) q[2] and q[1], written as $\overline{q[2] \vee q[1]}$ is equal to one (1) if and only if the value of q[2] is zero (0) and the value of q[1] is zero (0). Similarly, $\overline{q[2] \vee q[1]}$ is equal to zero (0) if and only if either the value of q[2] is one (1) or the value of q[1] is one (1) or both of them are one (1). Usually using a truth table with logic operation is to represent all possible combinations of inputs and the corresponding outputs. Thus, the rules in (2.8) for the **NOR** operation of a bit that takes two inputs of a bit and generates one single output of a bit may also be expressed in the form of a truth table that appears in Table 2.9.

2.6.1 Quantum Program of Implementing NOR Operation

We use one **CCNOT** gate that has three quantum input bits and three quantum output bits to implement **NOR** operation of a *classical* bit that acquires two inputs of a classical bit and generates one output of a classical bit. We apply the two control bits C_1 and C_2 of the **CCNOT** gate to encode two inputs q[2] and q[1] of a classical bit in **NOR** operation of a *classical* bit and make use of the target bit T of the **CCNOT** gate to store one output $\overline{q[2] \vee q[1]} = \overline{q[2]} \wedge \overline{q[1]}$ of a classical bit in **NOR** operation of a *classical* bit. Using the form of a truth table that appears in Table 2.10 may express the rule of applying one **CCNOT** gate to complete **NOR** operation. Its graph representation appears in Fig. 2.19. In Fig. 2.19, the first control bit (the top first wire) C_1 and the second control bit (the second wire) C_2 of the **CCNOT** gate respectively encode the first input q[2] and the second input q[1] of a classical bit in

Table 2.10 The truth table of using one *CCNOT* gate to implement **NOR** operation

Input			Output		
C_1	C_2	T	C_1	C_2	$T = \overline{q[2] \vee q[1]} = \overline{q[2]} \wedge \overline{q[1]}$
0	0	0	0	0	1
0	1	0	0	1	0
1	0	0	1	0	0
1	1	0	1	1	0

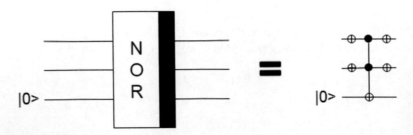

Fig. 2.19 The quantum circuit of implementing **NOR** operation of a *classical* bit

NOR operation of a classical bit. In Fig. 2.19, the target bit (the bottom wire) T of the *CCNOT* gate is to store one output $\overline{q[2] \vee q[1]} = \overline{q[2]} \wedge \overline{q[1]}$ in **NOR** operation of a classical bit.

The initial state of the target bit T in the *CCNOT* gate in Fig. 2.19 is set to $|0\rangle$. Implementing **NOR** operation of a *classical* bit takes two inputs q[2] and q[1] of a classical bit and generates one output $\overline{q[2] \vee q[1]} = \overline{q[2]} \wedge \overline{q[1]}$ of a classical bit. It is equivalent to complete **AND** operation of a classical bit that acquires two inputs $\overline{q[2]}$ and $\overline{q[1]}$ of a classical bit and produces one output $(\overline{q[2]} \wedge \overline{q[1]})$. Hence, in Fig. 2.19, we apply two *NOT* gates to operate the two control bits C_1 and C_2 of the *CCNOT* gate that encode two inputs q[2] and q[1] of a classical bit and to yield their negations $\overline{q[2]}$ and $\overline{q[1]}$. Next, in Fig. 2.19, we make use of one *CCNOT* gate to obtain their negations $\overline{q[2]}$ and $\overline{q[1]}$ as the input and to implement **AND** operation of a classical bit. From Table 2.10, two inputs q[2] and q[1] of a classical bit in **NOR** operation of a *classical* bit that is encoded by the two control bits C_1 and C_2 of the *CCNOT* gate in Fig. 2.19 are not changed. Therefore, we again use two *NOT* gates to operate the two control bits C_1 and C_2 of the *CCNOT* gate in Fig. 2.19 and to produce the result $\overline{\overline{q[2]}} = $ q[2] and $\overline{\overline{q[1]}} = $ q[1]. This implies that using *NOT* gate twice to the first control bit C_1 and the second control bit C_2 of the *CCNOT* gate in Fig. 2.19 does nothing to them.

In Listing 2.7, the program in the backend *ibmqx4* with five quantum bits in **IBM**'s quantum computer is the *seventh* example of the *second* chapter. We describe how to write a quantum program to implement one **NOR** operation with one *CCNOT* gate and four *NOT* gates.

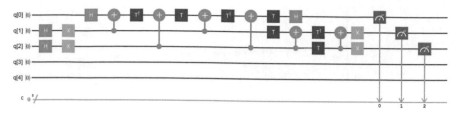

Fig. 2.20   The corresponding quantum circuit of the program in Listing 2.7

Figure 2.20 is the corresponding quantum circuit of the program in Listing 2.7. The statement "OPENQASM 2.0;" on line one of Listing 2.7 is to indicate that the program is written with version 2.0 of Open QASM. Next, the statement "include "qelib1.inc";" on line two of Listing 2.7 is to continue parsing the file "qelib1.inc" as if the contents of the file were pasted at the location of the include statement, where the file "qelib1.inc" is **Quantum Experience (QE) Standard Header** and the path is specified relative to the current workingdirectory.

Listing 2.7 The program of applying one *CCNOT* gate and four *NOT* gates to implement **NOR** operation.

```
1.   OPENQASM 2.0;
2.   include "qelib1.inc";
3.   qreg q[5];
4.   creg c[5];
5.   h q[1];
6.   h q[2];
7.   x q[1];
8.   x q[2];
9.   h q[0];
10.  cx q[1], q[0];
11.  tdg q[0];
12.  cx q[2], q[0];
13.  t q[0];
14.  cx q[1], q[0];
15.  tdg q[0];
16.  cx q[2], q[0];
17.  t q[0];
18.  t q[1];
19.  h q[0];
20.  cx q[2], q[1];
21.  tdg q[1];
22.  t q[2];
```

```
23.   cx q[2], q[1];
24.   x q[1];
25.   x q[2];
26.   measure q[0] → c[0];
27.   measure q[1] → c[1];
28.   measure q[2] → c[2];
```

Next, the statement "qreg q[5];" on line three of Listing 2.7 is to declare that in the program there are five quantum bits. In the left top of Fig. 2.20, five quantum bits are subsequently q[0], q[1], q[2], q[3] and q[4]. The initial value of each quantum bit is set to $|0\rangle$. We use three quantum bits q[2], q[1] and q[0] to encode respectively the first control bit, the second control bit and the target bit. This is to say that we use quantum bits q[2] and q[1] to encode two inputs of a classical bit in **NOR** operation of a *classical* bit and make use of quantum bit q[0] to store the result of **NOR** operation of a *classical* bit. For the convenience of our explanation, $q[k]^0$ for $0 \le k \le 4$ is to represent the value of q[k] to be zero (0) and $q[k]^1$ for $0 \le k \le 4$ is to represent the value of q[k] to be one (1). Similarly, for the convenience of our explanation, an initial state vector of implementing **NOR** operation of a classical bit is as follows:

$$|D_0\rangle = \left|q[2]^0\right\rangle\left|q[1]^0\right\rangle\left|q[0]^0\right\rangle = |0\rangle|0\rangle|0\rangle = |000\rangle.$$

Next, the statement "creg c[5];" on line four of Listing 2.7 is to declare that there are five classical bits in the program. In the left bottom of Fig. 2.20, five classical bits are subsequently c[0], c[1], c[2], c[3] and c[4]. The initial value of each classical bit is set to zero (0). Classical bit c[4] is the most significant bit and classical bit c[0] is the least significant bit.

Next, the two statements "h q[1];" and "h q[2];"on line five through line six of Listing 2.7 implement two Hadamard gates of the *first* time slot of the quantum circuit in Fig. 2.20. This is to say that the statement "h q[1];" converts q[1] from one state $|0\rangle$ to another state $\frac{1}{\sqrt{2}}(|0\rangle + |1\rangle)$ (its superposition) and the statement "h q[2];" converts q[2] from one state $|0\rangle$ to another state $\frac{1}{\sqrt{2}}(|0\rangle + |1\rangle)$ (its superposition). Because there is no gate to act on quantum bit q[0], its state is not changed. Therefore, after applying the two statements "h q[1];" and "h q[2];"on line five through line six of Listing 2.7 implements two Hadamard gates in the *first* time slot of the quantum circuit in Fig. 2.20, the following new state vector is

$$|D_1\rangle = \frac{1}{2}(|0\rangle|0\rangle|0\rangle + |0\rangle|1\rangle|0\rangle + |1\rangle|0\rangle|0\rangle + |1\rangle|1\rangle|0\rangle)$$

$$= \frac{1}{2}(|000\rangle + |010\rangle + |100\rangle + |110\rangle).$$

In the new state vector $|D_1\rangle$, four combinational states of quantum bits q[2] and q[1] with that the amplitude of each combinational state is $\frac{1}{2}$ encode all of the possible inputs for **NOR** operation of a classical bit. The initial state of quantum bit q[0] in four combinational states of quantum bits q[2] and q[1] is $|0\rangle$ and it stores the result for **NOR** operation of a classical bit.

Next, the two statements "x q[1];" and "x q[2];" on line seven through line eight of Listing 2.7 complete two **NOT** gates of the *second* time slot of the quantum circuit in Fig. 2.20. They take the new state vector $|D_1\rangle = \frac{1}{2}(|000\rangle + |010\rangle + |100\rangle + |110\rangle)$ as the input in the *second* time slot of Fig. 2.20. This indicates that in the new state vector $|D_1\rangle$ the state $(|0\rangle + |1\rangle)$ of q[2] is converted into the state $(|1\rangle + |0\rangle)$ and the state $(|0\rangle + |1\rangle)$ of q[1] is converted into the state $(|1\rangle + |0\rangle)$. There is no gate to act on quantum bit q[0], so its state is not changed. Thus, after two **NOT** gates in the *second* time slot of the quantum circuit in Fig. 2.20 are implemented by means of using the two statements "x q[1];" and "x q[2];" on line seven through line eight of Listing 2.7, the following new state vector is obtained:

$$|D_2\rangle = \frac{1}{2}(|1\rangle|1\rangle|0\rangle + |1\rangle|0\rangle|0\rangle + |0\rangle|1\rangle|0\rangle + |0\rangle|0\rangle|0\rangle)$$

$$= \frac{1}{2}(|110\rangle + |100\rangle + |010\rangle + |000\rangle).$$

The next 12 time slots in the quantum circuit of Fig. 2.20 implement **NOR** operation $(\overline{q[2] \vee q[1]} = \overline{q[2]} \wedge \overline{q[1]})$ of a classical bit that is equivalent to complete **AND** operation of a classical bit with two inputs $\overline{q[2]}$ and $\overline{q[1]}$ by means of implementing one **CCNOT** gate. From line nine through line twenty-three in Listing 2.7, there are fifteen statements. The front eight statements are "h q[0];", "cx q[1], q[0];", "tdg q[0];", "cx q[2], q[0];", "t q[0];", "cx q[1], q[0];", "tdg q[0];" and "cx q[2], q[0];". The rear seven statements are "t q[0];", "t q[1];", "h q[0];", "cx q[2], q[1];", "tdg q[1];", "t q[2];" and "cx q[2], q[1];". They implement each quantum gate from the *third* time slot through the *fourteenth* time slot in Fig. 2.20. They take the new state vector $|D_2\rangle = \frac{1}{2}(|110\rangle + |100\rangle + |010\rangle + |000\rangle)$ as the input in the *third* time slot and complete **NOR** operation $(\overline{q[2] \vee q[1]} = \overline{q[2]} \wedge \overline{q[1]})$ of a classical bit. This gives that the following new state vector is

$$|D_{17}\rangle = \frac{1}{2}(|111\rangle + |100\rangle + |010\rangle + |000\rangle).$$

Next, the two statements "x q[1];" and "x q[2];" on line twenty-four through line twenty-five of Listing 2.7 implement two **NOT** gates of the *fifteenth* time slot of the quantum circuit in Fig. 2.20. They take the new state vector $|D_{17}\rangle = \frac{1}{2}(|111\rangle + |100\rangle + |010\rangle + |000\rangle)$ as the input in the *fifteenth* time slot of Fig. 2.20. This indicates that in the new state vector $|D_{17}\rangle$ the state $(|111\rangle)$ is converted into the state $(|001\rangle)$, the state $(|100\rangle)$ is converted into the state $(|010\rangle)$, the state $(|010\rangle)$ is converted into the state $(|100\rangle)$ and the state $(|000\rangle)$ is converted into the state $(|110\rangle)$. Because there is no gate to act on quantum bit q[0], its state is not changed. Therefore, after two **NOT** gates in the *fifteenth* time slot of the quantum circuit in Fig. 2.20 were implemented by means of using the two statements "x q[1];" and "x q[2];" on line twenty-four through line twenty-five of Listing 2.7, the following new state vector is obtained:

$$|D_{18}\rangle = \frac{1}{2}(|001\rangle + |010\rangle + |100\rangle + |110\rangle).$$

Next, three measurements from the *sixteenth* time slot through the *eighteenth* time slot in Fig. 2.20 were implemented by the three statements "measure q[0] → c[0];", "measure q[1] → c[1];" and "measure q[2] → c[2];" on line twenty-six through line twenty-eight of Listing 2.7. They are to measure the first quantum bit q[0], the second quantum bit q[1] and the third quantum bit q[2]. They record the measurement outcome by overwriting the first classical bit c[0], the second classical bit c[1] and the third classical bit c[2]. In the backend *ibmqx4* with five quantum bits in **IBM**'s quantum computers, we apply the command "simulate" to execute the program in Listing 2.7. The measured result appears in Fig. 2.21. From Fig. 2.21, we obtain the answer 00001 (c[4] = 0, c[3] = 0, c[2] = 0 = q[2] = |0⟩, c[1] = 0 = q[1] = |0⟩ and c[0] = 1 = q[0] = |1⟩) with the probability 0.270. In **NOR** operation of a classical bit the value of the first input (the first control bit) q[2] is equal to zero (0) and the value of the second input (the second control bit) q[1] is equal to zero (0). Therefore, the value of the output (the target bit) q[0] is equal to one (1) with the probability 0.270.

Or we obtain the answer 00010 (c[4] = 0, c[3] = 0, c[2] = 0 = q[2] = |0⟩, c[1] = 1 = q[1] = |1⟩ and c[0] = 0 = q[0] = |0⟩) with the probability 0.270. In **NOR** operation of a classical bit the value of the first input (the first control bit) q[2] is equal to zero (0) and the value of the second input (the second control bit) q[1] is equal to one (1). Hence, the value of the output (the target bit) q[0] is equal to zero (0) with the probability 0.270. Or we obtain the answer 00100 (c[4] = 0, c[3] = 0, c[2] = 1 = q[2] = |1⟩, c[1] = 0 = q[1] = |0⟩ and c[0] = 0 = q[0] = |0⟩) with the probability 0.260. In **NOR** operation of a classical bit, the value of the first input (the first control bit) q[2] is equal to one (1) and the value of the second input (the second control bit) q[1] is equal to zero (0). Thus, the value of the output (the target bit) q[0] is equal to zero (0) with the probability 0.260. Or we obtain the answer 00110 (c[4] = 0, c[3] = 0, c[2] = 1 = q[2] = |1⟩, c[1] = 1 = q[1] = |1⟩ and c[0] = 0 = q[0] = |0⟩) with the probability 0.200. In **NOR** operation of a classical bit the value of the first input (the first control bit) q[2] is equal to one (1) and the value of the second input (the second control bit) q[1] is also equal to one (1). Therefore, the value of the output (the target bit) q[0] is equal to zero (0) with the probability 0.200.

Fig. 2.21 After the measurement to the program in Listing 2.7 is completed, we obtain the answer 00001 with the probability 0.270 or the answer 00010 with the probability 0.270 or the answer 00100 with the probability 0.260 or the answer 00110 with the probability 0.200

2.7 Introduction for Exclusive-OR Operation

The **Exclusive-OR (XOR)** operation of a bit takes two inputs of a bit and generates single output of a bit. If the value of the first input is the same as that of the second input, then it produces a result (output) of zero (0). However, if the value of the first input and the value of the second input are both different, then it generates an output of one (1). A symbol "$\oplus$" is used to represent the **XOR** operation. Hence, the **XOR** operation of a bit that gets two inputs of a bit is the following four possible combinational results:

$$0 \oplus 0 = 0$$
$$0 \oplus 1 = 1$$
$$1 \oplus 0 = 1$$
$$1 \oplus 1 = 0 \tag{2.9}$$

The value of a Boolean variable (a bit) is only one (1) or zero (0). Therefore, **XOR** operation of two Boolean variables (two inputs of a bit) q[2] and q[1], written as q[2] $\oplus$ q[1] is equal to one (1) if and only if the value of q[2] and the value of q[1] are different. Similarly, q[2] $\oplus$ q[1] is equal to zero (0) if and only if the value of q[2] and the value of q[1] are the same. Often using a truth table with logic operation is to represent all possible combinations of inputs and the corresponding outputs. Therefore, using the form of a truth table that appears in Table 2.11 may express the rules in (2.9) for the **XOR** operation of a bit that obtains two inputs of a bit and yields one single output of a bit.

2.7.1 Quantum Program of Implementing XOR Operation

We apply one **CNOT** gate that has two quantum input bits and two quantum output bits to implement **XOR** operation of a *classical* bit that takes two inputs of a classical bit and produces one output of a classical bit. We make use of the control bit C_1 and the target bit T of the **CNOT** gate to encode two inputs q[2] and q[1] of a classical bit

Table 2.11 The truth table for the **XOR** operation of a bit that acquires two inputs of a bit and produces one single output of a bit

Input		Output
q[2]	q[1]	q[2] $\oplus$ q[1]
0	0	0
0	1	1
1	0	1
1	1	0

in **XOR** operation of a *classical* bit. We use the target bit T of the $CNOT$ gate to store one output $q[2] \oplus q[1]$ of a classical bit in **XOR** operation of a *classical* bit. Using the form of a truth table that appears in Table 2.12 may express the rule of using one $CNOT$ gate to implement **XOR** operation. Its graph representation appears in Fig. 2.22.

In Fig. 2.22, the first control bit (the top wire) C_1 and the target bit (the bottom wire) T of the $CNOT$ gate respectively encode the first input $q[2]$ and the second input $q[1]$ of a classical bit in **XOR** operation of a classical bit in Table 2.11. In Fig. 2.22, the target bit (the bottom wire) T of the $CNOT$ gate also stores one output $q[2] \oplus q[1]$ of a classical bit in **XOR** operation of a classical bit in Table 2.11.

Implementing **XOR** operation of a *classical* bit acquires two inputs $q[2]$ and $q[1]$ of a classical bit and yields one output $q[2] \oplus q[1]$ of a classical bit. It is equivalent to implement one $CNOT$ gate that its control bit and its target bit encode two inputs $q[2]$ and $q[1]$ of a classical bit and its target bit stores one output $q[2] \oplus q[1]$. Therefore, in Fig. 2.22, we use one $CNOT$ gate to implement **XOR** operation of a classical bit.

In Listing 2.8, the program in the backend *ibmqx4* with five quantum bits in **IBM**'s quantum computer is the *eighth* example of the *second* chapter in which we illustrate how to write a quantum program to complete **XOR** operation of a classical bit by means of using one $CNOT$ gate of two quantum bits. Figure 2.23 is the corresponding quantum circuit of the program in Listing 2.8. The statement "OPENQASM 2.0;" on line one of Listing 2.8 is to point out that the program is written with version 2.0 of Open QASM. Then, the statement "include "qelib1.inc";" on line two of Listing 2.8 is to continue parsing the file "qelib1.inc" as if the contents of the file were pasted at the location of the include statement, where the file "qelib1.inc" is **Quantum**

Table 2.12 The truth table of using one $CNOT$ gate to implement **XOR** operation

Input		Output	
C_1	T	C_1	$T = q[2] \oplus q[1]$
0	0	0	0
0	1	0	1
1	0	1	1
1	1	1	0

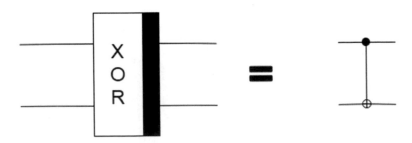

Fig. 2.22 The quantum circuit of implementing **XOR** operation of a *classical* bit

Fig. 2.23 The corresponding quantum circuit of the program in Listing 2.8

Experience (QE) Standard Header and the path is specified relative to the current working directory.

Listing 2.8 The program of using one **CNOT** gate to implement **XOR** operation.

```
1.  OPENQASM 2.0;
2.  include "qelib1.inc";
3.  qreg q[5];
4.  creg c[5];
5.  h q[1];
6.  h q[2];
7.  cx q[2], q[1];
8.  measure q[1] → c[1];
9.  measure q[2] → c[2];
```

Next, the statement "qreg q[5];" on line three of Listing 2.8 is to declare that in the program there are five quantum bits. In the left top of Fig. 2.23, five quantum bits are subsequently q[0], q[1], q[2], q[3] and q[4]. The initial value of each quantum bit is set to $|0\rangle$. We make use of two quantum bits q[2] and q[1] to encode respectively the control bit and the target bit of one **CNOT** gate. This implies that we apply quantum bits q[2] and q[1] to encode two inputs of a classical bit in **XOR** operation of a *classical* bit and use quantum bit q[1] to store the result of **XOR** operation of a *classical* bit. For the convenience of our explanation, $q[k]^0$ for $0 \leq k \leq 4$ is to represent the value of q[k] to be zero (0) and $q[k]^1$ for $0 \leq k \leq 4$ is to represent the value of q[k] to be one (1). Similarly, for the convenience of our explanation, an initial state vector of implementing **XOR** operation of a classical bit is as follows:

$$|E_0\rangle = \left|q[2]^0\right\rangle\left|q[1]^0\right\rangle = |0\rangle|0\rangle = |00\rangle.$$

Next, the statement "creg c[5];" on line four of Listing 2.8 is to declare that there are five classical bits in the program. In the left bottom of Fig. 2.23, five classical bits are respectively c[0], c[1], c[2], c[3] and c[4]. The initial value of each classical bit is set to zero (0). Classical bit c[4] is the most significant bit and classical bit c[0] is the least significant bit.

Then, the two statements "h q[1];" and "h q[2];"on line five through line six of Listing 2.8 implement two Hadamard gates of the *first* time slot of the quantum circuit in Fig. 2.23. This indicates that the statement "h q[1];" converts q[1] from one state $|0\rangle$ to another state $\frac{1}{\sqrt{2}}(|0\rangle + |1\rangle)$ (its superposition) and the statement "h q[2];" converts q[2] from one state $|0\rangle$ to another state $\frac{1}{\sqrt{2}}(|0\rangle + |1\rangle)$ (its superposition). Hence, after the two statements "h q[1];" and "h q[2];" on line five through line six of Listing 2.8 implement two Hadamard gates in the *first* time slot of the quantum circuit in Fig. 2.23 are implemented, the following new state vector is

$$|E_1\rangle = \frac{1}{2}(|0\rangle|0\rangle + |0\rangle|1\rangle + |1\rangle|0\rangle + |1\rangle|1\rangle)$$
$$= \frac{1}{2}(|00\rangle + |01\rangle + |10\rangle + |11\rangle).$$

In the new state vector $|E_1\rangle$, four combinational states of quantum bits q[2] and q[1] with that the amplitude of each combinational state is $\frac{1}{2}$ encode all of the possible inputs in **XOR** operation of a classical bit in Table 2.11. Quantum bit q[1] stores the result for **XOR** operation of a classical bit in Table 2.11.

Next, the statement "cx q[2], q[1];" on line seven of Listing 2.8 complete one **CNOT** gates of the *second* time slot of the quantum circuit in Fig. 2.23. They take the new state vector $|E_1\rangle = \frac{1}{2}(|00\rangle + |01\rangle + |10\rangle + |11\rangle)$ as the input in the *second* time slot of Fig. 2.23. This is to say that in the new state vector $|E_1\rangle$ the state ($|00\rangle$) of quantum bits q[2] and q[1] is not changed and the state ($|01\rangle$) of quantum bits q[2] and q[1] is also not changed because the value of the control bit q[2] is equal to zero (0). However, the value of the control bit q[2] is equal to one (1) and the statement "cx q[2], q[1];" flips the target bit q[1]. Therefore, the state ($|10\rangle$) of quantum bits q[2] and q[1] is converted into the state ($|11\rangle$) and the state ($|11\rangle$) of quantum bits q[2] and q[1] is converted into the state ($|10\rangle$). Therefore, after one **CNOT** gate in the *second* time slot of the quantum circuit in Fig. 2.23 is implemented by means of applying the statement "cx q[2], q[1];" on line seven of Listing 2.8, the following new state vector is obtained:

$$|E_2\rangle = \frac{1}{2}(|0\rangle|0\rangle + |0\rangle|1\rangle + |1\rangle|1\rangle + |1\rangle|0\rangle)$$
$$= \frac{1}{2}(|00\rangle + |01\rangle + |11\rangle + |10\rangle).$$

Next, two measurements from the *third* time slot through the *fourth* time slot in Fig. 2.23 were implemented by the two statements "measure q[1] → c[1];" and "measure q[2] → c[2];" on line eight through line nine of Listing 2.8. They are to measure the second quantum bit q[1] and the third quantum bit q[2]. They record the measurement outcome by overwriting the second classical bit c[1] and the third classical bit c[2]. In the backend *ibmqx4* with five quantum bits in **IBM**'s quantum

Fig. 2.24 After the measurement to the program in Listing 2.8 is completed, we obtain the answer 00010 with the probability 0.260 or the answer 00110 with the probability 0.260 or the answer 00100 with the probability 0.250 or the answer 00000 with the probability 0.230

computers, we use the command "simulate" to run the program in Listing 2.8. The measured result appears in Fig. 2.24. From Fig. 2.24, we get the answer 00010 ($c[4] = 0$, $c[3] = 0$, $c[2] = 0 = q[2] = |0\rangle$, $c[1] = 1 = q[1] = |1\rangle$ and $c[0] = 0$) with the probability 0.260. Because in **XOR** operation of a classical bit the value of the first input (the control bit) $q[2]$ is equal to zero (0) and the value of the second input (the target bit) $q[1]$ is equal to one (1), the value of the output (the target bit) $q[1]$ is equal to one (1) with the probability 0.260.

Or we obtain the answer 00110 ($c[4] = 0$, $c[3] = 0$, $c[2] = 1 = q[2] = |1\rangle$, $c[1] = 1 = q[1] = |1\rangle$ and $c[0] = 0$) with the probability 0.260. Since in **XOR** operation of a classical bit the value of the first input (the control bit) $q[2]$ is equal to one (1) and the value of the second input (the target bit) $q[1]$ is equal to zero (0), the value of the output (the target bit) $q[1]$ is equal to one (1) with the probability 0.260. Or we acquire the answer 00100 ($c[4] = 0$, $c[3] = 0$, $c[2] = 1 = q[2] = |1\rangle$, $c[1] = 0 = q[1] = |0\rangle$ and $c[0] = 0$) with the probability 0.250. Since in **XOR** operation of a classical bit the value of the first input (the control bit) $q[2]$ is equal to one (1) and the value of the second input (the target bit) $q[1]$ is equal to one (1), the value of the output (the target bit) $q[1]$ is equal to zero (0) with the probability 0.250. Or we get the answer 00000 ($c[4] = 0$, $c[3] = 0$, $c[2] = 0 = q[2] = |0\rangle$, $c[1] = 0 = q[1] = |0\rangle$ and $c[0] = 0$) with the probability 0.230. In **XOR** operation of a classical bit the value of the first input (the control bit) $q[2]$ is equal to zero (0) and the value of the second input (the target bit) $q[1]$ is also equal to zero (0). Therefore, the value of the output (the target bit) $q[1]$ is equal to zero (0) with the probability 0.230.

2.8 Introduction of Exclusive-NOR Operation

The one's complement of the **Exclusive-OR (XOR)** operation of a bit that acquires two inputs of a bit and yields one single output of a bit is known as the **Exclusive-NOR (XNOR)** operation of a bit. The **Exclusive-NOR (XNOR)** operation of a bit obtains two inputs of a bit and produces one single output of a bit. If the value of the first input is the same as that of the second input, then it generates a result (output) of one (1). However, if the value of the first input and the value of the second input are both different, then it produces an output of zero (0). A symbol "$\overline{\oplus}$" is applied to represent the **XNOR** operation of a bit. Therefore, the **XNOR** operation of a bit

Table 2.13 The truth table for the **XNOR** operation of a bit that acquires two inputs of a bit and generates one single output of a bit

Input		Output
q[2]	q[1]	$\overline{q[2] \oplus q[1]}$
0	0	1
0	1	0
1	0	0
1	1	1

that takes two inputs of a bit is the following four possible combinational results:

$$\overline{0 \oplus 0} = 1$$

$$\overline{0 \oplus 1} = 0$$

$$\overline{1 \oplus 0} = 0$$

$$\overline{1 \oplus 1} = 1 \tag{2.10}$$

The value of a Boolean variable (a bit) is just zero (0) or one (1). Hence, **XNOR** operation of two Boolean variables (two inputs of a bit) q[2] and q[1], written as $\overline{q[2] \oplus q[1]}$ is equal to one (1) if and only if the value of q[2] and the value of q[1] are the same. Similarly, $\overline{q[2] \oplus q[1]}$ is equal to zero (0) if and only if the value of q[2] and the value of q[1] are different. Usually applied a truth table with logic operation is to represent all possible combinations of inputs and the corresponding outputs. Hence, the rules in (2.10) for the **XNOR** operation of a bit that gets two inputs of a bit and produces one single output of a bit may also be expressed in the form of a truth table that appears in Table 2.13.

2.8.1 Quantum Program of Implementing XNOR Operation

We make use of one **CNOT** gate and one **NOT** gate to implement **XNOR** operation of a *classical* bit that acquires two inputs of a classical bit and generates one output of a classical bit. We use the control bit C_1 and the target bit T of the **CNOT** gate to encode two inputs q[2] and q[1] of a classical bit in **XNOR** operation of a *classical* bit. We also use the target bit T of the **CNOT** gate to store one output $q[2] \oplus q[1]$ of a classical bit in **XNOR** operation of a *classical* bit. Using the form of a truth table that appears in Table 2.14 may express the rule of applying one **CNOT** gate and one **NOT** gate to complete **XNOR** operation. Its graph representation appears in Fig. 2.25.

In Fig. 2.25, the first control bit (the top wire) C_1 and the target bit (the bottom wire) T of the **CNOT** gate respectively encode the first input q[2] and the second input q[1]of a classical bit in **XNOR** operation of a classical bit in Table 2.13. In

Table 2.14 The truth table of using one **CNOT** gate and one **NOT** gate to implement **XNOR** operation

Input		Output	
C_1	T	C_1	$T = \overline{q[2] \oplus q[1]}$
0	0	0	1
0	1	0	0
1	0	1	0
1	1	1	1

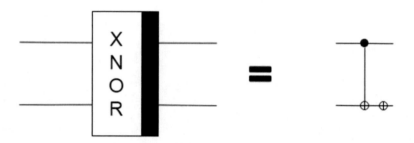

Fig. 2.25 The quantum circuit of implementing **XNOR** operation of a *classical* bit

Fig. 2.25, the target bit (the bottom wire) T of the **CNOT** gate also stores one output $\overline{q[2] \oplus q[1]}$ of a classical bit in **XNOR** operation of a classical bit in Table 2.13.

Implementing **XNOR** operation of a *classical* bit takes two inputs q[2] and q[1] of a classical bit and produces one output $\overline{q[2] \oplus q[1]}$ of a classical bit. It is equivalent to implement one **CNOT** gate and one **NOT** gate in which the control bit and the target bit encode two inputs q[2] and q[1] of a classical bit and its target bit also stores one output $\overline{q[2] \oplus q[1]}$. Hence, in Fig. 2.25, we first apply one **CNOT** gate to generate an output q[2] $\oplus$ q[1] of **XOR** operation that is stored in the target bit (the bottom wire) T. Next, we use one **NOT** gate to produce the negation of **XOR** operation (q[2]$\oplus$q[1]) that is to complete **XNOR** operation $\overline{q[2] \oplus q[1]}$ that is stored in the target bit T.

In Listing 2.9, the program in the backend *ibmqx4* with five quantum bits in **IBM**'s quantum computer is the *ninth* example of the *second* chapter in which we describe how to write a quantum program to complete **XNOR** operation of a classical bit by means of applying one **CNOT** gate and one **NOT** gate. Figure 2.26 is the corresponding quantum circuit of the program in Listing 2.9. The statement "OPEN-QASM 2.0;" on line one of Listing 2.9 is to indicate that the program is written with version 2.0 of Open QASM. Next, the statement "include "qelib1.inc";" on line two of Listing 2.9 is to continue parsing the file "qelib1.inc" as if the contents of the file were pasted at the location of the include statement, where the file "qelib1.inc" is **Quantum Experience (QE) Standard Header** and the path is specified relative to the current working directory.

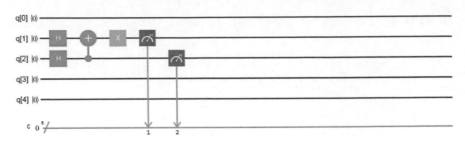

Fig. 2.26 The corresponding quantum circuit of the program in Listing 2.9

Listing 2.9 The program of using one *CNOT* gate and one *NOT* gate to implement **XNOR** operation.

```
1.   OPENQASM 2.0;
2.   include "qelib1.inc";
3.   qreg q[5];
4.   creg c[5];
5.   h q[1];
6.   h q[2];
7.   cx q[2], q[1];
8.   x q[1];
9.   measure q[1] → c[1];
10.  measure q[2] → c[2];
```

Next, the statement "qreg q[5];" on line three of Listing 2.9 is to declare that in the program there are five quantum bits. In the left top of Fig. 2.26, five quantum bits are subsequently q[0], q[1], q[2], q[3] and q[4]. The initial value of each quantum bit is set to $|0\rangle$. We use two quantum bits q[2] and q[1] to encode respectively the control bit and the target bit of one *CNOT* gate. This implies that we apply quantum bits q[2] and q[1] to encode two inputs of a classical bit in **XNOR** operation of a *classical* bit and use quantum bit q[1] to store the result of **XNOR** operation of a *classical* bit. For the convenience of our explanation, $q[k]^0$ for $0 \le k \le 4$ is to represent the value of q[k] to be zero (0) and $q[k]^1$ for $0 \le k \le 4$ is to represent the value of q[k] to be one (1). Similarly, for the convenience of our explanation, an initial state vector of implementing **XNOR** operation of a classical bit is as follows:

$$|F_0\rangle = \left|q[2]^0\right\rangle \left|q[1]^0\right\rangle = |0\rangle |0\rangle = |00\rangle.$$

Then, the statement "creg c[5];" on line four of Listing 2.9 is to declare that there are five classical bits in the program. In the left bottom of Fig. 2.26, five classical bits are respectively c[0], c[1], c[2], c[3] and c[4]. The initial value of each classical bit is set to zero (0). Classical bit c[4] is the most significant bit and classical bit c[0] is the least significant bit.

Next, the two statements "h q[1];" and "h q[2];"on line five through line six of Listing 2.9 implement two Hadamard gates of the *first* time slot of the quantum circuit in Fig. 2.26. This is to say that the statement "h q[1];" converts q[1] from one state $|0\rangle$ to another state $\frac{1}{\sqrt{2}}(|0\rangle + |1\rangle)$ (its superposition) and the statement "h q[2];" converts q[2] from one state $|0\rangle$ to another state $\frac{1}{\sqrt{2}}(|0\rangle + |1\rangle)$ (its superposition). Hence, after the two statements "h q[1];" and "h q[2];"on line five through line six of Listing 2.9 implement two Hadamard gates in the *first* time slot of the quantum circuit in Fig. 2.26, the following new state vector is

$$|F_1\rangle = \frac{1}{2}(|0\rangle|0\rangle + |0\rangle|1\rangle + |1\rangle|0\rangle + |1\rangle|1\rangle)$$

$$= \frac{1}{2}(|00\rangle + |01\rangle + |10\rangle + |11\rangle).$$

In the new state vector $|F_1\rangle$, four combinational states of quantum bits q[2] and q[1] with that the amplitude of each combinational state is $\frac{1}{2}$ encode all of the possible inputs in **XNOR** operation of a classical bit in Table 2.13. Quantum bit q[1] stores the result for **XNOR** operation of a classical bit in Table 2.13.

Next, the statement "cx q[2], q[1];" on line seven of Listing 2.9 implements one *CNOT* gates of the *second* time slot of the quantum circuit in Fig. 2.26. They take the new state vector $|F_1\rangle = \frac{1}{2}(|00\rangle + |01\rangle + |10\rangle + |11\rangle)$ as the input in the *second* time slot of Fig. 2.26. Inthe input state vector $|F_1\rangle$, because the value of the control bit q[2] is equal to zero (0), the value of the target bit q[1] is not changed. Hence, the state ($|00\rangle$) of quantum bits q[2] and q[1] is not changed and the state ($|01\rangle$) of quantum bits q[2] and q[1] is also not changed. Because the value of the control bit q[2] is equal to one (1), the statement "cx q[2], q[1];" flips the value of the target bit q[1]. Therefore, the state ($|10\rangle$) of quantum bits q[2] and q[1] is converted into the state ($|11\rangle$) and the state ($|11\rangle$) of quantum bits q[2] and q[1] is converted into the state ($|10\rangle$). Hence, after one *CNOT* gate in the *second* time slot of the quantum circuit in Fig. 2.26 is implemented by means of using the statement "cx q[2], q[1];" on line seven of Listing 2.9, we obtain the following new state vector

$$|F_2\rangle = \frac{1}{2}(|0\rangle|0\rangle + |0\rangle|1\rangle + |1\rangle|1\rangle + |1\rangle|0\rangle)$$

$$= \frac{1}{2}(|00\rangle + |01\rangle + |11\rangle + |10\rangle).$$

Next, the statement "x q[1];" on line eight of Listing 2.9 implements one *NOT* gate in the *third* time slot of the quantum circuit in Fig. 2.26. It takes the new state vector $|F_2\rangle = \frac{1}{2}(|00\rangle + |01\rangle + |11\rangle + |10\rangle)$ as the input in the *third* time slot of the quantum circuit in Fig. 2.26. This is to say that there is no quantum gate to act on quantum bit q[2] in the input state vector $|F_2\rangle$ and the value of quantum bit q[2] is not changed. Because the statement "x q[1];" flips the value of quantum bit q[1] in the input state vector $|F_2\rangle$, the states $(|00\rangle)$, $(|01\rangle)$, $(|11\rangle)$ and $(|10\rangle)$ of quantum bits q[2] and q[1] are subsequently converted into the new states $(|01\rangle)$, $(|00\rangle)$, $(|10\rangle)$ and $(|11\rangle)$. Therefore, after one *NOT* gate in the *third* time slot of the quantum circuit in Fig. 2.26 is implemented by means of applying the statement "x q[1];" on line eight of Listing 2.9, the following new state vector is obtained:

$$|F_3\rangle = \frac{1}{2}(|0\rangle|1\rangle + |0\rangle|0\rangle + |1\rangle|0\rangle + |1\rangle|1\rangle)$$

$$= \frac{1}{2}(|01\rangle + |00\rangle + |10\rangle + |11\rangle).$$

Next, two measurements from the *fourth* time slot through the *fifth* time slot of the quantum circuit in Fig. 2.26 were implemented by the two statements "measure q[1] $\rightarrow$ c[1];" and "measure q[2] $\rightarrow$ c[2];" on line nine through line ten of Listing 2.9. They are to measure the second quantum bit q[1] and the third quantum bit q[2]. They record the measurement outcome by overwriting the second classical bit c[1] and the third classical bit c[2]. In the backend *ibmqx4* with five quantum bits in **IBM**'s quantum computers, we make use of the command "simulate" to execute the program in Listing 2.9. The measured result appears in Fig. 2.27. From Fig. 2.27, we obtain the answer 00000 (c[4] = 0, c[3] = 0, c[2] = 0 = q[2] = $|0\rangle$, c[1] = 0 = q[1] = $|0\rangle$ and c[0] = 0) with the probability 0.370. Since in **XNOR** operation of a classical bit the value of the first input (the control bit) q[2] is equal to zero (0) and the value of the second input (the target bit) q[1] is equal to one (1), the value of the output (the target bit) q[1] is equal to zero (0) with the probability 0.370. Or we get the answer 00100 (c[4] = 0, c[3] = 0, c[2] = 1 = q[2] = $|1\rangle$, c[1] = 0 = q[1] = $|0\rangle$ and c[0] = 0) with the probability 0.260. Because in **XNOR** operation of a classical bit the value of the first input (the control bit) q[2] is equal to one (1) and the value of the second input (the target bit) q[1] is equal to zero (0), the value of the output (the target bit) q[1] is equal to zero (0) with the probability 0.260.

Or we acquire the answer 00010 (c[4] = 0, c[3] = 0, c[2] = 0 = q[2] = $|0\rangle$, c[1] = 1 = q[1] = $|1\rangle$ and c[0] = 0) with the probability 0.230. Because in **XNOR** operation of a classical bit the value of the first input (the control bit) q[2] is equal to zero (0) and the value of the second input (the target bit) q[1] is equal to zero (0), the value of

Fig. 2.27 After the measurement to the program in Listing 2.9 is completed, we obtain the answer 00000 with the probability 0.370 or the answer 00100 with the probability 0.260 or the answer 00010 with the probability 0.230 or the answer 00110 with the probability 0.140

the output (the target bit) q[1] is equal to one (1) with the probability 0.230. Or we obtain the answer 00110 (c[4] = 0, c[3] = 0, c[2] = 1 = q[2] = |1⟩, c[1] = 1 = q[1] = |1⟩ and c[0] = 0) with the probability 0.140. In **XNOR** operation of a classical bit, the value of the first input (the control bit) q[2] is equal to one (1) and the value of the second input (the target bit) q[1] is also equal to one (1). Therefore, the value of the output (the target bit) q[1] is equal to one (1) with the probability 0.140.

2.9 Summary

In this chapter we offered an illustration to how logic operations consisting of **NOT, AND, NAND, OR, NOR, Exclusive-OR (XOR)** and **Exclusive-NOR (XNOR)** on bits were implemented by means of using quantum bits and quantum gates in **IBM**'s quantum computers. We introduced the first program in Listing 2.1 and the second program in Listing 2.2 to explain how the one's complement (the **NOT** operation) of a bit and the one's complement (the **NOT** operation) of two bits were implemented by means of using quantum bits and the *X* gates (the *NOT* gates) in **IBM**'s quantum computers. Next, we described the third program in Listing 2.3 to show how decomposing **CCNOT** gate into six **CNOT** gates and nine gates of one bit in **IBM**'s quantum computers implemented one *CCNOT* gate.

Then, we introduced the fourth program in Listing 2.4 to reveal how the **AND** operation of a bit was implemented by means of using one *CCNOT* gate and three quantum bits in **IBM**'s quantum computers. We also illustrated the fifth program in Listing 2.5 to explain how applying one *CCNOT* gate and three quantum bits in **IBM**'s quantum computers implemented the **NAND** operation of a bit. Next, we described the sixth program in Listing 2.6 to show how using one *CCNOT* gate, four *NOT* gates (four *X* gates) and three quantum bits in **IBM**'s quantum computers implemented the **OR** operation of a bit.

We then illustrated the seventh program in Listing 2.7 to reveal how the **NOR** operation of a bit was implemented by means of applying one *CCNOT* gate, four *NOT* gates (four *X* gates) and three quantum bits in **IBM**'s quantum computers. We also introduced the eighth program in Listing 2.8 to explain how using one *CNOT* gate and two quantum bits in **IBM**'s quantum computers implemented the **XOR** operation of a bit. Next, we described the ninth program in Listing 2.9 to show how the **XNOR** operation of a bit was implemented by means of applying one *CNOT* gate, one *NOT* gate (one *X* gate) and two quantum bits in **IBM**'s quantum computers.

2.10 Bibliographical Notes

The textbooks written by these authors in Mano (1979, 1993), Chang and Vasilakos (2014) is a good illustration to logic operations including **NOT, AND, NAND, OR, NOR, Exclusive-OR (XOR)** and **Exclusive-NOR (XNOR)** on bits. A good introduction of decomposing **CCNOT** gate into six **CNOT** gates and nine gates of one bit can be found in the textbook (Nielsen and Chuang 2000) and in the famous article (Shende and Markov 2009). The famous menu in **IBM Q** (2016) is a good guide of writing nine quantum programs from Listing 2.1 to Listing 2.9. A good illustration to Boolean's functions discussed in exercises in Sect. 2.11 is Mano (1979, 1993), Brown and Vranesic (2007), Chang and Vasilakos (2014).

2.11 Exercises

2.1 The unary operator "−" denotes logical operation **NOT** and the binary operator "∨" denotes logical operation **OR**. For a logical operation, $\bar{x} \vee y$, x and y are Boolean variables that are subsequently the first input and the second input. Its truth table appears in Table 2.15. Please write a quantum program to implement the function of the logical operation, $\bar{x} \vee y$.

2.2 The unary operator "−" denotes logical operation **NOT** and the binary operator "∨" denotes logical operation **OR**. For a logical operation, $x \vee \bar{y}$, x and y are Boolean variables that are respectively its first input and its second input. Its truth table appears in Table 2.16. Please write a quantum program to implement the function of the logical operation, $x \vee \bar{y}$.

Table 2.15 The truth table to a logical operation $\bar{x} \vee y$

The first input (x)	The second input (y)	$\bar{x} \vee y$
0	0	1
0	1	1
1	0	0
1	1	1

Table 2.16 The truth table to a logical operation $x \vee \bar{y}$

The first input (x)	The second input (y)	$x \vee \bar{y}$
0	0	1
0	1	0
1	0	1
1	1	1

2.3 The unary operator "−" denotes logical operation **NOT** and the binary operator "∧" denotes logical operation **AND**. For a logical operation, $\bar{y} \wedge (x \vee \bar{x}) = \bar{y} \wedge 1 = \bar{y}$, x and y are Boolean variables that are subsequently the first input and the second input. Its truth table appears in Table 2.17. Please write a quantum program to implement the function of the logical operation, $\bar{y} \wedge (x \vee \bar{x}) = \bar{y} \wedge 1 = \bar{y}$.

2.4 The unary operator "−" denotes logical operation **NOT** and the binary operator "∧" denotes logical operation **AND**. For a logical operation, $\bar{x} \wedge (y \vee \bar{y}) = \bar{x} \wedge 1 = \bar{x}$, x and y are Boolean variables that are respectively the first input and the second input. Its truth table appears in Table 2.18. Please write a quantum program to implement the function of the logical operation, $\bar{x} \wedge (y \vee \bar{y}) = \bar{x} \wedge 1 = \bar{x}$.

2.5 The unary operator "−" denotes logical operation **NOT** and the binary operator "∧" denotes logical operation **AND**. For a logical operation, $\bar{x} \wedge y$, x and y are Boolean variables that are subsequently the first input and the second input. Its truth table appears in Table 2.19. Please write a quantum program to implement the function of the logical operation, $\bar{x} \wedge y$.

2.6 The unary operator "−" denotes logical operation **NOT** and the binary operator "∧" denotes logical operation **AND**. For a logical operation, $x \wedge \bar{y}$, x and y are

Table 2.17 The truth table to a logical operation $\bar{y} \wedge (x \vee \bar{x}) = \bar{y} \wedge 1 = \bar{y}$

The first input (x)	The second input (y)	$\bar{y} \wedge (x \vee \bar{x}) = \bar{y} \wedge 1 = \bar{y}$
0	0	1
0	1	0
1	0	1
1	1	0

Table 2.18 The truth table to a logical operation $\bar{x} \wedge (y \vee \bar{y}) = \bar{x} \wedge 1 = \bar{x}$

The first input (x)	The second input (y)	$\bar{x} \wedge (y \vee \bar{y}) = \bar{x} \wedge 1 = \bar{x}$
0	0	1
0	1	1
1	0	0
1	1	0

Table 2.19 The truth table to a logical operation $\bar{x} \wedge y$

The first input (x)	The second input (y)	$\bar{x} \wedge y$
0	0	0
0	1	1
1	0	0
1	1	0

Table 2.20 The truth table to a logical operation $x \wedge \overline{y}$

The first input (x)	The second input (y)	$x \wedge \overline{y}$
0	0	0
0	1	0
1	0	1
1	1	0

Boolean variables that are respectively its first input and its second input. Its truth table appears in Table 2.20. Please write a quantum program to implement the function of the logical operation, $x \wedge \overline{y}$.

References

Brown, S., Vranesic, Z.: Fundamentals of Digital Logic With Verilog Design. McGraw-Hill, New York (2007). ISBN: 978-0077211646

Chang, W.-L., Vasilakos, A.V.: Molecular Computing: Towards a Novel Computing Architecture for Complex Problem Solving. Springer, Berlin (2014). ISBN-10: 3319380982 and ISBN-13: 978-3319380988

Mano, M.M.: Digital Logic and Computer Design. Prentice-Hall, USA (1979). ISBN: 0-13-214510-3

Mano, M.M.: Computer System Architecture. Prentice Hall, USA (1993). ISBN: 978-0131755635

Nielsen, M.A., Chuang, I.L.: Quantum Computation and Quantum Information. Cambridge University Press, New York (2000). ISBN-10: 9781107002173 and ISBN-13: 978-1107002173

Shende, V.V., Markov, I.L.: On the CNOT-cost of TOFFOLI gates. Quantum Inf. Comput. **9**(5), 461–486 (2009)

The IBM Quantum Experience. https://www.research.ibm.com/quantum/. Accessed Nov 2016

Chapter 3
Quantum Search Algorithm and Its Applications

Because in **IBM**'s quantum computers, they only provide quantum gates of single quantum bit and two quantum bits, quantum gates of three quantum bits and many quantum bits must manually be decomposed into quantum gates of single quantum bit and two quantum bits. A good quantum algorithm of solving any given problem with the size of the input of n bits must have a constant successful probability of measuring its answer(s) that is close to one as soon as possible. In this chapter, we first illustrate how to decompose quantum gates of three quantum bits and many quantum bits into quantum gates of single quantum bit and two quantum bits. Next, we introduce how to write quantum programs with version 2.0 of Open QASM to implement decomposition among various kinds of quantum gates. A *quantum search algorithm* that is sometimes known as *Grover's algorithm* to find an item in unsorted databases with 2^n items that satisfies any given condition can give a quadratic speed-up and is the best one known. Hence, we then describe how to write quantum programs with version 2.0 of Open QASM to implement the quantum search algorithm in order to solve various applications.

3.1 Introduction to the Search Problem

It is assumed that a set X is equal to $\{x_1 x_2 \ldots x_n | \forall x_d \in \{0, 1\}$ for $1 \leq d \leq n\}$. From the set X the minimum element is $x_1^0 x_2^0 \ldots x_{n-1}^0 x_n^0$ with n bits and the maximum element is $x_1^1 x_2^1 \ldots x_{n-1}^1 x_n^1$ with n bits. For convenience of presentation, in the set X the decimal value of the minimum element with n bits is 0 and the decimal value of the maximum element with n bits is $2^n - 1$. We regard the set X as an unsorted database containing 2^n items (elements) with each item has n bits.

A search problem is to that from the set X that is $\{x_1 x_2 \ldots x_n - 1 x_n | \forall x_d \in \{0, 1\}$ for $1 \leq d \leq n\}$ and is also an unsorted database with 2^n items (elements) M items (elements) satisfy any given condition and we would like to find one of M solutions,

© The Author(s), under exclusive license to Springer Nature Switzerland AG 2021
W.-L. Chang and A. V. Vasilakos, *Fundamentals of Quantum Programming in IBM's Quantum Computers*, Studies in Big Data 81,
https://doi.org/10.1007/978-3-030-63583-1_3

where $1 \leq M \leq 2^n$. A common formulation of the search problem is as follows. For any given oracular function $O_f \{x_1 x_2 \ldots x_{n-1} x_n | \forall x_d \in \{0, 1\}$ for $1 \leq d \leq n\} \rightarrow \{0, 1\}$, there are M inputs of n bits from its domain, say λ_M, that satisfies the condition $O_f (\lambda_M) = 1$, whereas for all other inputs of n bits from the same domain, ω, for $0 \leq \omega \leq 2^n - 1$ and $\omega \neq \lambda_M$, $O_f (\omega) = 0$. The search problem is to find one of M solutions.

The most efficient classical algorithm for the search problem is to check whether the items (elements) in the domain one by one satisfy $O_f(\lambda_M) = 1$ or not. If an item (element) satisfies the required condition that is $O_f(\lambda_M) = 1$, then the most efficient classical algorithm is terminated. Otherwise, it continues to examine whether next item (element) satisfies $O_f(\lambda_M) = 1$ or not until the answer is found. The number of solutions that is one is the worst case in the search problem. For the worst case in the search problem, the best time complexity of finding the desired answer (item) is $O(1)$, the average time complexity of finding the desired answer (item) is $O\left(\frac{2^n+1}{2}\right)$ and the worst time complexity of finding the desired answer (item) is $O(2^n)$.

3.2 Introduction to the Satisfiability Problem

Let us consider an example $F(x_1, x_2) = (x_2 \vee x_1) \wedge (\overline{x_2} \vee \overline{x_1}) \wedge (x_1)$. The two variables x_2 and x_1 are two Boolean variables and their values could be 0 or 1. We suppose that 0 is "false" and 1 is "true". A symbol "$\vee$" is the "logical or" operation and a symbol "$\wedge$" is the "logical and" operation. Therefore, a Boolean formula $x_2 \vee x_1$ is 0 only if both x_2 and x_1 are 0; a Boolean formula $x_2 \wedge x_1$ is 1 only if both x_2 and x_1 are 1. We regard the Boolean formula $x_2 \vee x_1$ as one clause and we regard the Boolean formula $x_2 \wedge x_1$ as another clause.

We give $\overline{x_1}$ to represent the "negation" of x_1 and we give $\overline{x_2}$ to represent the "negation" of x_2. A Boolean formula $\overline{x_1}$ is 1 if x_1 is 0 and $\overline{x_1}$ is 0 if x_1 is 1. A Boolean formula $\overline{x_2}$ is 1 if x_2 is 0 and $\overline{x_2}$ is 0 if x_2 is 1. Of course, we also regard the Boolean formula $\overline{x_1}$ and the Boolean formula $\overline{x_2}$ as two different clauses. The satisfiability problem that is a **NP-complete** problem is to find Boolean values of x_2 and x_1 to make the formula $F(x_1, x_2)$ to be true that is equal to 1.

In this example, the answer is $x_2 = 0$ and $x_1 = 1$. In the formula $F(x_1, x_2)$, it actually includes three *clauses*: the first clause is "$(x_2 \vee x_1)$", the second clause is "$(\overline{x_2} \vee \overline{x_1})$" and the third clause is "(x_1)". A clause is a formula of the form $x_1 \vee x_2 \vee \ldots x_{n-1} \vee x_n$, where each variable x_k for $1 \leq k \leq n$ is a Boolean variable or its negation. Because the quantum program of implementing this example uses more quantum bits that exceed five quantum bits in the backend *ibmqx4* with five quantum bits in **IBM**'s quantum computers, we just use this example to explain what the satisfiability problem is. Next, we give Definition 3.1 to introduce the satisfiability problem.

Definition 3.1 In general, a satisfiability problem contains a Boolean formula of the form $C_1 \wedge C_2 \cdots \wedge C_m$, where each clause C_j for $1 \leq j \leq m$ is a formula of the form $x_1 \vee x_2 \vee \ldots x_{n-1} \vee x_n$ for each Boolean variable x_k to $1 \leq k \leq n$. Next, the question is to find values of each Boolean variable so that the whole formula has the value 1. This is the same as finding values of each Boolean variable that make each clause have the value 1.

From Definition 3.1, for a satisfiability problem with n Boolean variables and m clauses, we regard m clauses as any given oracular function O_f $(x_1, x_2, \ldots, x_{n-1}, x_n)$ and regard 2^n inputs of n Boolean variables as its domain $\{x_1 x_2 \ldots x_{n-1} x_n \ \forall x_d \in \{0, 1\} \ \text{for} \ 1 \le d \le n\}$. The satisfiability problem with n Boolean variables and m clauses is to find inputs of n bits (n Boolean variables) from its domain so that the whole formula O_f $(x_1, x_2, \ldots, x_{n-1}, x_n)$ has the value 1. This is to say that a satisfiability problem with n Boolean variables and m clauses is actually a kind of search problems.

3.2.1 Flowchart of Solving the Satisfiability Problem

From Definition 3.1, a satisfiability problem contains a Boolean formula of the form $C_1 \wedge C_2 \ldots \wedge C_m$, where each clause C_j for $1 \le j \le m$ is a formula of the form $x_1 \vee x_2 \vee \ldots x_{n-1} \vee x_n$ for each Boolean variable x_k to $1 \le k \le n$. For solving the satisfiability problem with n Boolean variables and m clause, we need to use auxiliary Boolean variables $r_{j, k}$ for $1 \le j \le m$ and $0 \le k \le n$ and auxiliary Boolean variables s_j for $0 \le j \le m$. Because we use auxiliary Boolean variables $r_{j, 0}$ for $1 \le j \le m$ as the first operand of the first logical or operation ("$\vee$") in each clause, the initial value of each auxiliary Boolean variable $r_{j, 0}$ for $1 \le j \le m$ is set to zero (0). This is to say that this setting does not change the correct result of the first logical or operation in each clause. We use **CCNOT** gates and **NOT** gates to implement the logical or operations in each clause and we apply auxiliary Boolean variables $r_{j, k}$ for $1 \le j$ $\le m$ and $1 \le k \le n$ to store the result of implementing the logical or operations in each clause. This indicates that each auxiliary Boolean variable $r_{j, k}$ for $1 \le j \le m$ and $1 \le k \le n$ is actually the target bit of a **CCNOT** gate of implementing a logical or operation. Therefore, the initial value of each auxiliary Boolean variable $r_{j, k}$ for $1 \le j \le m$ and $1 \le k \le n$ is set to one (1).

We use an auxiliary Boolean variable s_0 as the first operand of the first logical and operation ("$\wedge$") in a Boolean formula of the form $C_1 \wedge C_2 \ldots \wedge C_m$. The initial value of the auxiliary Boolean variable s_0 is set to one (1). This implies that this setting does not change the correct result of the first logical and operation in $C_1 \wedge C_2 \ldots \wedge C_m$. We use **CCNOT** gates to implement the logical and operations in $C_1 \wedge C_2 \ldots \wedge C_m$ and we apply auxiliary Boolean variables s_j for $1 \le j \le m$ to store the result of implementing the logical and operations in $C_1 \wedge C_2 \ldots \wedge C_m$. This is to say that each auxiliary Boolean variable s_j for $1 \le j \le m$ is actually the target bit of a **CCNOT** gate of implementing a logical and operation. Thus, the initial value of each auxiliary Boolean variable s_j for $1 \le j \le m$ is set to zero (0). For the convenience of our presentation, we assume that $|C_j|$ is the number of Boolean variable in the jth clause C_j.

Figure 3.1 is to flowchart of solving the satisfiability problem with n Boolean variables and m clauses. In Fig. 3.1, in statement S_1, it sets the index variable j of the first loop to one (1). Next, in statement S_2, it executes the conditional judgement of the first loop. If the value of j is less than or equal to the value of m, then *next executed* instruction is statement S_3. Otherwise, in statement S_9, it executes an *End* instruction to terminate the task that is to find values of each Boolean variable so that the whole formula has the value 1 and this is the same as finding values of each Boolean variable that make each clause have the value 1.

In statement S_3, it sets the index variable k of the second loop to one (1). Next, in statement S_4, it executes the conditional judgement of the *second* loop. If the value of k is less than or equal to the number of Boolean variables in the jth clause C_j, then next executed instruction is statement S_5. Otherwise, next executed instruction is statement S_7. In statement S_5, it implements a logical or operation "$r_{j,k} \leftarrow r_{j,k-1} \vee x_k$" that is the kth logical or operation in the jth clause C_j. Boolean variable $r_{j,\,k-1}$ is the first operand of the logical or operation and stores the result of the previous logical or operation. Boolean variable x_k is the second operand of the logical or operation. Boolean variable $r_{j,\,k}$ stores the result of implementing the kth logical or operation

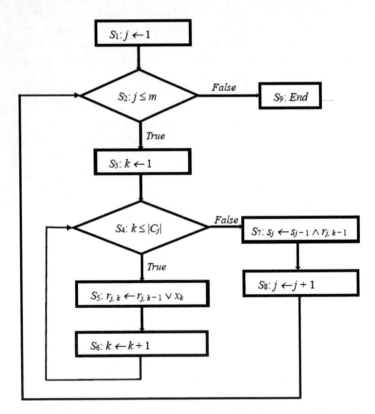

Fig. 3.1 Flowchart of solving the satisfiability problem with n Boolean variables and m clauses

in the jth clause C_j. Next, in statement S_6, it increases the value of the index variable k to the second loop. Repeat to execute statement S_4 through statement S_6 until in statement S_4 the conditional judgement becomes a *false* value.

When the value of the index variable k in the second loop is greater than the number of Boolean variables in the jth clause C_j, next executed instruction is statement S_7. In statement S_7, it executes a logical and operation "$s_j \leftarrow s_{j-1} \wedge r_{j,k-1}$" in $C_1 \wedge C_2 \ldots \wedge C_m$. Boolean variable s_{j-1} is the first operand of the logical and operation and stores the result of the previous logical and operation. Because the value of k is equal to $|C_j| + 1$, the value of $(k-1)$ is equal to $|C_j|$. Boolean variable $r_{j,k-1}$ is the second operand of the logical and operation and stores the result of implementing a formula of the form $x_1 \vee x_2 \vee \ldots x_n - 1 \vee x_n$ in the jth clause C_j. Boolean variable s_j stores the result of implementing the jth logical and operation in. Next, in statement S_8, it increases the value of the index variable j to the first loop. Repeat to execute statement S_2 through statement S_8 until in statement S_2 the conditional judgement becomes a *false* value. Because from Definition 3.1 each clause C_j for $1 \leq j \leq m$ is a formula of the form $x_1 \vee x_2 \vee \ldots x_{n-1} \vee x_n$, from Fig. 3.1 the total number of logical and operation and logical or operation is m logical and operations and $\left(\sum_{j=1}^m |C_j| \right) = (m \times n)$ logical or operations. This is the cost of implementing m clauses of one time for one of 2^n inputs to n Boolean variables. Therefore, the cost of implementing m clauses of 2^n times for 2^n inputs of n Boolean variables is $(2^n \times m)$ logical and operations and $(2^n \times m \times n)$ logical or operations.

3.2.2 Data Dependence Analysis for the Satisfiability Problem

A data dependence arises from two statements that access or modify the same resource. *Data dependence analysis* is to judge whether it is safe to *reorder* or *parallelize* statements. In a satisfiability problem with n Boolean variables and m clauses, it consists of 2^n inputs that are 2^n combinational states of n Boolean variables. The first input is $x_1^0 x_2^0 \ldots x_{n-1}^0 x_n^0$, the second input is $x_1^0 x_2^0 \ldots x_{n-1}^0 x_n^1$ and so on with that the last input is $x_1^1 x_2^1 \ldots x_{n-1}^1 x_n^1$. Each input needs to complete those operations in Fig. 3.1. Each input needs to use $(m \times (n+1))$ auxiliary Boolean variables $r_{j,k}$ for $1 \leq j \leq m$ and $0 \leq k \leq n$ and $(m+1)$ auxiliary Boolean variables s_j for $0 \leq j \leq m$. Since 2^n inputs of n Boolean variables implement those operations in Fig. 3.1 not to access or modify the same input and the same auxiliary Boolean variables, we can *parallelize* them without any error.

Let us consider another example that is $F(x_1, x_2) = x_1 \wedge x_2$, where two variables x_1 and x_2 are two Boolean variables and their values could be 0 or 1. In the formula $F(x_1, x_2) = x_1 \wedge x_2$, the first clause contains (x_1) and the second clause includes (x_2). The satisfiability problem for the Boolean formula $F(x_1, x_2) = x_1 \wedge x_2$ with two Boolean variable x_1 and x_2 is to find values of each Boolean variable so that the whole formula has the value 1. This is the same as finding values of each Boolean variable that make each clause have the value 1.

We regard the satisfiability problem for the Boolean formula $F(x_1, x_2) = x_1 \wedge x_2$ with two Boolean variable x_1 and x_2 as a search problem in which any given oracular function O_f is the Boolean formula $F(x_1, x_2) = x_1 \wedge x_2$, its domain is $\{x_1 x_2 \; \forall x_d \in \{0, 1\} \text{ for } 1 \leq d \leq 2\}$ and its range is $\{0, 1\}$. In the given oracular function $O_f = F(x_1, x_2) = x_1 \wedge x_2$ of the search problem, there are M inputs of *two* bits from its domain, say $\lambda_M = x_1 x_2$, that satisfies the condition $O_f(\lambda_M) = O_f(x_1 x_2) = F(x_1, x_2) = x_1 \wedge x_2 = 1$. Whereas for all other inputs of two bits from the same domain, $\omega = x_1 x_2$, for $0 \leq \omega \leq 2^2 - 1$ and $\omega \neq \lambda_M$, $O_f(\omega) = O_f(x_1 x_2) = F(x_1, x_2) = x_1 \wedge x_2 = 0$. The search problem is to find one of M solutions that is to find values of each Boolean variable so that the whole formula has the value 1. This is the same as finding values of each Boolean variable that make each clause have the value 1.

From the domain $\{x_1 x_2 | \forall x_d \in \{0, 1\} \text{ for } 1 \leq d \leq 2\}$ of the Boolean formula $F(x_1, x_2) = x_1 \wedge x_2$, there are four inputs $x_1^0 x_2^0, x_1^0 x_2^1, x_1^1 x_2^0$ and $x_1^1 x_2^1$. Because it contains two clauses in which each clause only consists of one Boolean variable, each input needs to complete "$r_{1,1} \leftarrow r_{1,0}^0 \vee x_1$", "$s_1 \leftarrow s_0^0 \wedge r_{1,1}$", "$r_{2,1} \leftarrow r_{2,0}^0 \vee x_2$" and "$s_2 \leftarrow s_1 \wedge r_{2,1}$". The result of implementing a logical or operation "$r_{1,1} \leftarrow r_{1,0}^0 \vee x_1$" is actually equal to the value of Boolean variable x_1. This is to say that Boolean variable $r_{1,1}$ stores the value of Boolean variable x_1. Next, a logical and operation "$s_1 \leftarrow s_0^1 \wedge r_{1,1}$" is equivalent to another logical and operation "$s_1 \leftarrow s_0^1 \wedge x_1$". Because the result of implementing "$s_1 \leftarrow s_0^1 \wedge x_1$" is actually equal to the value of Boolean variable x_1, Boolean variable s_1 stores the value of Boolean variable x_1.

Next, the result of implementing a logical or operation "$r_{2,1} \leftarrow r_{2,0}^0 \vee x_2$" is actually equal to the value of Boolean variable x_2. This indicates that Boolean variable $r_{2,1}$ stores the value of Boolean variable x_2. Next, a logical and operation "$s_2 \leftarrow s_1 \wedge r_{2,1}$" is equivalent to another logical and operation "$s_2 \leftarrow x_1 \wedge x_2$". This is to say that the result of implementing "$r_{1,1} \leftarrow r_{1,0}^0 \vee x_1$", "$s_1 \leftarrow s_0^1 \wedge r_{1,1}$", "$r_{2,1} \leftarrow r_{2,0}^0 \vee x_2$" and "$s_2 \leftarrow s_1 \wedge r_{2,1,1}$" is the same as that of implementing "$s_2 \leftarrow s_1 \wedge x_2$". Therefore, four results of implementing $F(x_1^0, x_2^0) = x_1^0 \wedge x_2^0$, $F(x_1^0, x_2^1) = x_1^0 \wedge x_2^1$, $F(x_1^1, x_2^0) = x_1^1 \wedge x_2^0$ and $F(x_1^1, x_2^1) = x_1^1 \wedge x_2^1$ are respectively s_2^0 (false), s_2^0 (false), s_2^0 (false) and s_2^1 (true). Because 2^2 inputs of two Boolean variables implement those instructions above not to access or modify the same input and the same auxiliary Boolean variable, we can *parallelize* them without any error.

3.2.3 Solution Space of Solving an Instance of the Satisfiability Problem

For the given oracular function $O_f = F(x_1, x_2) = x_1 \wedge x_2$ of the search problem, its domain is $\{x_1 x_2 | \forall x_\alpha \in \{0, 1\} \text{ for } 1 \leq d \leq 2\}$ and its range is $\{0, 1\}$. We regard its domain as its solution space in which there are four possible choices that satisfy $O_f = F(x_1, x_2) = x_1 \wedge x_2 = 1$. We use a basis $\{(1, 0, 0, 0), (0, 1, 0, 0), (0, 0, 1, 0), (0, 0, 0, 1)\}$ of the four-dimensional Hilbert space to construct solution space

$\{x_1x_2 \ \forall x_d \in \{0, 1\}$ for $1 \le d \le 2\}$. We make use of $(1, 0, 0, 0)$ to encode Boolean variable x_1^0 and Boolean variable x_2^0. Next, we use $(0, 1, 0, 0)$ to encode Boolean variable x_1^1 and Boolean variable x_2^0. We apply $(0, 0, 1, 0)$ to encode Boolean variable x_1^0 and Boolean variable x_2^1. Finally, we use $(0, 0, 0, 1)$ to encode Boolean variable x_1^1 and Boolean variable x_2^1.

We use a linear combination of each element in the basis that is $\frac{1}{\sqrt{2^2}}(1, 0, 0, 0) +$ $\frac{1}{\sqrt{2^2}} \times (0, 1, 0, 0) + \frac{1}{\sqrt{2^2}} \times (0, 0, 1, 0) + \frac{1}{\sqrt{2^2}} \times (0, 0, 0, 1) = \left(\frac{1}{\sqrt{2^2}}, \frac{1}{\sqrt{2^2}}, \frac{1}{\sqrt{2^2}}, \frac{1}{\sqrt{2^2}}\right)$ to construct solution space $\{x_1x_2 | \forall x_d \in \{0, 1\}$ for $1 \le d \le 2\}$. The amplitude of each possible choice is all $\frac{1}{\sqrt{2^2}}$ and the sum to the square of the absolute value of each amplitude is one. Because the length of the vector is one, it is a unit vector. This is to say that we use a unit vector to encode all of the possible choices that satisfy $O_f = F(x_1, x_2) = x_1 \wedge x_2$. We call the square of the absolute value of each amplitude as the cost (the successful probability) of that choice that satisfies the given oracular function $O_f = F(x_1, x_2) = x_1 \wedge x_2$. The cost (the successful probability) of the answer(s) is close to one as soon as possible.

3.2.4 Implementing Solution Space to an Instance of the Satisfiability Problem

In Listing 3.1, the program in the backend *ibmqx4* with five quantum bits in **IBM**'s quantum computer is to solve an instance of the satisfiability problem with $F(x_1, x_2) = x_1 \wedge x_2$ in which we illustrate how to write a quantum program to find values of each Boolean variable so that the whole formula has the value 1. Figure 3.2 is the quantum circuit of constructing solution space to an instance of the satisfiability problem with $F(x_1, x_2) = x_1 \wedge x_2$. The statement "OPENQASM 2.0;" on line one of Listing 3.1 is to indicate that the program is written with version 2.0 of Open QASM. Next, the statement "include "qelib1.inc";" on line two of Listing 3.1 is to continue parsing the file "qelib1.inc" as if the contents of the file were pasted at the location

Fig. 3.2 The quantum circuit of constructing solution space to an instance of the satisfiability problem with $F(x_1, x_2) = x_1 \wedge x_2$

of the include statement, where the file "qelib1.inc" is **Quantum Experience (QE) Standard Header** and the path is specified relative to the current working directory.

Listing 3.1 The program of solving an instance of the satisfiability problem with $F(x_1, x_2) = x_1 \wedge x_2$.

```
1.  OPENQASM 2.0;
2.  include "qelib1.inc";
3.  qreg q[5];
4.  creg c[5];
5.  x q[0];
6.  h q[3];
7.  h q[4];
8.  h q[0];
```

Next, the statement "qreg q[5];" on line three of Listing 3.1 is to declare that in the program there are five quantum bits. In the left top of Fig. 3.2, five quantum bits are subsequently q[0], q[1], q[2], q[3] and q[4]. The initial value of each quantum bit is set to $|0>$. We use quantum bit q[3] to encode Boolean variable x_1. We make use of quantum bit q[4] to encode Boolean variable x_2. We apply quantum bit q[2] to encode auxiliary Boolean variable s_2. We use quantum bit q[0] as an auxiliary working bit. We do not use quantum bit q[1].

For the convenience of our explanation, $q[k]^0$ for $0 \le k \le 4$ is to represent the value 0 of q[k] and $q[k]^1$ for $0 \le k \le 4$ is to represent the value 1 of q[k]. Similarly, for the convenience of our explanation, an initial state vector of constructing solution space to an instance of the satisfiability problem with $F(x_1, x_2) = x_1 \wedge x_2$ is as follows:

$$|\Phi_0\rangle = |q[4]^0\rangle|q[3]^0\rangle|q[2]^0\rangle|q[1]^0\rangle|q[0]^0\rangle = |0\rangle|0\rangle|0\rangle|0\rangle|0\rangle = |00000\rangle.$$

Then, the statement "creg c[5];" on line four of Listing 3.1 is to declare that there are five classical bits in the program. In the left bottom of Fig. 3.2, five classical bits are respectively c[0], c[1], c[2], c[3] and c[4]. The initial value of each classical bit is set to 0. Classical bit c[4] is the most significant bit and classical bit c[0] is the least significant bit.

Next, the three statements "x q[0];", "h q[3];" and "h q[4];" on line five through seven of Listing 3.1 is to implement one X gate (one *NOT* gate) and two Hadamard gates of the *first* time slot of the quantum circuit in Fig. 3.2. The statement "x q[0];" actually completes $\begin{pmatrix} 0 & 1 \\ 1 & 0 \end{pmatrix} \times \begin{pmatrix} 1 \\ 0 \end{pmatrix} = \begin{pmatrix} 0 \\ 1 \end{pmatrix} = (|1\rangle)$. This indicates that the statement "x q[0];" on line five of Listing 3.1 inverts into $|q[0]^1\rangle(|1\rangle)$. The two statements "h q[3];" and "h q[4];" both actually complete

$$\begin{pmatrix} \frac{1}{\sqrt{2}} & \frac{1}{\sqrt{2}} \\ \frac{1}{\sqrt{2}} & -\frac{1}{\sqrt{2}} \end{pmatrix} \times \begin{pmatrix} 1 \\ 0 \end{pmatrix} = \begin{pmatrix} \frac{1}{\sqrt{2}} \\ \frac{1}{\sqrt{2}} \end{pmatrix} \frac{1}{\sqrt{2}} \begin{pmatrix} 1 \\ 1 \end{pmatrix} = \frac{1}{\sqrt{2}} \left(\begin{pmatrix} 1 \\ 0 \end{pmatrix} + \begin{pmatrix} 0 \\ 1 \end{pmatrix} \right) = \frac{1}{\sqrt{2}}(|0\rangle + |1\rangle).$$

This is to say that converting q[3] from one state $|0\rangle$ to another state $\frac{1}{\sqrt{2}}(|0\rangle + |1\rangle)$ (its superposition) and converting q[4] from one state $|0\rangle$ to another state (its superposition) are completed. Therefore, the superposition of the two quantum bits q[4] and q[3] is $\left(\frac{1}{\sqrt{2}}(|0\rangle + |1\rangle)\right)\left(\frac{1}{\sqrt{2}}(|0\rangle + |1\rangle)\right) = \frac{1}{2}(|0\rangle|0\rangle + |0\rangle|1\rangle + |1\rangle|0\rangle + |1\rangle|1\rangle) = \frac{1}{2}(|00\rangle + |01\rangle + |10\rangle + |11\rangle)$. Because in the *first* time slot of the quantum circuit in Fig. 3.2 there is no quantum gate to act on quantum bits q[2] and q[1], their current states $|q[2]^0\rangle$ and $|q[1]^0\rangle$ are not changed. This is to say that we obtain the following new state vector.

$$|\Phi_1\rangle = \left(\frac{1}{\sqrt{2}}(|q[4]^0\rangle + |q[4]^1\rangle)\right)\left(\frac{1}{\sqrt{2}}(|q[3]^0\rangle + |q[3]^1\rangle)\right)(|q[2]^0\rangle|q[1]^0\rangle|q[0]^1\rangle)$$

$$= \frac{1}{2}(|q[4]^0\rangle|q[3]^0\rangle + |q[4]^0\rangle|q[3]^1\rangle + |q[4]^1\rangle|q[3]^0\rangle + |q[4]^1\rangle|q[3]^1\rangle)$$

$$(|q[2]^0\rangle\,|q[1]^0\rangle|q[0]^1\rangle)$$

$$= \frac{1}{2}(|0\rangle|0\rangle + |0\rangle|1\rangle + |1\rangle|0\rangle + |1\rangle|1\rangle)(|0\rangle|0\rangle|1\rangle).$$

Next, the statement "h q[0];" on line *eight* of Listing 3.1 is to implement one Hadamard gate of the *second* time slot of the quantum circuit in Fig. 3.2. The statement "h q[0];" actually completes $\begin{pmatrix} \frac{1}{\sqrt{2}} & \frac{1}{\sqrt{2}} \\ \frac{1}{\sqrt{2}} & -\frac{1}{\sqrt{2}} \end{pmatrix} \times \begin{pmatrix} 0 \\ 1 \end{pmatrix} = \begin{pmatrix} \frac{1}{\sqrt{2}} \\ -\frac{1}{\sqrt{2}} \end{pmatrix} = \frac{1}{\sqrt{2}}\begin{pmatrix} 1 \\ -1 \end{pmatrix} = \frac{1}{\sqrt{2}}\left(\begin{pmatrix} 1 \\ 0 \end{pmatrix} - \begin{pmatrix} 0 \\ 1 \end{pmatrix}\right) = \frac{1}{\sqrt{2}}(|0\rangle - |1\rangle)$. This indicates that converting q[0] from one state $|1\rangle$ to another state $\frac{1}{\sqrt{2}}(|0\rangle - |1\rangle)$ (its superposition) is completed. Because in the *second* time slot of the quantum circuit in Fig. 3.2 there is no quantum gate to act on quantum bits q[4] through q[1], their current states are not changed. This indicates that we obtain the following new state vector

$$|\Phi_2\rangle = \left(\frac{1}{2}(|q[4]^0\rangle|q[3]^0\rangle + |q[4]^0\rangle|q[3]^1\rangle + |q[4]^1\rangle|q[3]^0\rangle + |q[4]^1\rangle|q[3]^1\rangle)\right)$$

$$(|q[2]^0\rangle|q[1]^0\rangle)\left(\frac{1}{\sqrt{2}}(|q[0]^0\rangle - |q[0]^1\rangle)\right)$$

$$= \left(\frac{1}{2}(|0\rangle|0\rangle + |0\rangle|1\rangle + |1\rangle|0\rangle + |1\rangle|1\rangle)\right)$$

$$(|0\rangle|0\rangle)\left(\frac{1}{\sqrt{2}}(|0\rangle - |1\rangle)\right).$$

In the new state vector $|\Phi_2\rangle$, state $|q[4]^0\rangle|q[3]^0\rangle$ encodes Boolean variable x_1^0 and Boolean variable x_2^0. State $|q[4]^0\rangle|q[3]^1\rangle$ encodes Boolean variable x_1^0 and Boolean variables x_2^0. State $|q[4]^1\rangle|q[3]^0\rangle$ encodes Boolean variable x_1^0 and Boolean variable x_2^1. State $|q[4]^1\rangle|q[3]^1\rangle$ encodes Boolean variable x_1^1 and Boolean variable x_2^1. The amplitude of each choice is $\frac{1}{\sqrt{2^2}}$ and the cost (the successful possibility) of becoming the answer(s) to each choice is the same and is equal to $\frac{1}{2^2} = 1/4$.

3.2.5   The Oracle to an Instance of the Satisfiability Problem

The Oracle is to have the ability to *recognize* solutions to the given oracular function $O_f = F(x_1, x_2) = x_1 \wedge x_2$ of the satisfiability problem. The Oracle is to multiply the probability amplitude of the answer(s) by -1 and leaves any other amplitude unchanged. The Oracle of solving the satisfiability problem with the given oracular function $O_f = F(x_1, x_2) = x_1 \wedge x_2$ is a $(2^2 \times 2^2)$ matrix B that is equal to

$$\begin{pmatrix} 1 & 0 & 0 & 0 \\ 0 & 1 & 0 & 0 \\ 0 & 0 & 1 & 0 \\ 0 & 0 & 0 & -1 \end{pmatrix}_{4 \times 4}$$

We assume that a $(2^2 \times 2^2)$ matrix B^+ is the conjugate transpose of B. Because the transpose of B is equal to B and each element in the transpose of B is a real, the conjugate transpose of B is also equal to B. Hence, we obtain $B^+ = B$. Because B and B^+ are almost a $(2^2 \times 2^2)$ identity matrix, $B \times B_m^+ = I_{2^2 \times 2^2}$, and $B^+ \times B = I_{2^2 \times 2^2}$. Thus, we obtain $B \times B^+ = B^+ \times B$. This is to say that it is a unitary matrix (operator) to solve the satisfiability problem with the given oracular function $O_f = F(x_1, x_2) = x_1 \wedge x_2$. Implementing the given oracular function $O_f = F(x_1, x_2) = x_1 \wedge x_2$ in the satisfiability problem is equivalent to implement

the Oracle that is $\begin{pmatrix} 1 & 0 & 0 & 0 \\ 0 & 1 & 0 & 0 \\ 0 & 0 & 1 & 0 \\ 0 & 0 & 0 & -1 \end{pmatrix}_{4 \times 4} \times$

$$\begin{pmatrix} \frac{1}{\sqrt{2^2}} \\ \frac{1}{\sqrt{2^2}} \\ \frac{1}{\sqrt{2^2}} \\ \frac{1}{\sqrt{2^2}} \end{pmatrix}_{4 \times 1} = \begin{pmatrix} \frac{1}{\sqrt{2^2}} \\ \frac{1}{\sqrt{2^2}} \\ \frac{1}{\sqrt{2^2}} \\ \frac{-1}{\sqrt{2^2}} \end{pmatrix}_{4 \times 1} = \frac{1}{\sqrt{2^2}} \times \begin{pmatrix} 1 \\ 0 \\ 0 \\ 0 \end{pmatrix}_{4 \times 1} + \frac{1}{\sqrt{2^2}} \times \begin{pmatrix} 0 \\ 1 \\ 0 \\ 0 \end{pmatrix}_{4 \times 1}$$

$$+ \frac{1}{\sqrt{2^2}} \times \begin{pmatrix} 0 \\ 0 \\ 1 \\ 0 \end{pmatrix}_{4 \times 1} + \frac{-1}{\sqrt{2^2}} \times \begin{pmatrix} 0 \\ 0 \\ 0 \\ 1 \end{pmatrix}_{4 \times 1}$$

$$= \frac{1}{\sqrt{2^2}} |00\rangle + \frac{1}{\sqrt{2^2}} |01\rangle + \frac{1}{\sqrt{2^2}} |10\rangle + \frac{-1}{\sqrt{2^2}} |11\rangle.$$

Four computational basis vectors $\begin{pmatrix} 1 \\ 0 \\ 0 \\ 0 \end{pmatrix}_{4\times 1}, \begin{pmatrix} 0 \\ 1 \\ 0 \\ 0 \end{pmatrix}_{4\times 1}, \begin{pmatrix} 0 \\ 0 \\ 1 \\ 0 \end{pmatrix}_{4\times 1}$ and

$\begin{pmatrix} 0 \\ 0 \\ 0 \\ 1 \end{pmatrix}_{4\times 1}$ encode four states $|00\rangle$, $|01\rangle$, $|10\rangle$ and $|11\rangle$ and their current amplitudes

are respectively $\left(\frac{1}{\sqrt{2^2}}\right)$, $\left(\frac{1}{\sqrt{2^2}}\right)$, $\left(\frac{1}{\sqrt{2^2}}\right)$ and $\left(\frac{-1}{\sqrt{2^2}}\right)$. State $|00\rangle\left(|q[4]^0\rangle|q[3]^0\rangle\right)$ with the amplitude $\left(\frac{1}{\sqrt{2^2}}\right)$ encodes Boolean variable x_1^0 and Boolean variable x_2^0. State $|01\rangle\left(|q[4]^0\rangle|q[3]^1\rangle\right)$ with the amplitude $\left(\frac{1}{\sqrt{2^2}}\right)$ encodes Boolean variable x_1^1 and Boolean variable x_2^0. State $|10\rangle\left(|q[4]^1\rangle|q[3]^0\rangle\right)$ with the amplitude $\left(\frac{1}{\sqrt{2^2}}\right)$ encodes Boolean variable x_1^0 and Boolean variable x_2^1. State $|11\rangle\left(|q[4]^1\rangle|q[3]^1\rangle\right)$ with the amplitude $\left(\frac{-1}{\sqrt{2^2}}\right)$ encodes Boolean variable x_1^1 and Boolean variable x_2^1. This is to say that the Oracle multiplies the probability amplitude of the answer with Boolean variable x_1^1 and Boolean variable x_2^1 by -1 and leaves any other amplitude unchanged.

3.2.6 Implementing the Oracle to an Instance of the Satisfiability Problem

We use one **CCNOT** gate to implement the given oracular function $O_f = F(x_1, x_2) = x_1 \wedge x_2$ of the satisfiability problem. We use quantum bit q[3] to encode Boolean variable x_1, we use quantum bit q[4] to encode Boolean variable x_2 and we use quantum bit q[2] to encode Boolean variable s_2. So, quantum bits q[3], q[4], q[2] are respectively the first control bit, the second control bit and the target bit of the **CCNOT** gate. Because we use the **CCNOT** gate to implement a logical and operation, the initial value to quantum bit q[2] is set to $|0\rangle$.

From line *nine* through line *twenty-three* in Listing 3.1, there are the fifteen statements. The front eight statements are subsequently "h q[2];", "cx q[4],q[2];", "tdg q[2];", "cx q[3],q[2];", "t q[2];", "cx q[4],q[2];", "tdg q[2];" and "cx q[3],q[2];". The resting seven statements are "t q[4];", "t q[2];", "cx q[3],q[4];", "h q[2];", "t q[3];", "tdg q[4];" and "cx q[3], q[4];". They implement the **CCNOT** gate that completes the given oracular function $O_f = F(x_1, x_2) = x_1 \wedge x_2$.

Listing 3.1 Continued…

// We use the following *fifteen* statements to implement a ***CCNOT*** gate.

```
 9.   h q[2];
10.   cx q[4],q[2];
11.   tdg q[2];
12.   cx q[3],q[2];
13.   t q[2];
14.   cx q[4],q[2];
15.   tdg q[2];
16.   cx q[3],q[2];
17.   t q[4];
18.   t q[2];
19.   cx q[3],q[4];
20.   h q[2];
21.   t q[3];
22.   tdg q[4];
23.   cx q[3], q[4];
```

Figure 3.3 is the quantum circuit of implementing the Oracle to an instance of the satisfiability problem with $F(x_1, x_2) = x_1 \wedge x_2$. They take the state vector $|\Phi_2\rangle = \left(\frac{1}{2}\left(|q[4]^0\rangle|q[3]^0\rangle + |q[4]^0\rangle|q[3]^1\rangle + |q[4]^1\rangle|q[3]^0\rangle + |q[4]^1\rangle|q[3]^1\rangle\right)\right)$ $\left(|q[2]^0\rangle|q[1]^0\rangle\right)\left(\frac{1}{\sqrt{2}}\left(|q[0]^0\rangle - |q[0]^1\rangle\right)\right)$ as their input. After they actually implement six ***CNOT*** gates, two Hadamard gates, three T^+ gates and four T gates from the *first* time slot through the *eleventh* time slot in Fig. 3.3, we obtain the following new state vector.

$$|\Phi_3\rangle = \left(\frac{1}{2}\left(|q[4]^0\rangle|q[3]^0\rangle|q[2]^0\rangle + |q[4]^0\rangle|q[3]^1\rangle|q[2]^0\rangle + |q[4]^1\rangle|q[3]^0\rangle|q[2]^0\rangle\right.\right.$$

$$\left.\left. + |q[4]^1\rangle|q[3]^1\rangle|q[2]^1\rangle\right)\right)\left(|q[1]^0\rangle\right)\left(\frac{1}{\sqrt{2}}\left(|q[0]^0\rangle - |q[0]^1\rangle\right)\right)\right)$$

$$= \left(\frac{1}{2}\left(|0\rangle|0\rangle|0\rangle + |0\rangle|1\rangle|0\rangle + |1\rangle|0\rangle|0\rangle + |1\rangle|1\rangle|1\rangle\right)\right)$$

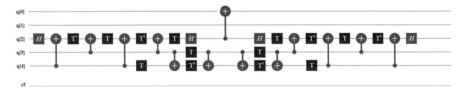

Fig. 3.3 The quantum circuit of implementing the Oracle to an instance of the satisfiability problem with $F(x_1, x_2) = x_1 \wedge x_2$

$$(|0\rangle)\left(\frac{1}{\sqrt{2}}(|0\rangle - |1\rangle)\right).$$

Next, from line *twenty-four* in Listing 3.1, the statement "cx q[2],q[0];" takes the new state vector $|\Phi_3\rangle$ as its input. It multiplies the probability amplitude of the answer. $|q[4]^1\rangle|q[3]^1\rangle$ encoding $x_1^1 x_2^1$ by -1 and leaves any other amplitude unchanged.

> **Listing 3.1 Continued…**
> // The Oracle multiplies the probability amplitude of the answer $x_1^1 x_2^1$ by -1 and.
>
> // leaves any other amplitude unchanged.
> 24. cx q[2],q[0];

This is to say that after the statement "cx q[2],q[0];" implements the **CNOT** gate in the *twelfth* time slot in Fig. 3.3, we obtain the following new state vector.

$$|\Phi_4\rangle = \left(\frac{1}{2}(|q[4]^0\rangle|q[3]^0\rangle|q[2]^0\rangle + |q[4]^0\rangle|q[3]^1\rangle|q[2]^0\rangle + |q[4]^1\rangle|q[3]^0\rangle|q[2]^0\rangle\right.$$
$$\left. + (-1)|q[4]^1\rangle|q[3]^1\rangle|q[2]^1\rangle))(|q[1]^0\rangle)\left(\frac{1}{\sqrt{2}}(|q[0]^0\rangle - |q[0]^1\rangle)\right)\right)$$
$$= \left(\frac{1}{2}(|0\rangle|0\rangle|0\rangle + |0\rangle|1\rangle|0\rangle + |1\rangle|0\rangle|0\rangle + (-1)|1\rangle|1\rangle|1\rangle)\right)$$
$$(|0\rangle)\left(\frac{1}{\sqrt{2}}(|0\rangle - |1\rangle)\right).$$

Because quantum operations are reversible by nature, executing the reversed order of implementing the **CCNOT** gate can restore the auxiliary quantum bits to their initial states. From line *twenty-five* through line *thirty-nine* in Listing 3.1, there are the *fifteen* statements. They are "cx q[3],q[4];", "tdg q[4];", "t q[3];", "h q[2];", "cx q[3],q[4];", "t q[2];", "t q[4];", "cx q[3],q[2];", "tdg q[2];", "cx q[4],q[2];", "t q[2];", "cx q[3],q[2];", "tdg q[2];", "cx q[4],q[2];" and "h q[2];". They run the reversed order of implementing the **CCNOT** gate that completes the given oracular function $O_f = F(x_1, x_2) = x_1 \wedge x_2$. They take the new state vector $|\Phi_4\rangle$ as their input. After they actually complete six **CNOT** gates, two Hadamard gates, three T^+ gates and four T gates from the *thirteenth* time slot through the *last* time slot in Fig. 3.3, we obtain the following new state vector

$$|\Phi_5\rangle = \left(\frac{1}{2}(|q[4]^0\rangle|q[3]^0\rangle + |q[4]^0\rangle|q[3]^1\rangle + |q[4]^1\rangle|q[3]^0\rangle\right.$$
$$\left. + (-1)|g[4]^1\rangle|q[3]^1\rangle))(|q[2]^0\rangle|q[1]^0\rangle)\left(\frac{1}{\sqrt{2}}(|q[0]^0\rangle - |q[0]^1\rangle)\right)\right)$$

$$= \left(\frac{1}{2}(|0\rangle|0\rangle + |0\rangle|1\rangle + |1\rangle|0\rangle + (-1)|1\rangle|1\rangle)\right)(|0\rangle|0\rangle)\left(\frac{1}{\sqrt{2}}(|0\rangle - |1\rangle)\right).$$

Listing 3.1 Continued…

// Because quantum operations are reversible by nature, executing the reversed.

// order of implementing the **CCNOT** gate can restore the auxiliary quantum bits.

// to their initial states.

```
25.  cx q[3],q[4];
26.  tdg q[4];
27.  t q[3];
28.  h q[2];
29.  cx q[3],q[4];
30.  t q[2];
31.  t q[4];
32.  cx q[3],q[2];
33.  tdg q[2];
34.  cx q[4],q[2];
35.  t q[2];
36.  cx q[3],q[2];
37.  tdg q[2];
38.  cx q[4],q[2];
39.  h q[2];
```

In the state vector $|\Phi_2\rangle$, the amplitude of each element in solution space $\{x_1 x_2 \forall x_d \in \{0, 1\}$ for $1 \le d \le 2\}$ is $(1/2)$. In the state vector $|\Phi_5\rangle$, the amplitude to three elements $x_1^0 x_2^0$, $x_1^0 x_2^1$, $x_1^1 x_2^0$ in solution space is all $(1/2)$ and the amplitude to the element $x_1^1 x_2^1$ in solution space is $(-1/2)$. This indicates that *thirty-one* state-

ments from line *nine* through *thirty-nine* in Listing 3.1 complete $\begin{pmatrix} 1 & 0 & 0 & 0 \\ 0 & 1 & 0 & 0 \\ 0 & 0 & 1 & 0 \\ 0 & 0 & 0 & -1 \end{pmatrix}_{4\times4}$

$\times \begin{pmatrix} \frac{1}{\sqrt{2^2}} \\ \frac{1}{\sqrt{2^2}} \\ \frac{1}{\sqrt{2^2}} \\ \frac{1}{\sqrt{2^2}} \end{pmatrix}_{4\times1}$ that is to complete the Oracle of solving the satisfiability problem

with the given oracular function $O_f = F(x_1, x_2) = x_1 \wedge x_2$.

3.2.7 The Grover Diffusion Operator to Amplify the Amplitude of the Answers in the Satisfiability Problem

We assume that a $(2^n \times 1)$ vector $|u\rangle$ is $\begin{pmatrix} u_{1,1} \\ u_{2,1} \\ \vdots \\ u_{2^n,1} \end{pmatrix}_{2^n \times 1}$ and another $(2^n \times 1)$

vector $|v\rangle$ is $\begin{pmatrix} v_{1,1} \\ v_{2,1} \\ \vdots \\ v_{2^n,1} \end{pmatrix}_{2^n \times 1}$. The transpose of $|v\rangle$ is a (1×2^n) vector that is

$\begin{pmatrix} v_{1,1} & v_{1,2} & \cdots & v_{1,2^n} \end{pmatrix}_{1 \times 2^n}$. Definition 3.2 introduces the outer product $|u\rangle\langle v|$ of two vectors $|u\rangle$ and $|v\rangle$.

Definition 3.2 The outer product $|u\rangle\langle v|$ of two vectors $|u\rangle$ and $|v\rangle$ is a $(2^n \times 2^n)$ matrix W that is

$$\begin{pmatrix} u_{1,1} \\ u_{2,1} \\ \vdots \\ u_{2^n,1} \end{pmatrix}_{2^n \times 1} \times \begin{pmatrix} v_{1,1} & v_{1,2} & \cdots & v_{1,2^n} \end{pmatrix}_{1 \times 2^n}$$

$$= \begin{pmatrix} u_{1,1} \times v_{1,1} & \cdots & u_{1,1} \times v_{1,2^n} \\ \vdots & \vdots & \vdots \\ u_{2^n,1} \times v_{1,1} & \cdots & u_{2^n,1} \times v_{1,2^n} \end{pmatrix}_{2^n \times 2^n}.$$

We assume that $H^{\otimes n}$ represents n Hadamard gates and $|0^{\otimes n}\rangle$ represents n quantum bits in which the value of each quantum bit is equal to $|0\rangle$. After we use a unitary operator $H^{\otimes n}$ to operate n quantum bits, $|0^{\otimes n}\rangle$, the state $|0\rangle$ of each quantum bit is converted into its superposition $\frac{1}{\sqrt{2}}(|0\rangle + |1\rangle)$. This is to say that the superposition to n quantum bits with $|0^{\otimes n}\rangle$ is $|\psi\rangle = \left(\otimes_{k=1}^{n} \left(\frac{1}{\sqrt{2}}(|0\rangle + |1\rangle) \right) \right) = \frac{1}{\sqrt{2^n}} \sum_{k=0}^{2^n-1} |k\rangle =$

$$\frac{1}{\sqrt{2^n}} \left(\begin{pmatrix} 1 \\ 0 \\ \vdots \\ 0 \end{pmatrix}_{2^n \times 1} + \cdots + \begin{pmatrix} 0 \\ 0 \\ \vdots \\ 1 \end{pmatrix}_{2^n \times 1} \right) = \begin{pmatrix} \frac{1}{\sqrt{2^n}} \\ \frac{1}{\sqrt{2^n}} \\ \vdots \\ \frac{1}{\sqrt{2^n}} \end{pmatrix}_{2^n \times 1}.$$ The state vector $|\psi\rangle$ is

the uniform superposition of states, is a $(2^n \times 1)$ vector and its length is one. This indicates that it is a unit vector.

The matrix D that is a $(2^n \times 2^n)$ matrix defines the Grover diffusion operator D as follows:

$$D_{a,b} = \frac{2}{2^n} \text{ if } a \neq b \text{ and } D_{a,a} = \frac{2}{2^n} - 1.$$

This diffusion transform, D, can be implemented as $D = 2|\psi\rangle\langle\psi| - I_{2^n \times 2^n} = H^{\otimes n}(2|0^{\otimes n}\rangle\langle 0^{\otimes n}| - I_{2^n \times 2^n})H^{\otimes n}$. The rotation matrix R that is a $(2^n \times 2^n)$ matrix defines a phase shifter operator, $(2|0^{\otimes n}\rangle\langle 0^{\otimes n}| - I_{2^n \times 2^n})$, as follows:

$$R_{a,b} = 0 \text{ if } a \neq b; R_{a,a} = 1 \text{ if } a = 0; R_{a,a} = -1 \text{ if } a \neq 0.$$

This implies that the phase shifter operator, $(2|0^{\otimes n}\rangle\langle 0^{\otimes n}| - I_{2^n \times 2^n})$, negates all the states except for $|0\rangle$. It turns out that a quantum circuit with a phase shift operator, $2|0^{\otimes n}\rangle\langle 0^{\otimes n}| - I_{2^n \times 2^n}$, that negates all the states except for $|0\rangle$ sandwiched between $H^{\otimes n}$ gates can implement the Grover diffusion operator D. We use Lemma 3.1 to show that the Grover diffusion operator, $D = 2|\psi\rangle\langle\psi| - I_{2^n \times 2^n} = H^{\otimes n}(2|0^{\otimes n}\rangle\langle 0^{\otimes n}| - I_{2^n \times 2^n})H^{\otimes n}$, is a unitary operator.

Lemma 3.1 *The Grover diffusion operator,* $D = 2|\psi\rangle\langle\psi| - I_{2^n \times 2^n} = H^{\otimes n}(2|0^{\otimes n}\rangle\langle 0^{\otimes n}| - I_{2^n \times 2^n})H^{\otimes n}$, *is a unitary operator.*

Proof The outer product $|\psi\rangle\langle\psi|$ of the $(2^n \times 1)$ vector $|\psi\rangle$ with itself leads to a $(2^n \times 2^n)$ matrix V that is

$$\begin{pmatrix} \frac{1}{\sqrt{2^n}} \\ \vdots \\ \frac{1}{\sqrt{2^n}} \end{pmatrix}_{2^n \times 1} \times \begin{pmatrix} \frac{1}{\sqrt{2^n}} & \cdots & \frac{1}{\sqrt{2^n}} \end{pmatrix}_{2^n \times 1} = \begin{pmatrix} \frac{1}{2^n} & \cdots & \frac{1}{2^n} \\ \vdots & \vdots & \vdots \\ \frac{1}{2^n} & \cdots & \frac{1}{2^n} \end{pmatrix}_{2^n \times 2^n}.$$

Subtracting the identity matrix $I_{2^n \times 2^n}$ from the double of the $(2^n \times 2^n)$ matrix V, we

obtain a new $(2^n \times 2^n)$ matrix D that is $\begin{pmatrix} \frac{2}{2^n} & \cdots & \frac{2}{2^n} \\ \vdots & \vdots & \vdots \\ \frac{2}{2^n} & \cdots & \frac{2}{2^n} \end{pmatrix}_{2^n \times 2^n} - \begin{pmatrix} 1 & \cdots & 0 \\ \vdots & \vdots & \vdots \\ 0 & \cdots & 1 \end{pmatrix}_{2^n \times 2^n} =$

$$\begin{pmatrix} \frac{2}{2^n} - 1 & \cdots & \frac{2}{2^n} \\ \vdots & \vdots & \vdots \\ \frac{2}{2^n} & \cdots & \frac{2}{2^n} - 1 \end{pmatrix}_{2^n \times 2^n}.$$

We assume that a $(2^n \times 2^n)$ matrix D^+ is the conjugate transpose of D. We assume that N is equal to $(1/2^n)$. Because the transpose of D is equal to D and each element in the transpose of D is a real, the conjugate transpose of D is equal to D. Hence, we obtain $D^+ = D$. Since $(D \times D^+)_{a,a} = \left(\frac{2}{N} - 1\right)^2 + \left(\frac{2}{N}\right)^2 \times (N - 1) = 1$ and $(D \times D^+)_{a,b} = \left(\frac{2}{N}1\right) \times \left(\frac{2}{N}\right) + \left(\frac{2}{N}\right) \times \left(\frac{2}{N} - 1\right) + \left(\frac{2}{N}\right)^2 \times (N - 2) = 0$, we obtain $D \times D^+ = I_{2^n \times 2^n}$. Because $D^+ = D$ and $D \times D^+ = I_{2^n \times 2^n}$, we obtain $D^+ \times D = D \times D^+ = I_{2^n \times 2^n}$. Because $|\psi\rangle = H^{\otimes n}(|0^{\otimes n}\rangle)$ and $I_{2^n \times 2^n} = H^{\otimes n} I_{2^n \times 2^n} H^{\otimes n}$, we obtain $D = 2|\psi\rangle\langle\psi| - I_{2^n \times 2^n} = 2 H^{\otimes n}|0^{\otimes n}\rangle\langle 0^{\otimes n}| H^{\otimes n} - I_{2^n \times 2^n} = 2 H^{\otimes n}|0^{\otimes n}\rangle\langle 0^{\otimes n}| H^{\otimes n} - H^{\otimes n} I_{2^n \times 2^n} H^{\otimes n} = H^{\otimes n}(2|0^{\otimes n}\rangle\langle 0^{\otimes n}| - I_{2^n \times 2^n})H^{\otimes n}$. From the statements above, it is at once inferred that the Grover diffusion operator, $D = 2 |\psi\rangle\langle\psi| - I_{2^n \times 2^n} = H^{\otimes n} (2|0^{\otimes n}\rangle\langle 0^{\otimes n}| - I_{2^n \times 2^n}) H^{\otimes n}$ is a unitary operator. ∎

3.2.8 Implementing the Grover Diffusion Operator to Amplify the Amplitude of the Answer in an Instance of the Satisfiability Problem

The new state vector $|\Phi_5\rangle$ is $\left(\frac{1}{2}(|q[4]^0\rangle|q[3]^0\rangle + |q[4]^0\rangle|q[3]^1\rangle + |q[4]^1\rangle|q[3]^0\rangle + (-1)|q[4]^1\rangle|q[3]^1\rangle)\right)$ $\left(|q[2]^0\rangle|q[1]^0\rangle\right)\left(\frac{1}{\sqrt{2}}(|q[0]^0\rangle - |q[0]^1\rangle)\right)$. It consists of two subsystem. The first subsystem is $\left(\frac{1}{2}(|q[4]^0\rangle|q[3]^0\rangle + |q[4]^0\rangle|q[3]^1\rangle + |q[4]^1\rangle|q[3]^0\rangle + (-1)|q[4]^1\rangle|q[3]^1\rangle)\right)$ and the second subsystem is $\left(|q[2]^0\rangle|q[1]^0\rangle\right)\left(\frac{1}{\sqrt{2}}(|q[0]^0\rangle - |q[0]^1\rangle)\right)$. The two subsystems are independent each other. Amplifying the amplitude of each answer in the satisfiability problem with the given oracular function $O_f = F(x_1, x_2) = x_1 \wedge x_2$ just needs to consider the first subsystem in the new state vector $|\Phi_5\rangle$. Because for the satisfiability problem with the given oracular function $O_f = F(x_1, x_2) = x_1 \wedge x_2$ the $(2^2 \times 1)$ vector $\begin{pmatrix} \frac{1}{\sqrt{2^2}} \\ \frac{1}{\sqrt{2^2}} \\ \frac{1}{\sqrt{2^2}} \\ \frac{-1}{\sqrt{2^2}} \end{pmatrix}_{4\times 1}$ encodes the first

subsystem of the new state vector $|\Phi_5\rangle$ and $\begin{pmatrix} \frac{2}{2^2}-1 & \frac{2}{2^2} & \frac{2}{2^2} & \frac{2}{2^2} \\ \frac{2}{2^2} & \frac{2}{2^2}-1 & \frac{2}{2^2} & \frac{2}{2^2} \\ \frac{2}{2^2} & \frac{2}{2^2} & \frac{2}{2^2}-1 & \frac{2}{2^2} \\ \frac{2}{2^2} & \frac{2}{2^2} & \frac{2}{2^2} & \frac{2}{2^2}-1 \end{pmatrix}_{2^2 \times 2^2}$

is a $(2^2 \times 2^2)$ diffusion operator, amplifying the amplitude of the answer is to complete $\begin{pmatrix} \frac{2}{2^2}-1 & \frac{2}{2^2} & \frac{2}{2^2} & \frac{2}{2^2} \\ \frac{2}{2^2} & \frac{2}{2^2}-1 & \frac{2}{2^2} & \frac{2}{2^2} \\ \frac{2}{2^2} & \frac{2}{2^2} & \frac{2}{2^2}-1 & \frac{2}{2^2} \\ \frac{2}{2^2} & \frac{2}{2^2} & \frac{2}{2^2} & \frac{2}{2^2}-1 \end{pmatrix}_{2^2 \times 2^2} \begin{pmatrix} \frac{1}{\sqrt{2^2}} \\ \frac{1}{\sqrt{2^2}} \\ \frac{1}{\sqrt{2^2}} \\ \frac{-1}{\sqrt{2^2}} \end{pmatrix}_{4\times 1} = \begin{pmatrix} 0 \\ 0 \\ 0 \\ 1 \end{pmatrix}_{4 \times 1}$.

This is to say that the amplitude of the answer $x_1^1 x_2^1$ is one and the amplitude of other three possible choices $x_1^0 x_2^0$, $x_1^0 x_2^1$ and $x_1^1 x_2^0$ is all zero.

The quantum circuit in Fig. 3.4 implements the Grover diffusion operator, $H^{\otimes 2}(2|0^{\otimes 2}\rangle\langle 0^{\otimes 2}| - I_{2^2 \times 2^2}) H^{\otimes 2}$. The statements "h q[3];" and "h q[4];" from line *forty* through line *forty-one* in Listing 3.1 complete the first $H^{\otimes 2}$ gate in the diffusion operator, $H^{\otimes 2} (2|0^{\otimes 2}\rangle\langle 0^{\otimes 2}| - I_{2^2 \times 2^2}) H^{\otimes 2}$. They takes $\left(\frac{1}{2}(|q[4]^0\rangle|q[3]^0\rangle + |q[4]^0\rangle|q[3]^1\rangle + |q[4]^1\rangle|q[3]^0\rangle + (-1)|q[4]^1\rangle|q[3]^1\rangle)\right)$ as their input and complete two Hadamard gates in the first time slot of Fig. 3.4. State $\left(\frac{1}{2}|q[4]^0\rangle|q[3]^0\rangle\right)$ is converted into state

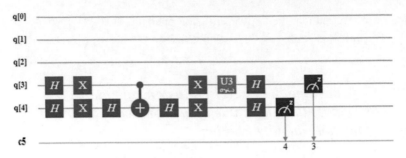

Fig. 3.4 The quantum circuit of implementing the Grover diffusion operator, $H^{\otimes 2}$ $(2|0^{\otimes 2}\rangle\langle 0^{\otimes 2}| - I_{2^2 \times 2^2})H^{\otimes 2}$, to an instance of the satisfiability problem with $F(x_1, x_2) = x_1 \wedge x_2$

$\frac{1}{4}(|q[4]^0\rangle|q[3]^0\rangle + |q[4]^0\rangle|q[3]^1\rangle + |q[4]^1\rangle|q[3]^0\rangle + |q[4]^1\rangle|q[3]^1\rangle)$.   State $(\frac{1}{2}|q[4]^0\rangle|q[3]^1\rangle)$ is converted into state $\frac{1}{4}(|q[4]^0\rangle|q[3]^0\rangle - |q[4]^0\rangle|q[3]^1\rangle$ $+|q[4]^1\rangle|q[3]^0\rangle - |q[4]^1\rangle|q[3]^1\rangle)$.   State $(\frac{1}{2}|q[4]^1\rangle|q[3]^0\rangle)$ is converted into state $\frac{1}{4}(|q[4]^0\rangle|q[3]^0\rangle + |q[4]^0\rangle|q[3]^1\rangle$ $-|q[4]^1\rangle|q[3]^0\rangle - |q[4]^1\rangle|q[3]^1\rangle)$.   State $(-\frac{1}{2}|q[4]^1\rangle|q[3]^1\rangle)$ is converted into state $-\frac{1}{4}(|q[4]^0\rangle|q[3]^0\rangle - |q[4]^0\rangle|q[3]^1\rangle$ $-|q[4]^1\rangle|q[3]^0\rangle + |q[4]^1\rangle|q[3]^1\rangle)$. This is to say that we obtain the following new state vector.

$$|\Phi_6\rangle = \left(\frac{1}{2}(|q[4]^0\rangle|q[3]^0\rangle + |q[4]^0\rangle|q[3]^1\rangle + |q[4]^1\rangle|q[3]^0\rangle + (-1)|q[4]^1\rangle|q[3]^1\rangle)\right).$$

Listing 3.1 Continued...

//We complete the amplitude amplification of the answer.

40.   h q[3];
41.   h q[4];

Next, from line forty-two through forty-three in Listing 3.1 the two statements "x q[3]" and "x q[4]" implement two **NOT** gates in the second time slot of Fig. 3.4. They take $(\frac{1}{2}(|q[4]^0\rangle|q[3]^0\rangle + |q[4]^0\rangle|q[3]^1\rangle + |q[4]^1\rangle|q[3]^0\rangle + (-1)|q[4]^1\rangle|q[3]^1\rangle))$ in the new state vector $|\Phi_6 >$ as their input. State $(\frac{1}{2}|q[4]^0\rangle|q[3]^0\rangle)$ is converted into state $\frac{1}{2}(|q[4]^1\rangle|q[3]^1\rangle)$ State $\frac{1}{2}(|q[4]^0\rangle|q[3]^1\rangle)$ is converted into state $\frac{1}{2}(|q[4]^1\rangle|q[3]^0\rangle)$. State $\frac{1}{2}(|q[4]^1\rangle|q[3]^0\rangle)$ is converted into state $\frac{1}{2}(|q[4]^0\rangle|q[3]^1\rangle)$. State $-\frac{1}{2}(|q[4]^1\rangle|q[3]^1\rangle)$ is converted into state $-\frac{1}{2}(|q[4]^0\rangle|q[3]^0\rangle)$. This indicates that we obtain the following new state vector

$$|\Phi_7\rangle = \left(\frac{1}{2}(-1)|q[4]^0\rangle|q[3]^0\rangle + |q[4]^0\rangle|q[3]^1\rangle + |q[4]^1\rangle|q[3]^0\rangle + |q[4]^1\rangle|q[3]^1\rangle\right).$$

Next, from line forty-four in Listing 3.1 the statement "h q[4];" implements one Hadamard gate in the third time slot of Fig. 3.4. They take $\left(\frac{1}{2}(-1)\left|q[4]^0\right\rangle\left|q[3]^0\right\rangle + \left|q[4]^0\right\rangle\left|q[3]^1\right\rangle + \left|q[4]^1\right\rangle\left|q[3]^0\right\rangle + \left|q[4]^1\right\rangle\left|q[3]^1\right\rangle\right)$ in the new state vector $\left|\Phi_7\right\rangle$ as their input. State $\left(-\frac{1}{2}\left|q[4]^0\right\rangle\left|q[3]^0\right\rangle\right)$ is converted into state $-\frac{1}{2\sqrt{2}}\left(\left|q[4]^0\right\rangle\left|q[3]^0\right\rangle + \left|q[4]^1\right\rangle\left|q[3]^0\right\rangle\right)$. State $\left(\frac{1}{2}\left|q[4]^0\right\rangle\left|q[3]^1\right\rangle\right)$ is converted into state $\frac{1}{2\sqrt{2}}\left(\left|q[4]^0\right\rangle\left|q[3]^1\right\rangle + \left|q[4]^1\right\rangle\left|q[3]^1\right\rangle\right)$. State $\left(\frac{1}{2}\left|q[4]^1\right\rangle\left|q[3]^0\right\rangle\right)$ is converted into state $\frac{1}{2\sqrt{2}}\left(\left|q[4]^0\right\rangle\left|q[3]^0\right\rangle - \left|q[4]^1\right\rangle\left|q[3]^0\right\rangle\right)$. State $\left(\frac{1}{2}\left|q[4]^1\right\rangle\left|q[3]^1\right\rangle\right)$ is converted into state $\frac{1}{2\sqrt{2}}\left(\left|q[4]^0\right\rangle\left|q[3]^1\right\rangle - \left|q[4]^1\right\rangle\left|q[3]^1\right\rangle\right)$. This is to say that we obtain the following newstate vector

$$\left|\Phi_8\right\rangle = \left(\frac{1}{\sqrt{2}}\left(\left|q[4]^0\right\rangle\left|q[3]^1\right\rangle - \left|q[4]^1\right\rangle\left|q[3]^0\right\rangle\right)\right).$$

Next, from the line forty-five in Listing 3.1 the statement "cx q[3],q[4];" implements one **CNOT** gate in the fourth time slot of Fig. 3.4. They take $\frac{1}{\sqrt{2}}\left(\left|q[4]^0\right\rangle\left|q[3]^1\right\rangle - \left|q[4]^1\right\rangle\left|q[3]^0\right\rangle\right)$ in the new state vector $\left|\Phi_8\right\rangle$ as their input. State $\left(\frac{1}{\sqrt{2}}\left|q[4]^0\right\rangle\left|q[3]^1\right\rangle\right)$ is converted into state $\left(\frac{1}{\sqrt{2}}\left|q[4]^1\right\rangle\left|q[3]^1\right\rangle\right)$. State $\left(-\frac{1}{\sqrt{2}}\left|q[4]^1\right\rangle\left|q[3]^0\right\rangle\right)$ is converted into state $\left(-\frac{1}{\sqrt{2}}\left|q[4]^1\right\rangle\left|q[3]^0\right\rangle\right)$. This indicates that we obtain the following new state vector

$$\left|\Phi_9\right\rangle = \left(\frac{1}{\sqrt{2}}\left(\left|q[4]^1\right\rangle\left|q[3]^1\right\rangle - \left|q[4]^1\right\rangle\left|q[3]^0\right\rangle\right)\right).$$

Next, from the line forty-six in Listing 3.1 the statement "h q[4];" implements one Hadamard gate in the fifth time slot of Fig. 3.4. They take
$\left(\frac{1}{\sqrt{2}}\left(|q[4]^1\rangle|q[3]^1\rangle - |q[4]^1\rangle|q[3]^0\rangle\right)\right)$ in the new state vector $|\Phi_9\rangle$ as their input. State
$\left(\frac{1}{\sqrt{2}}|q[4]^1\rangle|q[3]^1\rangle\right)$ is converted into state $\frac{1}{2}\left(|q[4]^0\rangle|q[3]^1\rangle - |q[4]^1\rangle|q[3]^1\rangle\right)$. State
$\left(-\frac{1}{\sqrt{2}}|q[4]^1\rangle|q[3]^0\rangle\right)$ is converted into state $-\frac{1}{2}\left(|q[4]^0\rangle|q[3]^0\rangle - |q[4]^1\rangle|q[3]^0\rangle\right)$. This
indicates that we obtain the following new state vector

$$|\Phi_{10}\rangle = \frac{1}{2}\left((-1)|q[4]^0\rangle|q[3]^0\rangle + |q[4]^0\rangle|q[3]^1\rangle\right.$$
$$\left. + |q[4]^1\rangle|q[3]^0\rangle + (-1)|q[4]^1\rangle|q[3]^1\rangle\right).$$

Listing 3.1 Continued…

46. h q[4];

Next, from the line forty-seven through line forty-eight in Listing 3.1 the
two statements "x q[4];" and "x q[3];" implements two **X** (**NOT**) gates in
the *sixth* time slot of Fig. 3.4. They take $\frac{1}{2}\left((-1)|q[4]^0\rangle|q[3]^0\rangle + |q[4]^0\rangle|q[3]^1\rangle\right.$
$\left. + |q[4]^1\rangle|q[3]^0\rangle + (-1)|q[4]^1\rangle|q[3]^1\rangle\right)$. in the new state vector $|\Phi_{10}\rangle$ as their
input. State $\left(-\frac{1}{2}|q[4]^0\rangle|q[3]^0\rangle\right)$ is converted into state $\left(-\frac{1}{2}|q[4]^1\rangle|q[3]^1\rangle\right)$. State
$\left(\frac{1}{2}|q[4]^0\rangle|q[3]^1\rangle\right)$ is converted into state $\left(\frac{1}{2}|q[4]^1\rangle|q[3]^0\rangle\right)$. State $\left(\frac{1}{2}|q[4]^1\rangle|q[3]^0\rangle\right)$ is
converted into state $\left(\frac{1}{2}|q[4]^0\rangle|q[3]^1\rangle\right)$. State $\left(-\frac{1}{2}|q[4]^1\rangle|q[3]^1\rangle\right)$ is converted into state
$\left(-\frac{1}{2}|q[4]^0\rangle|q[3]^0\rangle\right)$. This is to say that we obtain the following new state vector.

$$|\Phi_{11}\rangle = \frac{1}{2}\left((-1)|q[4]^0\rangle|q[3]^0\rangle + |q[4]^0\rangle|q[3]^1\rangle\right.$$
$$\left. + |q[4]^1\rangle|q[3]^0\rangle + (-1)|q[4]^1\rangle|q[3]^1\rangle\right).$$

Listing 3.1 Continued…

47. x q[4];
48. x q[3];

Next, from line forty-nine in Listing 3.1, the statement "u3(2*pi,0*pi,0*pi) q[3];"
completes one u3(2*pi,0*pi,0*pi) gate that is a (2×2) matrix $\begin{pmatrix} -1 & 0 \\ 0 & -1 \end{pmatrix}_{2\times2}$ in
the seventh time slot of Fig. 3.4. It takes $\frac{1}{2}\left((-1)|q[4]^0\rangle|q[3]^0\rangle + |q[4]^0\rangle|q[3]^1\rangle\right.$
$\left. + |q[4]^1\rangle|q[3]^0\rangle + (-1)|q[4]^1\rangle|q[3]^1\rangle\right)$ in the new state vector $|\Phi_{11}\rangle$ as its input. State

$-\frac{1}{2}|q[4]^0\rangle|q[3]^0\rangle$ receives one phase (-1). State $\frac{1}{2}|q[4]^0\rangle|q[3]^1\rangle$ receives one phase (-1). State $\frac{1}{2}|q[4]^1\rangle|q[3]^0\rangle$ receives one phase (-1). State $\frac{1}{2}|q[4]^1\rangle|q[3]^1\rangle$ receives one phase (-1). This indicates that we obtain the following new state vector

$$|\Phi_{12}\rangle = \frac{1}{2}((-1 \times -1)|q[4]^0\rangle|q[3]^0\rangle + (-1)|q[4]^0\rangle|q[3]^1\rangle + (-1)|q[4]^1\rangle|q[3]^0\rangle$$
$$+(-1 \times -1)|q[4]^1\rangle|q[3]^1\rangle)$$
$$= \frac{1}{2}(|q[4]^0\rangle|q[3]^0\rangle + (-1)|q[4]^0\rangle|q[3]^1\rangle$$
$$+(-1)|q[4]^1\rangle|q[3]^0\rangle + |q[4]^1\rangle|q[3]^1\rangle).$$

Listing 3.1 Continued…

49. u3(2*pi,0*pi,0*pi) q[3];

Those quantum gates from the *second* time slot through the seventh time slot of Fig. 3.4 completes the phase shifter, $(2 - |0^{\otimes 2}\rangle\langle 0^{\otimes 2}|I_{2^2 \times 2^2})$, in the Grover diffusion operator to an instance of the satisfiability problem with $F(x_1, x_2) = x_1 \wedge x_2$. Next, from the line fifty through line fifty-one in Listing 3.1 the two statements "h q[4];" and "h q[3];" implement two Hadamard gates in the eighth time slot of Fig. 3.4. They complete the second $H^{\otimes 2}$ gate in the diffusion operator, $H^{\otimes 2} (2|0^{\otimes 2}\rangle\langle 0^{\otimes 2}| - I_{2^2 \times 2^2}) H^{\otimes 2}$. They take $\frac{1}{2}(|q[4]^0\rangle|q[3]^0\rangle + (-1)|q[4]^0\rangle|q[3]^1\rangle$ $+(-1)|q[4]^1\rangle|q[3]^0\rangle + |q[4]^1\rangle|q[3]^1\rangle)$ in the new state vector $|\Phi_{12}\rangle$ as their input. State $(\frac{1}{2}|q[4]^0\rangle|q[3]^0\rangle)$ is converted into state $(\frac{1}{4}(|q[4]^0\rangle|q[3]^0\rangle+|q[4]^0\rangle|q[3]^1\rangle$ $+|q[4]^1\rangle|q[3]^0\rangle+|q[4]^1\rangle|q[3]^1\rangle))$. State $(-\frac{1}{2}|q[4]^0\rangle|q[3]^1\rangle)$ is converted into state $(-\frac{1}{4}(|q[4]^0\rangle|q[3]^0\rangle - |q[4]^0\rangle|q[3]^1\rangle +|q[4]^1\rangle|q[3]^0\rangle - |q[4]^1\rangle|q[3]^1\rangle))$. State $(-\frac{1}{2}|q[4]^1\rangle|q[3]^0\rangle)$ is converted into state $(-\frac{1}{4}(|q[4]^0\rangle|q[3]^0\rangle + |q[4]^0\rangle|q[3]^1\rangle -|q[4]^1\rangle|q[3]^0\rangle - |q[4]^1\rangle|q[3]^1\rangle))$. State $(\frac{1}{2}|q[4]^1\rangle|q[3]^1\rangle)$ is converted into state $(\frac{1}{4}(|q[4]^0\rangle|q[3]^0\rangle - |q[4]^0\rangle|q[3]^1\rangle -|q[4]^1\rangle|q[3]^0\rangle + |q[4]^1\rangle|q[3]^1\rangle))$. This is to say that we obtain the following new state vector

$$|\Phi_{13}\rangle=|q[4]^1\rangle|q[3]^1\rangle.$$

Listing 3.1 Continued…

50. h q[4];
51. h q[3];

Fig. 3.5   After the measurement to solve an instance of the satisfiability problem with $F(x_1, x_2) = x_1 \wedge x_2$ is completed, we obtain the answer 11,000 with the probability 1 (100%)

Next, from line fifty-two in Listing 3.1 the statement "measure q[4] -> c[4];" is to measure the fifth quantum bit q[4] and to record the measurement outcome by overwriting the fifth classical bit c[4]. From line fifty-three in Listing 3.1 the statement "measure q[3] -> c[3];" is to measure the fourth quantum bit q[3] and to record the measurement outcome by overwriting the fourth classical bit c[3]. They complete the measurement from the ninth time slot through the tenth time slot of Fig. 3.4.

Listing 3.1 Continued...
// We complete the measurement of the answer.
52.   measure q[4] -> c[4];
53.   measure q[3] -> c[3];

In the backend *ibmqx4* with five quantum bits in **IBM**'s quantum computers, we use the command "simulate" to execute the program in Listing 3.1 . The measured result appears in Fig. 3.5. From Fig. 3.5, we obtain the answer 11,000 (c[4] = 1 = q[4] = $|1\rangle$, c[3] = 1 = q[3] = $|1\rangle$, c[2] = 0, c[1] = 0 and c[0] = 0) with the probability 1 (100%). This is to say that with the possibility 1 (100%) we obtain that the value of quantum bit q[3] is $|1\rangle$ and the value of quantum bit q[4] is $|1\rangle$. For solving an instance of the satisfiability problem with $F(x_1, x_2) = x_1 \wedge x_2$ we use quantum bit q[3] to encode Boolean variable x_1 and use quantum bit q[4] to encode Boolean variable x_2. Therefore, the answer is to that the value of Boolean variable x_1 is 1 (one) and the value of Boolean variable x_2 is 1 (one).

3.2.9   The Quantum Search Algorithm to the Satisfiability Problem

A satisfiability problem has n Boolean variables and any given oracular function $O_f(x_1 x_2 \ldots x_{n-1} x_n)$. Any given oracular function $O_f(x_1 x_2 \ldots x_{n-1} x_n)$ is a Boolean formula of the form $C_1 \wedge C_2 \ldots \wedge C_m$. Each clause C_j for $1 \leq j \leq m$ is a formula of the form $x_1 \vee x_2 \vee \ldots x_{n-1} \vee x_n$ for each Boolean variable x_k to $1 \leq k \leq n$. The question is to find values of each Boolean variable so that any given oracular function $O_f(x_1 \ x_2 \ \ldots, \ x_{n-1} \ x_n)$ (the whole formula) has the value 1. We use the quantum search algorithm to find one of M solutions to the question, where $0 \leq M \leq 2^n$.

The quantum circuit in Fig. 3.6 is to implement the quantum search algorithm to solve an instance of the satisfiability problem with n Boolean variables and m clauses. The first quantum register in the left top of Fig. 3.6 is $\left(\left|0^{\otimes n}\right\rangle\right)$. This is to say that the initial value of each quantum bit is $|0\rangle$. The second quantum register in the left bottom of Fig. 3.6 has $(m \times n + 2 \times m + 1)$ quantum bits and is an auxiliary quantum register.

The initial value of each quantum bit in the second quantum register is $|0\rangle$ or $|1\rangle$ that is dependent on implementing a logical or operation or a logical and operation. The third quantum register in the left bottom of Fig. 3.6 is $(\langle 1|)$.

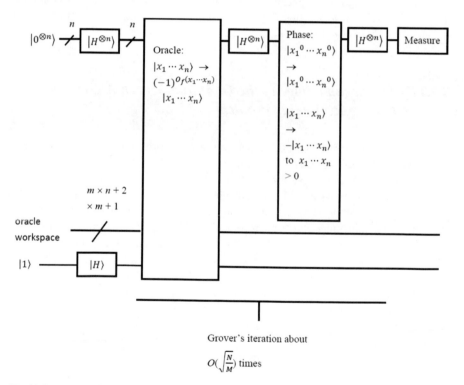

Grover's iteration about

$$O\left(\sqrt{\frac{N}{M}}\right) \text{ times}$$

Fig. 3.6 Circuit of implementing the quantum search algorithm to solve an instance of the satisfiability problem with n Boolean variables and m clauses

3.2.10   The First Stage of the Quantum Search Algorithm to the Satisfiability Problem

In Fig. 3.6, the first stage of the quantum search problem to solve an instance of the satisfiability problem with n Boolean variables and m clauses is to use n Hadamard gates $\left(\left|H^{\otimes n}\right\rangle\right)$ to operate the first quantum register $\left(\left|0^{\otimes n}\right\rangle\right)$. This indicates that it generates the superposition of n quantum bits that is $\left(\otimes_{k=1}^{n}\left(\frac{\left|x_k^0\right\rangle+\left|x_k^1\right\rangle}{\sqrt{2}}\right)\right) = $ $\frac{1}{\sqrt{2^n}}\left(\sum_{\{x_1 x_2 \ldots x_n | \forall x_d \in \{0,1\} \text{for} 1 \le d \le n\}} |x_1 x_2 \ldots x_n\rangle\right)$. Solution space of solving an instance of the satisfiability problem with n Boolean variables and m clauses is $\{x_1\, x_2 \ldots x_{n-1}\, x_n \mid \forall\, x_d \in \{0, 1\} \text{ for } 1 \le d \le n\}$. In the first stage of the quantum search algorithm, state $\left(\left|x_1^0 x_2^0 \cdots x_n^0\right\rangle\right)$ with the amplitude $\left(\frac{1}{\sqrt{2^n}}\right)$ encodes the *first* element x_1^0 $x_2^0 \ldots x_n^0$ in solution space and so on with that state $\left(\left|x_1^1 x_2^1 \cdots x_n^1\right\rangle\right)$ with the amplitude $\left(\frac{1}{\sqrt{2^n}}\right)$ encodes the *last* element $x_1^1\, x_2^1 \ldots x_n^1$ in solution space. In the first stage of the quantum search algorithm, it uses one Hadamard gate to operate the third quantum register $(|1\rangle)$. This is to say that it generates the superposition of the third quantum register $(|1\rangle)$ that is $\left(\frac{1}{\sqrt{2}}(|0\rangle|1\rangle)\right)$.

3.2.11   The Second Stage of the Quantum Search Algorithm to the Satisfiability Problem

In Fig. 3.6, the *second* stage of the quantum search algorithm to solve an instance of the satisfiability problem with n Boolean variables and m clauses is to complete the Oracle. The Oracle is to have the ability to *recognize* solutions in the satisfiability problem with n Boolean variables and m clauses. The Oracle is to multiply the probability amplitude of the answer(s) by -1 and leaves any other amplitude unchanged. Any given oracular function $O_f(x_1\, x_2 \ldots x_{n-1}\, x_n)$ is a Boolean formula of the form $C_1 \wedge C_2 \ldots \wedge C_m$. Each clause C_j for $1 \le j \le m$ is a formula of the form $x_1 \vee x_2 \vee \ldots$ $x_{n-1} \vee x_n$ for each Boolean variable x_k to $1 \le k \le n$. Therefore, for implementing any given oracular function $O_f(x_1\, x_2 \ldots x_{n-1}\, x_n)$, we need to complete $(m \times n)$ logical or operations and (m) logical and operations.

In oracle workspace in the second stage of the quantum search algorithm, we use auxiliary quantum bits $\left|r_{j,k}\right\rangle$ for $1 \le j \le m$ and $0 \le k \le n$ to encode auxiliary Boolean variables $r_{j,k}$ for $1 \le j \le m$ and $0 \le k \le n$. We use auxiliary quantum bits $\left|s_j\right\rangle$ for $0 \le j$ $\le m$ to encode auxiliary Boolean variables s_j for $0 \le j \le m$. Since we use auxiliary quantum bits $\left|r_{j,0}\right\rangle$ for $1 \le j \le m$ as the first operand of the first logical or operation ("$\vee$") in each clause, the initial value of each auxiliary quantum bit $\left|r_{j,0}\right\rangle$ for $1 \le j$ $\le m$ is set to $|0\rangle$. This implies that this setting does not change the correct result of the first logical or operation in each clause. We use a **CCNOT** gate and four **NOT** gates to implement each logical or operation in each clause. We apply auxiliary quantum bits $\left|r_{j,k}\right\rangle$ for $1 \le j \le m$ and $1 \le k \le n$ to store the result of implementing the logical or operations in each clause. This is to say that each auxiliary quantum bit $\left|r_{j,k}\right\rangle$ for $1 \leq j \leq m$ and $1 \leq k \leq n$ is actually the target bit of a **CCNOT** gate of implementing a logical or operation. Thus, the initial value of each auxiliary quantum bit $\left|r_{j,k}\right\rangle$ for $1 \leq j \leq m$ and $1 \leq k \leq n$ is set to $|1\rangle$.

We use an auxiliary quantum bit $|s_0\rangle$ as the first operand of the first logical and operation ("$\wedge$") in any given oracular function $O_f(x_1 x_2 \ldots x_{n-1} x_n)$ with a Boolean formula of the form $C_1 \wedge C_2 \ldots \wedge C_m$. The initial value of the auxiliary quantum bit $|s_0\rangle$ is set to $|1\rangle$. This is to say that this setting does not change the correct result of the first logical and operation in any given oracular function $O_f(x_1 x_2 \ldots x_{n-1} x_n)$ with a Boolean formula of the form $C_1 \wedge C_2 \ldots \wedge C_m$. We use a **CCNOT** gate to implement each logical and operation in any given oracular function $O_f(x_1 x_2 \ldots x_{n-1} x_n)$ with a Boolean formula of the form $C_1 \wedge C_2 \ldots \wedge C_m$. We apply auxiliary quantum bits $\left|s_j\right\rangle$ for $1 \leq j \leq m$ to store the result of implementing the logical and operations in any given oracular function $O_f(x_1 x_2 \ldots x_{n-1} x_n)$ with a Boolean formula of the form $C_1 \wedge C_2 \ldots \wedge C_m$. This indicates that each auxiliary quantum bit $\left|s_j\right\rangle$ for $1 \leq j \leq m$ is actually the target bit of a **CCNOT** gate of implementing a logical and operation. Therefore, the initial value of each auxiliary quantum bit $\left|s_j\right\rangle$ for $1 \leq j \leq m$ is set to $|0\rangle$.

A **CCNOT** gate and four **NOT** gate can implement a logical or operation. A **CCNOT** gate can implement a logical and operation. From Fig. 3.1, implementing any given oracular function $O_f(x_1 x_2 \ldots x_{n-1} x_n)$ is to complete $(m \times n)$ logical or operations and (m) logical and operations. This is to say that implementing any given oracular function $O_f(x_1 x_2 \ldots x_{n-1} x_n)$ is to complete $(m \times n + m)$ **CCNOT** gates and $(4 \times m \times n)$ **NOT** gates. Quantum bit $|s_m\rangle$ is to store the result of implementing any given oracular function $O_f(x_1 x_2 \ldots x_{n-1} x_n)$. If the value to quantum bit $|s_m\rangle$ is equal to $|1\rangle$, then any given oracular function $O_f(x_1 x_2 \ldots x_{n-1} x_n)$ has the value 1 (one). Otherwise, it has the value 0 (zero).

We use one **CNOT** gate $\left|\frac{|0\rangle - |1\rangle}{\sqrt{2}} \oplus s_m\right\rangle$ to multiply the probability amplitude of the answer(s) by -1 and to leave any other amplitude unchanged, where quantum bit $\left|\left(\frac{|0\rangle - |1\rangle}{\sqrt{2}}\right)\right\rangle$ is the target bit of the **CNOT** gate and quantum bit $(|s_m\rangle)$ is the control bit of the **CNOT** gate. When the value of the control bit $(|s_m\rangle)$ is equal to $(|1\rangle)$, the target bit becomes $\left(\frac{|1\rangle - |0\rangle}{\sqrt{2}}\right) = (-1)\left(\frac{|0\rangle - |1\rangle}{\sqrt{2}}\right)$. This is to multiply the probability amplitude of the answer(s) by -1. When the value of the control bit $(|s_m\rangle)$ is equal to $(|0\rangle)$, the target bit still is $\left(\frac{|0\rangle - |1\rangle}{\sqrt{2}}\right)$. This is to leave any other amplitude unchanged.

Because quantum operations are reversible by nature, executing the reversed order of implementing any given oracular function $O_f(x_1 x_2 \ldots x_{n-1} x_n)$ can restore the auxiliary quantum bits to their initial states. This is to say that the second stage of the quantum search algorithm to solve the satisfiability problem with n Boolean variables and m clauses converts $\frac{1}{\sqrt{2^n}}\left(\sum_{\{x_1 x_2 \ldots x_n | \forall x_d \in \{0,1\} \text{for} 1 \leq d \leq n\}} |x_1 x_2 \ldots x_n\rangle\right)$ into $(-1)^{O_f(x_1 \cdots x_n)} \frac{1}{\sqrt{2^n}}\left(\sum_{\{x_1 x_2 \ldots x_n | \forall x_d \in \{0,1\} \text{for} 1 \leq d \leq n\}} |x_1 x_2 \ldots x_n\rangle\right)$. The cost of completing the Oracle in the second stage of the quantum search algorithm in Fig. 3.6 is to implement $(2 \times (m \times n + m))$ **CCNOT** gates, $(8 \times m \times n)$ **NOT** gates and one **CNOT** gate.

3.2.12   The Third Stage of the Quantum Search Algorithm to the Satisfiability Problem

In Fig. 3.6, the *third* stage of the quantum search algorithm to solve an instance of the satisfiability problem with n Boolean variables and m clauses is to complete the Grover diffusion operator. Implementing the Grover diffusion operator is equivalent to implement $H^{\otimes n} \left(2|0^{\otimes n}\rangle\langle 0^{\otimes n}| I_{2^n \times 2^n}\right) H^{\otimes n}$. A phase shifter operator, $\left(2|0^{\otimes n}\rangle\langle 0^{\otimes n}| - I_{2^n \times 2^n}\right)$ negates all the states except for $(|x_1^0 \ldots x_n^0\rangle)$. In Fig. 3.6, the *third* stage of the quantum search problem is to use that the phase shift operator, $2|0^{\otimes n}\rangle\langle 0^{\otimes n}| - I_{2^n \times 2^n}$, negates all the states except for $\left(|x_1^0 \cdots x_n^0\rangle\right)$ sandwiched between $H^{\otimes n}$ gates. This indicates that the *third* stage of the quantum search algorithm to solve an instance of the satisfiability problem with n Boolean variables and m clauses is to increase the amplitude of the answer(s) and to decrease the amplitude of the non-answer(s).

For solving an instance of the satisfiability problem with n Boolean variables and m clauses, we regard the Oracle and the Grover diffusion operator from the second stage through the third stage of the quantum search algorithm in Fig. 3.6 as a subroutine. We call the subroutine as the *Grover iteration*. After repeat to execute the Grover iteration of $O\left(\sqrt{\frac{N}{M}}\right)$ times, the successful probability of measuring the answer(s) is at least $(1/2)$. When the value of (M/N) is equal to $(1/4)$, the successful probability of measuring the answer(s) is one (100%) with the Grover iteration of one time. This is the *best* case of the quantum search algorithm to solve an instance of the satisfiability problem with n Boolean variables and m clauses. When the value of M is equal to one, the successful probability of measuring the answer(s) is at least $(1/2)$ with the Grover iteration of $O(\sqrt{N})$ times. This is the *worst* case of the quantum search algorithm to solve an instance of the satisfiability problem with n Boolean variables and m clauses. This indicates that the quantum search algorithm to solve an instance of the satisfiability problem with n Boolean variables and m clauses only gives a quadratic speed-up.

3.3   Introduction to the Maximal Clique Problem

We assume that a graph $G = (V, E)$ has n vertices and θ edges, where V is a set of vertices in G and E is a set of edges in G. For a graph $G = (V, E)$ with n vertices and θ edges, its *complementary* graph is to contain all edges missing in the *original* graph $G = (V, E)$ with n vertices and θ edges. Therefore, we assume that its *complementary* graph $\overline{G} = (V, \overline{E})$ has n vertices and m edges in which each edge in $\overline{E}$ is out of E,

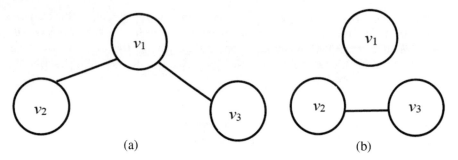

Fig. 3.7 a The graph has three vertices and two edges. **b** Its *complementary* graph has the same vertices and one edge missing in the *original* graph

where V is a set of vertices in $\overline{G}$ and $\overline{E}$ is a set of edges in $\overline{G}$. This is to say that a graph $G = (V, E)$ and its *complementary* graph $\overline{G} = (V, \overline{E})$ has the same vertices and its *complementary* graph $\overline{G} = (V, \overline{E})$ contains all edges missing in the *original* graph $G = (V, E)$. In Fig. 3.7a, the graph has three vertices $\{v_1, v_2, v_3\}$ and two edges $\{(v_1, v_2), (v_1, v_3)\}$. In Fig. 3.7b, its *complementary* graph has the same three vertices $\{v_1, v_2, v_3\}$ and one edge $\{(v_2, v_3)\}$ missing in the *original* graph in Fig. 3.7a.

Mathematically, a *clique* for a graph $G = (V, E)$ with n vertices and θ edges is defined as a set of vertices in which every vertex is connected to every other vertex by an edge. This is to say that mathematically, a *clique* for a graph $G = (V, E)$ with n vertices and θ edges is a *complete* subgraph to G. For the graph with three vertices $\{v_1, v_2, v_3\}$ and two edges $\{(v_1, v_2), (v_1, v_3)\}$ in Fig. 3.7a, there are eight subsets of vertices. Eight subsets of vertices are subsequently $\{v_1, v_2, v_3\}$, $\{v_1, v_2\}$, $\{v_1, v_3\}$, $\{v_2, v_3\}$, $\{v_1\}$, $\{v_2\}$, $\{v_3\}$ and an empty subset of vertices that is $\{\}$.

Any two vertices connected in the *complementary* graph with three vertices $\{v_1, v_2, v_3\}$ and one edge $\{(v_2, v_3)\}$ in Fig. 3.7b are disconnected in the *original* graph with three vertices $\{v_1, v_2, v_3\}$ and two edges $\{(v_1, v_2), (v_1, v_3)\}$ in Fig. 3.7a. This is to say that any two vertices connected in the *complementary* graph cannot be members of the same clique. Because the edge (v_2, v_3) in the *complementary* graph in Fig. 3.7b connects the two vertices v_2 and v_3, eight subsets of vertices containing the two vertices v_2 and v_3 are not a clique. A subset of vertices $\{v_1, v_2, v_3\}$ contains the two vertices v_2 and v_3, so it is not a clique. Another subset of vertices $\{v_2, v_3\}$ consists of the two vertices v_2 and v_3 and therefore it is not a clique. This indicates that other six subsets of vertices $\{v_1, v_2\}$, $\{v_1, v_3\}$, $\{v_1\}$, $\{v_2\}$, $\{v_3\}$ and an empty subset of vertices that is $\{\}$ are all a clique.

The number of vertex to clique $\{v_1, v_2\}$ and clique $\{v_1, v_3\}$ is both two. The number of vertex to clique $\{v_1\}$, clique $\{v_2\}$ and clique $\{v_3\}$ is all one. The number of vertex to clique $\{\}$ that is an empty subset of vertices is zero. The *maximal* clique problem that is a **NP-complete** problem is to find a *maximum-sized* clique in the graph with three vertices $\{v_1, v_2, v_3\}$ and two edges $\{(v_1, v_2), (v_1, v_3)\}$ in Fig. 3.7a. Therefore, clique $\{v_1, v_2\}$ and clique $\{v_1, v_3\}$ are two *maximum-sized* cliques to solve the *maximal* clique problem for the graph with three vertices $\{v_1, v_2, v_3\}$ and two edges $\{(v_1, v_2),$

$(v_1, v_3)\}$ in Fig. 3.7a. Because the quantum program of implementing this example uses more quantum bits that exceed five quantum bits in the backend *ibmqx4* with five quantum bits in **IBM**'s quantum computers, we just use this example to explain what the maximal clique problem is. Next, we give Definition 3.3 to describe the maximal clique problem.

Definition 3.3 Mathematically, a *clique* for a graph $G = (V, E)$ with n vertices and θ edges is defined as a set of vertices in which every vertex is connected to every other vertex by an edge. The maximal clique problem to a graph $G = (V, E)$ with n vertices and θ edges is to find a *maximum-sized* clique in the graph $G = (V, E)$ with n vertices and θ edges.

From Definition 3.3, all of the possible solutions to the clique problem of graph G with n vertices and θ edges consist of 2^n possible choices. Every possible choice corresponds to a subset of vertices (a possible clique in G). Hence, we assume that a set X contains 2^n possible choices and the set X is equal to $\{x_1 x_2 \ldots x_{n-1} x_n | \forall x_d \in \{0, 1\}$ for $1 \leq d \leq n\}$. This is to say that the length of each element in $\{x_1 x_2 \ldots x_{n-1} x_n | \forall x_d \in \{0, 1\}$ for $1 \leq d \leq n\}$ is n bits and every element stands for one of 2^n possible choices.

For the sake of presentation, we suppose that x_d^0 denotes the fact that the value of x_d is zero and x_d^1 denotes the fact that the value of x_d is one. If an element $x_1 x_2 \ldots x_{n-1} x_n$ in $\{x_1 x_2 \ldots x_{n-1} x_n | \forall x_d \in \{0, 1\}$ for $1 \leq d \leq n\}$ is a clique and the value of x_d for $1 \leq d \leq n$ is one, then x_d^1 represents that the dth vertex is within the clique. If an element $x_1 x_2 \ldots x_{n-1} x_n$ in $\{x_1 x_2 \ldots x_{n-1} x_n | \forall x_d \in \{0, 1\}$ for $1 \leq d \leq n\}$ is a clique and the value of x_d for $1 \leq d \leq n$ is zero, then x_d^0 stands for that the dth vertex is not within the clique.

From $\{x_1 x_2 \ldots x_{n-1} x_n | \forall x_d \in \{0, 1\}$ for $1 \leq d \leq n\}$, the first element $x_1^0 x_2^0 \ldots x_{n-1}^0 x_n^0$ encodes an empty set of vertices that is $\{\}$. The second element $x_1^0 x_2^0 \ldots x_{n-1}^0 x_n^1$ encodes a set of vertices $\{v_n\}$. The third element $x_1^0 x_2^0 \ldots x_{n-1}^1 x_n^0$ encodes a set of vertices $\{v_{n-1}\}$ and so on with that the last element $x_1^1 x_2^1 \ldots x_{n-1}^1 x_n^1$ encodes a set of vertices $\{v_1 v_2 \ldots v_{n-1} v_n\}$. We regard $\{x_1 x_2 \ldots x_{n-1} x_n | \forall x_d \in \{0, 1\}$ for $1 \leq d \leq n\}$ as an unsorted database containing 2^n possible choices (2^n possible cliques) to the maximal clique problem of graph G with n vertices and θ edges.

3.3.1   Flowchart of Recognize Cliques to the Maximal Clique Problem

From Definition 3.3, solving the clique problem for a graph $G = (V, E)$ with n vertices and θ edges and its complementary graph $\overline{G} = (V, \overline{E})$ with n vertices and m edges in which each edge is out of E is to find a subset V^1 of vertices with size r that satisfies V^1 to be a maximum-sized clique in G. This indicates that all of the possible solutions are 2^n subset of vertices in which each subset of vertices corresponds to a possible clique. Any two vertices connected in the *complementary* graph $\overline{G} = (V,$

$\overline{E}$) with n vertices and m edges are disconnected in the *original* graph with $G = (V, E)$ with n vertices and θ edges. This is to say that any two vertices connected in the *complementary* graph cannot be members of the same clique.

If any one of 2^n possible choices (cliques) does not include any edge in the *complementary* graph $\overline{G}$, then it is a clique in the original graph G. Otherwise, it is not a clique in the original graph G. Boolean variables x_i and x_j encode vertices v_i and v_j for $1 \leq i \leq n$ and $1 \leq j \leq n$. In the *complementary* graph $\overline{G}$, the kth edge is $e_k = (v_i, v_j)$ to $1 \leq k \leq m$. The requested condition of deciding whether any one of 2^n subsets of vertices does not include the kth edge $e_k = (v_i, v_j)$ or not is to satisfy a formula of the form $\left(\overline{x_i \wedge x_j}\right)$ that is the true value, where $\left(\overline{x_i \wedge x_j}\right)$ is one **NAND** gate. We regard a formula of the form $\left(\overline{x_i \wedge x_j}\right)$ as a clause. When the value of Boolean variable x_i is 1 (one) and the value of Boolean variable x_j is 1 (one), the output (result) of implementing $\left(\overline{x_i \wedge x_j}\right)$ is 0 (zero). This indicates that each possible choice containing the two vertices v_i and v_j is not a clique. When the value of Boolean variable x_i is 1 (one) and the value of Boolean variable x_j is 0 (zero), the output (result) of implementing $\left(\overline{x_i \wedge x_j}\right)$ is 1 (one). This is to say that each possible choice consisting of vertex v_i and not containing vertex v_j is perhaps a clique. When the value of Boolean variable x_i is 0 (zero) and the value of Boolean variable x_j is 1 (one), the output (result) of implementing $\left(\overline{x_i \wedge x_j}\right)$ is 1 (one). This implies that each possible choice consisting of vertex v_j and not containing vertex v_i is perhaps a clique. When the value of Boolean variable x_i is 0 (zero) and the value of Boolean variable x_j is 0 (zero), the output (result) of implementing $\left(\overline{x_i \wedge x_j}\right)$ is 1 (one). This indicates that each possible choice not containing vertex v_i and vertex v_j is perhaps a clique.

The requested condition of checking whether any one of 2^n subsets of vertices does not include m edges in the complementary graph or not is to satisfy a formula of the form $(\wedge_m^{k=1}\left(\overline{x_i \wedge x_j}\right))$ that is the true value. We regard a formula of the form $(\wedge_m^{k=1}\left(\overline{x_i \wedge x_j}\right))$ as a Boolean formula of the form $C_1 \wedge C_2 \ldots \wedge C_m$, where each clause C_j for $1 \leq j \leq m$ is a formula of the form $\left(\overline{x_i \wedge x_j}\right)$ to Boolean variables x_i and x_j for $1 \leq i \leq n$ and $1 \leq j \leq n$. Any one of 2^n subsets of vertices is a clique if finding values of each Boolean variable satisfy the whole formula has the value 1 (one). This is the same as finding values of each Boolean variable that make each clause have the value 1 (one).

Recognizing clique(s) is equivalent to implement a formula of the form $(\wedge_m^{k=1}\left(\overline{x_i \wedge x_j}\right))$. Therefore, we need to make use of auxiliary Boolean variables r_k for $1 \leq k \leq m$ and auxiliary Boolean variables s_k for $0 \leq k \leq m$. We use **CCNOT** gates to implement the only **NAND** gate $\left(\overline{x_i \wedge x_j}\right)$ in each clause and we apply auxiliary Boolean variables r_k for $1 \leq k \leq m$ to store the result of implementing the only **NAND** gate $\left(\overline{x_i \wedge x_j}\right)$ in each clause. This is to say that each auxiliary Boolean variable r_k for $1 \leq k \leq m$ is actually the target bit of a **CCNOT** gate of implementing a **NAND** gate $(\overline{x_i \wedge x_j})$. Hence, the initial value of each auxiliary Boolean variable r_k for $1 \leq k \leq m$ is set to one (1).

We make use of an auxiliary Boolean variable s_0 as the first operand of the first logical and operation ("$\wedge$") in a Boolean formula of the form $(\wedge_m^{k=1}\left(\overline{x_i \wedge x_j}\right))$. The initial value of the auxiliary Boolean variable s_0 is set to one (1). This is to say that this setting does not change the correct result of the first logical and operation in $(\wedge_m^{k=1}(\overline{x_i \wedge x_j}))$. We apply **CCNOT** gates to implement the logical and operations in $(\wedge_m^{k=1}(\overline{x_i \wedge x_j}))$ and we use auxiliary Boolean variables s_k for $1 \leq k \leq m$ to store the result of implementing the logical and operations in $(\wedge_m^{k=1}(\overline{x_i \wedge x_j}))$. This indicates that each auxiliary Boolean variable s_k for $1 \leq k \leq m$ is actually the target bit of a **CCNOT** gate of implementing a logical and operation. Therefore, the initial value of each auxiliary Boolean variable s_k for $1 \leq k \leq m$ is set to zero (0).

Figure 3.8 is to flowchart of recognizing cliques to the maximal clique problem for a graph $G = (V, E)$ with n vertices and θ edges and its complementary graph $\overline{G} = (V, \overline{E})$ with n vertices and m edges. In Fig. 3.8, in statement S_1, it sets the index variable k of the first loop to one (1). Next, in statement S_2, it executes the conditional judgement of the first loop. If the value of k is less than or equal to the value of m, then *next executed* instruction is statement S_3. Otherwise, in statement S_6, it executes an *End* instruction to terminate the task that is to find values of each Boolean variable so that the whole formula has the value 1 and this is the same as finding values of each Boolean variable that make each clause have the value 1.

In statement S_3, it implements a **NAND** gate "$r_k \leftarrow (\overline{x_i \wedge x_j})$". Boolean variables x_i and x_j respectively encode vertex v_i and vertex v_j that are connected by the kth edge in the complementary graph $\overline{G} = (V, \overline{E})$ with n vertices and m edges. Boolean

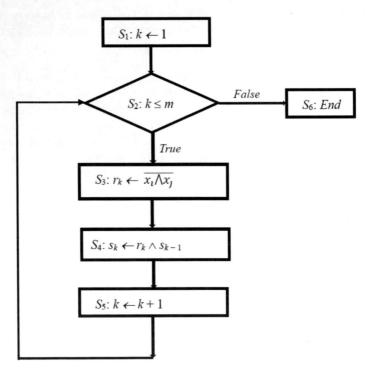

Fig. 3.8 Flowchart of recognizing cliques to the maximal clique problem for a graph $G = (V, E)$ with n vertices and θ edges and its complementary graph $\overline{G} = (V, \overline{E})$ with n vertices and m edges

variable r_k stores the result of implementing $((\overline{x_i \wedge x_j}))$ (the kth **NAND** gate). Next, in statement S_4, it executes a logical and operation "$s_k \leftarrow r_k \wedge s_{k-1}$" that is the kth clause in $(\wedge_m^{k=1}(\overline{x_i \wedge x_j}))$. Boolean variable r_k stores the result of implementing the kth **NAND** gate and is the first operand of the logical and operation. Boolean variable s_{k-1} is the second operand of the logical and operation and stores the result of the previous logical and operation. Boolean variable s_k for $1 \leq k \leq m$ stores the result of implementing $r_k \wedge s_{k-1}$ that is the kth clause that is the kth **AND** gate. Next, in statement S_5, it increases the value of the index variable k to the first loop. Repeat to execute statement S_2 through statement S_5 until in statement S_2 the conditional judgement becomes a *false* value. From Fig. 3.8, the total number of **NAND** gate is m **NAND** gates. The total number of logical and operation is m **AND** gates (logical and operations). Therefore, the cost of recognizing clique(s) is to implement m **NAND** gates and m **AND** gates.

3.3.2 Flowchart of Computing the Number of Vertex in Each Clique to the Maximal Clique Problem

For a graph $G = (V, E)$ with n vertices and θ edges and its complementary graph $\overline{G} = (V, \overline{E})$ with n vertices and m edge, solution space of the maximal clique problem is 2^n subsets of vertices. We use $\{x_1 x_2 \ldots x_{n-1} x_n | \forall x_d \in \{0, 1\}$ for $1 \leq d \leq n\}$ to encode 2^n subsets of vertices. After each element encoding one subset of vertices in $\{x_1 x_2 \ldots x_{n-1} x_n | \forall x_d \in \{0, 1\}$ for $1 \leq d \leq n\}$ completes each operation in Fig. 3.8, Boolean variable s_m stores the result of deciding whether it is a clique or not. If the value of Boolean variable s_m is equal to 1 (one), then it is a clique. Otherwise, it is not a clique.

For computing the number of vertex, we need auxiliary Boolean variables $z_{i+1,j}$ and $z_{i+1,j+1}$ for $0 \leq i \leq n-1$ and $0 \leq j \leq i$. Auxiliary Boolean variables $z_{i+1,j}$ and $z_{i+1,j+1}$ for $0 \leq i \leq n-1$ and $0 \leq j \leq i$ are set to the initial value 0 (zero). Boolean variable $z_{i+1,j+1}$ for $0 \leq i \leq n-1$ and $0 \leq j \leq i$ is to store the number of vertex in a clique after figuring out the influence of Boolean variable x_{i+1} that encodes the $(i+1)$th vertex to the number of ones (vertices). If the value of Boolean variable $z_{i+1,j+1}$ for $0 \leq i \leq n-1$ and $0 \leq j \leq i$ is equal to 1 (one), then this indicates that there are $(j+1)$ ones (vertices) in the clique. Boolean variable $z_{i+1,j}$ for $0 \leq i \leq n-1$ and $0 \leq j \leq i$ is to store the number of vertex in a clique after figuring out the influence of Boolean variable x_{i+1} that encodes the $(i+1)$th vertex to the number of ones (vertices). If the value of Boolean variable $z_{i+1,j}$ for $0 \leq i \leq n-1$ and $0 \leq j \leq i$ is equal to 1 (one), then this is to say that there are j ones (vertices) in the clique.

In a clique, Boolean variable x_1 encodes the *first* vertex v_1. If the value of Boolean variable x_1 is equal to 1 (one), then the first vertex v_1 is within the clique and it increases the number of vertex to the clique. If the value of Boolean variable x_1 is equal to 0 (zero), then the first vertex v_1 is not within the clique and it reserves the number of vertex to the clique. Therefore, the influence of Boolean variable x_1 to increase the number of vertex to a clique is to satisfy the formula $(s_m \wedge x_1)$ that is the true value. Similarly, the influence of Boolean variable x_1 to reserve the number of vertex to a clique is to satisfy the formula $(s_m \wedge \overline{x_1})$ that is the true value.

In a clique, Boolean variable x_{i+1} encodes the $(i+1)$th vertex v_{i+1} for $1 \le i \le n - 1$. If the value of Boolean variable x_{i+1} is equal to 1 (one), then the $(i+1)$th vertex v_{i+1} is within the clique and it increases the number of vertex to the clique. If the value of Boolean variable x_{i+1} is equal to 0 (zero), then the $(i+1)$th vertex v_{i+1} is not within the clique and it reserves the number of vertex to the clique. Therefore, the influence of Boolean variable x_{i+1} to increase the number of vertex to a clique that has currently has j vertices is to satisfy the formula $(x_{i+1} \wedge z_{i,j})$ that is the true value. Similarly, the influence of Boolean variable x_{i+1} to reserve the number of vertex to a clique that has currently has j vertices is to satisfy the formula $((\overline{x_{i+1}}) \wedge z_{i,j})$ that is the true value.

Figure 3.9 is the logical flowchart of counting the number of vertex in each clique to the maximal clique problem for a graph $G = (V, E)$ with n vertices and θ edges and its complementary graph $\overline{G} = (V, \overline{E})$ with n vertices and m edges. In Fig. 3.9, in statement S_1, it implements a logical and operation "$z_{1,1} \leftarrow s_m \wedge x_1$" that is one **AND** gate. Boolean variable $z_{1,1}$ stores the result of implementing one **AND** gate $(s_m \wedge x_1)$. If the value of Boolean variable $z_{1,1}$ is equal to 1 (one), then it increases the number of vertex so that the number of vertex in each clique with the first vertex v_1 is one. Next, in statement S_2, it implements a logical and operation "$z_{1,0} \leftarrow s_m \wedge \overline{x_1}$" that is one **AND** gate $(s_m \wedge \overline{x_1})$. Boolean variable $z_{1,0}$ stores the result of implementing one **AND** gate $(s_m \wedge \overline{x_1})$. If the value of Boolean variable $z_{1,0}$ is equal to 1 (one), then it reserves the number of vertex so that the number of vertex in each clique *without* the first vertex v_1 is zero.

Next, in statement S_3, it sets the index variable i of the first loop to one. Next, in statement S_4, it executes the conditional judgement of the first loop. If the value of i is less than or equal to the value of $(n - 1)$, then *next executed* instruction is statement S_5. Otherwise, in statement S_{11}, it executes an *End* instruction to terminate the task that is to count the number of vertex in each clique. In statement S_5, it sets the index variable j of the second loop to the value of the index variable i in the first loop. Next, in statement S_6, it executes the conditional judgement of the *second* loop. If the value of j is greater than or equal to zero, then next executed instruction is statement S_7. Otherwise, next executed instruction is statement S_{10}.

In statement S_7, it implements a logical and operation "$z_{i+1, j+1} \leftarrow x_{i+1} \wedge z_{i,j}$" that is one **AND** gate. Boolean variable x_{i+1} encodes the $(i+1)$th vertex and is the first operand of the logical and operation. Boolean variable $z_{i,j}$ is the second operand of the logical and operation. Boolean variable $z_{i,j}$ stores the number of vertex in a clique after figuring out the influence of Boolean variable x_i that encodes the ith vertex to the number of ones (vertices). If the value of Boolean variable $z_{i,j}$ is equal to 1 (one), then this indicates that there are j ones (vertices) in the clique. Boolean variable $z_{i+1, j+1}$ stores the result of implementing the logical and operation "$z_{i+1, j+1} \leftarrow x_{i+1} \wedge z_{i,j}$". This is to say that Boolean variable $z_{i+1, j+1}$ stores the number of vertex in a clique after figuring out the influence of Boolean variable x_{i+1} that encodes the $(i+1)$th vertex to the number of ones (vertices). If the value of Boolean variable

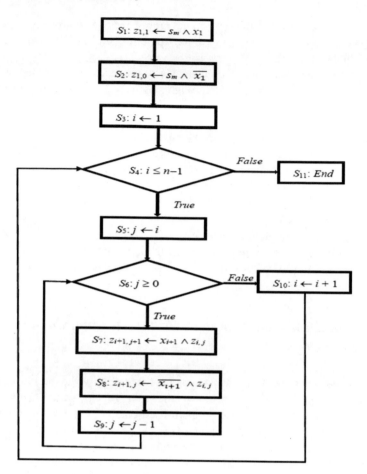

Fig. 3.9 Flowchart of computing the number of vertex in each clique to the maximal clique problem for a graph $G = (V, E)$ with n vertices and θ edges and its complementary graph $\overline{G} = (V, \overline{E})$ with n vertices and m edges

$z_{i+1, j+1}$ is equal to 1 (one), then this implies that there are $(j + 1)$ ones (vertices) in the clique.

Next, in statement S_8, it implements a logical and operation "$z_{i+1, j} \leftarrow \overline{x_{i+1}} \wedge z_{i, j}$" that is one **AND** gate. Boolean variable x_{i+1} encodes the $(i + 1)$th vertex and its negation $\overline{x_{i+1}}$ is the first operand of the logical and operation. Boolean variable $z_{i, j}$ is the second operand of the logical and operation. Boolean variable $z_{i, j}$ stores the number of vertex in a clique after figuring out the influence of Boolean variable x_i that encodes the ith vertex to the number of ones (vertices). If the value of Boolean variable $z_{i, j}$ is equal to 1 (one), then this is to say that there are j ones (vertices) in the clique. Boolean variable $z_{i+1, j}$ stores the result of implementing the logical and operation "$z_{i+1, j} \leftarrow \overline{x_{i+1}} \wedge z_{i, j}$". This indicates that Boolean variable $z_{i+1, j}$ stores the number of vertex in a clique after figuring out the influence of Boolean variable

x_{i+1} that encodes the $(i + 1)$th vertex to the number of ones (vertices). If the value of Boolean variable $z_{i+1,j}$ is equal to 1 (one), then this is to say that there are j ones (vertices) in the clique.

Next, in statement S_9, it decreases the value of the index variable j in the second loop. Repeat to execute statement S_6 through statement S_9 until in statement S_6 the conditional judgement becomes a *false* value. Next, in statement S_{10}, it increases the value of the index variable i in the first loop. Repeat to execute statement S_4 through statement S_{10} until in statement S_4 the conditional judgement becomes a *false* value. When in statement S_4 the conditional judgement becomes a *false* value, next executed statement is statement S_{11}. In statement S_{11}, it executes an *End* instruction to terminate the task that is to count the number of vertex in each clique. The cost of one time to complete each operation in Fig. 3.9 is to implement $(n \times (n + 1))$ **AND** gates and $(\frac{n \times (n+1)}{2})$ **NOT** gates. This is to say that the cost of counting the number of vertex to a clique is to implement $(n \times (n + 1))$ **AND** gates and $(\frac{n \times (n+1)}{2})$ **NOT** gates. Therefore, for counting the number of vertex in all cliques, the cost is to implement $(2^n \times n \times (n + 1))$ **AND** gates and $(2^n \times \frac{n \times (n+1)}{2})$ **NOT** gates.

3.3.3 Data Dependence Analysis for the Maximal Clique Problem

A data dependence arises from two statements that read or write the same memory. *Data dependence analysis* is to decide whether to *reorder* or *parallelize* statements is safe or not. In a maximal clique problem for a graph $G = (V, E)$ with n vertices and θ edges and its complementary graph $\overline{G} = (V, \overline{E})$ with n vertices and m edges, it contains 2^n subsets of vertices (2^n possible choices). We use a set $\{x_1 x_2 \ldots x_{n-1} x_n |$ $\forall x_d \in \{0, 1\}$ for $1 \leq d \leq n\}$ to encode 2^n subsets of vertices. In the set $\{x_1 x_2 \ldots x_{n-1} x_n | \forall x_d \in \{0, 1\}$ for $1 \leq d \leq n\}$, the first element $x_1^0 x_2^0 \ldots x_{n-1}^0 x_n^0$ encodes an empty subset without any vertex. The second element $x_1^0 x_2^0 \ldots x_{n-1}^0 x_n^1$ encodes $\{v_n\}$. The third element $x_1^0 x_2^0 \ldots x_{n-1}^1 x_n^0$ encodes $\{v_{n-1}\}$ and so on with that the last element $x_1^1 x_2^1 \ldots x_{n-1}^1 x_n^1$ encodes $\{v_1 v_2 \ldots v_{n-1} v_n\}$ that contains each vertex. Each element needs to implement those operations in Figs. 3.8 and 3.9. Each element needs to use m auxiliary Boolean variables r_k for $1 \leq k \leq m$, $(m + 1)$ auxiliary Boolean variables s_k for $0 \leq k \leq m$ and $((n \times (n + 3))/2)$ auxiliary Boolean variables $z_{i+1,j}$ and $z_{i+1,j+1}$ for $0 \leq i \leq n - 1$ and $0 \leq j \leq i$. Because 2^n subsets of vertices (2^n inputs) implement those instructions from Fig. 3.8 through Fig. 3.9 not to access or modify the same input and the same auxiliary Boolean variables, we can *parallelize* them without any error.

Let us consider another graph in Fig. 3.10a and its complementary graph in Fig. 3.10b. In Fig. 3.10a, the graph has two vertices $\{v_1, v_2\}$ and one edge $\{(v_1, v_2)\}$. In Fig. 3.10b, its *complementary* graph has the same two vertices $\{v_1, v_2\}$ and no edge missing in the *original* graph in Fig. 3.10a. For the graph in Fig. 3.10a with two vertices $\{v_1, v_2\}$ and one edge $\{(v_1, v_2)\}$ and its complementary graph in

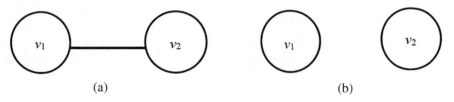

Fig. 3.10 a The graph has two vertices and one edge. **b** Its *complementary* graph has the same vertices and no edge missing in the *original* graph

Fig. 3.10b with the same two vertices $\{v_1, v_2\}$ and no edge, solving the maximal clique problem is to find a subset V^1 of vertices with size r that satisfies V^1 to be a maximum-sized clique. This is to say that the value of n is equal to two, the value of θ is equal to one and the value of m is equal to zero.

We regard the maximal clique problem for the graph with two vertices $\{v_1, v_2\}$ and one edge $\{(v_1, v_2)\}$ in Fig. 3.10a as a search problem. Any given oracular function O_f is to implement those instructions from Fig. 3.8 to Fig. 3.9 to recognize the maximal clique(s). Its domain is $\{x_1 \, x_2 | \, \forall \, x_d \in \{0, 1\}$ for $1 \le d \le 2\}$ to encode 2^2 subsets of vertices. In the domain $\{x_1 \, x_2 | \, \forall \, x_d \in \{0, 1\}$ for $1 \le d \le 2\}$, the first element x_1^0 x_2^0 encodes an empty subset without vertex. The second element $x_1^0 \, x_2^1$ encodes $\{v_2\}$. The third element $x_1^1 \, x_2^0$ encodes $\{v_1\}$. The fourth element $x_1^1 \, x_2^1$ encodes $\{v_1, v_2\}$.

From the domain $\{x_1 \, x_2 | \, \forall \, x_d \in \{0, 1\}$ for $1 \le d \le 2\}$, there are four inputs x_1^0 x_2^0, $x_1^0 \, x_2^1$, $x_1^1 \, x_2^0$ and $x_1^1 \, x_2^1$. Because the value of m is equal to zero, there is no edge in its complementary graph in Fig. 3.10b. Therefore, each input does not need to implement those instructions in Fig. 3.8. Each input is a clique. This indicates that an empty subset $\{\}$, $\{v_2\}$, $\{v_1\}$ and $\{v_1, v_2\}$ are all cliques. Next, for computing the number of vertex in each clique, each input needs to implement "$z_{1,1} \leftarrow s_0^1 \wedge x_1$", "$z_{1,0} \leftarrow s_0^1 \wedge \overline{x_1}$", "$z_{2,2} \leftarrow x_2 \wedge z_{1,1}$", "$z_{2,1} \leftarrow \overline{x_2} \wedge z_{1,1}$", "$z_{2,1} \leftarrow x_2 \wedge z_{1,0}$" and "$z_{2,0} \leftarrow \overline{x_2} \wedge z_{1,0}$" in Fig. 3.9. After each input completes six instructions above, the input $x_1^1 \, x_2^1$ has the result $z_{2,2}^1$ and other inputs $x_1^0 \, x_2^0$, $x_1^0 \, x_2^1$ and $x_1^1 \, x_2^0$ have the same result $z_{2,2}^0$. Because Boolean variable $z_{2,2}^1$ indicates that the input $x_1^1 \, x_2^1$ encodes $\{v_1, v_2\}$ to be the maximal clique, the answer is to $\{v_1, v_2\}$ and the number of vertex in the answer is two. Because 2^2 subsets of vertices implement six instructions above not to access or modify the same input and the same auxiliary Boolean variable, we can *parallelize* them without any error.

3.3.4   Solution Space of Solving an Instance of the Maximal Clique Problem

In the graph with two vertices $\{v_1, v_2\}$ and one edge $\{(v_1, v_2)\}$ in Fig. 3.10a, an empty subset $\{\}$, $\{v_2\}$, $\{v_1\}$ and $\{v_1, v_2\}$ are all cliques. The maximal clique problem to the graph in Fig. 3.10a is to find a maximum-sized clique in which the number of vertex is two. Implementing the instruction "$z_{1,1} \leftarrow s_0^1 \wedge x_1$" is equivalent to implement the instruction "$z_{1,1} \leftarrow x_1$" in which Boolean variable $z_{1,1}$ actually stores the value of x_1. Therefore, implementing the instruction "$z_{2,2} \leftarrow x_2 \wedge z_{1,1}$" is equivalent to implement the instruction "$z_{2,2} \leftarrow x_2 \wedge x_1$". So, any given oracular function O_f to recognize a maximal-sized clique for the graph in Fig. 3.10a is to implement the instruction "$z_{2,2} \leftarrow x_2 \wedge x_1$". Its domain is $\{x_1 \, x_2 | \forall \, x_d \in \{0, 1\} \text{ for } 1 \leq d \leq 2\}$ and its range is $\{0, 1\}$.

We regard its domain as its solution space in which there are four possible choices that satisfy $O_f = F(x_1, x_2) = x_1 \wedge x_2 = 1$. We make use of a basis $\{(1, 0, 0, 0), (0, 1, 0, 0), (0, 0, 1, 0), (0, 0, 0, 1)\}$ of the four-dimensional Hilbert space to construct solution space $\{x_1 \, x_2 | \forall \, x_d \in \{0, 1\} \text{ for } 1 \leq d \leq 2\}$. We use $(1, 0, 0, 0)$ to encode Boolean variable x_1^0 and Boolean variable x_2^0 that represent an empty clique $\{\}$ without any vertex. Next, we apply $(0, 1, 0, 0)$ to encode Boolean variable x_1^1 and Boolean variable x_2^0 that represent a clique $\{v_1\}$. We make use of $(0, 0, 1, 0)$ to encode Boolean variable x_1^0 and Boolean variable x_2^1 that represent a clique $\{v_2\}$. Finally, we apply $(0, 0, 0, 1)$ to encode Boolean variable x_1^1 and Boolean variable x_2^1 that represent the maximal clique $\{v_1, v_2\}$.

We use a linear combination of each element in the basis that is $\frac{1}{\sqrt{2^2}}(1, 0, 0, 0) + \frac{1}{\sqrt{2^2}} \times (0, 1, 0, 0) + \frac{1}{\sqrt{2^2}} \times (0, 0, 1, 0) + \frac{1}{\sqrt{2^2}} \times (0, 0, 0, 1) = (\frac{1}{\sqrt{2^2}}, \frac{1}{\sqrt{2^2}}, \frac{1}{\sqrt{2^2}}, \frac{1}{\sqrt{2^2}})$ to construct solution space $\{x_1 \, x_2 | \forall \, x_d \in \{0, 1\} \text{ for } 1 \leq d \leq 2\}$. The amplitude of each possible choice is all $\frac{1}{\sqrt{2^2}}$ and the sum to the square of the absolute value of each amplitude is one. The length of the vector is one, so it is a unit vector. This indicates that we make use of a unit vector to encode all of the possible choices that satisfy $O_f = F(x_1, x_2) = x_1 \wedge x_2$. We call the square of the absolute value of each amplitude as the cost (the successful probability) of that choice that satisfies the given oracular function $O_f = F(x_1, x_2) = x_1 \wedge x_2$. The cost (the successful probability) of the answer(s) is close to one as soon as possible.

3.3.5 Implementing Solution Space of Solving an Instance of the Maximal Clique Problem

In Listing 3.2, the program in the backend *ibmqx4* with five quantum bits in **IBM**'s quantum computer is to solve an instance of the maximal clique problem to the graph with two vertices $\{v_1, v_2\}$ and one edge $\{(v_1, v_2)\}$ in Fig. 3.10a. Because the given oracular function of recognizing the maximal clique(s) is $O_f = F(x_1, x_2) = x_1 \wedge x_2$, in Listing 3.2, we introduce how to write a quantum program to find values of each Boolean variable so that the whole formula has the value 1. Figure 3.11 is the quantum circuit of constructing solution space to the question. The statement "OPENQASM 2.0;" on line one of Listing 3.2 is to point out that the program is written with version 2.0 of Open QASM. Next, the statement "include "qelib1.inc";" on line two of Listing 3.2 is to continue parsing the file "qelib1.inc" as if the contents of the file were pasted at the location of the include statement, where the file "qelib1.inc" is

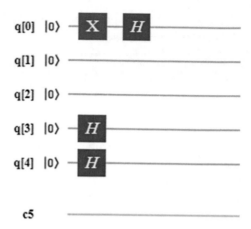

Fig. 3.11 The quantum circuit of constructing solution space to an instance of the maximal clique problem to the graph with two vertices $\{v_1, v_2\}$ and one edge $\{(v_1, v_2)\}$ in Fig. 3.10a

Quantum Experience (QE) Standard Header and the path is specified relative to the current working directory.

Listing 3.2 The program of solving an instance of the maximal clique problem to the graph with two vertices $\{v_1, v_2\}$ and one edge $\{(v_1, v_2)\}$ in Fig. 3.10a.

```
1.  OPENQASM 2.0;
2.  include "qelib1.inc";
3.  qreg q[5];
4.  creg c[5];
5.  x q[0];
6.  h q[3];
7.  h q[4];
8.  h q[0];
```

Next, the statement "qreg q[5];" on line three of Listing 3.2 is to declare that in the program there are five quantum bits. In the left top of Fig. 3.11, five quantum bits are respectively q[0], q[1], q[2], q[3] and q[4]. The initial value of each quantum bit is set to $|0\rangle$. We apply quantum bit q[3] to encode Boolean variable x_1. We use quantum bit q[4] to encode Boolean variable x_2. We make use of quantum bit q[2] to encode auxiliary Boolean variable $z_{2,2}$. We apply quantum bit q[0] as an auxiliary working bit. We do not use quantum bit q[1].

For the convenience of our explanation, $q[k]^0$ for $0 \le k \le 4$ is to represent the value 0 of q[k] and $q[k]^1$ for $0 \le k \le 4$ is to represent the value 1 of q[k]. Similarly, for the convenience of our explanation, an initial state vector of constructing solution space to an instance of the maximal clique problem in the graph with two vertices $\{v_1, v_2\}$ and one edge $\{(v_1, v_2)\}$ in Fig. 3.10a is as follows:

$$|\alpha_0\rangle = |q[4]^0\rangle|q[3]^0\rangle|q[2]^0\rangle|q[1]^0\rangle|q[0]^0\rangle = |0\rangle|0\rangle|0\rangle|0\rangle|0\rangle = |00000\rangle.$$

Next, the statement "creg c[5];" on line four of Listing 3.2 is to declare that there are five classical bits in the program. In the left bottom of Fig. 3.11, five classical bits are subsequently c[0], c[1], c[2], c[3] and c[4]. The initial value of each classical bit is set to 0. Classical bit c[4] is the most significant bit and classical bit c[0] is the least significant bit.

Next, the three statements "x q[0];", "h q[3];" and "h q[4];" on line five through seven of Listing 3.2 is to complete one X gate (one NOT gate) and two Hadamard gates of the *first* time slot of the quantum circuit in Fig. 3.11. The statement "x q[0];" actually implements $\begin{pmatrix} 0 & 1 \\ 1 & 0 \end{pmatrix} \times \begin{pmatrix} 1 \\ 0 \end{pmatrix} = \begin{pmatrix} 0 \\ 1 \end{pmatrix} = (|1\rangle)$. This is to say that the statement "x q[0];" on line five of Listing 3.2 inverts $|q[0]^0\rangle(|0\rangle)$ into $|q[0]^1\rangle(|1\rangle)$.

The two statements "h q[3];" and "h q[4];" both actually run $\begin{pmatrix} \frac{1}{\sqrt{2}} & \frac{1}{\sqrt{2}} \\ \frac{1}{\sqrt{2}} & -\frac{1}{\sqrt{2}} \end{pmatrix} \times \begin{pmatrix} 1 \\ 0 \end{pmatrix}$

$= \begin{pmatrix} \frac{1}{\sqrt{2}} \\ \frac{1}{\sqrt{2}} \end{pmatrix} = \frac{1}{\sqrt{2}} \begin{pmatrix} 1 \\ 1 \end{pmatrix} = \frac{1}{\sqrt{2}} \left(\begin{pmatrix} 1 \\ 0 \end{pmatrix} + \begin{pmatrix} 0 \\ 1 \end{pmatrix} \right) = \frac{1}{\sqrt{2}} (|0\rangle + |1\rangle)$. This implies that

converting q[3] from one state $|0>$ to another state $\frac{1}{\sqrt{2}} (|0\rangle + |1\rangle)$ (its superposition) and converting q[4] from one state $|0>$ to another state $\frac{1}{\sqrt{2}}(|0\rangle + |1\rangle)$ (its superposition) are implemented. Thus, the superposition of the two quantum bits q[4] and q[3] is $\left(\frac{1}{\sqrt{2}}(|0\rangle + |1\rangle) \right) \cdot \left(\frac{1}{\sqrt{2}}(|0\rangle + |1\rangle) \right) = \frac{1}{2}(|0\rangle|0\rangle + |0\rangle|1\rangle + |1\rangle|0\rangle + |1\rangle|1\rangle) = \frac{1}{2}(|00\rangle + |01\rangle + |10\rangle + |11\rangle)$. Since in the *first* time slot of the quantum circuit in Fig. 3.11 there is no quantum gate to act on quantum bits q[2] and q[1], their current states $|q[2]^0\rangle$ and $|q[1]^0\rangle$ are not changed. This indicates that we obtain the following new state vector

$$|\alpha_1\rangle = \left(\frac{1}{\sqrt{2}}(|q[4]^0\rangle + |q[4]^1\rangle) \right) \left(\frac{1}{\sqrt{2}}(|q[3]^0\rangle + |q[3]^1\rangle) \right)$$
$$(|q[2]^0\rangle|q[1]^0\rangle|q[0]^1\rangle)$$
$$= \frac{1}{2}(|q[4]^0\rangle|q[3]^0\rangle + |q[4]^0\rangle|q[3]^1\rangle + |q[4]^1\rangle|q[3]^0\rangle + |q[4]^1\rangle|q[3]^1\rangle)$$
$$(|q[2]^0\rangle |q[1]^0\rangle|q[0]^1\rangle)$$
$$= \frac{1}{2}(|0\rangle|0\rangle + |0\rangle|1\rangle + |1\rangle|0\rangle + |1\rangle|1\rangle)(|0\rangle|0\rangle|1\rangle).$$

Then, the statement "h q[0];" on line *eight* of Listing 3.2 is to execute one Hadamard gate of the *second* time slot of the quantum circuit in Fig. 3.11. The statement "h q[0];" actually implements $\begin{pmatrix} \frac{1}{\sqrt{2}} & \frac{1}{\sqrt{2}} \\ \frac{1}{\sqrt{2}} & -\frac{1}{\sqrt{2}} \end{pmatrix} \times \begin{pmatrix} 0 \\ 1 \end{pmatrix} = \begin{pmatrix} \frac{1}{\sqrt{2}} \\ -\frac{1}{\sqrt{2}} \end{pmatrix} = \frac{1}{\sqrt{2}}$

$\begin{pmatrix} 1 \\ -1 \end{pmatrix} = \frac{1}{\sqrt{2}} \left(\begin{pmatrix} 1 \\ 0 \end{pmatrix} - \begin{pmatrix} 0 \\ 1 \end{pmatrix} \right) = \frac{1}{\sqrt{2}}(|0> - |1>)$. This implies that converting q[0]

from one state $|1>$ to another state $\frac{1}{\sqrt{2}}(|0> - |1>)$ (its superposition) is performed. Because in the *second* time slot of the quantum circuit in Fig. 3.11 there is no quantum gate to act on quantum bits q[4] through q[1], their current states are not changed. This is to say that we obtain the following new state vector

$$|\alpha_2\rangle = \left(\frac{1}{2}(|q[4]^0\rangle|q[3]^0\rangle + |q[4]^0\rangle|q[3]^1\rangle + |q[4]^1\rangle|q[3]^0\rangle + |q[4]^1\rangle|q[3]^1\rangle) \right)$$

$$\left(|q[2]^0\rangle|q[1]^0\rangle\right)\left(\frac{1}{\sqrt{2}}(|q[0]^0\rangle - |q[0]^1\rangle) \right)$$

$$= \left(\frac{1}{2}(|0\rangle|0\rangle + |0\rangle|1\rangle + |1\rangle|0\rangle + |1\rangle|1\rangle) \right)(|0\rangle|0\rangle)\left(\frac{1}{\sqrt{2}}(|0\rangle - |1\rangle) \right).$$

In the new state vector $|\alpha_2\rangle$, state $|q[4]^0\rangle|q[3]^0\rangle$ encodes Boolean variable x_1^0 and Boolean variable x_2^0 that represent a possible choice without any vertex. State $|q[4]^0\rangle|q[3]^1\rangle$ encodes Boolean variable x_1^1 and Boolean variable x_2^0 that represent a possible choice with the *first* vertex v_1. State $|q[4]^1\rangle|q[3]^0\rangle$ encodes Boolean variable x_1^0 and Boolean variable x_2^1 that represent a possible choice with the *second* vertex v_2. State $|q[4]^1\rangle|q[3]^1\rangle$ encodes Boolean variable x_1^1 and Boolean variable x_2^1 that represent a possible choice with the two vertices v_1 and v_2. The amplitude of each possible choice is $\frac{1}{\sqrt{2^2}}$ and the cost (the successful possibility) of becoming the answer(s) to each possible choice is the same and is equal to $\frac{1}{2^2} = 1/4$.

3.3.6 The Oracle to an Instance of the Maximal Clique Problem

For solving the maximal clique problem of the graph with two vertices $\{v_1, v_2\}$ and one edge $\{(v_1, v_2)\}$ in Fig. 3.10a, the Oracle is to have the ability to *recognize* the maximal-sized clique(s). The Oracle is to multiply the probability amplitude of the maximal-sized clique(s) by -1 and leaves any other amplitude unchanged. Since the given oracular function of recognizing the maximal-sized clique(s) is $O_f = F(x_1, x_2) = x_1 \wedge x_2$, the Oracle to solve an instance of the maximal clique problem is a $(2^2 \times$

$2^2)$ matrix B that is equal to $\begin{pmatrix} 1 & 0 & 0 & 0 \\ 0 & 1 & 0 & 0 \\ 0 & 0 & 1 & 0 \\ 0 & 0 & 0 & -1 \end{pmatrix}_{4\times4}$.

We suppose that a $(2^2 \times 2^2)$ matrix B^+ is the conjugate transpose of matrix B. The transpose of matrix B is equal to *itself* (matrix B) and each element in the transpose of matrix B is a real, so the conjugate transpose of matrix B is also equal to itself (matrix B). Therefore, we obtain $B^+ = B$. Because matrix B and its conjugate transpose B^+ are almost a $(2^2 \times 2^2)$ identity matrix, $B \times B^+ = I_{2^2 \times 2^2}$, and $B^+ \times B = I_{2^2 \times 2^2}$. Hence, we get $B \times B^+ = B^+ \times B$. This indicates that it is a unitary matrix (operator) to solve an instance of the maximal clique problem in the graph with two vertices $\{v_1, v_2\}$ and one edge $\{(v_1, v_2)\}$ with the given oracular function $O_f = F(x_1, x_2) = x_1 \wedge x_2$. Implementing the given oracular function $O_f = F(x_1, x_2) = x_1 \wedge x_2$ that recognizes solution(s) in an instance of the maximal clique problem is equivalent to implement the Oracle that is

$$\begin{pmatrix} 1 & 0 & 0 & 0 \\ 0 & 1 & 0 & 0 \\ 0 & 0 & 1 & 0 \\ 0 & 0 & 0 & -1 \end{pmatrix}_{4\times 4} \times \begin{pmatrix} \frac{1}{\sqrt{2^2}} \\ \frac{1}{\sqrt{2^2}} \\ \frac{1}{\sqrt{2^2}} \\ \frac{1}{\sqrt{2^2}} \end{pmatrix}_{4\times 1}$$

$$= \begin{pmatrix} \frac{1}{\sqrt{2^2}} \\ \frac{1}{\sqrt{2^2}} \\ \frac{1}{\sqrt{2^2}} \\ \frac{-1}{\sqrt{2^2}} \end{pmatrix}_{4\times 1} = \frac{1}{\sqrt{2^2}} \times \begin{pmatrix} 1 \\ 0 \\ 0 \\ 0 \end{pmatrix}_{4\times 1} + \frac{1}{\sqrt{2^2}}$$

$$\times \begin{pmatrix} 0 \\ 1 \\ 0 \\ 0 \end{pmatrix}_{4\times 1} + \frac{1}{\sqrt{2^2}} \times \begin{pmatrix} 0 \\ 0 \\ 1 \\ 0 \end{pmatrix}_{4\times 1} + \frac{-1}{\sqrt{2^2}} \times \begin{pmatrix} 0 \\ 0 \\ 0 \\ 1 \end{pmatrix}_{4\times 1}$$

$$= \frac{1}{\sqrt{2^2}}|00\rangle + \frac{1}{\sqrt{2^2}}|01\rangle + \frac{1}{\sqrt{2^2}}|10\rangle + \frac{-1}{\sqrt{2^2}}|11\rangle.$$

Four computational basis vectors $\begin{pmatrix} 1 \\ 0 \\ 0 \\ 0 \end{pmatrix}_{4\times 1}$, $\begin{pmatrix} 0 \\ 1 \\ 0 \\ 0 \end{pmatrix}_{4\times 1}$, $\begin{pmatrix} 0 \\ 0 \\ 1 \\ 0 \end{pmatrix}_{4\times 1}$ and

$\begin{pmatrix} 0 \\ 0 \\ 0 \\ 1 \end{pmatrix}_{4\times 1}$ encode four states $|00\rangle(|q[4]^0\rangle|q[3]^0\rangle)$, $|01\rangle(|q[4]^0\rangle|q[3]^1\rangle)$, $|10\rangle(|q[4]^1\rangle|q[3]^0\rangle)$ and $|11\rangle(|q[4]^1\rangle|q[3]^1\rangle)$ and their current amplitudes are respectively $\left(\frac{1}{\sqrt{2^2}}\right), \left(\frac{1}{\sqrt{2^2}}\right), \left(\frac{1}{\sqrt{2^2}}\right)$ and $\left(\frac{-1}{\sqrt{2^2}}\right)$. State $|00\rangle(|q[4]^0\rangle|q[3]^0\rangle)$ with the amplitude $\left(\frac{1}{\sqrt{2^2}}\right)$ encodes Boolean variable x_1^0 and Boolean variable x_2^0 that represent a possible choice without any vertex. State $|01\rangle(|q[4]^0\rangle|q[3]^1\rangle)$ with the amplitude

$\left(\frac{1}{\sqrt{2^2}}\right)$ encodes Boolean variable x_1^1 and Boolean variable x_2^0 that represent a possible choice with the first vertex v_1. State $|10\rangle\left(|q[4]^1\rangle|q[3]^0\rangle\right)$ with the amplitude $\left(\frac{1}{\sqrt{2^2}}\right)$ encodes Boolean variable x_1^0 and Boolean variable x_2^1 that represent a possible choice with the second vertex v_2. State $|11\rangle\left(|q[4]^1\rangle|q[3]^1\rangle\right)$ with the amplitude $\left(\frac{-1}{\sqrt{2^2}}\right)$ encodes Boolean variable x_1^1 and Boolean variable x_2^1 that represent the maximal-sized clique with the two vertices v_1 and v_2. This indicates that the Oracle multiplies the probability amplitude of the maximal-sized clique with the two vertices v_1 and v_2 by -1 and leaves any other amplitude unchanged.

3.3.7 Implementing the Oracle to an Instance of the Maximal Clique Problem

We apply one **CCNOT** gate to run the given oracular function $O_f = F(x_1, x_2) = x_1 \wedge x_2$ that recognizes the maximal-sized clique(s) to the maximal clique problem in the graph with two vertices $\{v_1, v_2\}$ and one edge $\{(v_1, v_2)\}$ in Fig. 3.10a. We make use of quantum bit q[3] to encode Boolean variable x_1, we use quantum bit q[4] to encode Boolean variable x_2 and we apply quantum bit q[2] to encode Boolean variable $z_{2,2}$. Therefore, quantum bits q[3], q[4], q[2] are subsequently the first control bit, the second control bit and the target bit of the **CCNOT** gate. We make use of the **CCNOT** gate to implement a logical and operation, so the initial value to quantum bit q[2] is set to $|0\rangle$.

From line *nine* through line *twenty-three* in Listing 3.2, there are the fifteen statements. They are subsequently "h q[2];", "cx q[4],q[2];", "tdg q[2];", "cx q[3],q[2];", "t q[2];", "cx q[4],q[2];", "tdg q[2];", "cx q[3],q[2];", "t q[4];", "t q[2];", "cx q[3],q[4];", "h q[2];", "t q[3];", "tdg q[4];" and "cx q[3], q[4];". They complete the **CCNOT** gate that implements the given oracular function $O_f = F(x_1, x_2) = x_1 \wedge x_2$ that recognizes the maximal-sized clique(s) to the maximal clique problem in the graph with two vertices $\{v_1, v_2\}$ and one edge $\{(v_1, v_2)\}$ in Fig. 3.10a. Figure 3.12 is the quantum circuit of implementing the given oracular function $O_f = F(x_1, x_2) = x_1 \wedge x_2$ that recognizes the maximal-sized clique(s) to the maximal clique problem in the graph with two vertices $\{v_1, v_2\}$ and one edge $\{(v_1, v_2)\}$ in Fig. 3.10a.

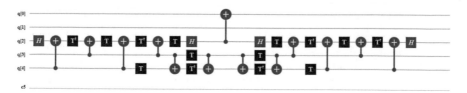

Fig. 3.12 The quantum circuit of implementing the given oracular function $O_f = F(x_1, x_2) = x_1 \wedge x_2$ that recognizes the maximal-sized clique(s) to the maximal clique problem in the graph with two vertices $\{v_1, v_2\}$ and one edge $\{(v_1, v_2)\}$ in Fig. 3.10a

Listing 3.2 Continued…

// We use the following *fifteen* statements to implement a **CCNOT** gate.

9.   h q[2];
10. cx q[4],q[2];
11. tdg q[2];
12. cx q[3],q[2];
13. t q[2];
14. cx q[4],q[2];
15. tdg q[2];
16. cx q[3],q[2];
17. t q[4];
18. t q[2];
19. cx q[3],q[4];
20. h q[2];
21. t q[3];
22. tdg q[4];
23. cx q[3], q[4];

They take the state vector $|\alpha_2\rangle = \left(\frac{1}{2}(|q[4]^0\rangle|q[3]^0\rangle + |q[4]^0\rangle|q[3]^1\rangle + |q[4]^1\rangle|q[3]^0\rangle + |q[4]^1\rangle|q[3]^1\rangle)\right)\left(|q[2]^0\rangle|q[1]^0\rangle\right)\left(\frac{1}{\sqrt{2}}(|q[0]^0\rangle - |q[0]^1\rangle)\right)$ as their input. After they actually perform six **CNOT** gates, two Hadamard gates, three T^+ gates and four **T** gates from the *first* time slot through the *eleventh* time slot in Fig. 3.12, we gain the following new state vector

$$|\alpha_3\rangle = \left(\frac{1}{2}(|q[4]^0\rangle|q[3]^0\rangle|q[2]^0\rangle + |q[4]^0\rangle|q[3]^1\rangle|q[2]^0\rangle + |q[4]^1\rangle|q[3]^0\rangle|q[2]^0\rangle + |q[4]^1\rangle|q[3]^1\rangle|q[2]^1\rangle)\right)\left(|q[1]^0\rangle\right)\left(\frac{1}{\sqrt{2}}(|q[0]^0\rangle - |q[0]^1\rangle)\right)$$

$$= \left(\frac{1}{2}(|0\rangle|0\rangle|0\rangle + |0\rangle|1\rangle|0\rangle + |1\rangle|0\rangle|0\rangle + |1\rangle|1\rangle|1\rangle)\right)$$

$$(|0\rangle)\left(\frac{1}{\sqrt{2}}(|0\rangle - |1\rangle)\right).$$

Then, from line *twenty-four* in Listing 3.2, the statement "cx q[2],q[0];" takes the new state vector $|\alpha_3\rangle$ as its input. It multiplies the probability amplitude of the answer. $(|q[4]^1\rangle|q[3]^1\rangle)$ encoding the maximal-sized clique $\{v_1, v_2\}$ by -1 and leaves any other amplitude unchanged. This implies that after the statement "cx q[2],q[0];" completes the **CNOT** gate in the *twelfth* time slot in Fig. 3.12, we get the following new state vector.

Listing 3.2 Continued...

// The Oracle multiplies the probability amplitude of the maximal-sized clique $\{v_{21},$

// $v\}$ by -1 and leaves any other amplitude unchanged.
24. cx q[2],q[0];

$$|\alpha_4\rangle = \left(\frac{1}{2}\left(|q[4]^0\rangle|q[3]^0\rangle|q[2]^0\rangle + |q[4]^0\rangle|q[3]^1\rangle|q[2]^0\rangle + |q[4]^1\rangle|q[3]^0\rangle|q[2]^0\rangle\right.\right.$$

$$\left.\left. + (-1)|q[4]^1\rangle|q[3]^1\rangle|q[2]^1\rangle\right)\right)\left(|q[1]^0\rangle\right)\left(\frac{1}{\sqrt{2}}\left(|q[0]^0\rangle - |q[0]^1\rangle\right)\right)\right)$$

$$= \left(\frac{1}{2}\left(|0\rangle|0\rangle|0\rangle + |0\rangle|1\rangle|0\rangle + |1\rangle|0\rangle|0\rangle + (-1)|1\rangle|1\rangle|1\rangle\right)\right)$$

$$(|0\rangle)\left(\frac{1}{\sqrt{2}}\right)(|0\rangle - |1\rangle).$$

Since quantum operations are reversible by nature, executing the reversed order of implementing the **CCNOT** gate can restore the auxiliary quantum bits to their initial states. From line *twenty-five* through line *thirty-nine* in Listing 3.2, there are the *fifteen* statements. They are "cx q[3],q[4];", "tdg q[4];", "t q[3];", "h q[2];", "cx q[3],q[4];", "t q[2];", "t q[4];", "cx q[3],q[2];", "tdg q[2];", "cx q[4],q[2];", "t q[2];", "cx q[3],q[2];", "tdg q[2];", "cx q[4],q[2];" and "h q[2];". They run the reversed order of implementing.

Listing 3.2 Continued...

// Because quantum operations are reversible by nature, executing the reversed.

// order of implementing the **CCNOT** gate can restore the auxiliary quantum bits.

// to their initial states.
25. cx q[3],q[4];
26. tdg q[4];
27. t q[3];
28. h q[2];
29. cx q[3],q[4];
30. t q[2];

```
31.  t q[4];
32.  cx q[3],q[2];
33.  tdg q[2];
34.  cx q[4],q[2];
35.  t q[2];
36.  cx q[3],q[2];
37.  tdg q[2];
38.  cx q[4],q[2];
39.  h q[2];
```

the **CCNOT** gate that performs the given oracular function $O_f = F(x_1, x_2) = x_1 \wedge x_2$ that recognizes the maximal-sized clique(s). They take the new state vector $|\alpha_4 >$ as their input. After they actually complete six **CNOT** gates, two Hadamard gates, three T^+ gates and four T gates from the *thirteenth* time slot through the *last* time slot in Fig. 3.12, we obtain the following new state vector.

$$|\alpha_5\rangle = \left(\frac{1}{2}(|q[4]^0\rangle|q[3]^0\rangle + |q[4]^0\rangle|q[3]^1\rangle + |q[4]^1\rangle|q[3]^0\rangle + (-1)|q[4]^1\rangle|q[3]^1\rangle) \right)$$

$$\left(|q[2]^0\rangle|q[1]^0\rangle \right) \left(\frac{1}{\sqrt{2}}(|g[0]^0\rangle - |q[0]^1\rangle) \right)$$

$$= \left(\frac{1}{2}(|0\rangle|0\rangle + |0\rangle|1\rangle + |1\rangle|0\rangle + (-1)|1\rangle|1\rangle) \right)(|0\rangle|0\rangle)\left(\frac{1}{\sqrt{2}}(|0\rangle - |1\rangle) \right).$$

In the state vector $|\alpha_5\rangle$, the amplitude of each element in solution space $\{x_1\, x_2|$ $\forall\, x_d \in \{0, 1\}$ for $1 \leq d \leq 2\}$ is $(1/2)$. In the state vector $|\alpha_5\rangle$, the amplitude to three elements $x_1^0\, x_2^0$ encoding an empty clique, $x_1^0\, x_2^1$ encoding a clique $\{v_2\}$, $x_1^1\, x_2^0$ encoding a clique $\{v_1\}$ in solution space is all $(1/2)$ and the amplitude to the element $x_1^1\, x_2^1$ encoding the maximal-sized clique $\{v_1, v_2\}$ in solution space is $(-1/2)$. This is to say that *thirty-one* statements from line *nine* through *thirty-nine* in Listing 3.2

complete $\begin{pmatrix} 1 & 0 & 0 & 0 \\ 0 & 1 & 0 & 0 \\ 0 & 0 & 1 & 0 \\ 0 & 0 & 0 & -1 \end{pmatrix}_{4\times4} \times \begin{pmatrix} \frac{1}{\sqrt{2^2}} \\ \frac{1}{\sqrt{2^2}} \\ \frac{1}{\sqrt{2^2}} \\ \frac{1}{\sqrt{2^2}} \end{pmatrix}_{4\times1}$ that is to implement the Oracle that

recognizes the maximal-sized clique(s) to solve the maximal clique problem in the graph with two vertices $\{v_1, v_2\}$ and one edge $\{(v_1, v_2)\}$ in Fig. 3.10a.

3.3.8 Implementing the Grover Diffusion Operator to Amplify the Amplitude of the Solution in an Instance of the Maximal Clique Problem

The new state vector $|\alpha_5\rangle$ is $\frac{1}{2}(|q[4]^0\rangle|q[3]^0\rangle + |q[4]^0\rangle|q[3]^1\rangle + |q[4]^1\rangle|q[3]^0\rangle + (-1)|q[4]^1\rangle|q[3]^1\rangle)) \; (|q[2]^0\rangle|q[1]^0\rangle)(\frac{1}{\sqrt{2}}(|q[0]^0\rangle - |q[0]^1\rangle))$.
It includes two independent subsystem. The first subsystem is $(\frac{1}{2}(|q[4]^0\rangle|q[3]^0\rangle + |q[4]^0\rangle|q[3]^1\rangle + |q[4]^1\rangle|q[3]^0\rangle + (-1)|q[4]^1\rangle|q[3]^1\rangle))$ and the second subsystem is $(|q[2]^0\rangle|q[1]^0\rangle)(\frac{1}{\sqrt{2}}(|q[0]^0\rangle - |q[0]^1\rangle))$. Amplifying the amplitude of each solution in the maximal clique problem in the graph with two vertices $\{v_1, v_2\}$ and one edge $\{(v_1, v_2)\}$ in Fig. 3.10a only needs to consider the first subsystem in the new state vector $|\alpha_5\rangle$. Since for an instance of the maximal

clique problem the $(2^2 \times 1)$ vector $\begin{pmatrix} \frac{1}{\sqrt{2^2}} \\ \frac{1}{\sqrt{2^2}} \\ \frac{1}{\sqrt{2^2}} \\ \frac{-1}{\sqrt{2^2}} \end{pmatrix}_{4 \times 1}$ encodes the first subsystem of

the new state vector $|\alpha_5\rangle$ and $\begin{pmatrix} \frac{2}{2^2}-1 & \frac{2}{2^2} & \frac{2}{2^2} & \frac{2}{2^2} \\ \frac{2}{2^2} & \frac{2}{2^2}-1 & \frac{2}{2^2} & \frac{2}{2^2} \\ \frac{2}{2^2} & \frac{2}{2^2} & \frac{2}{2^2}-1 & \frac{2}{2^2} \\ \frac{2}{2^2} & \frac{2}{2^2} & \frac{2}{2^2} & \frac{2}{2^2}-1 \end{pmatrix}_{2^2 \times 2^2}$ is a $(2^2 \times$

$2^2)$ diffusion operator, amplifying the amplitude of the solution is to implement

$\begin{pmatrix} \frac{2}{2^2}-1 & \frac{2}{2^2} & \frac{2}{2^2} & \frac{2}{2^2} \\ \frac{2}{2^2} & \frac{2}{2^2}-1 & \frac{2}{2^2} & \frac{2}{2^2} \\ \frac{2}{2^2} & \frac{2}{2^2} & \frac{2}{2^2}-1 & \frac{2}{2^2} \\ \frac{2}{2^2} & \frac{2}{2^2} & \frac{2}{2^2} & \frac{2}{2^2}-1 \end{pmatrix}_{2^2 \times 2^2} \times \begin{pmatrix} \frac{1}{\sqrt{2^2}} \\ \frac{1}{\sqrt{2^2}} \\ \frac{1}{\sqrt{2^2}} \\ \frac{-1}{\sqrt{2^2}} \end{pmatrix}_{4 \times 1} = \begin{pmatrix} 0 \\ 0 \\ 0 \\ 1 \end{pmatrix}_{4 \times 1}$. This indi-

cates that the amplitude of the solution that encodes the maximal-sized clique $\{v_1, v_2\}$ is one and the amplitude of other three possible choices is all zero.

The quantum circuit in Fig. 3.13 implements the Grover diffusion operator, $H^{\otimes 2}(2|0^{\otimes 2}\rangle\langle 0^{\otimes 2}| - I_{2^2 \times 2^2}) \, H^{\otimes 2}$. From line *forty* through line *fifty-one* in Listing 3.2, there are the *twelve* statements. They are subsequently "h q[3];", "h q[4];", "x q[3];", "x q[4];", "h q[4];", "cx q[3],q[4];", "h q[4];", "x q[4];", "x q[3];", "u3(2*pi,0*pi,0*pi) q[3];", "h q[4];" and "h q[3];".

Listing 3.2 Continued...

//We complete the amplitude amplification of the answer.
40. h q[3];
41. h q[4];

// We complete phase shifters.

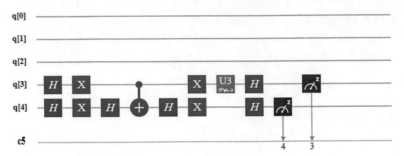

Fig. 3.13 The quantum circuit of implementing the Grover diffusion operator, $H^{\otimes 2} \left(2\left|0^{\otimes 2}\right\rangle\left\langle0^{\otimes 2}\right| - I_{2^2 \times 2^2}\right) H^{\otimes 2}$, to an instance of the maximal clique problem in the graph with two vertices $\{v_1, v_2\}$ and one edge $\{(v_1, v_2)\}$ in Fig. 3.10a

```
42.  xq[3];
43.  xq[4];
44.  h q[4];
45.  cx q[3],q[4];
46.  h q[4];
47.  xq[4];
48.  xq[3];
49.  u3(2*pi,0*pi,0*pi) q[3];
50.  h q[4];
51.  h q[3];
```

They take the new state vector $\frac{1}{2}\left(\left|q[4]^0\right\rangle\left|q[3]^0\right\rangle + \left|q[4]^0\right\rangle\left|q[3]^1\right\rangle + \left|q[4]^1\right\rangle\left|q[3]^0\right\rangle + (-1)\left|q[4]^1\right\rangle\left|q[3]^1\right\rangle\right)$ as their input. They complete the diffusion operator, $H^{\otimes 2} \left(2\left|0^{\otimes 2}\right\rangle\left\langle0^{\otimes 2}\right| - I_{2^2 \times 2^2}\right) H^{\otimes 2}$ from the first time slot through the eighth time slot in Fig. 3.13. This is to say that we obtain the following new state vector.

$$|\alpha_6\rangle = \left|q[4]^1\right\rangle\left|q[3]^1\right\rangle.$$

Next, from line fifty-two in Listing 3.2the statement "measure q[4] -> c[4];" is to measure the *fifth* quantum bit q[4] and to record the measurement outcome by overwriting the *fifth* classical bit c[4]. From line fifty-three in Listing 3.2 the statement "measure q[3] -> c[3];" is to measure the *fourth* quantum bit q[3] and to record the measurement outcome by overwriting the *fourth* classical bit c[3]. They implement the measurement from the ninth time slot through the tenth time slot of Fig. 3.13.

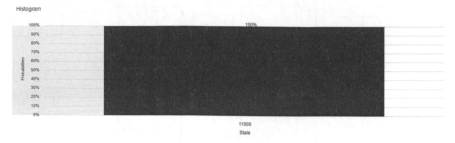

Fig. 3.14 After the measurement to solve an instance of the maximal clique problem in the graph with two vertices $\{v_1, v_2\}$ and one edge $\{(v_1, v_2)\}$ in Fig. 3.10a is completed, we obtain the answer 11,000 with the probability 1 (100%)

Listing 3.2 Continued…

// We complete the measurement of the answer.
```
52.  measure q[4] -> c[4];
53.  measure q[3] -> c[3];
```

In the backend *ibmqx4* with five quantum bits in **IBM**'s quantum computers, we apply the command "simulate" to execute the program in Listing 3.2. The measured result appears in Fig. 3.14. From Fig. 3.14, we get the answer 11,000 ($c[4] = 1 = q[4] = |1\rangle$, $c[3] = 1 = q[3] = |1\rangle$, $c[2] = 0$, $c[1] = 0$ and $c[0] = 0$) with the probability 1 (100%). This implies that with the possibility 1 (100%) we gain that the value of quantum bit $q[3]$ is equal to $|1\rangle$ and the value of quantum bit $q[4]$ is equal to $|1\rangle$. Therefore, the maximal-sized clique is to $\{v_1, v_2\}$.

3.3.9 The Quantum Search Algorithm to the Maximal Clique Problem

A maximal clique problem to a graph $G = (V, E)$ with n vertices and θ edges and its *complementary* graph $\overline{G} = (V, \overline{E})$ with n vertices and m edges in which each edge in $\overline{E}$ is out of E is to find the maximal-sized clique(s) among 2^n subsets of vertices. Any given oracular function $O_f(x_1 x_2 \ldots x_{n-1} x_n)$ is to recognize the maximal-sized clique(s) among 2^n subsets of vertices. It implements flowchart of recognizing cliques in Fig. 3.8. Next, it implements flowchart of computing the number of vertex in each clique in Fig. 3.9. We make use of the quantum search algorithm to find one of M solutions among 2^n subsets of vertices, where $0 \leq M \leq 2^n$.

The quantum circuit in Fig. 3.15 is to complete the quantum search algorithm to solve an instance of the maximal clique problem in a graph $G = (V, E)$ with n vertices and θ edges. The first quantum register in the left top of Fig. 3.15 is $(|0^{\otimes n}\rangle)$. This implies that the initial value of each quantum bit is $|0\rangle$. The second quantum register

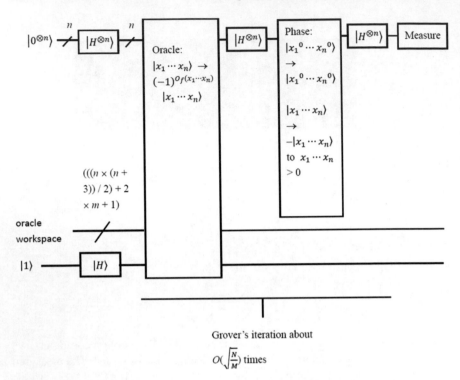

Grover's iteration about

$$O\left(\sqrt{\frac{N}{M}}\right) \text{ times}$$

Fig. 3.15 Circuit of implementing the quantum search algorithm to solve an instance of the maximal problem in a graph $G = (V, E)$ with n vertices and θ edges

in the left bottom of Fig. 3.15 has $(((n \times (n + 3))/2) + 2 \times m + 1)$ quantum bits and is an auxiliary quantum register. The initial value of each quantum bit in the second quantum register is $|0\rangle$ or $|1\rangle$ that is dependent on implementing **NAND** operations or **AND** operations. The third quantum register in the left bottom of Fig. 3.15 is $(|1\rangle)$.

3.3.10 The First Stage of the Quantum Search Algorithm to the Maximal Clique Problem

In Fig. 3.15, the first stage of the quantum search problem to solve an instance of the maximal clique problem in a graph $G = (V, E)$ with n vertices and θ edges is to apply n Hadamard gates $(|H^{\otimes n}\rangle)$ to operate the first quantum register $(|0^{\otimes n}\rangle)$. This is to say that it produces the superposition of n quantum bits to be $(\otimes_{k=1}^{n}\left(\frac{|x_k^0\rangle + |x_k^1\rangle}{\sqrt{2}}\right)) = \frac{1}{\sqrt{2^n}}(\sum_{\{x_1 x_2 \ldots x_n | \forall x_d \in \{0,1\} \text{for} 1 \leq d \leq n\}} |x_1 x_2 \ldots x_n\rangle)$. The superposition $\frac{1}{\sqrt{2^n}}(\sum_{\{x_1 x_2 \ldots x_n | \forall x_d \in \{0,1\} \text{for} 1 \leq d \leq n\}} |x_1 x_2 \ldots x_n\rangle)$ encodes solution space $\{x_1 x_2 \ldots x_{n-1} x_n | \forall x_d \in \{0, 1\} \text{ for } 1 \leq d \leq n\}$ to solve an instance of the maximal clique problem in a graph $G = (V, E)$ with n vertices and θ edges. This indicates that state $(|x_1^0 x_2^0 \ldots x_n^0\rangle)$ with the amplitude $(\frac{1}{\sqrt{2^n}})$ encodes an empty subset (choice) of vertex, state $(|x_1^0 x_2^0 \ldots x_n^1\rangle)$ with the amplitude $(\frac{1}{\sqrt{2^n}})$ encodes $\{v_n\}$ and so on with that state $(|x_1^1 x_2^1 \ldots x_n^1\rangle)$ with the amplitude $(\frac{1}{\sqrt{2^n}})$ encodes $\{ v_1 \; v_2 \cdots v_n \}$. In the first stage of the quantum search algorithm, it uses another Hadamard gate to operate the third quantum register $(|1\rangle)$. This implies that it yields the superposition of the third quantum register $(|1\rangle)$ that is $\left(\frac{1}{\sqrt{2}}(|0\rangle - |1\rangle)\right)$.

3.3.11 The Second Stage of the Quantum Search Algorithm to the Maximal Clique Problem

In Fig. 3.15, the *second* stage of the quantum search algorithm to solve an instance of the maximal clique problem in a graph $G = (V, E)$ with n vertices and θ edges and its *complementary* graph $\overline{G} = (V, \overline{E})$ with n vertices and m edges in which each edge in $\overline{E}$ is out of E is to implement the Oracle. The Oracle is to have the ability to *recognize* solutions that are the maximal-sized cliques among 2^n subsets (possible choices) of vertices. The Oracle is to multiply the probability amplitude of the maximal-sized clique(s) by -1 and leaves any other amplitude unchanged.

The *first* main task of the Oracle is to that recognizing clique(s) among 2^n possible choices (subsets) of vertices is equivalent to implement a formula of the form $(\wedge_m^{k=1}(\overline{x_i \wedge x_j}))$. This is to say that for completing the *first* main task of the Oracle we need to use $(2 \times m + 1)$ auxiliary quantum bits and to complete m **NAND** operations and m **AND** operations that is equivalent to implement $(2 \times m)$ **CCNOT** gates.

Next, the *second* main task of the Oracle is to that computing the number of vertex in each clique is equivalent to decide that the influence of each vertex is to increase the number of vertex to each clique or to reserve the number of vertex to each clique. This indicates that for performing the *second* main task of the Oracle we need to use $((n \times (n + 3))/2)$ auxiliary quantum bits and to implement $(n \times (n + 1))$ **AND** operations and $(\frac{n \times (n+1)}{2})$ **NOT** operations that is equivalent to complete $(n \times (n + 1))$ **CCNOT** gates and $(\frac{n \times (n+1)}{2})$ **NOT** gates.

In oracle workspace in the second stage of the quantum search algorithm, we make use of auxiliary quantum bits $|r_k\rangle$ for $1 \leq k \leq m$ to encode auxiliary Boolean variables r_k for $1 \leq k \leq m$. We apply auxiliary quantum bits $|s_k\rangle$ for $0 \leq k \leq m$ to encode auxiliary Boolean variables s_k for $0 \leq k \leq m$. We use auxiliary quantum bits $|z_{i+1,j}\rangle$ and $|z_{i+1,j+1}\rangle$ for $0 \leq i \leq n - 1$ and $0 \leq j \leq i$ to encode auxiliary Boolean variables $z_{i+1,j}$ and $z_{i+1,j+1}$ for $0 \leq i \leq n - 1$ and $0 \leq j \leq i$.

We use **CCNOT** gates to implement each **NAND** gate $(\overline{x_i \wedge x_j})$ in a formula of the form $(\wedge_m^{k=1}(\overline{x_i \wedge x_j}))$ and we apply auxiliary quantum bits $|r_k\rangle$ for $1 \leq k \leq m$ to store the result of implementing each **NAND** gate $(\overline{x_i \wedge x_j})$. This is to say that each auxiliary quantum bit $|r_k\rangle$ for $1 \leq k \leq m$ is actually the target bit of a **CCNOT** gate of implementing a **NAND** gate $(\overline{x_i \wedge x_j})$. Therefore, the initial value of each auxiliary quantum bit $|r_k\rangle$ for $1 \leq k \leq m$ is set to one $|1\rangle$.

We use an auxiliary quantum bit $|S_0\rangle$ as the first operand of the first logical and operation ("$\wedge$") in a formula of the form $(\wedge_m^{k=1}(\overline{x_i \wedge x_j}))$. The initial value of the auxiliary quantum bit $|S_0\rangle$ is set to $|1\rangle$. This indicates that this setting does not change the correct result of the first logical and operation. We make use of a **CCNOT** gate to implement each logical and operation in a formula of the form $(\wedge_m^{k=1}(\overline{x_i \wedge x_j}))$. We apply auxiliary quantum bits $|S_k\rangle$ for $1 \leq k \leq m$ to store the result of implementing each logical and operation. This implies that each auxiliary quantum bit $|s_k>$ for $1 \leq k \leq m$ is actually the target bit of a **CCNOT** gate of implementing a logical and operation. Thus, the initial value of each auxiliary quantum bit $|s_k>$ for $1 \leq k \leq m$ is set to $|0>$.

We use a **CCNOT** gate to implement $(s_m \wedge x_1)$ that is to compute the influence of the first vertex to increase the number of vertex in each clique. We apply a **CCNOT** gate and two **NOT** gates to implement $(s_m \wedge \overline{x_1})$ that is to compute the influence of the first vertex to reserve the number of vertex in each clique. We make use of auxiliary quantum bits $|z_{1,1}\rangle$ to store the result of implementing $(s_m \wedge x_1)$ and apply auxiliary quantum bits $|z_{1,0}\rangle$ to store the result of implementing $(s_m \wedge \overline{x_1})$. This is to say that the two auxiliary quantum bits $|z_{1,1}\rangle$ and $|z_{1,0}\rangle$ are actually the target bits of two **CCNOT** gates of implementing two logical and operations. Therefore, the initial value of the two auxiliary quantum bits $|z_{1,1}\rangle$ and $|z_{1,0}\rangle$ is set to $|0\rangle$.

We use a **CCNOT** gate to implement $(x_i + 1 \wedge z_{i,j})$ that is to figure the influence of the $(i + 1)$th vertex v_{i+1} for $1 \leq i \leq n - 1$ to increase the number of vertex in each clique. We apply a **CCNOT** gate and two **NOT** gates to implement $((\overline{x_{i+1}}) \wedge z_{i,j})$ that is to deal with the influence of the $(i + 1)$th vertex v_{i+1} for $1 \leq i \leq n - 1$ to reserve the number of vertex in each clique. We make use of auxiliary quantum bits $|z_{i+1,j+1}\rangle$ and $|z_{i+1,j}\rangle$ for $0 \leq i \leq n - 1$ and $0 \leq j \leq i$ to store the result of implementing them. This is to say that auxiliary quantum bit $|z_{i+1,j+1}\rangle$ and $|z_{i+1,j}\rangle$ for $0 \leq i \leq n - 1$ and $0 \leq j \leq i$ are the target bits of the corresponding **CCNOT** gates of implementing logical and operations. Hence, the initial value of auxiliary quantum bit $|z_{i+1,j+1}\rangle$ and $|z_{i+1,j}\rangle$ for $0 \leq i \leq n - 1$ and $0 \leq j \leq i$ is set to $|0>$. Quantum bit $|z_{n,j}\rangle$ for $n \geq j \geq 0$ is to store the result that has j ones after computing the influence of n vertices to the number of vertex in each clique. If the value of quantum bit $|z_{n,j}\rangle$ for $n \geq j \geq 0$ is equal to $|1\rangle$, then it has j vertices.

We use one **CNOT** gate $\left|\frac{|0\rangle - |1\rangle}{\sqrt{2}} \oplus z_{n,j}\right\rangle$ to multiply the probability amplitude of the maximal-sized clique(s) by $- 1$ and to leave any other amplitude unchanged, where quantum bit $\left(\frac{|0\rangle - |1\rangle}{\sqrt{2}}\right)$ is the target bit of the **CNOT** gate and quantum bit $\left(|z_{n,j}\rangle\right)$ is the control bit of the **CNOT** gate. When the value of the control bit $\left(|z_{n,j}\rangle\right)$ is equal to $(|1>)$, the target bit becomes $\left(\frac{|1\rangle - |0\rangle}{\sqrt{2}}\right) = (-1)\left(\frac{|0\rangle - |1\rangle}{\sqrt{2}}\right)$. This is to multiply the probability amplitude of the answer(s) by $- 1$. When the value of the control bit $\left(|Z_{n,j}\rangle\right)$ is equal to $(|0>)$, the target bit still is $\left(\frac{|0\rangle - |1\rangle}{\sqrt{2}}\right)$. This is to leave any other amplitude unchanged.

Quantum operations are reversible by nature, so executing the reversed order of implementing the two main tasks of the Oracle can restore the auxiliary quantum bits to their initial states. This indicates that the second stage of the quantum search algorithm to solve an instance of the maximal clique problem in a graph $G = (V, E)$ with n vertices and θ edges converts $\frac{1}{\sqrt{2^n}}(\sum_{\{x_1 x_2 \cdots x_n | \forall x_d \in \{0,1\} \text{for} 1 \leq d \leq n\}} |x_1 x_2 \ldots x_n\rangle)$ into $(-1)^{O_f(x_1 \cdots x_n)} \frac{1}{\sqrt{2^n}}(\sum_{\{x_1 x_2 \ldots x_n | \forall x_d \in \{0,1\} \text{ for } 1 \leq d \leq n\}} |x_1 x_2 \ldots x_n\rangle)$. The cost of completing the Oracle in the second stage of the quantum search algorithm in Fig. 3.15 is to implement $(2 \times ((2 \times m) + (n \times (n + 1))))$ **CCNOT** gates, $(n \times (n + 1))$ **NOT** gates and one **CNOT** gate.

Because the number of the vertices to the maximal clique(s) is from n through zero, we first check whether there is/are clique(s) with n vertices. The condition is to implement one **CNOT** gate $\left| \frac{|0\rangle - |1\rangle}{\sqrt{2}} \oplus z_{n,n} \right)$ to multiply the probability amplitude of the answer(s) with n vertices by -1 and to leave any other amplitude unchanged. If the answer(s) with n vertices are found, then we terminate the execution of the program. Otherwise, we need to continue to test whether there is/are clique(s) with $(n - 1)$ vertices, $(n - 2)$ vertices and so on with that has one vertex until the answer(s) are found.

3.3.12 The Third Stage of the Quantum Search Algorithm to the Maximal Clique Problem

In Fig. 3.15, the *third* stage of the quantum search algorithm to solve an instance of the maximal clique problem in a graph $G = (V, E)$ with n vertices and θ edges is to perform the Grover diffusion operator that is a $(2^n \times 2^n)$ matrix D in which $D_{a,b} = \frac{2}{2^n}$ if $a \neq b$ and $D_{a,a} = \frac{2}{2^n} - 1$. Executing the Grover diffusion operator is equivalent to implement $H^{\otimes n} (2 |0^{\otimes n}\rangle\langle 0^{\otimes n}| - I_{2^n \times 2^n}) H^{\otimes n}$. A phase shifter operator, $(2 |0^{\otimes n}\rangle\langle 0^{\otimes n}| - I_{2^n \times 2^n})$ negates all the states except for $(|x_1^0 \ldots x_n^0\rangle)$. In Fig. 3.15, the *third* stage of the quantum search problem is to apply that the phase shift operator, $2 |0^{\otimes n}\rangle\langle 0^{\otimes n}| - I_{2^n \times 2^n}$, negates all the states except for $(|x_1^0 \ldots x_n^0\rangle)$ sandwiched between $H^{\otimes n}$ gates. This is to say that the *third* stage of the quantum search algorithm to solve an instance of the maximal clique problem in a graph $G = (V, E)$ with n vertices and θ edges is to increase the amplitude of the maximal-sized clique(s) and to decrease the amplitude of the non-answer(s).

For solving an instance of the maximal clique problem in a graph $G = (V, E)$ with n vertices and θ edges, we regard the Oracle and the Grover diffusion operator from the second stage through the third stage of the quantum search algorithm in Fig. 3.15 as a subroutine. We call the subroutine as the *Grover iteration*. The number of the vertices to the maximal clique(s) is from n through zero, but in advanced it is *unknown*. Therefore, we first test whether there is/are clique(s) with n vertices. The condition is to implement one **CNOT** gate $\left| \frac{|0\rangle - |1\rangle}{\sqrt{2}} \oplus z_{n,n} \right)$ to multiply the probability amplitude of the answer(s) with n vertices by -1 and to leave any other amplitude unchanged. If there is/are clique(s) with n vertices, then after repeat to execute the Grover iteration of $O\left(\sqrt{\frac{N}{M}}\right)$ times, the successful probability of measuring the answer(s) with n vertices is at least (1/2). Otherwise, the measurement has a failed result and we need to continue to check whether there is/are clique(s) with $(n-1)$ vertices, $(n-2)$ vertices and so on with that has one vertex until the answer(s) is/are found.

When the value of (M/N) is equal to (1/4), the successful probability of measuring the answer(s) is one (100%) with the Grover iteration of one time. This is the *best* case of the quantum search algorithm to solve an instance of the maximal clique problem in a graph $G = (V, E)$ with n vertices and θ edges. When the value of M is equal to one, the successful probability of measuring the answer(s) is at least (1/2) with the Grover iteration of $O(\sqrt{N})$ times. This is the *worst* case of the quantum search algorithm to solve an instance of the maximal clique problem in a graph $G = (V, E)$ with n vertices and θ edges. This implies that the quantum search algorithm to solve an instance of the maximal clique problem in a graph $G = (V, E)$ with n vertices and θ edges only gives a quadratic speed-up.

3.4 Summary

In this chapter, we gave a formal illustration of the search problem. We provided the satisfiability problem and the maximal clique problem as two examples of the search problem. First, we provided flowcharts of solving the satisfiability problem and the maximal clique problem. We also introduced data dependence analysis of solving the satisfiability problem and the maximal clique problem. We illustrated solution space of solving the satisfiability problem and the maximal clique problem. Next, we described the quantum circuits of implementing solution space for solving the satisfiability problem and the maximal clique problem. We also introduced the Oracle and provided the quantum circuits of implementing the Oracle to solve the satisfiability problem and the maximal clique problem. We then illustrated the Grover diffusion operator to amplify the amplitude of the answers to solve the satisfiability problem and the maximal clique problem. We also introduced the quantum circuits of implementing the Grover diffusion operator to amplify the amplitude of the answers in the satisfiability problem and in the maximal clique problem. Finally, we offered the two quantum search algorithms to solve the satisfiability problem and the maximal clique problem.

3.5 Bibliographical Notes

In this chapter, a more detailed introduction to the Search Problem can be found in the recommended books that are (Imre and Balazs 2007; Lipton and Regan 2014; Nielsen and Chuang 2000; Silva 2018). A complete description for the satisfiability problem with m clauses and m Boolean variables can be found in the famous article

(Cook 1971) and in the famous textbook (Garey and Johnson 1979). A complete quantum algorithm for solving an instance of the satisfiability problem with m clauses and m Boolean variables can be found in the famous article (Chang et al. 2018).

A complete illustration to the maximal clique problem to a graph $G = (V, E)$ with n vertices and θ edges and its complementary graph $\overline{G} = (V, \overline{E})$ has n vertices and m edges in which each edge in $\overline{E}$ is out of E can be found in the famous article (Karp 1975) and in the famous textbook (Garey and Johnson 1979). A complete quantum algorithm for solving an instance of the maximal clique problem in a graph $G = (V, E)$ with n vertices and θ edges and its complementary graph $\overline{G} = (V, \overline{E})$ has n vertices and m edges in which each edge in $\overline{E}$ is out of E can be found in the famous article (Chang et al. 2018).

A complete description of quantum search algorithm can be found in the famous article (Grover 1996) and in the famous textbooks (Imre and Balazs 2007; Lipton and Regan 2014; Nielsen and Chuang 2000; Silva 2018). A detailed introduction (quantum circuit) of implementing the Grover diffusion operator can be found in the famous articles (Coles et al. 2018; Mandviwalla et al. 2018) and in the famous textbooks (Imre and Balazs 2007; Lipton and Regan 2014; Nielsen and Chuang 2000; Silva 2018).

3.6　Exercises

3.1. The unary operator "-" denotes logical operation **NOT** and the binary operator "$\wedge$" denotes logical operation **AND**. We regard the satisfiability problem for the Boolean formula $F(x_1, x_2) = \overline{x_1} \wedge x_2$ with two Boolean variables x_1 and x_2 as a search problem in which any given oracular function O_f is the Boolean formula $F(x_1, x_2) = \overline{x_1} \wedge x_2$. Its domain is $\{x_1\ x_2 |\ \forall\ x_d \in \{0, 1\}$ for $1 \leq d \leq 2\}$, its range is $\{0, 1\}$ and the logical operation $\overline{x_1}$ is the negation of Boolean variable x_1. In the given oracular function $O_f = F(x_1, x_2) = \overline{x_1} \wedge x_2$ of the search problem, there are M inputs of *two* bits from its domain, say $\lambda_M = x_1 x_2$, that satisfies the condition $O_f(\lambda_M) = O_f(x_1\ x_2) = F(x_1, x_2) = \overline{x_1} \wedge x_2 = 1$. Please use quantum search algorithms to write a quantum program with what possibility to find the answer $(\lambda_M = x_1\ x_2)$ that satisfies $O_f(\lambda_M) = O_f(x_1\ x_2) = F(x_1, x_2) = \overline{x_1} \wedge x_2 = 1$.

3.2. The binary operator "$\wedge$" denotes logical operation **AND**. The unary operator "-" denotes logical operation **NOT**. Any given oracular function O_f is the Boolean formula $F(x_1, x_2) = x_1 \wedge \overline{x_2}$ with two Boolean variables x_1 and x_2. Its domain is $\{x_1\ x_2 |\ \forall\ x_d \in \{0, 1\}$ for $1 \leq d \leq 2\}$, its range is $\{0, 1\}$ and the logical operation $\overline{x_2}$ is the negation of Boolean variable x_2. Please make use of quantum search algorithms to write a quantum program with what possibility to compute each value of two Boolean variables x_1 and x_2 that satisfies the given oracular function $O_f(x_1\ x_2) = F(x_1, x_2) = x_1 \wedge \overline{x_2} = 1$.

3.3 The binary operator "·" denotes logical operation **NAND**. Any given oracular function O_f is the Boolean formula $F(x_1, x_2) = x_1 \wedge \overline{x_2}$ with two Boolean variables x_1 and x_2. Its domain is $\{x_1 \, x_2 | \, \forall \, x_d \in \{0, 1\} \text{ for } 1 \le d \le 2\}$ and its range is $\{0, 1\}$. Please apply quantum search algorithms to write a quantum program with what possibility to calculate each value of two Boolean variables x_1 and x_2 that satisfies the given oracular function $O_f(x_1 \, x_2) = F(x_1, x_2) = x_1 \wedge \overline{x_2} = 1$.

3.4. The binary operator "$\wedge$" denotes logical operation **AND**. Any given oracular function O_f is the Boolean formula $F(x_1, x_2) = x_1 \wedge x_2$ with two Boolean variables x_1 and x_2. Its domain is $\{x_1 \, x_2 | \, \forall \, x_d \in \{0, 1\} \text{ for } 1 \le d \le 2\}$ and its range is $\{0, 1\}$. Please use quantum search algorithms to write a quantum program with what possibility to determine each value of two Boolean variables x_1 and x_2 that satisfies the given oracular function $O_f(x_1 \, x_2) = F(x_1, x_2) = x_1 \wedge x_2 = 1$.

References

Chang, W.-L., Ren, T.-T., Luo, J.n., Feng, M., Guo, M., Lin, K.W.: Quantum algorithms for bio-molecular solutions of the satisfiability problem on a quantum machine. IEEE Trans. Nanobiosci. 7(3), 215–222 (2008)

Chang, W.-L., Yu, Q., Li, Z., Chen, J., Peng, X., Feng, M.: Quantum speedup in solving the maximal-clique problem. Phys. Rev. A **97**, 032344 (2018)

Coles, P.J., Eidenbenz, S., Pakin, S., Adedoyin, A., Ambrosiano, J., Anisimov, P., Casper, W., Chennupati, G., Coffrin, C., Djidjev, H., Gunter, D., Karra, S., Lemons, N., Lin, S., Lokhov, A., Malyzhenkov, A., Mascarenas, D., Mniszewski, S., Nadiga, B., O'Malley, D., Oyen, D., Prasad, L., Roberts, R., Romero, P., Santhi, N., Sinitsyn, N., Swart, P., Vuffray, M., Wendelberger, J., Yoon, B., Zamora, R., Zhu, W.: Quantum algorithm implementations for beginners (2018). https://arxiv.org/abs/1804.03719

Cook, S.: The complexity of theorem proving procedures. In: Proceedings of Third Annual ACM Symposium on Theory of Computing, pp. 151–158 (1971)

Garey, M.R., Johnson, D.S.: Computer and Intractability: A Guide to the Theory of NP-Completeness. W. H. Freeman Company (1979). ISBN-13: 978-0716710448

Grover, L.K.: A fast quantum mechanical algorithm for database search. In: Proceedings of the Twenty-Eighth Annual ACM Symposium on Theory of Computing, pp. 212–219 (1996)

Imre, S., Balazs, F.: Quantum Computation and Communications: An Engineering Approach. Wiley, UK (2007). ISBN-10: 047086902X and ISBN-13: 978-0470869024, 2005

Karp, R.: On the computational complexity of combinatorial problems. Networks 5(45), 68 (1975)

Lipton, R.J., Regan, K.W.: Quantum Algorithms via Linear Algebra: A Primer. The MIT Press (2014). ISBN 978-0-262-02839-4

Mandviwalla, A., Ohshiro, K. Ji, B.: Implementing Grover's algorithm on the IBM quantum computers. In: 2018 IEEE International Conference on Big Data (Big Data), pp. 1–7 (2018). https://doi.org/10.1109/bigdata.2018.8622457

Nielsen, M.A., Chuang, I.L.: Quantum Computation and Quantum Information. Cambridge University Press, New York, NY (2000). ISBN-10: 9781107002173 and ISBN-13: 978-1107002173

Silva, V.: Practical Quantum Computing for Developers: Programming Quantum Rigs in the Cloud using Python, Quantum Assembly Language and IBM Q Experience. Apress, December 13, 2018. ISBN-10: 1484242173 and ISBN-13: 978-1484242179

Chapter 4
Quantum Fourier Transform and Its Applications

When in 1822 Fourier published his most famous article (work), people originally used his transform in thermodynamics. Now, the most common use for the Fourier transform is in signal processing. A signal is given in the time domain: as a function mapping time to amplitude. The Fourier transform allows us to express the signal as a weighted sum of phase-shifted sinusoids of varying frequencies. The phases and the weights associated with frequencies characterize the signal in the frequency domain. The Fourier transform establishes the bridge between signal representation in time and frequency domains.

The success of the Fourier transform is due to its discrete version and we call it as the *Discrete Fourier Transform* (or *DFT*). The discrete Fourier transform has a computationally very efficient implementation in the form of the *Fast Fourier Transform* (or **FFT**). In this chapter, we introduce complex roots of *unity* and study their properties. Next, we illustrate the discrete Fourier transform and the *inverse* operation of the discrete Fourier transform. Next, we introduce the *Quantum Fourier Transform* (or *QFT*) that is the quantum version of the Fourier transform and is analogous to the fast Fourier transform. Next, we describe how to write a quantum program for implementing the discrete Fourier transform and the *inverse* operation of the discrete Fourier transform.

4.1 Introduction to Complex Roots of Unity

The definition of the exponential of a complex number is

$$e^{\sqrt{-1}\times\theta} = \cos(\theta) + \sqrt{-1} \times \sin(\theta). \tag{4.1}$$

The value of θ in Eq. (4.1) is a real number. The distance between any point (a, b) at the complex plane and the origin $(0, 0)$ of the complex plane is

$$\sqrt{(a-0)^2 + (b-0)^2} = \sqrt{a^2 + b^2}. \tag{4.2}$$

The value of a is a real number and the value of b is also a real number. Because according to Eq. (4.2) the distance between the exponential of a complex number $e^{\sqrt{-1} \times \theta}$ in Eq. (4.1) and the origin of the complex plane is $\sqrt{(\cos(\theta) - 0)^2 + (\sin(\theta) - 0)^2} = \sqrt{(\cos(\theta))^2 + (\sin(\theta))^2} = 1$, each $e^{\sqrt{-1} \times \theta}$ is equally spaced around the circle of unit radius centered at the origin of the complex plane.

A *complex nth root of unity* is a complex number ω such that

$$\omega^n = 1. \tag{4.3}$$

There are actually n complex nth roots of unity: $e^{\sqrt{-1} \times 2 \times \pi \times k/n}$ to $k = 0$, 1, 2, ..., $n - 2$, $n - 1$. The distance between each complex nth root of unity, $e^{\sqrt{-1} \times 2 \times \pi \times k/n}$, for $k = 0, 1, 2, ..., n - 2, n - 1$ and the origin of the complex plane is $\sqrt{(\cos(2 \times \pi \times k/n) - 0)^2 + (\sin(2 \times \pi \times k/n) - 0)^2} = \sqrt{(\cos(2 \times \pi \times k/n))^2 + (\sin(2 \times \pi \times k/n))^2} = 1$. Therefore, each $e^{\sqrt{-1} \times 2 \times \pi \times k/n}$ for $k = 0, 1, 2, ..., n - 2, n - 1$ that is a complex nth root of unity is equally spaced around the circle of unit radius centered at the origin of the complex plane.

The principal nth root of unity is

$$\omega_n = e^{\sqrt{-1} \times 2 \times \pi/n}. \tag{4.4}$$

Because all of the other complex nth roots of unity are $e^{\sqrt{-1} \times 2 \times \pi \times k/n} = (e^{\sqrt{-1} \times 2 \times \pi/n})^k = \omega_n^k$ for $k = 0, 2, 3, ..., n - 2, n - 1$, they are powers of ω_n in Eq. (4.4). Therefore, n complex nth roots of unity are subsequently

$$\omega_n^0, \omega_n^1, \omega_n^2, \ldots, \omega_n^{n-2}, \omega_n^{n-1}. \tag{4.5}$$

Figure 4.1 shows that in (4.5) $\omega_n^0, \omega_n^1, \omega_n^2, \ldots, \omega_n^{n-2}, \omega_n^{n-1}$ that are the n complex n-th roots of unity are equally spaced around the circle of unit radius centered at the origin of the complex plane. This is to say that $\omega_n^0, \omega_n^1, \omega_n^2, \ldots, \omega_n^{n-2}, \omega_n^{n-1}$ lie in the circumference of the circle of unit radius centered at the origin of the complex plane.

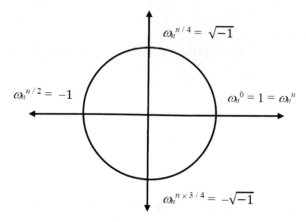

Fig. 4.1 The values of $\omega_n^0, \omega_n^1, \omega_n^2, \ldots, \omega_n^{n-2}, \omega_n^{n-1}$ in the complex plane

4.2 Illustration of an Abelian Group for n Complex nth Roots of Unity Together with a Binary Operation of Multiplication

A **group** $(S, *)$ is a set S together with a binary operation $*$ denoted on S for which the following four properties hold. The first property is to that for all $a, b \in S$ we have $a * b \in S$. This is to say that the group $(S, *)$ has **closure**. The second property is to that there is an element $e \in S$, called the **identity** of the group $(S, *)$, such that $e * a = a * e = a$ for all $a \in S$. The third property is to that for all $a, b, c \in S$ we have $(a * b) * c = a * (b * c)$. This indicates that the group $(S, *)$ has **associativity**. The fourth property is to that for each $a \in S$, there exists a unique element b, called the **inverse** of a, such that $a * b = b * a = e$. If a group $(S, *)$ satisfies the **commutative law** $a * c = c * a$ for all $a, c \in S$, then it is an **abelian group**.

We assume that a set S is $\{\omega_n^0, \omega_n^1, \omega_n^2, \ldots, \omega_n^{n-2}, \omega_n^{n-1}\}$ that is a set of n complex nth roots of unity. We also assume that a binary operation $\times$ is multiplication denoted on S. We use **Lemma 4.1** to demonstrate that $(S = \{\omega_n^0, \omega_n^1, \omega_n^2, \ldots, \omega_n^{n-2}, \omega_n^{n-1}\}, \times)$ is an Abelian group.

Lemma 4.1 $(S = \{\omega_n^0, \omega_n^1, \omega_n^2, \ldots, \omega_n^{n-2}, \omega_n^{n-1}\}, \times)$ *is an abelian group.*

Proof For all $\omega_n^j, \omega_n^k \in S$ to $0 \leq j \leq (n-1)$ and $0 \leq k \leq (n-1)$, we have $\omega_n^j \times \omega_n^k = \omega_n^{j+k}$. If the value of $(j+k)$ is from 0 (zero) through $(n-1)$, then $\omega_n^{j+k} \in S$. If the value of $(j+k)$ is from n through $(2 \times n - 2)$, then $\omega_n^{j+k} = \omega_n^{n+p} = \omega_n^n \times \omega_n^p = 1 \times \omega_n^p = \omega_n^p \in S$ to $0 \leq p \leq (n-2)$. This is to say that it satisfies **closure**. In S, there is an element ω_n^0 that is equal to one such that each element $\omega_n^k \in S$ to $0 \leq k \leq (n-1)$ satisfies $\omega_n^0 \times \omega_n^k = 1 \times \omega_n^k = \omega_n^k$ and $\omega_n^k \times \omega_n^0 = \omega_n^k \times 1 = \omega_n^k$. This indicates that $\omega_n^0 \times \omega_n^k = \omega_n^k \times \omega_n^0 = \omega_n^k$. Hence, it has an **identity** ω_n^0.

For all $\omega_n^j, \omega_n^k, \omega_n^l \in S$ to $0 \le j \le (n-1)$ and $0 \le k \le (n-1)$ and $0 \le l \le (n-1)$, we have $(\omega_n^j \times \omega_n^k) \times \omega_n^l = \omega_n^{(j+k)+l} = \omega_n^{j+(k+l)} = \omega_n^j \times \omega_n^{k+l} = \omega_n^j \times (\omega_n^k \times \omega_n^l)$. This implies that it satisfies **associativity**. For each element $\omega_n^k \in S$ to $0 \le k \le (n-1)$, we have $\omega_n^k \times \omega_n^{n-k} = \omega_n^{n-k} \times \omega_n^k = \omega_n^n = 1 = \omega_n^0$. This is to say that ω_n^{n-k} is the **inverse** of ω_n^k. For all $\omega_n^j, \omega_n^k \in S$ to $0 \le j \le (n-1)$ and $0 \le k \le (n-1)$, we have $\omega_n^j \times \omega_n^k = \omega_n^{j+k} = \omega_n^{k+j} = \omega_n^k \times \omega_n^j$. This is to say that it satisfies the **commutative law**. Therefore, from the statements above, we infer at once that $(S = \{\omega_n^0, \omega_n^1, \omega_n^2, \dots, \omega_n^{n-2}, \omega_n^{n-1}\}, \times)$ is an abelian group. ∎

4.3 Description to Properties of Complex nth Roots of Unity

We use the following lemmas to introduce essential properties of the complex nth roots of unity.

Lemma 4.2 *For any integers $n \ge 0$, $k \ge 0$, and $j > 0$,*

$$\omega_{j \times n}^{j \times k} = \omega_n^k. \tag{4.6}$$

Proof According to Eq. (4.4), we have $\omega_{j \times n}^{j \times k} = \left(e^{\sqrt{-1} \times 2 \times \pi / j \times n}\right)^{j \times k} = \left(e^{\sqrt{-1} \times 2 \times \pi / n}\right)^k = \omega_n^k$. Therefore, we at once infer that $\omega_{j \times n}^{j \times k} = \omega_n^k$ for any integers $n \ge 0$, $k \ge 0$, and $j > 0$. ∎

Lemma 4.3 *For any even integer $n > 0$,*

$$\omega_n^{n/2} = \omega_2 = -1. \tag{4.7}$$

Proof In light of Eq. (4.4), we have $\omega_n^{n/2} = \left(e^{\sqrt{-1} \times 2 \times \pi / n}\right)^{n/2} = \left(e^{\sqrt{-1} \times 2 \times \pi / 2}\right) = \omega_2 = -1$. Hence, we at once derive that $\omega_n^{n/2} = \omega_2 = -1$ for any *even* integer $n > 0$. ∎

Lemma 4.4 *For any positive even integer n, the squares of the n complex nth roots of unity are the $(n/2)$ complex $(n/2)$th roots of unity.*

Proof The complex nth roots of unity are ω_n^k for any nonnegative integer k. The complex $(n/2)$th roots of unity are $\omega_{n/2}^k$ for any nonnegative integer k. Because the squares of the complex nth roots of unity are $(\omega_n^k)^2$, we have $(\omega_n^k)^2 = \left(e^{\sqrt{-1} \times 2 \times \pi \times k / n}\right)^2 = e^{\sqrt{-1} \times 2 \times \pi \times k \times 2 / (\frac{n}{2}) \times 2} = e^{\sqrt{-1} \times 2 \times \pi \times k / (\frac{n}{2})} = \omega_{n/2}^k$. This is to say that the squares of the complex nth roots of unity are the complex $(n/2)$th roots of unity.

From **Lemma 4.3**, we have $\omega_n^{n/2} = -1$. This implies that $\omega_n^{k+(n/2)} = \omega_n^k \times \omega_n^{n/2}$ $= \omega_n^k \times (-1) = -\omega_n^k$. Thus, we obtain $(\omega_n^{k+(n/2)})^2 = (-\omega_n^k)^2 = (\omega_n^k)^2$. This indicates that ω_n^k and $\omega_n^{k+(n/2)}$ have the same square. This is also to say that if we square all of the complex nth roots of unity, then we actually obtain each complex $(n/2)$th roots of unity twice. Therefore, from the statements above, we at once infer that for any *positive even* integer n, the squares of the n complex nth roots of unity are the $(n/2)$ complex $(n/2)$th roots of unity. ∎

Lemma 4.5 *For any integer $n \geq 1$ and nonzero integer k not divisible by n,*

$$\sum_{j=0}^{n-1} (\omega_n^k)^j = 0. \tag{4.8}$$

Proof The common ratio to $\sum_{j=0}^{n-1} (\omega_n^k)^j$ in Eq. (4.8) is ω_n^k. Because nonzero integer k is not divisible by n, the value of ω_n^k is not equal to one. Therefore, we have $\sum_{j=0}^{n-1} (\omega_n^k)^j =$ $((\omega_n^k)^n - 1)/(\omega_n^k - 1) = ((\omega_n^n)^k - 1)/(\omega_n^k - 1) = ((1)^k - 1)/(\omega_n^k - 1) = (1 - 1)/(\omega_n^k - 1) = (0)/(\omega_n^k - 1) = 0.$ ∎

4.4 Introduction to the Discrete Fourier Transform and the Inverse Discrete Fourier Transform

We assume that a $(n \times 1)$ vector $a = (a_0, a_1, a_2, ..., a_{n-1})^T$ has that each coordinate a_k to $0 \leq k \leq (n-1)$ is a complex number, where the $(n \times 1)$ vector $a = (a_0, a_1, a_2, ..., a_{n-1})^T$ is the transpose of the $(1 \times n)$ vector $(a_0, a_1, a_2, ..., a_{n-1})$. Also we suppose that another vector $y = (y_0, y_1, y_2, ..., y_{n-1})^T$ has that each coordinate y_k to $0 \leq k \leq (n-1)$ is a complex number, where the $(n \times 1)$ vector $y = (y_0, y_1, y_2, ..., y_{n-1})^T$ is the transpose of the $(1 \times n)$ vector $(y_0, y_1, y_2, ..., y_{n-1})$. The matrix **DFT** of the discrete Fourier transform is as follows

$$\left(\frac{1}{\sqrt{n}}\right) \times \begin{pmatrix} \omega_n^{0\times0} & \omega_n^{0\times1} & \omega_n^{0\times2} & \omega_n^{0\times3} & \cdots & \omega_n^{0\times(n-1)} \\ \omega_n^{1\times0} & \omega_n^{1\times1} & \omega_n^{1\times2} & \omega_n^{1\times3} & \cdots & \omega_n^{1\times(n-1)} \\ \omega_n^{2\times0} & \omega_n^{2\times1} & \omega_n^{2\times2} & \omega_n^{2\times3} & \cdots & \omega_n^{2\times(n-1)} \\ \omega_n^{3\times0} & \omega_n^{3\times1} & \omega_n^{3\times2} & \omega_n^{3\times3} & \cdots & \omega_n^{3\times(n-1)} \\ \vdots & \vdots & \vdots & \vdots & \vdots & \vdots \\ \omega_n^{(n-1)\times0} & \omega_n^{(n-1)\times1} & \omega_n^{(n-1)\times2} & \omega_n^{(n-1)\times3} & \cdots & \omega_n^{(n-1)\times(n-1)} \end{pmatrix}_{n\times n} \tag{4.9}$$

The discrete Fourier transform of the vector $a = (a_0, a_1, a_2, \ldots, a_{n-1})^T$ is denoted by the vector $y = (y_0, y_1, y_2, \ldots, y_{n-1})^T = \mathbf{DFT} \times a$. The Fourier coefficient y_k for $0 \leq k \leq (n-1)$ in the vector y is as follows

$$y_k = \left(\frac{1}{\sqrt{n}}\right) \times \sum_{l=0}^{n-1} a_l \times \omega_n^{k \times l} \tag{4.10}$$

The matrix **IDFT** of the inverse discrete Fourier transform is as follows

$$\left(\frac{1}{\sqrt{n}}\right) \times \begin{pmatrix} \omega_n^{-0 \times 0} & \omega_n^{-0 \times 1} & \omega_n^{-0 \times 2} & \omega_n^{-0 \times 3} & \cdots & \omega_n^{-0 \times (n-1)} \\ \omega_n^{-1 \times 0} & \omega_n^{-1 \times 1} & \omega_n^{-1 \times 2} & \omega_n^{-1 \times 3} & \cdots & \omega_n^{-1 \times (n-1)} \\ \omega_n^{-2 \times 0} & \omega_n^{-2 \times 1} & \omega_n^{-2 \times 2} & \omega_n^{-2 \times 3} & \cdots & \omega_n^{-2 \times (n-1)} \\ \omega_n^{-3 \times 0} & \omega_n^{-3 \times 1} & \omega_n^{-3 \times 2} & \omega_n^{-3 \times 3} & \cdots & \omega_n^{-3 \times (n-1)} \\ \vdots & \vdots & \vdots & \vdots & \vdots & \vdots \\ \omega_n^{-(n-1) \times 0} & \omega_n^{-(n-1) \times 1} & \omega_n^{-(n-1) \times 2} & \omega_n^{-(n-1) \times 3} & \cdots & \omega_n^{-(n-1) \times (n-1)} \end{pmatrix}_{n \times n} \tag{4.11}$$

The inverse discrete Fourier transform of the vector $y = (y_0, y_1, y_2, \ldots, y_{n-1})^T$ is denoted by the vector $a = (a_0, a_1, a_2, \ldots, a_{n-1})^T = \mathbf{IDFT} \times y$. The inverse Fourier coefficient a_k for $0 \leq k \leq (n-1)$ in the vector a is as follows

$$a_k = \left(\frac{1}{\sqrt{n}}\right) \times \sum_{l=0}^{n-1} y_l \times \omega_n^{-k \times l}. \tag{4.12}$$

The matrix **IDFT** of the inverse discrete Fourier transform in Eq. (4.11) is actually to the conjugate transpose of the matrix **DFT** of the discrete Fourier transform in Eq. (4.9). Similarly, the matrix **DFT** of the discrete Fourier transform in Eq. (4.9) is actually to the conjugate transpose of the matrix **IDFT** of the inverse discrete Fourier transform in Eq. (4.11). We use the following two lemmas to show that the matrix **DFT** of the discrete Fourier transform in Eq. (4.9) is a unitary matrix (a unitary operator) and the matrix **IDFT** of the inverse discrete Fourier transform in Eq. (4.11) is also a unitary matrix (a unitary operator).

Lemma 4.6 *The matrix* **DFT** *of the discrete Fourier transform in* Eq. (4.9) *is a unitary matrix (a unitary operator).*

Proof For $0 \leq j \leq (n-1)$ and $0 \leq k \leq (n-1)$, the (k, j) entry of the matrix **DFT** of the discrete Fourier transform in Eq. (4.9) is $\left(\frac{1}{\sqrt{n}} \times \omega_n^{k \times j}\right)$. Similarly, for $0 \leq j \leq (n-1)$ and $0 \leq k \leq (n-1)$, the (k, j) entry of the matrix **IDFT** of the inverse discrete Fourier transform in Eq. (4.11) is $\left(\frac{1}{\sqrt{n}} \times \omega_n^{-k \times j}\right)$. We have the (j, l) entry of **DFT** $\times$ **IDFT** is equal to $(1/n) \times \left(\sum_{k=0}^{n-1} \left(\omega_n^{j \times k}\right) \times \left(\omega_n^{-k \times l}\right)\right) = (1/n) \times \left(\sum_{k=0}^{n-1} \left(\omega_n^{(j-l) \times k}\right)\right)$.

This summation is equal to 1 if the value of j is equal to the value of l, and it is 0 otherwise by **Lemma 4.5** that is to the summation lemma. This is to say that **DFT** $\times$ **IDFT** is an $(n \times n)$ *identity* matrix.

Similarly, the (j, l) entry of **IDFT** $\times$ **DFT** is equal to $(1/n) \times \left(\sum_{k=0}^{n-1} \left(\omega_n^{-j \times k} \right) \times \left(\omega_n^{k \times l} \right) \right) = (1/n) \times \left(\sum_{k=0}^{n-1} \left(\omega_n^{(l-j) \times k} \right) \right)$. This summation is equal to 1 if $j = l$, and it is 0 otherwise by **Lemma 4.5** that is to the summation lemma. This indicates that **IDFT** $\times$ **DFT** is an $(n \times n)$ *identity* matrix. Therefore, from the statements above, we at once infer that the matrix **DFT** of the discrete Fourier transform in Eq. (4.9) is a unitary matrix (a unitary operator). ∎

Lemma 4.7 *The matrix* **IDFT** *of the inverse discrete Fourier transform in* Eq. (4.11) *is also a unitary matrix (a unitary operator).*

Proof For $0 \leq j \leq (n-1)$ and $0 \leq k \leq (n-1)$, the (k, j) entry of the matrix **IDFT** of the inverse discrete Fourier transform in Eq. (4.11) is $\left(\frac{1}{\sqrt{n}} \times \omega_n^{-k \times j} \right)$. Similarly, for $0 \leq j \leq (n-1)$ and $0 \leq k \leq (n-1)$, the (k, j) entry of the matrix **DFT** of the discrete Fourier transform in Eq. (4.9) is $\left(\frac{1}{\sqrt{n}} \times \omega_n^{k \times j} \right)$. We have the (j, l) entry of **IDFT** $\times$ **DFT** is equal to $(1/n) \times \left(\sum_{k=0}^{n-1} \left(\omega_n^{-j \times k} \right) \times \left(\omega_n^{k \times l} \right) \right) = (1/n) \times \left(\sum_{k=0}^{n-1} \left(\omega_n^{(l-j) \times k} \right) \right)$. This summation is equal to 1 if the value of j is equal to the value of l, and it is 0 otherwise by **Lemma 4.5** that is to the summation lemma. This implies that **IDFT** $\times$ **DFT** is an $(n \times n)$ *identity* matrix.

Similarly, the (j, l) entry of **DFT** $\times$ **IDFT** is equal to $(1/n) \times \left(\sum_{k=0}^{n-1} \left(\omega_n^{j \times k} \right) \times \left(\omega_n^{-k \times l} \right) \right) = (1/n) \times \left(\sum_{k=0}^{n-1} \left(\omega_n^{(j-l) \times k} \right) \right)$. This summation is equal to 1 if $j = l$, and it is 0 otherwise by **Lemma 4.5** that is to the summation lemma. This is to say that **DFT** $\times$ **IDFT** is an $(n \times n)$ *identity* matrix. Hence, from the statements above, we at once derive that the matrix **IDFT** of the inverse discrete Fourier transform in Eq. (4.11) is a unitary matrix (a unitary operator). ∎

4.5 The Quantum Fourier Transform of Implementing the Discrete Fourier Transform

An n *dimensional orthonormal* basis α of a Hilbert space is as follows

$$\alpha = \{(1, 0, 0, \ldots, 0)^T, (0, 1, 0, \ldots, 0)^T, (0, 0, 1, \ldots, 0)^T, \ldots, (0, 0, 0, \ldots, 1)^T\}.$$
(4.13)

Without losing generality, we assume that n is a power of two (2). The length of each element in the n dimensional *orthonormal* basis α is unity. The inner product of any two elements in the n dimensional *orthonormal* basis α is zero. Each element in

α is a computational basis vector. We use computational basis state $|0\rangle$ to encodes the first computational basis vector $(1, 0, 0, \ldots, 0)^T$. We apply computational basis state $|1\rangle$ to encode the second computational basis vector $(0, 1, 0, \ldots, 0)^T$. We make use of computational basis state $|2\rangle$ to encode the third computational basis vector $(0, 0, 1, \ldots, 0)^T$ and so on with that we use computational basis state $|n-1\rangle$ to encode the last computational basis vector $(0, 0, 0, \ldots, 1)^T$. This is to say that the basis $\{|0\rangle, |1\rangle, |2\rangle, \ldots, |n-1\rangle\}$ encodes a n *dimensional orthonormal* basis α of a Hilbert space.

A superposition $|\beta\rangle$ is a linear combination of each computational basis vector (or each computational basis state) in α such that the superposition $|\beta\rangle$ is as follows

$$\left| \beta \right\rangle = \sum_{a=0}^{n-1} \beta_a \times \left| a \right\rangle = \sum_{a=0}^{n-1} \beta_a |a\rangle . \tag{4.14}$$

In the superposition $|\beta\rangle$ in Eq. (4.14), β_a for $0 \le a \le (n-1)$ is the amplitude of each computational basis state $|a\rangle$ such that $\sum_{a=0}^{n-1} |\beta_a|^2 = 1$.

The matrix **QFT** of the quantum Fourier transform is the same as the matrix **DFT** of the discrete Fourier transform in Eq. (4.9). Because **QFT** and **DFT** are the same matrix and **DFT** is a unitary matrix (a unitary operator) from **Lemma 4.6**, **QFT** is also a unitary matrix (a unitary operator). The quantum Fourier transform is a computationally very efficient implementation to the discrete Fourier transform in the form of tracing back **QFT** to its tensor product decomposition. The quantum Fourier transform transforms a superposition $|\beta\rangle = \sum_{a=0}^{n-1} \beta_a |a\rangle$ in Eq. (4.14) to the following new superposition

$$|y\rangle = \mathbf{QFT}(|\beta\rangle) = \mathbf{QFT}\left(\sum_{a=0}^{n-1} \beta_a |a\rangle \right) = \left(\frac{1}{\sqrt{n}} \right) \times \left(\sum_{k=0}^{n-1} \sum_{a=0}^{n-1} \beta_a \times \omega_n^{k \times a} |k\rangle \right). \tag{4.15}$$

In the new superposition $|y\rangle$ in Eq. (4.15), the Fourier coefficient y_k for $0 \le k \le (n-1)$ is as follows

$$y_k = \left(\frac{1}{\sqrt{n}} \right) \times \left(\sum_{a=0}^{n-1} \beta_a \times \omega_n^{k \times a} \right). \tag{4.16}$$

Applying **QFT** (the quantum Fourier transform) to computational basis state $|a\rangle$ for $0 \le a \le (n-1)$ generates

$$\mathbf{QFT} \times (|a\rangle) = \mathbf{QFT}(|a\rangle) = \left(\frac{1}{\sqrt{n}} \right) \times \left(\sum_{k=0}^{n-1} \omega_n^{k \times a} |k\rangle \right). \tag{4.17}$$

In the following, we take $n = 2^N$, where N is the number of quantum bit in a quantum computer and the basis $\{|0\rangle, |1\rangle, |2\rangle, \ldots, |2^N - 1\rangle\}$ is the computational basis for the quantum computer. We mention here that $(\omega_n^{k \times a}) = (\omega_{2^N}^{k \times a}) = (e^{(\sqrt{-1} \times 2 \times \pi \times k \times a)/2^N})$. It is very useful to write computational basis state $|a\rangle$ for $0 \leq a \leq (2^N - 1)$ using the binary representation $a = a_1 a_2 a_3 \ldots a_N = a_1 \times 2^{N-1} + a_2 \times 2^{N-2} + a_3 \times 2^{N-3} + \ldots + a_N \times 2^0$. It is also very useful to write the notation $0.a_1 a_2 a_3 \ldots a_N$ to represent the binary fraction $a_1 \times 2^{-1} + a_2 \times 2^{-2} + a_3 \times 2^{-3} + \ldots + a_N \times 2^{-N}$. The following very useful product representation is to tensor product decomposition of the quantum Fourier transform:

$$\mathbf{QFT}(|a\rangle) = \mathbf{QFT}(|a_1 a_2 a_3 \ldots a_N\rangle)$$

$$= \left(\frac{|0\rangle + e^{\sqrt{-1} \times 2 \times \pi \times 0.a_N} |1\rangle}{\sqrt{2}} \right)$$

$$\otimes \left(\frac{|0\rangle + e^{\sqrt{-1} \times 2 \times \pi \times 0.a_{N-1} a_N} |1\rangle}{\sqrt{2}} \right)$$

$$\otimes \left(\frac{|0\rangle + e^{\sqrt{-1} \times 2 \times \pi \times 0.a_{N-2} a_{N-1} a_N} |1\rangle}{\sqrt{2}} \right)$$

$$\otimes \ldots \otimes \left(\frac{|0\rangle + e^{\sqrt{-1} \times 2 \times \pi \times 0.a_1 a_2 \ldots a_N} |1\rangle}{\sqrt{2}} \right) \tag{4.18}$$

The product representation (4.18) is actually to the *definition* of the quantum Fourier transform. We use the following lemma to show the equivalence of the product representation (4.18) and the definition (4.17) to the quantum Fourier transform.

Lemma 4.8 *The product representation* (4.18) *is equivalent to the definition* (4.17) *for the quantum Fourier transform and is to the* definition *of the quantum Fourier transform.*

Proof Because we take $n = 2^N$ to apply **QFT** (the quantum Fourier transform) to computational basis state $|a\rangle$ for $0 \leq a \leq (2^N - 1)$, the definition (4.17) of the quantum Fourier transform is as follows

$$\mathbf{QFT}(|a\rangle) = \left(\frac{1}{\sqrt{2^N}} \right) \times \left(\sum_{k=0}^{2^N-1} \omega_{2^N}^{k \times a} |k\rangle \right)$$

$$= \left(\frac{1}{\sqrt{2^N}} \right) \times \left(\sum_{k=0}^{2^N-1} e^{(\sqrt{-1} \times 2 \times \pi \times k \times a)/2^N} |k\rangle \right). \tag{4.19}$$

We write computational basis state $|k\rangle$ for $0 \le k \le (2^N - 1)$ using the binary representation $k = k_1 k_2 k_3 \ldots k_N = k_1 \times 2^{N-1} + k_2 \times 2^{N-2} + k_3 \times 2^{N-3} + \ldots + k_N \times 2^0$. We also write the notation $0. k_1 k_2 k_3 \ldots k_N$ to represent the binary fraction $k_1 \times 2^{-1} + k_2 \times 2^{-2} + k_3 \times 2^{-3} + \ldots + k_N \times 2^{-N}$. The computation of the division $k/2^N$ is to $k/2^N = k_1 k_2 k_3 \ldots k_N/2^N = (k_1 \times 2^{N-1} + k_2 \times 2^{N-2} + k_3 \times 2^{N-3} + \ldots + k_N \times 2^0)/2^N = k_1 \times 2^{-1} + k_2 \times 2^{-2} + k_3 \times 2^{-3} + \ldots + k_N \times 2^{-N} = \left(\sum_{l=1}^{N} k_l \times 2^{-l} \right)$. The computation of the sum $\sum_{k=0}^{2^N-1}$ is to complete the sum of 2^N items. The computation of the sum $\sum_{k_1=0}^{1} \cdots \sum_{k_N=0}^{1}$ also is to complete the sum of 2^N items. If they process the same items, then $\left(\sum_{k=0}^{2^N-1} \right) = \left(\sum_{k_1=0}^{1} \cdots \sum_{k_N=0}^{1} \right)$. Therefore, we rewrite the definition (4.19) of the quantum Fourier transform as follows

$$\mathbf{QFT}(|a\rangle) = \left(\frac{1}{\sqrt{2^N}} \right) \times \left(\sum_{k_1=0}^{1} \cdots \sum_{k_N=0}^{1} e^{\left(\sqrt{-1} \times 2 \times \pi \times a \times \left(\sum_{l=1}^{N} k_l \times 2^{-l} \right) \right)} |k_1 \cdots k_N\rangle \right). \tag{4.20}$$

The product representation $(\otimes_{l=1}^{N} e^{\left(\sqrt{-1} \times 2 \times \pi \times a \times k_l \times 2^{-l} \right)} |k_l\rangle)$ is equal to $\left(e^{\left(\sqrt{-1} \times 2 \times \pi \times a \times k_1 \times 2^{-1} \right)} |k_1\rangle \right) \otimes \left(e^{\left(\sqrt{-1} \times 2 \times \pi \times a \times k_2 \times 2^{-2} \right)} |k_2\rangle \right) \otimes \ldots \otimes \left(e^{\left(\sqrt{-1} \times 2 \times \pi \times a \times k_N \times 2^{-N} \right)} |k_N\rangle \right) =$

$\left(e^{\left(\sqrt{-1} \times 2 \times \pi \times a \times k_1 \times 2^{-1} \right)} \times \cdots \times e^{\left(\sqrt{-1} \times 2 \times \pi \times a \times k_N \times 2^{-N} \right)} |k_1 \cdots k_N\rangle \right) =$

$\left(e^{\left(\sqrt{-1} \times 2 \times \pi \times a \times (k_1 \times 2^{-1} + \cdots + k_N \times 2^{-N}) \right)} |k_1 \cdots k_N\rangle \right)$

$$= \left(e^{\left(\sqrt{-1} \times 2 \times \pi \times a \times \left(\sum_{l=1}^{N} k_l \times 2^{-l} \right) \right)} |k_1 \cdots k_N\rangle \right).$$ Hence, we rewrite the definition (4.20)

of the quantum Fourier transform as follows

$$\mathbf{QFT}(|a\rangle) = \left(\frac{1}{\sqrt{2^N}} \right) \times \left(\sum_{k_1=0}^{1} \cdots \sum_{k_N=0}^{1} \otimes_{l=1}^{N} e^{\left(\sqrt{-1} \times 2 \times \pi \times a \times k_l \times 2^{-l} \right)} |k_l\rangle \right) \tag{4.21}$$

$$= \left(\frac{1}{\sqrt{2^N}} \right) \times \left(\otimes_{l=1}^{N} \left(\sum_{k_l=0}^{1} e^{\left(\sqrt{-1} \times 2 \times \pi \times a \times k_l \times 2^{-l} \right)} |k_l\rangle \right) \right) \tag{4.22}$$

$$= \left(\frac{1}{\sqrt{2^N}} \right) \times \left(\otimes_{l=1}^{N} \left(e^{\left(\sqrt{-1} \times 2 \times \pi \times a \times 0 \times 2^{-l} \right)} |0\rangle + e^{\left(\sqrt{-1} \times 2 \times \pi \times a \times 1 \times 2^{-l} \right)} |1\rangle \right) \right) \tag{4.23}$$

$$= \left(\frac{1}{\sqrt{2^N}}\right) \times \left(\otimes_{l=1}^{N}\left(|0\rangle + e^{(\sqrt{-1}\times 2\times\pi\times a\times 1\times 2^{-l})}|1\rangle\right)\right). \tag{4.24}$$

The computation of the division $(a \times 2^{-l} = a/2^l)$ for $1 \le l \le N$ is to complete the left shift of l position to the decimal point ".". This is to say that for $1 \le l \le N$ we obtain $a \times 2^{-l} = a/2^l = a_1 \cdots$ $a_{N-l}.a_{N-l+1} \; a_{N-l+2} \; \cdots \; a_N$. Therefore, we obtain $e^{(\sqrt{-1}\times 2\times\pi\times a\times 1\times 2^{-l})}$ $= \; e^{(\sqrt{-1}\times 2\times\pi\times a_1\cdots a_{N-l}.a_{N-l+1}\cdots a_N)} \; = \; e^{(\sqrt{-1}\times 2\times\pi\times(a_1\cdots a_{N-l}+0.a_{N-l+1}\cdots a_N))}$ $= \; \left(e^{(\sqrt{-1}\times 2\times\pi\times(a_1\cdots a_{N-l}))} \times e^{(\sqrt{-1}\times 2\times\pi\times(0.a_{N-l+1}\cdots a_N))}\right) \; =$ $\left(1 \times e^{(\sqrt{-1}\times 2\times\pi\times(0.a_{N-l+1}\cdots a_N))}\right) = \left(e^{(\sqrt{-1}\times 2\times\pi\times(0.a_{N-l+1}\cdots a_N))}\right)$. Next, we rewrite the definition (4.24) of the quantum Fourier transform as follows

$$\mathbf{QFT}(|a\rangle) = \mathbf{QFT}(|a_1 a_2 a_3 \ldots a_N\rangle)$$

$$= \left(\frac{|0\rangle + e^{\sqrt{-1}\times 2\times\pi\times 0.a_N}|1\rangle}{\sqrt{2}}\right) \otimes \left(\frac{|0\rangle + e^{\sqrt{-1}\times 2\times\pi\times 0.a_{N-1}a_N}|1\rangle}{\sqrt{2}}\right)$$

$$\otimes \left(\frac{|0\rangle + e^{\sqrt{-1}\times 2\times\pi\times 0.a_{N-2}a_{N-1}a_N}|1\rangle}{\sqrt{2}}\right)$$

$$\otimes \ldots \otimes \left(\frac{|0\rangle + e^{\sqrt{-1}\times 2\times\pi\times 0.a_1 a_2\cdots a_N}|1\rangle}{\sqrt{2}}\right). \tag{4.25}$$

Therefore, from the statements above we at once infer that the product representation (4.18) is equivalent to the definition (4.17) for the quantum Fourier transform and is to the *definition* of the quantum Fourier transform. ■

4.6 Quantum Circuits of Implementing the Quantum Fourier Transform

"If C is true, then do D". This type of controlled operations is one of the most useful in deriving a very efficient circuit for the quantum Fourier transform. A (2×2) matrix $R_{k,c}$ is as follows

$$R_{k,c} = \begin{bmatrix} 1 & 0 \\ 0 & e^{\sqrt{-1}\times 2\times\pi/2^k} \end{bmatrix} = \begin{bmatrix} 1 & 0 \\ 0 & \omega_{2^k}^1 \end{bmatrix}. \tag{4.26}$$

A (2×2) matrix $\overline{R_{k,c}}$ that is the conjugate transpose of the (2×2) matrix $R_{k,c}$ is as follows

$$\overline{R_{k,c}} = \begin{bmatrix} 1 & 0 \\ 0 & e^{\sqrt{-1}\times 2\times \pi/2^k} \end{bmatrix} = \begin{bmatrix} 1 & 0 \\ 0 & e^{-\sqrt{-1}\times 2\times \pi/2^k} \end{bmatrix} = \begin{bmatrix} 1 & 0 \\ 0 & \omega_{2^2}^{-1} \end{bmatrix}. \qquad (4.27)$$

Because $R_{k,c} \times \overline{R_{k,c}} = (\begin{bmatrix} 1 & 0 \\ 0 & \omega_{2^k}^1 \end{bmatrix} \times \begin{bmatrix} 1 & 0 \\ 0 & \omega_{2^k}^{-1} \end{bmatrix}) = (\begin{bmatrix} 1 & 0 \\ 0 & \omega_{2^k}^1 \times \omega_{2^k}^{-1} \end{bmatrix}) =$

$(\begin{bmatrix} 1 & 0 \\ 0 & \omega_{2^k}^{1-1} \end{bmatrix}) = (\begin{bmatrix} 1 & 0 \\ 0 & \omega_{2^k}^0 \end{bmatrix}) = (\begin{bmatrix} 1 & 0 \\ 0 & 1 \end{bmatrix})$ and $\overline{R_{k,c}} \times R_{k,c} = (\begin{bmatrix} 1 & 0 \\ 0 & \omega_{2^k}^{-1} \end{bmatrix} \times \begin{bmatrix} 1 & 0 \\ 0 & \omega_{2^k}^1 \end{bmatrix}) =$

$(\begin{bmatrix} 1 & 0 \\ 0 & \omega_{2^k}^{-1} \times \omega_{2^k}^1 \end{bmatrix}) = (\begin{bmatrix} 1 & 0 \\ 0 & \omega_{2^k}^{-1+1} \end{bmatrix}) = (\begin{bmatrix} 1 & 0 \\ 0 & \omega_{2^k}^0 \end{bmatrix}) = (\begin{bmatrix} 1 & 0 \\ 0 & 1 \end{bmatrix})$, matrix $R_{k,c}$ is a unitary

matrix (a unitary operator) and matrix $\overline{R_{k,c}}$ is a unitary matrix (a unitary operator). A
controlled-$R_{k,c}$ operation is an operation of two quantum bits that contain a control
bit and a target bit. If the control bit is set to $|1\rangle$, then matrix $R_{k,c}$ is applied to the
target bit. Otherwise, matrix $R_{k,c}$ does not change the target bit.

Similarly, a *controlled-$\overline{R_{k,c}}$* operation is an operation of two quantum bits that
include a control bit and a target bit. If the control bit is set to $|1\rangle$, then matrix $\overline{R_{k,c}}$
is applied to the target bit. Otherwise, matrix $\overline{R_{k,c}}$ does not change the target bit.
The matrix representation of the *controlled-$R_{k,c}$* operation and the *controlled-$\overline{R_{k,c}}$*
operation is

$$CR_{k,c} = \begin{bmatrix} 1 & 0 & 0 & 0 \\ 0 & 1 & 0 & 0 \\ 0 & 0 & 1 & 0 \\ 0 & 0 & 0 & e^{\sqrt{-1}\times 2\times\pi/2^k} \end{bmatrix} \text{ and } C\overline{R_{k,c}} = \begin{bmatrix} 1 & 0 & 0 & 0 \\ 0 & 1 & 0 & 0 \\ 0 & 0 & 1 & 0 \\ 0 & 0 & 0 & e^{-\sqrt{-1}\times 2\times\pi/2^k} \end{bmatrix}. \qquad (4.28)$$

Matrix $C\overline{R_{k,c}}$ in Eq. (4.28) is the conjugate transpose of matrix $CR_{k,c}$ in Eq. (4.28).

Because $CR_{k,c} \times C\overline{R_{k,c}} = \begin{bmatrix} 1 & 0 & 0 & 0 \\ 0 & 1 & 0 & 0 \\ 0 & 0 & 1 & 0 \\ 0 & 0 & 0 & e^{\sqrt{-1}\times 2\times\pi/2^k\times(1-1)} \end{bmatrix} = \begin{bmatrix} 1 & 0 & 0 & 0 \\ 0 & 1 & 0 & 0 \\ 0 & 0 & 1 & 0 \\ 0 & 0 & 0 & 1 \end{bmatrix}$ and $C\overline{R_{k,c}}$

$\times CR_{k,c} = \begin{bmatrix} 1 & 0 & 0 & 0 \\ 0 & 1 & 0 & 0 \\ 0 & 0 & 1 & 0 \\ 0 & 0 & 0 & e^{\sqrt{-1}\times 2\times\pi/2^k\times(-1+1)} \end{bmatrix} = \begin{bmatrix} 1 & 0 & 0 & 0 \\ 0 & 1 & 0 & 0 \\ 0 & 0 & 1 & 0 \\ 0 & 0 & 0 & 1 \end{bmatrix}$, $CR_{k,c} \times C\overline{R_{k,c}}$ is an

identity matrix and $C\overline{R_{k,c}} \times CR_{k,c}$ also is an identity matrix. Therefore, $CR_{k,c} \times C_i\overline{R_{k,c}} = C\overline{R_{k,c}} \times CR_{k,c}$ and matrix $CR_{k,c}$ and matrix $C\overline{R_{k,c}}$ are both a unitary matrix
(a unitary operator). Figure 4.2 shows the circuit representation of the *controlled-$R_{k,c}$*
and Fig. 4.3 shows the circuit representation of the *controlled-$\overline{R_{k,c}}$* operation.

A (4×4) matrix *swap* and its conjugate transpose $\overline{swap}$ are respectively

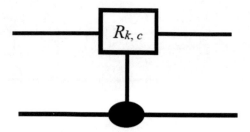

Fig. 4.2 Controlled-$R_{k,\,c}$ operation with that the bottom line is the control bit and the top line is the target bit

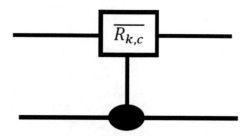

Fig. 4.3 Controlled-$\overline{R_{k,c}}$ operation with that the bottom line is the control bit and the top line is the target bit

$$swap = \begin{bmatrix} 1 & 0 & 0 & 0 \\ 0 & 0 & 1 & 0 \\ 0 & 1 & 0 & 0 \\ 0 & 0 & 0 & 1 \end{bmatrix} \text{ and } \overline{swap} = \begin{bmatrix} 1 & 0 & 0 & 0 \\ 0 & 0 & 1 & 0 \\ 0 & 1 & 0 & 0 \\ 0 & 0 & 0 & 1 \end{bmatrix}. \tag{4.29}$$

Because $swap \times \overline{swap} = \begin{bmatrix} 1 & 0 & 0 & 0 \\ 0 & 1 & 0 & 0 \\ 0 & 0 & 1 & 0 \\ 0 & 0 & 0 & 1 \end{bmatrix} = \overline{swap} \times swap$, matrix $swap$ and matrix

$\overline{swap}$ are both a unitary matrix (a unitary operator). Matrix $swap$ is to matrix representation of a swap gate of two quantum bits. The functionality of the swap gate is to exchange the information contained in two quantum bits. The left picture in Fig. 4.4 is the circuit representation of a swap gate with two quantum bits and the right picture in Fig. 4.4 is to the circuit representation of implementing the swap gate by means of using three **CNOT** gates.

An initial state vector is $|\beta_0\rangle = |A\rangle \otimes |B\rangle$ for the left circuit and the right circuit in Fig. 4.4, where $A \in \{0, 1\}$ and $B \in \{0, 1\}$. The output of the swap gate in the left circuit of Fig. 4.4 is $|\beta_1\rangle = |B\rangle \otimes |A\rangle$. The output of the first **CNOT** gate with the control bit $|A\rangle$ and the target bit $|B\rangle$ in the *right* circuit of Fig. 4.4 is $|\theta_1\rangle = |A\rangle \otimes |A \oplus B\rangle$. Next, the output of the second **CNOT** gate with the control bit $|A \oplus B\rangle$ and

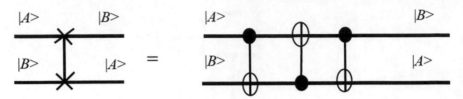

Fig. 4.4 Circuit Representation of a swap gate with two quantum bits

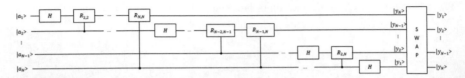

Fig. 4.5 Efficient circuit of implementing the quantum Fourier transform

the target bit $|A\rangle$ in the *right* circuit of Fig. 4.4 is $|\theta_2\rangle = |A \oplus (A \oplus B)\rangle \otimes |A \oplus B\rangle = |A \oplus A \oplus B\rangle \otimes |A \oplus B\rangle = |0 \oplus B\rangle \otimes |A \oplus B\rangle = |B\rangle \otimes |A \oplus B\rangle$. Finally, the output of the third **CNOT** gate with the control bit $|B\rangle$ and the target bit $|A \oplus B\rangle$ in the *right* circuit of Fig. 4.4 is $|\theta_3\rangle = |B\rangle \otimes |(A \oplus B) \oplus B\rangle = |B\rangle \otimes |A \oplus B \oplus B\rangle = |B\rangle \otimes |A \oplus 0\rangle = |B\rangle \otimes |A\rangle$. Because the two state vectors $|\beta_1\rangle$ and $|\theta_3\rangle$ are the same, the left circuit of Fig. 4.4 is equivalent to the right circuit of Fig. 4.4.

The product representation (4.18) makes it easy to infer a very efficient circuit of implementing the quantum Fourier transform. Such a circuit appears in Fig. 4.5. Consider what happens when the state $|a_1 \cdots a_N\rangle$ is an input for the pictured circuit in Fig. 4.5 that computes the quantum Fourier transform. Making use of the Hadamard gate to the first quantum bit generates the following state

$$\frac{1}{\sqrt{2}}\left(|0\rangle + e^{\sqrt{-1} \times 2 \times \pi \times 0.a_1}|1\rangle\right)|a_2 \cdots a_N\rangle \tag{4.30}$$

When $a_1 = 1$, we obtain $e^{\sqrt{-1} \times 2 \times \pi \times 0.a_1} = e^{\sqrt{-1} \times 2 \times \pi \times 0.1} = e^{\sqrt{-1} \times 2 \times \pi \times \left(\frac{1}{2}\right)} = e^{\sqrt{-1} \times \pi} = -1$. When $a_1 = 0$, we obtain $e^{\sqrt{-1} \times 2 \times \pi \times 0.a_1} = e^{\sqrt{-1} \times 2 \times \pi \times 0.0} = 1$. This satisfies the functionality of the Hadamard gate.

Next, using the controlled-$R_{2,2}$ gate produces the following state

$$\frac{1}{\sqrt{2}}\left(|0\rangle + e^{\sqrt{-1} \times 2 \times \pi \times 0.a_1} \times e^{\sqrt{-1} \times 2 \times \pi \times 0.0a_2}|1\rangle\right)|a_2 \cdots a_N\rangle$$

$$= \frac{1}{\sqrt{2}}\left(|0\rangle + e^{\sqrt{-1} \times 2 \times \pi \times (0.a_1 + 0.0a_2)}|1\rangle\right)|a_2 \cdots a_N\rangle$$

$$= \frac{1}{\sqrt{2}}\left(|0\rangle + e^{\sqrt{-1} \times 2 \times \pi \times 0.a_1a_2}|1\rangle\right)|a_2 \cdots a_N\rangle. \tag{4.31}$$

When $a_2 = 1$ is equivalent to the control bit $|a_2\rangle$ to be $|1\rangle$, it adds the phase $e^{\sqrt{-1}\times 2\times\pi\times(\frac{1}{4})} = e^{\sqrt{-1}\times 2\times\pi\times 0.01} = e^{\sqrt{-1}\times 2\times\pi\times 0.0a_2}$ to the coefficient of the first $|1\rangle$. When $a_2 = 0$ is equivalent to the control bit $|a_2\rangle$ to be $|0\rangle$, it does not change the coefficient of the first $|1\rangle$ because $e^{\sqrt{-1}\times 2\times\pi\times 0.0a_2} = e^{\sqrt{-1}\times 2\times\pi\times 0.00} = 1$. This satisfies the functionality of the controlled-$R_{2,2}$ gate.

Next, applying the controlled-$R_{3,3}$ gate generates the following state

$$\frac{1}{\sqrt{2}}\left(|0\rangle + e^{\sqrt{-1}\times 2\times\pi\times 0.a_1 a_2} \times e^{\sqrt{-1}\times 2\times\pi\times 0.00a_3}|1\rangle\right)|a_2\cdots a_N\rangle$$

$$= \frac{1}{\sqrt{2}}\left(|0\rangle + e^{\sqrt{-1}\times 2\times\pi\times(0.a_1 a_2 + 0.00a_3)}|1\rangle\right)|a_2\cdots a_N\rangle$$

$$= \frac{1}{\sqrt{2}}\left(|0\rangle + e^{\sqrt{-1}\times 2\times\pi\times 0.a_1 a_2 a_3}|1\rangle\right)|a_2\cdots a_N\rangle \tag{4.32}$$

When $a_3 = 1$ is equivalent to the control bit $|a_3\rangle$ to be $|1\rangle$, it adds the phase $e^{\sqrt{-1}\times 2\times\pi\times(\frac{1}{8})} = e^{\sqrt{-1}\times 2\times\pi\times 0.001} = e^{\sqrt{-1}\times 2\times\pi\times 0.00a_3}$ to the coefficient of the first $|1\rangle$. When $a_3 = 0$ is equivalent to the control bit $|a_3\rangle$ to be $|0\rangle$, it does not change the coefficient of the first $|1\rangle$ because $e^{\sqrt{-1}\times 2\times\pi\times 0.00a_3} = e^{\sqrt{-1}\times 2\times\pi\times 0.000} = 1$. This satisfies the functionality of the controlled-$R_{3,3}$ gate.

We continue making use of the controlled-$R_{4,4}$, $R_{5,5}$ through $R_{N,N}$ gates, they subsequently add the phase $(e^{\sqrt{-1}\times 2\times\pi\times 0.000a_4})$, the phase $(e^{\sqrt{-1}\times 2\times\pi\times 0.0000a_5})$ through the phase $(e^{\sqrt{-1}\times 2\times\pi\times 0.00\cdots a_N})$ to the coefficient of the first $|1\rangle$. At the end of this procedure, we obtain the following state

$$\frac{1}{\sqrt{2}}\left(\begin{array}{l}|0\rangle + e^{\sqrt{-1}\times 2\times\pi\times 0.a_1 a_2 a_3} \times e^{\sqrt{-1}\times 2\times\pi\times 0.000a_4} \\ \times e^{\sqrt{-1}\times 2\times\pi\times 0.0000a_5} \times \cdots \times e^{\sqrt{-1}\times 2\times\pi\times 0.00\cdots a_N}|1\rangle\end{array}\right)|a_2\cdots a_N\rangle$$

$$= \frac{1}{\sqrt{2}}\left(|0\rangle + e^{\sqrt{-1}\times 2\times\pi\times(0.a_1 a_2 a_3 + 0.000a_4 + 0.0000a_5 + 0.00\cdots a_N)}|1\rangle\right)|a_2\cdots a_N\rangle$$

$$= \frac{1}{\sqrt{2}}\left(|0\rangle + e^{\sqrt{-1}\times 2\times\pi\times 0.a_1 a_2 a_3 a_4 a_5 \cdots a_N}|1\rangle\right)|a_2\cdots a_N\rangle \tag{4.33}$$

Next, we complete a similar procedure on the second quantum bit. Applying the Hadamard gate to the second quantum bit produces the following state

$$\left(\frac{|0\rangle + e^{\sqrt{-1}\times 2\times\pi\times 0.a_1 a_2\cdots a_N}|1\rangle}{\sqrt{2}}\right)\left(\frac{|0\rangle + e^{\sqrt{-1}\times 2\times\pi\times 0.a_2}|1\rangle}{\sqrt{2}}\right)|a_3\cdots a_N\rangle. \tag{4.34}$$

Next, applying the controlled-$R_{2,3}$, $R_{3,4}$ through $R_{N-1,N}$ gates produces the following state

$$\left(\frac{|0\rangle + e^{\sqrt{-1}\times 2\times\pi\times 0.a_1 a_2\cdots a_N}|1\rangle}{\sqrt{2}}\right)\left(\frac{|0 + e^{\sqrt{-1}\times 2\times\pi\times 0.a_2\cdots a_N}|1\rangle}{\sqrt{2}}\right)|a_3\cdots a_N\rangle.$$

$$\tag{4.35}$$

We continue in this fashion for each quantum bit, giving a final state

$$\left(\frac{|0\rangle + e^{\sqrt{-1}\times 2\times\pi\times 0.a_1 a_2 \cdots a_N}|1\rangle}{\sqrt{2}}\right)\left(\frac{|0\rangle + e^{\sqrt{-1}\times 2\times\pi\times 0.a_2 \cdots a_N}|1\rangle}{\sqrt{2}}\right)$$
$$\cdots \left(\frac{|0\rangle + e^{\sqrt{-1}\times 2\times\pi\times 0.a_N}|1\rangle}{\sqrt{2}}\right). \tag{4.36}$$

Next, applying swap operations reverses the order of the quantum bits. After the swap operations, the state of the quantum bits is

$$\left(\frac{\left|0\right\rangle + e^{\sqrt{-1}\times 2\times\pi\times 0.a_N}|1\rangle}{\sqrt{2}}\right)\otimes\left(\frac{\left|0\right\rangle + e^{\sqrt{-1}\times 2\times\pi\times 0.a_{N-1}a_N}|1\rangle}{\sqrt{2}}\right)$$
$$\otimes\left(\frac{\left|0\right\rangle + e^{\sqrt{-1}\times 2\times\pi\times 0.a_{N-2}a_{N-1}a_N}|1\rangle}{\sqrt{2}}\right)\otimes\cdots\otimes\left(\frac{\left|0\right\rangle + e^{\sqrt{-1}\times 2\times\pi\times 0.a_1 a_2 \cdots a_N}|1\rangle}{\sqrt{2}}\right)$$
$$\tag{4.37}$$

Comparing with Eq. (4.18) and Eq. (4.37), we look at the desired output from the quantum Fourier transform. This construction also demonstrates that the quantum Fourier transform is unitary because each gate in the circuit of Fig. 4.5 is unitary.

4.7 Assessment of Time Complexity for Implementing the Quantum Fourier Transform

How many gates does the circuit of implementing the quantum Fourier transform in Fig. 4.5 use? We begin by means of doing a Hadamard gate and $(n - 1)$ conditional rotations on the first quantum bit. This requires a total of n gates. Next step (the second step) is to do a Hadamard gate and $(n - 2)$ conditional rotations on the second quantum bit. This requires a total of $n + (n - 1)$ gates. Next step (the third step) is to do a Hadamard gate and $(n - 3)$ conditional rotations on the third quantum bit. This requires a total of $n + (n - 1) + (n - 2)$ gates. Continuing in this way, this requires a total of $n + (n - 1) + (n - 2) + \ldots + 1 = n \times (n + 1)/2$ gates, plus at most $n/2$ swap gate s that reverse the order of the quantum bits. Because each swap gate can be implemented by means of using three controlled-NOT gates, this circuit in Fig. 4.5 provides a $O(n^2)$ algorithm for completing the quantum Fourier transform.

The fast Fourier transform is the best classical algorithm to compute the discrete Fourier transform on 2^n elements applying $O(n \times 2^n)$ gates. This is to say that it requires exponentially more operations to compute the discrete Fourier transform on a digital (classical) computer than it does to implement the quantum Fourier transform on a quantum computer. The Fourier transform is a critical step for many

applications in the real world. Computer speech recognition is one of the most important applications to the Fourier transform. The first step in phoneme recognition of
computer speech recognition is to Fourier transform the digitized sound. Can we
make use of the quantum Fourier transform to enhance the performance of these
Fourier transforms? Unfortunately, there is no known way to complete this. Because
measurement in a quantum computer cannot directly access the amplitudes, there is
no way of computing the Fourier transformed amplitudes of the original state. Worse
still, a problem is to that there is in general no way to efficiently prepare the original
state to be Fourier transformed.

4.8 The Inverse Quantum Fourier Transform of Implementing the Inverse Discrete Fourier Transform

The matrix **IQFT** of the inverse quantum Fourier transform is the same as the matrix
IDFT of the inverse discrete Fourier transform in Eq. (4.11). Because **IQFT** and
IDFT are the same matrix and **IDFT** is a unitary matrix (a unitary operator) from
Lemma 4.7, **IQFT** is also a unitary matrix (a unitary operator). The inverse quantum
Fourier transform is a computationally very efficient implementation to the inverse
discrete Fourier transform in the form of tracing back **IQFT** to its tensor product
decomposition. In the following, we take $n = 2^N$, where N is the number of quantum
bit in a quantum computer and the basis $\{|0\rangle, |1\rangle, |2\rangle, ..., |2^N - 1\rangle\}$ is the computa-
tional basis for the quantum computer. We mention here that $\left(\omega_n^{-k \times a}\right) = \left(\omega_{2^N}^{-k \times a}\right) = \left(e^{\left(-\sqrt{-1} \times 2 \times \pi \times k \times a\right)/2^N}\right)$.

Applying **IQFT** (the inverse quantum Fourier transform) to computational basis
state $|a\rangle$ for $0 \leq a \leq (2^N - 1)$ yields

$$\mathbf{IQFT} \times (|a\rangle) = \mathbf{IQFT}(|a\rangle)$$

$$= \left(\frac{1}{\sqrt{2^N}}\right) \times \left(\sum_{k=0}^{2^N-1} \omega_{2^N}^{-k \times a}|k\rangle\right)$$

$$= \left(\frac{1}{\sqrt{2^N}}\right) \times \left(\sum_{k=0}^{2^N-1} e^{\left(-\sqrt{-1} \times 2 \times \pi \times k \times a\right)/2^N}|k\rangle\right). \tag{4.38}$$

It is very useful to write computational basis state $|a\rangle$ for $0 \leq a \leq (2^N - 1)$ using
the binary representation $a = a_1 a_2 a_3 ... a_N = a_1 \times 2^{N-1} + a_2 \times 2^{N-2} + a_3 \times 2^{N-3} + ... + a_N \times 2^0$. It is also very useful to write the notation $0. a_1 a_2 a_3 ... a_N$
to represent the binary fraction $a_1 \times 2^{-1} + a_2 \times 2^{-2} + a_3 \times 2^{-3} + ... + a_N \times 2^{-N}$.
The following very useful product representation is to tensor product decomposition
of the inverse quantum Fourier transform:

$$\mathbf{IQFT}(|a\rangle) = \mathbf{IQFT}\big(|a_1 \ a_2 \ a_3 \ \dots \ a_N\rangle\big)$$

$$= \left(\frac{|0\rangle + e^{-\sqrt{-1}\times 2\times\pi\times 0.a_N}\,|1\rangle}{\sqrt{2}}\right) \otimes \left(\frac{|0\rangle + e^{-\sqrt{-1}\times 2\times n\times 0.a_{N-1}a_N}\,|1\rangle}{\sqrt{2}}\right)$$

$$\otimes \left(\frac{|0\rangle + e^{-\sqrt{-1}\times 2\times\pi\times 0.a_{N-2}a_{N-1}a_N}\,|1\rangle}{\sqrt{2}}\right)$$

$$\otimes \dots \otimes \left(\frac{|0\rangle + e^{-\sqrt{-1}\times 2\times\pi\times 0.a_1 a_2\cdots a_N}\,|1\rangle}{\sqrt{2}}\right) \tag{4.39}$$

The product representation (4.39) is actually to the *definition* of the inverse quantum Fourier transform. We make use of the following lemma to demonstrate the equivalence of the product representation (4.39) and the definition (4.38) to the inverse quantum Fourier transform.

Lemma 4.9 *The product representation* (4.39) *is equivalent to the definition* (4.38) *for the inverse quantum Fourier transform and is to the* definition *of the inverse quantum Fourier transform.*

Proof Because we take $n = 2^N$ to use **IQFT** (the inverse quantum Fourier transform) to computational basis state $|a\rangle$ for $0 \le a \le (2^N - 1)$, the definition (4.38) of the inverse quantum Fourier transform is

$$\mathbf{IQFT}(|a\rangle) = \left(\frac{1}{\sqrt{2^N}}\right)\left(\sum_{k=0}^{2^N-1} \omega_{2^N}^{-k\times a}\,|k\rangle\right) = \left(\frac{1}{\sqrt{2^N}}\right)\left(\sum_{k=0}^{2^N-1} e^{-\sqrt{-1}\times 2\times\pi\times k\times a)/2^N}\,|k\rangle\right) \tag{4.40}$$

We write computational basis state $|k\rangle$ for $0 \le k \le (2^N - 1)$ applying the binary representation $k = k_1 \ k_2 \ k_3 \ \dots \ k_N = k_1 \times 2^{N-1} + k_2 \times 2^{N-2} + k_3 \times 2^{N-3} + \dots + k_N \times 2^0$. We also write the notation $0.\,k_1 \ k_2 \ k_3 \ \dots \ k_N$ to represent the binary fraction $k_1 \times 2^{-1} + k_2 \times 2^{-2} + k_3 \times 2^{-3} + \dots + k_N \times 2^{-N}$. The computation of the division $k/2^N$ is to $k/2^N = k_1 \ k_2 \ k_3 \ \dots \ k_N/2^N = (k_1 \times 2^{N-1} + k_2 \times 2^{N-2} + k_3 \times 2^{N-3} + \dots + k_N \times 2^0)/2^N = k_1 \times 2^{-1} + k_2 \times 2^{-2} + k_3 \times 2^{-3} + \dots + k_N \times 2^{-N} = \left(\sum_{l=1}^{N} k_l \times 2^{-l}\right)$. The computation of the sum $\sum_{k=0}^{2^N-1}$ is to complete the sum of 2^N items. The computation of the sum $\sum_{k_1=0}^{1} \cdots \sum_{k_N=0}^{1}$ also is to complete the sum of 2^N items. If they process the same items, then $\left(\sum_{k=0}^{2^N-1}\right) = \left(\sum_{k_1=0}^{1} \cdots \sum_{k_N=0}^{1}\right)$. Hence, we rewrite the definition (4.40) of the inverse quantum Fourier transform as follows

$$\mathbf{IQFT}(|a\rangle) = \left(\frac{1}{\sqrt{2^N}}\right) \times \left(\sum_{k_1=0}^{1} \cdots \sum_{k_N=0}^{1} e^{\left(-\sqrt{-1}\times 2\times\pi\times a\times\left(\sum_{l=1}^{N} k_l\times 2^{-l}\right)\right)} |k_1 \ldots k_N\rangle\right)$$

(4.41)

The product representation $\left(\otimes_{l=1}^{N} e^{\left(-\sqrt{-1}\times 2\times\pi\times a\times k_l\times 2^{-l}\right)} |k_l\rangle\right)$ is equal to $\left(e^{\left(-\sqrt{-1}\times 2\times\pi\times a\times k_1\times 2^{-1}\right)} |k_1\rangle\right) \otimes \left(e^{\left(-\sqrt{-1}\times 2\times\pi\times a\times k_2\times 2^{-2}\right)} |k_2\rangle\right) \otimes \cdots \otimes$

$\left(e^{\left(-\sqrt{-1}\times 2\times\pi\times a\times k_N\times 2^{-N}\right)} |k_N\rangle\right) =$

$\left(e^{\left(-\sqrt{-1}\times 2\times\pi\times a\times k_1\times 2^{-1}\right)} \times \cdots \times e^{\left(-\sqrt{-1}\times 2\times\pi\times a\times k_N\times 2^{-N}\right)} |k_1 \cdots k_N\rangle\right) =$

$\left(e^{\left(-\sqrt{-1}\times 2\times\pi\times a\times\left(k_1\times 2^{-1}+\cdots+k_N\times 2^{-N}\right)\right)} |k_1 \cdots k_N\rangle\right) =$

$\left(e^{\left(-\sqrt{-1}\times 2\times\pi\times a\times\left(\sum_{l=1}^{N} k_l\times 2^{-l}\right)\right)} |k_1 \cdots k_N\rangle\right)$. Therefore, we rewrite the definition

(4.41) of the inverse quantum Fourier transform as follows

$$\mathbf{IQFT}(|a\rangle) = \left(\frac{1}{\sqrt{2^N}}\right) \times \left(\sum_{k_1=0}^{1} \cdots \sum_{k_N=0}^{1} \otimes_{l=1}^{N} e^{\left(-\sqrt{-1}\times 2\times\pi\times a\times k_l\times 2^{-l}\right)} |k_l\rangle\right) \quad (4.42)$$

$$= \left(\frac{1}{\sqrt{2^N}}\right) \times \left(\otimes_{l=1}^{N} \left(\sum_{k_l=0}^{1} e^{\left(-\sqrt{-1}\times 2\times\pi\times a\times k_l\times 2^{-l}\right)} |k_l\rangle\right)\right) \quad (4.43)$$

$$= \left(\frac{1}{\sqrt{2^N}}\right) \times \left(\otimes_{l=1}^{N} \left(e^{\left(-\sqrt{-1}\times 2\times\pi\times a\times 0\times 2^{-l}\right)} |0\rangle + e^{\left(-\sqrt{-1}\times 2\times\pi\times a\times 1\times 2^{-l}\right)} |1\rangle\right)\right) \quad (4.44)$$

$$= \left(\frac{1}{\sqrt{2^N}}\right) \times \left(\otimes_{l=1}^{N} \left(|0\rangle + e^{\left(-\sqrt{-1}\times 2\times\pi\times a\times 1\times 2^{-l}\right)} |1\rangle\right)\right). \quad (4.45)$$

The computation of the division ($a \times 2^{-l} = a/2^l$) for $1 \leq l \leq N$ is to complete the left shift of l position to the decimal point ".". This indicates that for $1 \leq l \leq N$ we obtain $a \times 2^{-l} = a/2^l = a_1$ $\cdots a_{N-l}\cdot a_{N-l+1} a_{N-l+2} \cdots a_N$. Thus, we obtain $e^{\left(-\sqrt{-1}\times 2\times\pi\times a\times 1\times 2^{-l}\right)}$ $= e^{\left(-\sqrt{-1}\times 2\times\pi\times a_1\cdots a_{N-l}.a_{N-l+1}\cdots a_N\right)} = e^{\left(-\sqrt{-1}\times 2\times\pi\times (a_1\cdots a_{N-l}+0.a_{N-l+1}\cdots a_N)\right)}$ $= \left(e^{\left(-\sqrt{-1}\times 2\times\pi\times (a_1\cdots a_{N-l})\right)} \times e^{\left(-\sqrt{-1}\times 2\times\pi\times (0.a_{N-l+1}\cdots a_N)\right)}\right) = (1 \times$ $e^{\left(-\sqrt{-1}\times 2\times\pi\times (0.a_{N-l+1}\cdots a_N)\right)}) = \left(e^{\left(-\sqrt{-1}\times 2\times\pi\times (0.a_{N-l+1}\cdots a_N)\right)}\right)$. Next, we rewrite the definition (4.45) of the inverse quantum Fourier transform as follows

$$\mathbf{IQFT}(|a\rangle) = \mathbf{IQFT}(|a_1 a_2 a_3 \ldots a_N\rangle)$$

$$= \left(\frac{|0\rangle + e^{-\sqrt{-1}\times 2\times\pi\times 0.a_N} |1\rangle}{\sqrt{2}}\right) \otimes \left(\frac{|0\rangle + e^{-\sqrt{-1}\times 2\times\pi\times 0.a_{N-1} a_N} |1\rangle}{\sqrt{2}}\right)$$

$$\otimes \left(\frac{\left| 0 \right\rangle + e^{-\sqrt{-1} \times 2 \times \pi \times 0.a_{N-2} a_{N-1} a_N} \left| 1 \right\rangle}{\sqrt{2}} \right)$$

$$\otimes \cdots \otimes \left(\frac{\left| 0 \right\rangle + e^{-\sqrt{-1} \times 2 \times \pi \times 0.a_1 a_2 \ldots a_N} \left| 1 \right\rangle}{\sqrt{2}} \right) \tag{4.46}$$

Therefore, from the statements above we at once derive that the product representation (4.39) is equivalent to the definition (4.38) for the inverse quantum Fourier transform and is to the *definition* of the inverse quantum Fourier transform. ∎

4.9   Quantum Circuits of Implementing the Inverse Quantum Fourier Transform

The product representation (4.39) makes it easy to design a very efficient circuit of implementing the inverse quantum Fourier transform. Such a circuit appears in Fig. 4.6. Consider what happens when the state $|a_1 \cdots a_N\rangle$ is an input for the pictured circuit in Fig. 4.6 that computes the inverse quantum Fourier transform. Applying the Hadamard gate to the first quantum bit yields the following state

$$\frac{1}{\sqrt{2}} \left(|0 + e^{-\sqrt{-1} \times 2 \times \pi \times 0.a_1} |1\rangle \right) |a_2 \cdots a_N\rangle \tag{4.47}$$

When $a_1 = 1$, we obtain $e^{-\sqrt{-1} \times 2 \times \pi \times 0.a_1} = e^{-\sqrt{-1} \times 2 \times \pi \times 0.1} = e^{-\sqrt{-1} \times 2 \times \pi \times \left(\frac{1}{2}\right)} = e^{-\sqrt{-1} \times \pi} = -1$. When $a_1 = 0$, we obtain $e^{-\sqrt{-1} \times 2 \times \pi \times 0.a_1} = e^{-\sqrt{-1} \times 2 \times \pi \times 0.0} = 1$. This satisfies the functionality of the Hadamard gate.

Next, using the controlled-$\overline{R_{2,2}}$ gate generates the following state

$$\frac{1}{\sqrt{2}} \left(|0\rangle + e^{-\sqrt{-1} \times 2 \times \pi \times 0.a_1} \times e^{-\sqrt{-1} \times 2 \times \pi \times 0.0a_2} |1\rangle \right) |a_2 \cdots a_N\rangle$$

$$= \frac{1}{\sqrt{2}} \left(|0\rangle + e^{-\sqrt{-1} \times 2 \times \pi \times (0.a_1 + 0.0a_2)} |1\rangle \right) |a_2 \cdots a_N\rangle$$

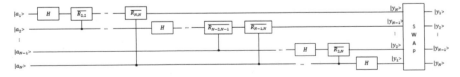

Fig. 4.6 Efficient circuit of implementing the inverse quantum Fourier transform

$$= \frac{1}{\sqrt{2}}\left(|0\rangle + e^{-\sqrt{-1}\times 2\times\pi\times 0.a_1 a_2}|1\rangle\right)|a_2\cdots a_N\rangle. \tag{4.48}$$

When $a_2 = 1$ is equivalent to the control bit $|a_2\rangle$ to be $|1\rangle$, it adds the phase $e^{-\sqrt{-1}\times 2\times\pi\times\left(\frac{1}{4}\right)} = e^{-\sqrt{-1}\times 2\times\pi\times 0.01} = e^{-\sqrt{-1}\times 2\times\pi\times 0.0a_2}$ to the coefficient of the first $|1\rangle$. When $a_2 = 0$ is equivalent to the control bit $|a_2\rangle$ to be $|0\rangle$, it does not change the coefficient of the first $|1\rangle$ because $e^{-\sqrt{-1}\times 2\times\pi\times 0.0a_2} = e^{-\sqrt{-1}\times 2\times\pi\times 0.00} = 1$. This satisfies the functionality of the controlled-$\overline{R_{2,2}}$ gate.

Next, making use of the controlled-$\overline{R_{3,3}}$ gate produces the following state

$$\frac{1}{\sqrt{2}}\left(|0\rangle + e^{-\sqrt{-1}\times 2\times\pi\times 0.a_1 a_2} \times e^{-\sqrt{-1}\times 2\times\pi\times 0.00a_3}|1\rangle\right)|a_2\cdots a_N\rangle$$

$$= \frac{1}{\sqrt{2}}\left(|0\rangle + e^{-\sqrt{-1}\times 2\times\pi\times(0.a_1 a_2 + 0.00a_3)}|1\rangle\right)|a_2\cdots a_N\rangle$$

$$= \frac{1}{\sqrt{2}}\left(|0\rangle + e^{-\sqrt{-1}\times 2\times\pi\times 0.a_1 a_2 a_3}|1\rangle\right)|a_2\cdots a_N\rangle. \tag{4.49}$$

When $a_3 = 1$ is equivalent to the control bit $|a_3\rangle$ to be $|1\rangle$, it adds the phase $e^{-\sqrt{-1}\times 2\times\pi\times\left(\frac{1}{8}\right)} = e^{-\sqrt{-1}\times 2\times\pi\times 0.001} = e^{-\sqrt{-1}\times 2\times\pi\times 0.00a_3}$ to the coefficient of the first $|1\rangle$. When $a_3 = 0$ is equivalent to the control bit $|a_3\rangle$ to be $|0\rangle$, it does not change the coefficient of the first $|1\rangle$ because $e^{-\sqrt{-1}\times 2\times\pi\times 0.00a_3} = e^{-\sqrt{-1}\times 2\times\pi\times 0.000} = 1$. This satisfies the functionality of the controlled-$\overline{R_{3,3}}$ gate.

We continue using the controlled-$\overline{R_{4,4}}$, $\overline{R_{5,5}}$ through $\overline{R_{N,N}}$ gates, they respectively add the phase $\left(e^{-\sqrt{-1}\times 2\times\pi\times 0.000a_4}\right)$, the phase $\left(e^{-\sqrt{-1}\times 2\times\pi\times 0.0000a_5}\right)$ through the phase $\left(e^{-\sqrt{-1}\times 2\times\pi\times 0.00\cdots a_N}\right)$ to the coefficient of the first $|1\rangle$. At the end of this procedure, we get the following state

$$\frac{1}{\sqrt{2}}\left(\begin{array}{c}|0\rangle + e^{-\sqrt{-1}\times 2\times\pi\times 0.a_1 a_2 a_3} \times e^{-\sqrt{-1}\times 2\times\pi\times 0.000a_4} \\ \times e^{-\sqrt{-1}\times 2\times\pi\times 0.0000a_5} \times \cdots \times e^{-\sqrt{-1}\times 2\times\pi\times 0.00\cdots a_N}|1\rangle\end{array}\right)|a_2\cdots a_N\rangle$$

$$= \frac{1}{\sqrt{2}}\left(|0\rangle + e^{-\sqrt{-1}\times 2\times\pi\times(0.a_1 a_2 a_3 + 0.000a_4 + 0.0000a_5 + 0.00\cdots a_N)}|1\rangle\right)|a_2\cdots a_N\rangle$$

$$= \frac{1}{\sqrt{2}}\left(|0\rangle + e^{-\sqrt{-1}\times 2\times\pi\times 0.a_1 a_2 a_3 a_4 a_5\cdots a_N}|1\rangle\right)|a_2\cdots a_N\rangle. \tag{4.50}$$

Next, we complete a similar procedure on the second quantum bit. Using the Hadamard gate to the second quantum bit yields the following state

$$\left(\frac{|0 + e^{-\sqrt{-1}\times 2\times\pi\times 0.a_1 a_2\cdots a_N}|1\rangle}{\sqrt{2}}\right)\left(\frac{|0 + e^{-\sqrt{-1}\times 2\times\pi\times 0.a_2}|1\rangle}{\sqrt{2}}\right)|a_3\cdots a_N\rangle. \tag{4.51}$$

Next, applying the controlled-$\overline{R_{2,3}}$, $\overline{R_{3,4}}$ through $\overline{R_{N-1,N}}$ gates produces the following state

$$
\left(\frac{|0\rangle + e^{-\sqrt{-1}\times 2\times\pi \times 0.a_1 a_2 \cdots a_N}|1\rangle}{\sqrt{2}} \right) \left(\frac{|0\rangle + e^{-\sqrt{-1}\times 2\times\pi \times 0.a_2 \cdots a_N}|1\rangle}{\sqrt{2}} \right) |a_3 \cdots a_N\rangle .
$$

$$(4.52)$$

We continue in this fashion for each quantum bit, giving a final state

$$
\left(\frac{|0\rangle + e^{-\sqrt{-1}\times 2\times\pi \times 0.a_1 a_2 \cdots a_N}|1\rangle}{\sqrt{2}} \right) \left(\frac{|0\rangle + e^{-\sqrt{-1}\times 2\times\pi \times 0.a_2 \cdots a_N}|1\rangle}{\sqrt{2}} \right)
$$
$$
\cdots \left(\frac{|0\rangle + e^{-\sqrt{-1}\times 2\times\pi \times 0.a_N}|1\rangle}{\sqrt{2}} \right) .
$$

$$(4.53)$$

Next, making use of swap operations reverses the order of the quantum bits. After the swap operations, the state of the quantum bits is

$$
\left(\frac{\left|0\right\rangle + e^{-\sqrt{-1}\times 2\times\pi \times 0.a_N}\left|1\right\rangle}{\sqrt{2}} \right) \otimes \left(\frac{\left|0\right\rangle + e^{-\sqrt{-1}\times 2\times\pi \times 0.a_{N-1}a_N}\left|1\right\rangle}{\sqrt{2}} \right)
$$
$$
\otimes \left(\frac{|0\rangle + e^{-\sqrt{-1}\times 2\times\pi \times 0.a_{N-2}a_{N-1}a_N}|1\rangle}{\sqrt{2}} \right)
$$
$$
\otimes \cdots \otimes \left(\frac{\left|0\right\rangle + e^{-\sqrt{-1}\times 2\times\pi \times 0.a_1 a_2 \cdots a_N}\left|1\right\rangle}{\sqrt{2}} \right) .
$$

$$(4.54)$$

Comparing with Eq. (4.39) and Eq. (4.54), we look at the desired output from the inverse quantum Fourier transform. This construction also shows that the inverse quantum Fourier transform is unitary because each gate in the circuit of Fig. 4.6 is unitary.

4.10 Assessment of Time Complexity for Implementing the Inverse Quantum Fourier Transform

How many gates does the circuit of implementing the inverse quantum Fourier transform in Fig. 4.6 apply? We start by means of doing a Hadamard gate and $(n - 1)$ conditional rotations on the first quantum bit. This requires a total of n gates. Next similar procedure is to implement a Hadamard gate and $(n - 2)$ conditional rotations on the second quantum bit. This requires a total of $n + (n - 1)$ gates. Next similar procedure is to do a Hadamard gate and $(n - 3)$ conditional rotations on the third quantum bit. This requires a total of $n + (n - 1) + (n - 2)$ gates. Continuing in this way, this requires a total of $n + (n - 1) + (n - 2) + \ldots + 1 = n \times (n + 1)/2$ gates, plus at most $n/2$ swap gates that reverse the order of the quantum bits.

Because applying three controlled-NOT gates can implement each swap gate, this circuit in Fig. 4.6 provides a $O(n^2)$ algorithm for performing the inverse quantum Fourier transform. The best classical algorithm of figuring out the inverse discrete Fourier transform on 2^n elements is the fast Fourier transform with $O(n \times 2^n)$ gates. This indicates that it requires exponentially more operations to deal with the inverse discrete Fourier transform on a digital (classical) computer than it does to implement the inverse quantum Fourier transform on a quantum computer.

4.11 Compute the Period and the Frequency of a Given Oracular Function

A given oracular function is $O_f: \{a_1 \, a_2 \, a_3 \, a_4 \mid \forall \, a_d \in \{0, 1\} \text{ for } 1 \leq d \leq 4\} \to \{\frac{1}{\sqrt{2^4}} \times e^{\sqrt{-1} \times 2 \times \pi \times \frac{1}{2} \times a_4} \mid a_4 \in \{0, 1\}\}$. The value $(\frac{1}{\sqrt{2^4}} \times e^{\sqrt{-1} \times 2 \times \pi \times \frac{1}{2} \times a_4})$ is the amplitude (the weight) of each input $a_1 \, a_2 \, a_3 \, a_4$ for the given oracular function O_f. The square of the absolute value of the amplitude (the weight) to an input $a_1 \, a_2 \, a_3 \, a_4$ is the possibility of obtaining the output of O_f that takes the input $a_1 \, a_2 \, a_3 \, a_4$ as its input value. The sum of the square of the absolute value to the amplitude (the weight) of each input $a_1 \, a_2 \, a_3 \, a_4$ is equal to one. Sixteen outputs of O_f that takes each input from $a_1^0 \, a_2^0 \, a_3^0 \, a_4^0$ through $a_1^1 \, a_2^1 \, a_3^1 \, a_4^1$ are subsequently $\left(\frac{1}{\sqrt{2^4}}\right), \left(-\frac{1}{\sqrt{2^4}}\right), \left(\frac{1}{\sqrt{2^4}}\right), \left(-\frac{1}{\sqrt{2^4}}\right), \left(\frac{1}{\sqrt{2^4}}\right),$ $\left(-\frac{1}{\sqrt{2^4}}\right), \left(\frac{1}{\sqrt{2^4}}\right), \left(-\frac{1}{\sqrt{2^4}}\right), \left(\frac{1}{\sqrt{2^4}}\right), \left(-\frac{1}{\sqrt{2^4}}\right), \left(\frac{1}{\sqrt{2^4}}\right), \left(-\frac{1}{\sqrt{2^4}}\right), \left(\frac{1}{\sqrt{2^4}}\right), \left(-\frac{1}{\sqrt{2^4}}\right), \left(\frac{1}{\sqrt{2^4}}\right)$ and $\left(-\frac{1}{\sqrt{2^4}}\right)$.

The period r of O_f is to satisfy $O_f(a_1 \, a_2 \, a_3 \, a_4) = O_f(a_1 \, a_2 \, a_3 \, a_4 + r)$ to any two inputs $a_1 \, a_2 \, a_3 \, a_4$ and $a_1 \, a_2 \, a_3 \, a_4 + r$. The frequency f of O_f is equal to the number of the period per sixteen outputs. This gives that $r \times f = 16$. Hidden patterns and information stored in a given oracular function O_f are to that its output rotates back to its starting value $\left(\frac{1}{\sqrt{2^4}}\right)$ eight times. This implies that the number of the period per sixteen outputs is eight and the frequency f of O_f is equal to eight. Therefore, this gives that the period r of O_f is $16 / 8 = 2$. When O_f takes two inputs $a_1 \, a_2 \, a_3$ a_4 and $a_1 \, a_2 \, a_3 \, a_4 + 2$ as its input values, it produces the same output to the two inputs. This is to say that $O_f(a_1 \, a_2 \, a_3 \, a_4) = O_f(a_1 \, a_2 \, a_3 \, a_4 + 2)$. For obtaining the frequency f and the period r of O_f, it needs to implement at least sixteen exponential computations to $e^{\sqrt{-1} \times 2 \times \pi \times \frac{1}{2} \times a_4}$ and multiplication of sixteen times.

4.11.1   The Period and the Frequency of Signals in a Given Oracular Function

However, we can use another way (the second way) to obtain the period r and the frequency f to a given oracular function O_f. In another way, the value $\left(\frac{1}{\sqrt{2^4}}\right)$ is the *magnitude* of the amplitude (the weight) to each input $a_1\ a_2\ a_3\ a_4$. The square of the absolute value for the *magnitude* of the amplitude (the weight) to an input $a_1\ a_2\ a_3\ a_4$ is the possibility of obtaining the output of O_f that takes the input $a_1\ a_2\ a_3\ a_4$ as its input value. The value $(2 \times \pi \times (1/2) \times a_4)$ is the *phase* among the amplitudes of these inputs. The phase can take any value from 0 degree to 360 degrees. The phase of each input from $a_1^0\ a_2^0\ a_3^0\ a_4^0$ through $a_1^1\ a_2^1\ a_3^1\ a_4^1$ is subsequently 0°, 180°, 0°, 180°, 0°, 180°, 0°, 180°, 0°, 180°, 0°, 180°, 0°, 180°, 0° and 180°. We think of the input domain of O_f as the time domain and the phases from its output as signals. Computing the period r and the frequency f of O_f is equivalent to determine the period r and the frequency f of signals in the time domain (the input domain).

Because the phase of the output of each input from $a_1^0\ a_2^0\ a_3^0\ a_4^0$ through $a_1^1\ a_2^1\ a_3^1\ a_4^1$ is subsequently 0°, 180°, 0°, 180°, 0°, 180°, 0°, 180°, 0°, 180°, 0°, 180°, 0°, 180°, 0° and 180°, we take the sixteen input values as the corresponding sixteen time units and the sixteen phases as the sixteen samples of signals. Each sample encodes an angle. The angle can take 0° or 180°. The sixteen input values from $a_1^0\ a_2^0\ a_3^0\ a_4^0$ through $a_1^1\ a_2^1\ a_3^1\ a_4^1$ corresponds to sixteen time units from zero through fifteen. We use Fig. 4.7 to explain the reason of why computing the period r and the frequency f of O_f is equivalent to determine the period r and the frequency f of signals in the time domain (the input domain). In Fig. 4.7, the horizontal axis is to represent the time domain in which it contains the input domain of O_f and the vertical axis is to represent signals in which it consists of the sixteen phases from its output. From Fig. 4.7, the signal rotates back to its starting value 0° *eight* times. This is to say that there are eight periods of signals per sixteen time units and the frequency f of signals

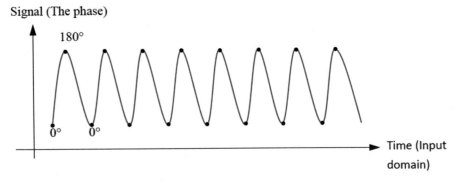

Fig. 4.7 Sampling sixteen points from sixteen phases of outputs of a given oracular function that is $O_f\colon \{a_1\ a_2\ a_3\ a_4 \mid \forall\ a_d \in \{0,1\}\ \text{for}\ 1 \le d \le 4\} \to \{\frac{1}{\sqrt{2^4}} \times e^{\sqrt{-1}\times 2\times\pi\times\frac{1}{2}\times a_4} \mid a_4 \in \{0,1\}\}$

is equal to eight. This gives that the period r of signals is equal to $(16/8) = (2)$. Because the magnitude of each output to O_f is the same and is equal to $\left(\frac{1}{\sqrt{2^4}}\right)$ and the signal (the phase of each output to O_f) rotates back to its starting value $0°$ *eight* times, its output rotates back to its starting value $\left(\frac{1}{\sqrt{2^4}}\right)$ *eight* times. This indicates that the number of the period per sixteen outputs is eight and the frequency f of O_f is equal to eight. Therefore, this gives that the period r of O_f is $16/8 = 2$.

4.11.2 Circle Notation of Representing the Period and the Frequency of Signals in a Given Oracular Function

Because in Fig. 4.7 in subsection 4.11.1 sampling sixteen points just encodes sixteen phases from sixteen output of O_f, we use circle notation (another concept) to explain how figuring out the period r and the frequency f of O_f is equivalent to determine the period r and the frequency f of signals in the time domain (the input domain). In a circle, the filled *radius* represents the *magnitude* of the amplitude to an input a_1 a_2 a_3 a_4 of O_f. This implies that the size (the shaded area) of the circle is directly proportional to the square of the absolute value of the magnitude of the amplitude to that input a_1 a_2 a_3 a_4 of O_f. This means that the size (the shaded area) of the circle is directly proportional to the possibility of obtaining the output of O_f that takes that input a_1 a_2 a_3 a_4 as its input value. A darker line drawn in the circle indicates that the phase (a positive angle) rotates the circle *counterclockwise* or the phase (a negative angle) rotates the circle *clockwise*. A number a_1 a_2 a_3 a_4 below the circle encodes an input to O_f. In Fig. 4.8, it contains sixteen outputs of a given oracular function that is O_f: $\{a_1 a_2 a_3 a_4 \mid \forall a_d \in \{0, 1\}$ for $1 \le d \le 4\} \to \{\frac{1}{\sqrt{2^4}} \times e^{\sqrt{-1} \times 2 \times \pi \times \frac{1}{2} \times a_4} \mid a_4 \in \{0, 1\}\}$. In Fig. 4.8, sixteen circles encode sampling sixteen points in Fig. 4.7 and sixteen numbers below each circle encode sixteen time units of the time domain of Fig. 4.7.

Because in Fig. 4.8 the shaded area of each circle is directly proportional to $\left(\frac{1}{\sqrt{2^4}}\right)^2$, the shaded area of each circle is the same. In Fig. 4.8, the radius of the left first circle is $\left(\frac{1}{\sqrt{2^4}}\right) = (1/4)$ that is the magnitude of output to $O_f(a_1^0 a_2^0 a_3^0 a_4^0)$. The darker line drawn in the left first circle indicates that the phase to output of $O_f(a_1^0 a_2^0 a_3^0 a_4^0)$ is a $0°$ and gives a $0°$ rotation to the left first circle. A number $a_1^0 a_2^0 a_3^0 a_4^0$

Fig. 4.8 Sixteen outputs of a given oracular function that is O_f: $\{a_1 a_2 a_3 a_4 \mid \forall a_d \in \{0, 1\}$ for $1 \le d \le 4\} \to \{\frac{1}{\sqrt{2^4}} \times e^{\sqrt{-1} \times 2 \times \pi \times \frac{1}{2} \times a_4} \mid a_4 \in \{0, 1\}\}$

below the left first circle encodes the input that is taken by $O_f(a_1^0 \, a_2^0 \, a_3^0 \, a_4^0)$. Similarly, in Fig. 4.8, the radius of the left second circle is $\left(\frac{1}{\sqrt{2^4}}\right) = (1/4)$ that is the magnitude of output to $O_f(a_1^0 \, a_2^0 \, a_3^0 \, a_4^1)$. The darker line drawn in the left second circle indicates that the phase to output of $O_f(a_1^0 \, a_2^0 \, a_3^0 \, a_4^1)$ is a $180°$ and gives a $180°$ rotation to the left second circle. A number $a_1^0 \, a_2^0 \, a_3^0 \, a_4^1$ below the left second circle encodes the input that is taken by $O_f(a_1^0 \, a_2^0 \, a_3^0 \, a_4^1)$. Similarly, in Fig. 4.8, the radius of the left third circle through the last circle is all $\left(\frac{1}{\sqrt{2^4}}\right) = (1/4)$ that is the magnitude of output to $O_f(a_1^0 \, a_2^0 \, a_3^1 \, a_4^0)$ through $O_f(a_1^1 \, a_2^1 \, a_3^1 \, a_4^1)$. The darker line drawn in the left third circle through the last circle indicates that the phase to output of $O_f(a_1^0 \, a_2^0 \, a_3^1 \, a_4^0)$ through $O_f(a_1^1 \, a_2^1 \, a_3^1 \, a_4^1)$ is subsequently $0°, 180°, 0°, 180°, 0°, 180°, 0°, 180°, 0°,$ $180°, 0°, 180°, 0°$ and $180°$. They subsequently give a $0°$ rotation and a $180°$ rotation of seven times to the left third circle through the last circle. Fourteen numbers $a_1^0 \, a_2^0$ $a_3^1 \, a_4^0$ through $a_1^1 \, a_2^1 \, a_3^1 \, a_4^1$ below the left third circle trough the last circle encode the fourteen inputs that are subsequently taken by $O_f(a_1^0 \, a_2^0 \, a_3^1 \, a_4^0)$ through $O_f(a_1^1 \, a_2^1 \, a_3^1 \, a_4^1)$.

From Fig. 4.8, hidden patterns and information stored in signals of sixteen circles (sampling sixteen points) are that the signal (the phase) rotates back to its starting value $0°$ *eight* times. This indicates that there are eight periods of signals per sixteen inputs (per sixteen time units) and the frequency f of signals is equal to eight. This gives that the period r of signals is equal to $(16/8) = (2)$. Because the signal (the phase of each output to O_f) rotates back to its starting value $0°$ *eight* times, its output rotates back to its starting value $\left(\frac{1}{\sqrt{2^4}}\right)$ *eight* times. This implies that the number of the period per sixteen outputs is eight and the frequency f of O_f is equal to eight. Therefore, this gives that the period r of O_f is $16/8 = 2$. So, when a given oracular function O_f takes two inputs $a_1 \, a_2 \, a_3 \, a_4$ and $a_1 \, a_2 \, a_3 \, a_4 + 2$ as its input values, it gives the same output to the two inputs. This indicates that $O_f(a_1 \, a_2 \, a_3 \, a_4) = O_f(a_1 \, a_2 \, a_3 \, a_4 + 2)$ to any two inputs $a_1 \, a_2 \, a_3 \, a_4$ and $a_1 \, a_2 \, a_3 \, a_4 + 2$. In this way, for obtaining the period r and the frequency f of O_f, it only needs to implement eight multiplications $(2 \times \pi \times (1/2) \times 1) \, (180°)$ and eight multiplications $(2 \times \pi \times (1/2) \times 0) \, (0°)$. The second way very significantly enhances the performance of the first way to determine the period r and the frequency f of O_f.

4.11.3   The First Stage of Quantum Programs for Finding the Period and the Frequency of Signals in a Given Oracular Function

In Listing 4.1, the program is in the backend that is *simulator* of Open QASM with *thirty-two* quantum bits in **IBM**'s quantum computer. The program is to determine the frequency f and the period r of O_f so that $O_f(a_1 \, a_2 \, a_3 \, a_4) = O_f(a_1 \, a_2 \, a_3 \, a_4 + r)$ to any two inputs $a_1 \, a_2 \, a_3 \, a_4$ and $a_1 \, a_2 \, a_3 \, a_4 + r$. In Listing 4.1, we introduce how to

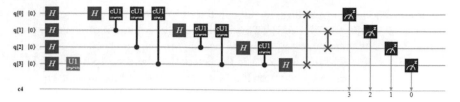

Fig. 4.9 The quantum circuit of computing the frequency f and the period r to a given oracular function O_f: $\{a_1\, a_2\, a_3\, a_4 \mid \forall\, a_d \in \{0, 1\}$ for $1 \le d \le 4\} \to \{\frac{1}{\sqrt{2^4}} \times e^{\sqrt{-1} \times 2 \times \pi \times \frac{1}{2} \times a_4} \mid a_4 \in \{0, 1\}\}$

use the inverse quantum Fourier transform to find the frequency f and the period r of O_f. Figure 4.9 is the quantum circuit of computing the frequency f and the period r of O_f. The statement "OPENQASM 2.0;" on line one of Listing 4.1 is to point out that the program is written with version 2.0 of Open QASM. Next, the statement "include "qelib1.inc";" on line two of Listing 4.1 is to continue parsing the file "qelib1.inc" as if the contents of the file were pasted at the location of the include statement, where the file "qelib1.inc" is **Quantum Experience (QE) Standard Header** and the path is specified relative to the current working directory.

Listing 4.1 The program of determining the frequency f and the period r to a given oracular function O_f: $\{a_1\, a_2\, a_3\, a_4 \mid \forall\, a_d \in \{0, 1\}$ for $1 \le d \le 4\} \to \{\frac{1}{\sqrt{2^4}} \times e^{\sqrt{-1} \times 2 \times \pi \times \frac{1}{2} \times a_4} \mid a_4 \in \{0, 1\}\}$.

```
1.  OPENQASM 2.0;
2.  include "qelib1.inc";
3.  qreg q[4];
4.  creg c[4];
```

Next, the statement "qreg q[4];" on line three of Listing 4.1 is to declare that in the program there are four quantum bits. In the left top of Fig. 4.9, four quantum bits are respectively q[0], q[1], q[2] and q[3]. The initial value of each quantum bit is set to state $|0\rangle$. We use four quantum bits q[0], q[1], q[2] and q[3] to encode the input domain $\{a_1\, a_2\, a_3\, a_4 \mid \forall\, a_d \in \{0, 1\}$ for $1 \le d \le 4\}$ of O_f. This is to say that quantum bit q[0] encodes bit a_1, quantum bit q[1] encodes bit a_2, quantum bit q[2] encodes bit a_3 and quantum bit q[3] encodes bit a_4.

For the convenience of our explanation, q[k]^0 for $0 \le k \le 3$ is to represent the value 0 of q[k] and q[k]^1 for $0 \le k \le 3$ is to represent the value 1 of q[k]. Similarly, for the convenience of our explanation, an initial state vector of determining the frequency f and the period r of O_f so that $O_f(a_1\, a_2\, a_3\, a_4) = O_f(a_1\, a_2\, a_3\, a_4 + r)$ to any two inputs $a_1\, a_2\, a_3\, a_4$ and $a_1\, a_2\, a_3\, a_4 + r$ is

$$\langle|\Omega_0\rangle = |q[0]^0\rangle |q[1]^0\rangle |q[2]^0\rangle |q[3]^0\rangle = |0\rangle |0\rangle |0\rangle |0\rangle = |0000\rangle.$$

The corresponding decimal value of the initial state vector $|\Omega_0\rangle$ is $2^3 \times q[0]^0 + 2^2$ $\times q[1]^0 + 2^1 \times q[2]^0 + 2^0 \times q[3]^0$. This implies that quantum bit $q[0]^0$ is the most significant bit and quantum bit $q[3]^0$ is the least significant bit. Then, the statement "creg c[4];" on line four of Listing 4.1 is to declare that there are four classical bits in the program. In the left bottom of Fig. 4.9, four classical bits are subsequently c[0], c[1], c[2] and c[3]. The initial value of each classical bit is set to zero (0). For the convenience of our explanation, $c[k]^0$ for $0 \le k \le 3$ is to represent the value 0 of c[k] and $c[k]^1$ for $0 \le k \le 3$ is to represent the value 1 of c[k]. The corresponding decimal value of the four initial classical bits $c[3]^0$ $c[2]^0$ $c[1]^0$ $c[0]^0$ is $2^3 \times c[3]^0 +$ $2^2 \times c[2]^0 + 2^1 \times c[1]^0 + 2^0 \times c[0]^0$. This indicates that classical bit $c[3]^0$ is the most significant bit and classical bit $c[0]^0$ is the least significant bit.

Next, the four statements "h q[0];", "h q[1];" "h q[2];" and "h q[3];" on line five through eight of Listing 4.1 are to implement four Hadamard gates of the *first* time slot

---

Listing 4.1 continued...

// We use the following *five* statements to implement a given oracular function O_f:

//$\{a_1\ a_2\ a_3\ a_4\ |\ \forall\ a_d \in \{0, 1\}$ for $1 \le d \le 4\} \rightarrow \{\ \frac{1}{\sqrt{2^4}} \times e^{\sqrt{-1} \times 2 \times \pi \times \frac{1}{2} \times a_4}\ |$ $a_4 \in \{0, 1\}\}$.

```
5.  h q[0];
6.  h q[1];
7.  h q[2];
8.  h q[3];
9.  u1(2*pi*1/2) q[3];
```

---

of the quantum circuit in Fig. 4.9. The four statements "h q[0];", "h q[1];" "h q[2];" and "h q[3];" take the initial state vector $|\Omega_0\rangle = |q[0]^0\rangle\ |q[1]^0\rangle\ |q[2]^0\rangle\ |q[3]^0\rangle$ as their input state vector. Because the initial state of each quantum bit is set to state $|0\rangle$, the four statements "h q[0];", "h q[1];" "h q[2];" and "h q[3];" actually implement

$$\begin{pmatrix} \frac{1}{\sqrt{2}} & \frac{1}{\sqrt{2}} \\ \frac{1}{\sqrt{2}} & -\frac{1}{\sqrt{2}} \end{pmatrix} \times \begin{pmatrix} 1 \\ 0 \end{pmatrix} = \begin{pmatrix} \frac{1}{\sqrt{2}} \\ \frac{1}{\sqrt{2}} \end{pmatrix} = \frac{1}{\sqrt{2}} \begin{pmatrix} 1 \\ 1 \end{pmatrix} = \frac{1}{\sqrt{2}} \left(\begin{pmatrix} 1 \\ 0 \end{pmatrix} + \begin{pmatrix} 0 \\ 1 \end{pmatrix} \right) = \frac{1}{\sqrt{2}} (|0\rangle + |1\rangle).$$

This is to say that the four statements convert four quantum bits q[0], q[1], q[2] and q[3] from one state $|0\rangle$ to another state $\frac{1}{\sqrt{2}} (|0\rangle + |1\rangle)$ (their superposition). Hence, the superposition of the four quantum bits q[0], q[1], q[2] and q[3] is $(\frac{1}{\sqrt{2}} (|0\rangle + |1\rangle))$ $(\frac{1}{\sqrt{2}} (|0\rangle + |1\rangle))$ $(\frac{1}{\sqrt{2}} (|0\rangle + |1\rangle))$ $(\frac{1}{\sqrt{2}} (|0\rangle + |1\rangle))$. This implies that we obtain the following new state vector

$$|\Omega_1\rangle = \left(\frac{1}{\sqrt{2}} (q[0]^0\rangle + |q[0]^1\rangle) \right) \left(\frac{1}{\sqrt{2}} (|q[1]^0\rangle + |q[1]^1\rangle) \right)$$
$$\left(\frac{1}{\sqrt{2}} (|q[2]^0\rangle + |q[2]^1\rangle) \right) \left(\frac{1}{\sqrt{2}} (|q[3]^0\rangle + |q[3]^1\rangle) \right).$$

Next, the statement "u1(2*pi*1/2) q[3];" on line 9 in Listing 4.1 actually imple-

ments one rotation gate $\begin{pmatrix} 1 & 0 \\ 0 & e^{\sqrt{-1} \times 2 \times \pi \times 1/2} \end{pmatrix}$ of the *second* time slot of the quantum

circuit in Fig. 4.9. The statement "u1(2*pi*1/2) q[3];" takes the new state vector
$(|\Omega_1\rangle)$ as its input state vector. Because the state of quantum bit q[3] is $(\frac{1}{\sqrt{2}}(|q[3]^0\rangle +$

$|q[3]^1\rangle))$, the statement "u1(2*pi*1/2) q[3];" actually completes $\begin{pmatrix} 1 & 0 \\ 0 & e^{\sqrt{-1} \times 2 \times \pi \times 1/2} \end{pmatrix}$

$$\times \begin{pmatrix} \frac{1}{\sqrt{2}} \\ \frac{1}{\sqrt{2}} \end{pmatrix} = \begin{pmatrix} \frac{1}{\sqrt{2}} \\ \frac{1}{\sqrt{2}} \times e^{\sqrt{-1} \times 2 \times \pi \times 1/2} \end{pmatrix} = \frac{1}{\sqrt{2}} \begin{pmatrix} 1 \\ e^{\sqrt{-1} \times 2 \times \pi \times 1/2} \end{pmatrix} = \frac{1}{\sqrt{2}} (\begin{pmatrix} 1 \\ 0 \end{pmatrix} +$$

$$\begin{pmatrix} 0 \\ e^{\sqrt{-1} \times 2 \times \pi \times 1/2} \end{pmatrix}) = \frac{1}{\sqrt{2}} (\begin{pmatrix} 1 \\ 0 \end{pmatrix} + e^{\sqrt{-1} \times 2 \times \pi \times 1/2} \begin{pmatrix} 0 \\ 1 \end{pmatrix}) = \frac{1}{\sqrt{2}} (|0\rangle + e^{\sqrt{-1} \times 2 \times \pi \times 1/2}$$

$|1\rangle) = \frac{1}{\sqrt{2}} (|0\rangle + e^{\sqrt{-1} \times \pi} |1\rangle) = \frac{1}{\sqrt{2}} (|0\rangle + (-1)|1\rangle)$. This indicates that the statement
"u1(2*pi*1/2) q[3];" adds the phase $e^{\sqrt{-1} \times 2 \times \pi \times 1/2} = e^{\sqrt{-1} \times \pi} = \cos(\pi) + \sqrt{-1} \times$
$\sin(\pi) = -1 + \sqrt{-1} \times 0 = -1$ to the coefficient of the state $|1\rangle$ in the superposition
of the quantum bit q[3] and does not change the coefficient of the state $|0\rangle$ in the
superposition of the quantum bit q[3]. Because in the *second* time slot of the quantum
circuit in Fig. 4.9 there is no quantum gate to act on quantum bits q[0] through q[2],
their current states are not changed. This is to say that we obtain the following new
state vector

$$|\Omega_2\rangle = \left(\frac{1}{\sqrt{2}}(|q[0]^0\rangle + |q[0]^1\rangle)\right)\left(\frac{1}{\sqrt{2}}(|q[1]^0\rangle + |q[1]^1\rangle)\right)\left(\frac{1}{\sqrt{2}}(|q[2]^0\rangle + |q[2]^1\rangle)\right)$$
$$\left(\frac{1}{\sqrt{2}}(|q[3]^0\rangle + e^{\sqrt{-1} \times 2 \times \pi \times 1/2}|q[3]^1\rangle)\right)$$
$$= \left(\frac{1}{\sqrt{2}}(|q[0]^0\rangle + |q[0]^1\rangle)\right)\left(\frac{1}{\sqrt{2}}(|q[1]^0\rangle + |q[1]^1\rangle)\right)\left(\frac{1}{\sqrt{2}}(|q[2]^0\rangle + |q[2]^1\rangle)\right)$$
$$\left(\frac{1}{\sqrt{2}}(|q[3]^0\rangle + (-1)|q[3]^1\rangle)\right)$$

In the new state vector $(|\Omega_2\rangle)$, the sixteen amplitudes from state $|q[0]^0\rangle |q[1]^0\rangle$
$|q[2]^0\rangle |q[3]^0\rangle$ through state $|q[0]^1\rangle |q[1]^1\rangle |q[2]^1\rangle |q[3]^1\rangle$ are subsequently $\left(\frac{1}{\sqrt{2^4}}\right)$,
$\left(-\frac{1}{\sqrt{2^4}}\right)$, $\left(\frac{1}{\sqrt{2^4}}\right)$, $\left(-\frac{1}{\sqrt{2^4}}\right)$, $\left(\frac{1}{\sqrt{2^4}}\right)$, $\left(-\frac{1}{\sqrt{2^4}}\right)$, $\left(\frac{1}{\sqrt{2^4}}\right)$, $\left(-\frac{1}{\sqrt{2^4}}\right)$, $\left(\frac{1}{\sqrt{2^4}}\right)$, $\left(-\frac{1}{\sqrt{2^4}}\right)$,
$\left(\frac{1}{\sqrt{2^4}}\right)$, $\left(-\frac{1}{\sqrt{2^4}}\right)$, $\left(\frac{1}{\sqrt{2^4}}\right)$, $\left(-\frac{1}{\sqrt{2^4}}\right)$, $\left(\frac{1}{\sqrt{2^4}}\right)$ and $\left(-\frac{1}{\sqrt{2^4}}\right)$. This is to say that in the
new state vector $(|\Omega_2\rangle)$, it uses the amplitude of each state to encode sixteen outputs
to a given oracular function O_f: $\{a_1 \, a_2 \, a_3 \, a_4 \,|\, \forall \, a_d \in \{0, 1\} \text{ for } 1 \leq d \leq 4\} \rightarrow \{\frac{1}{\sqrt{2^4}}$
$\times e^{\sqrt{-1} \times 2 \times \pi \times \frac{1}{2} \times a_4} \,| a_4 \in \{0, 1\}\}$. Hidden patterns and information stored in the new
state vector $(|\Omega_2\rangle)$ are to that the amplitude rotates back to its starting value $\left(\frac{1}{\sqrt{2^4}}\right)$
eight times.

Similarly, in the new state vector ($|\Omega_2\rangle$), it makes use of the magnitude $\left(\frac{1}{\sqrt{2^4}}\right) =$ (1/4) of the amplitude of each state as the radius of each circle in Fig. 4.8. Because in Fig. 4.8 the shaded area of each circle is directly proportional to $(1/4)^2$, the shaded area of each circle is the same. In the new state vector ($|\Omega_2\rangle$), the sixteen phases from state $|q[0]^0\rangle$ $|q[1]^0\rangle$ $|q[2]^0\rangle$ $|q[3]^0\rangle$ through state $|q[0]^1\rangle$ $|q[1]^1\rangle$ $|q[2]^1\rangle$ $|q[3]^1\rangle$ are subsequently 0°, 180°, 0°, 180°, 0°, 180°, 0°, 180°, 0°, 180°, 0°, 180°, 0°, 180°, 0° and 180°. They subsequently give a 0° rotation and a 180° rotation of eight times to the left first circle through the last circle in Fig. 4.8. In the new state vector ($|\Omega_2\rangle$), it uses the darker line drawn in the left first circle through the last circle in Fig. 4.8 to indicate a 0° rotation and a 180° rotation of eight times to the sixteen circles in Fig. 4.8. This is to say that hidden patterns and information stored in the new state vector ($|\Omega_2\rangle$) are to that the phase rotates back to its starting value 0° *eight* times.

4.11.4 The Inverse Quantum Fourier Transform in Quantum Programs of Finding the Period and the Frequency of Signals in a Given Oracular Function

Next, the *twelve* statements from line 10 through line 21 in Listing 4.1 implement the inverse quantum Fourier transform with four quantum bits. The statement "h q[0];"

Listing 4.1 continued...

```
// We use the following twelve statements to implement the inverse quantum
   // Fourier transform with four quantum bits.
10.  h q[0];
11.  cu1(-2*pi*1/4) q[1],q[0];
12.  cu1(-2*pi*1/8) q[2],q[0];
13.  cu1(-2*pi*1/16) q[3],q[0];
14.  h q[1];
15.  cu1(-2*pi*1/4) q[2],q[1];
16.  cu1(-2*pi*1/8) q[3],q[1];
```

on line 10 in Listing 4.1 implements one Hadamard gate in the third time slot of Fig. 4.9. It takes the new state vector $|\Omega_2\rangle$ as its input state vector. Because the current state of quantum bit q[0] in $|\Omega_2\rangle$ is ($\frac{1}{\sqrt{2}}$ ($|q[0]^0\rangle + |q[0]^1\rangle$)), the statement "h q[0];" on line 10 in Listing 4.1 actually implements $\begin{pmatrix} \frac{1}{\sqrt{2}} & \frac{1}{\sqrt{2}} \\ \frac{1}{\sqrt{2}} & -\frac{1}{\sqrt{2}} \end{pmatrix} \times \begin{pmatrix} \frac{1}{\sqrt{2}} \\ \frac{1}{\sqrt{2}} \end{pmatrix} = \begin{pmatrix} 1 \\ 0 \end{pmatrix}$ = $|0\rangle$. This indicates that the statement "h q[0];" converts quantum bit q[0] from one state ($\frac{1}{\sqrt{2}}$ ($|q[0]^0\rangle + |q[0]^1\rangle$)) (its superposition) to another state $|q[0]^0\rangle$. Because in the *third* time slot of the quantum circuit in Fig. 4.9 there is no quantum gate to act on quantum bits q[1] through q[3], their current states are not changed. This is to say that we obtain the following new state vector

$$|\Omega_3\rangle = \left(|q[0]^0\rangle\right)\left(\frac{1}{\sqrt{2}}(|q[1]^0\rangle + |q[1]^1\rangle)\right)\left(\frac{1}{\sqrt{2}}(|q[2]^0\rangle + |q[2]^1\rangle)\right)$$
$$\left(\frac{1}{\sqrt{2}}(|q[3]^0\rangle + (-1)|q[3]^1\rangle)\right).$$

Next, the statement "cu1(−2*pi*1/4) q[1],q[0];" on line 11 in Listing 4.1 is a

controlled rotation gate $\begin{bmatrix} 1 & 0 & 0 & 0 \\ 0 & 1 & 0 & 0 \\ 0 & 0 & 1 & 0 \\ 0 & 0 & 0 & e^{\sqrt{-1}\times -2\times\pi\times 1/4} \end{bmatrix}$. The control bit is quantum bit

q[1] and the target bit is quantum bit q[0]. If the control bit is $|1\rangle$ and the target bit is $|1\rangle$, then it adds the phase $e^{\sqrt{-1}\times -2\times\pi\times 1/4}$ to the coefficient of the state $|1\rangle$ of the target bit. Otherwise, it does not change the target bit. The statement "cu1(-2*pi*1/4) q[1],q[0];" on line 11 in Listing 4.1 takes the new state vector $|\Omega_3\rangle$ as its input state vector and implements one controlled rotation gate in the fourth time slot of Fig. 4.9. Because the state of the target bit q[0] is $(|q[0]^0\rangle)$, the statement "cu1(−2*pi*1/4) q[1],q[0];" does not change the state $(|q[0]^0\rangle)$ of the target bit q[0]. In the *fourth* time slot of the quantum circuit in Fig. 4.9 there is no quantum gate to act on quantum bits q[2] through q[3], their current states are not changed. Therefore, we obtain the following new state vector

$$|\Omega_4\rangle = \left(|q[0]^0\rangle\right)\left(\frac{1}{\sqrt{2}}(|q[1]^0\rangle + |q[1]^1\rangle)\right)$$
$$\left(\frac{1}{\sqrt{2}}(|q[2]^0\rangle + |q[2]^1\rangle)\right)\left(\frac{1}{\sqrt{2}}(|q[3]^0\rangle + (-1)\ |q[3]^1\rangle)\right).$$

Next, the statement "cu1(−2*pi*1/8) q[2],q[0];" on line 12 in Listing 4.1 is a

controlled rotation gate $\begin{bmatrix} 1 & 0 & 0 & 0 \\ 0 & 1 & 0 & 0 \\ 0 & 0 & 1 & 0 \\ 0 & 0 & 0 & e^{\sqrt{-1}\times -2\times\pi\times 1/8} \end{bmatrix}$. The control bit is quantum bit

q[2] and the target bit is quantum bit q[0]. If the control bit is $|1\rangle$ and the target bit is $|1\rangle$, then it adds the phase $e^{\sqrt{-1}\times -2\times\pi\times 1/8}$ to the coefficient of the state $|1\rangle$ of the target bit. Otherwise, it does not change the target bit. The statement "cu1(−2*pi*1/8) q[2],q[0];" on line 12 in Listing 4.1 takes the new state vector $|\Omega_4\rangle$ as its input state vector and implements one controlled rotation gate in the fifth time slot of Fig. 4.9. Because the state of the target bit q[0] is $(|q[0]^0\rangle)$, the statement "cu1(-2*pi*1/8) q[2],q[0];" does not change the state $(|q[0]^0\rangle)$ of the target bit q[0]. In the *fifth* time slot of the quantum circuit in Fig. 4.9 there is no quantum gate to act on quantum bits q[1] and q[3], their current states are not changed. This indicates that we obtain the following new state vector

$$|\Omega_5\rangle = \left(|q[0]^0\rangle\right)\left(\frac{1}{\sqrt{2}}(|q[1]^0\rangle+|q[1]^1\rangle)\right)\left(\frac{1}{\sqrt{2}}(|q[2]^0\rangle+|q[2]^1\rangle)\right)$$
$$\left(\frac{1}{\sqrt{2}}(|q[3]^0\rangle + (-1)\ |q[3]^1\rangle)\right).$$

Next, the statement "cu1($-2*$pi$*1/16$) q[3],q[0];" on line 13 in Listing 4.1 is a

controlled rotation gate $\begin{bmatrix} 1 & 0 & & 0 & 0 \\ 0 & 1 & & 0 & 0 \\ 0 & 0 & 1 & & 0 \\ 0 & 0 & 0 & e^{\sqrt{-1}\times-2\times\pi\times1/16} \end{bmatrix}$. The control bit is quantum bit

q[3] and the target bit is quantum bit q[0]. If the control bit is $|1\rangle$ and the target bit is $|1\rangle$, then it adds the phase $e^{\sqrt{-1}\times-2\times\pi\times1/16}$ to the coefficient of the state $|1\rangle$ of the target bit. Otherwise, it does not change the target bit. The statement "cu1($-2*$pi$*1/16$) q[3],q[0];" on line 13 in Listing 4.1 takes the new state vector $|\Omega_5\rangle$ as its input state vector and implements one controlled rotation gate in the sixth time slot of Fig. 4.9. Because the state of the target bit q[0] is ($|q[0]^0\rangle$), the statement "cu1($-2*$pi$*1/16$) q[3],q[0];" does not change the state ($|q[0]^0\rangle$) of the target bit q[0]. In the *sixth* time slot of the quantum circuit in Fig. 4.9 there is no quantum gate to act on quantum bits q[1] and q[2], their current states are not changed. This implies that we obtain the following new state vector

$$|\Omega_6\rangle = \left(|q[0]^0\rangle\right)\left(\frac{1}{\sqrt{2}}(|q[1]^0\rangle+|q[1]^1\rangle)\right)\left(\frac{1}{\sqrt{2}}(|q[2]^0\rangle+|q[2]^1\rangle)\right)$$
$$\left(\frac{1}{\sqrt{2}}(|q[3]^0\rangle + (-1)\ |q[3]^1\rangle)\right).$$

Next, the statement "h q[1];" on line 14 in Listing 4.1 completes one Hadamard gate in the seventh time slot of Fig. 4.9. It takes the new state vector $|\Omega_6\rangle$ as its input state vector. Because the current state of quantum bit q[1] in $|\Omega_6\rangle$ is ($\frac{1}{\sqrt{2}}$ ($|q[1]^0\rangle$ + $|q[1]^1\rangle$)), the statement "h q[1];" on line 14 in Listing 4.1 actually implements $\begin{pmatrix} \frac{1}{\sqrt{2}} & \frac{1}{\sqrt{2}} \\ \frac{1}{\sqrt{2}} & -\frac{1}{\sqrt{2}} \end{pmatrix} \times \begin{pmatrix} \frac{1}{\sqrt{2}} \\ \frac{1}{\sqrt{2}} \end{pmatrix} = \begin{pmatrix} 1 \\ 0 \end{pmatrix} = |0\rangle$. This implies that the statement "h q[1];" converts quantum bit q[1] from one state ($\frac{1}{\sqrt{2}}$ ($|q[1]^0\rangle$ + $|q[1]^1\rangle$)) (its superposition) to another state $|q[1]^0\rangle$. Because in the *seventh* time slot of the quantum circuit in Fig. 4.9 there is no quantum gate to act on quantum bits q[0], q[2] and q[3], their current states are not changed. This indicates that we obtain the following new state vector

$$|\Omega_7\rangle = \left(|q[0]^0\rangle\right)\left(|q[1]^0\rangle\right)\left(\frac{1}{\sqrt{2}}(|q[2]^0\rangle+|q[2]^1\rangle)\right)$$

$$\left(\frac{1}{\sqrt{2}}\left(|q[3]^0\rangle + (-1)|q[3]^1\rangle\right)\right)$$

Next, the statement "cu1(−2*pi*1/4) q[2],q[1];" on line 15 in Listing 4.1 is a controlled rotation gate $\begin{bmatrix} 1\ 0 & 0\ 0 \\ 0\ 1 & 0\ 0 \\ 0\ 0\ 1 & 0 \\ 0\ 0\ 0 & e^{\sqrt{-1}\times-2\times\pi\times1/4} \end{bmatrix}$. The control bit is quantum bit q[2] and the target bit is quantum bit q[1]. If the control bit is |1⟩ and the target bit is |1⟩, then it adds the phase $e^{\sqrt{-1}\times-2\times\pi\times1/4}$ to the coefficient of the state |1⟩ of the target bit. Otherwise, it does not change the target bit. The statement "cu1(−2*pi*1/4) q[2],q[1];" on line 15 in Listing 4.1 takes the new state vector |Ω7⟩ as its input state vector and implements one controlled rotation gate in the eighth time slot of Fig. 4.9. Because the state of the target bit q[1] is (|q[1]^0⟩), the statement "cu1(−2*pi*1/4) q[2],q[1];" does not change the state (|q[1]^0⟩) of the target bit q[1]. In the *eighth* time slot of the quantum circuit in Fig. 4.9 there is no quantum gate to act on quantum bits q[0] and q[3], their current states are not changed. Therefore, we obtain the following new state vector

$$|\Omega_8\rangle = \left(|q[0]^0\rangle\right)\left(|q[1]^0\rangle\right)\left(\frac{1}{\sqrt{2}}\left(|q[2]^0\rangle + |q[2]^1\rangle\right)\right)$$
$$\left(\frac{1}{\sqrt{2}}\left(|q[3]^0\rangle + (-1)|q[3]^1\rangle\right)\right).$$

Next, the statement "cu1(−2*pi*1/8) q[3],q[1];" on line 16 in Listing 4.1 is a controlled rotation gate $\begin{bmatrix} 1\ 0 & 0\ 0 \\ 0\ 1 & 0\ 0 \\ 0\ 0\ 1 & 0 \\ 0\ 0\ 0 & e^{\sqrt{-1}\times-2\times\pi\times1/8} \end{bmatrix}$. The control bit is quantum bit q[3] and the target bit is quantum bit q[1]. If the control bit is |1⟩ and the target bit is |1⟩, then it adds the phase $e^{\sqrt{-1}\times-2\times\pi\times1/8}$ to the coefficient of the state |1⟩ of the target bit. Otherwise, it does not change the target bit. The statement "cu1(−2*pi*1/8) q[3],q[1];" on line 16 in Listing 4.1 takes the new state vector |Ω8⟩ as its input state vector and implements one controlled rotation gate in the ninth time slot of Fig. 4.9. Because the state of the target bit q[1] is (|q[1]^0⟩), the statement "cu1(−2*pi*1/8) q[3],q[1];" does not change the state (|q[1]^0⟩) of the target bit q[1]. In the *ninth* time slot of the quantum circuit in Fig. 4.9 there is no quantum gate to act on quantum bits q[0] and q[2], their current states are not changed. This indicates that we obtain the following new state vector

$$|\Omega_9\rangle = \left(|q[0]^0\rangle\right)\left(|q[1]^0\rangle\right)\left(\frac{1}{\sqrt{2}}\left(|q[2]^0\rangle + |q[2]^1\rangle\right)\right)$$
$$\left(\frac{1}{\sqrt{2}}\left(|q[3]^0\rangle + (-1)|q[3]^1\rangle\right)\right)$$

Next, the statement "h q[2];" on line 17 in Listing 4.1 performs one Hadamard gate in the tenth time slot of Fig. 4.9. It takes the new state vector $|\Omega_9\rangle$ as its input state

vector. Because the current state of quantum bit q[2] in $|\Omega_9\rangle$ is $(\frac{1}{\sqrt{2}}(|q[2]^0\rangle + |q[2]^1\rangle))$, the statement "h q[2];" on line 17 in Listing 4.1 actually implements

$$\begin{pmatrix} \frac{1}{\sqrt{2}} & \frac{1}{\sqrt{2}} \\ \frac{1}{\sqrt{2}} & -\frac{1}{\sqrt{2}} \end{pmatrix} \times \begin{pmatrix} \frac{1}{\sqrt{2}} \\ \frac{1}{\sqrt{2}} \end{pmatrix} = \begin{pmatrix} 1 \\ 0 \end{pmatrix} = |0\rangle.$$ This is to say that the statement "h q[2];"

converts quantum bit q[2] from one state $(\frac{1}{\sqrt{2}}(|q[2]^0\rangle + |q[2]^1\rangle))$ (its superposition) to another state $|q[2]^0\rangle$. Because in the *tenth* time slot of the quantum circuit in Fig. 4.9 there is no quantum gate to act on quantum bits q[0], q[1] and q[3], their current states are not changed. This is to say that we obtain the following new state vector

$$|\Omega_{10}\rangle = \left(|q[0]^0\rangle\right)\left(|q[1]^0\rangle\right)\left(|q[2]^0\rangle\right)\left(\frac{1}{\sqrt{2}}(|q[3]^0\rangle + (-1)|q[3]^1\rangle)\right).$$

Next, the statement "cu1(−2*pi*1/4) q[3],q[2];" on line 18 in Listing 4.1 is a

controlled rotation gate $\begin{bmatrix} 1 & 0 & 0 & 0 \\ 0 & 1 & 0 & 0 \\ 0 & 0 & 1 & 0 \\ 0 & 0 & 0 & e^{\sqrt{-1} \times -2 \times \pi \times 1/4} \end{bmatrix}$. The control bit is quantum bit

q[3] and the target bit is quantum bit q[2]. If the control bit is $|1\rangle$ and the target bit is $|1\rangle$, then it adds the phase $e^{\sqrt{-1} \times -2 \times \pi \times 1/4}$ to the coefficient of the state $|1\rangle$ of the target bit. Otherwise, it does not change the target bit. The statement "cu1(−2*pi*1/4) q[3],q[2];" on line 18 in Listing 4.1 takes the new state vector $|\Omega_{10}\rangle$ as its input state vector and implements one controlled rotation gate in the eleventh time slot of the quantum circuit in Fig. 4.9. Because the state of the target bit q[2] is $(|q[2]^0\rangle)$, the statement "cu1(−2*pi*1/4) q[3],q[2];" does not change the state $(|q[2]^0\rangle)$ of the target bit q[2]. In the *eleventh* time slot of the quantum circuit in Fig. 4.9 there is no quantum gate to act on quantum bits q[0] and q[1], their current states are not changed. Therefore, we obtain the following new state vector

$$|\Omega_{11}\rangle = \left(|q[0]^0\rangle\right)\left(|q[1]^0\rangle\right)\left(|q[2]^0\rangle\right)\left(\frac{1}{\sqrt{2}}(|q[3]^0\rangle + (-1)|q[3]^1\rangle)\right).$$

Next, the statement "h q[3];" on line 19 in Listing 4.1 implements one Hadamard gate in the *twelfth* time slot of the quantum circuit in Fig. 4.9. It takes the new state vector $|\Omega_{11}\rangle$ as its input state vector. Because the current state of quantum bit q[3] in $|\Omega_{11}\rangle$ is $(\frac{1}{\sqrt{2}}(|q[3]^0\rangle + (-1)|q[3]^1\rangle))$, the statement "h q[3];" on line 19 in Listing 4.1 actually implements $\begin{pmatrix} \frac{1}{\sqrt{2}} & \frac{1}{\sqrt{2}} \\ \frac{1}{\sqrt{2}} & -\frac{1}{\sqrt{2}} \end{pmatrix} \times \begin{pmatrix} \frac{1}{\sqrt{2}} \\ -\frac{1}{\sqrt{2}} \end{pmatrix} = \begin{pmatrix} 0 \\ 1 \end{pmatrix} = |1\rangle$. This indicates that the statement "h q[3];" converts quantum bit q[3] from one state $(\frac{1}{\sqrt{2}}(|q[3]^0\rangle + (-1)$ $|q[3]^1\rangle))$ to another state $|q[3]^1\rangle$. Because in the *twelfth* time slot of the quantum circuit in Fig. 4.9 there is no quantum gate to act on quantum bits q[0], q[1] and q[2], their current states are not changed. This implies that we obtain the following new state vector

$$|\Omega_{12}\rangle = (|q[0]^0\rangle)(|q[1]^0\rangle)(|q[2]^0\rangle)(|q[3]^1\rangle).$$

Next, the statement "swap q[0],q[3];" on line 20 in Listing 4.1 is a swap gate $\begin{bmatrix} 1 & 0 & 0 & 0 \\ 0 & 0 & 1 & 0 \\ 0 & 1 & 0 & 0 \\ 0 & 0 & 0 & 1 \end{bmatrix}$ that is to exchange the information contained in the two quantum bits q[0] and q[3]. It takes the new state vector $|\Omega_{12}\rangle$ as its input state vector and implements one swap gate in the *thirteenth* time slot of the quantum circuit in Fig. 4.9. Because in the *thirteenth* time slot of the quantum circuit in Fig. 4.9 there is no quantum gate to act on quantum bits q[1] and q[2], their current states are not changed. Therefore, after the statement "swap q[0],q[3];" on line 20 in Listing 4.1, we obtain the following new state vector
$|\Omega_{13}\rangle = (|q[0]^1\rangle)(|q[1]^0\rangle)(|q[2]^0\rangle)(|q[3]^0\rangle)$.

Next, the statement "swap q[1],q[2];" on line 21 in Listing 4.1 is a swap gate $\begin{bmatrix} 1 & 0 & 0 & 0 \\ 0 & 0 & 1 & 0 \\ 0 & 1 & 0 & 0 \\ 0 & 0 & 0 & 1 \end{bmatrix}$ that is to exchange the information contained in the two quantum bits q[1] and q[2]. It takes the new state vector $|\Omega_{13}\rangle$ as its input state vector and implements one swap gate in the *fourteenth* time slot of the quantum circuit in Fig. 4.9. Because in the *fourteenth* time slot of the quantum circuit in Fig. 4.9 there is no quantum gate to act on quantum bits q[0] and q[3], their current states are not changed. Hence, after the statement "swap q[1],q[2];" on line 21 in Listing 4.1, we obtain the following new state vector

$$|\Omega_{14}\rangle = (|q[0]^1\rangle)(|q[1]^0\rangle)(|q[2]^0\rangle)(|q[3]^0\rangle).$$

4.11.5   Measurement in Quantum Programs to Read Out the Period and the Frequency of Signals in a Given Oracular Function

Quantum bit q[0] is the most significant bit and quantum bit q[3] is the least significant bit. Classical bit c[3] is the most significant bit and classical bit c[0] is the

Listing 4.1 continued...

```
22.   measure q[0] → c[3];
23.   measure q[1] → c[2];
24.   measure q[2] → c[1];
25.   measure q[3] → c[0];
```

least significant bit. Therefore, the four statements "measure q[0] → c[3];", "measure q[1] → c[2];", "measure q[2] → c[1];" and "measure q[3] → c[0];" from line 22 through line 25 in Listing 4.1 are to measure the most significant quantum bit q[0] through the least significant quantum bit q[3]. They record the measurement outcome by means of overwriting the most significant classical bit c[3] through the least significant classical bit c[0]. They complete the measurement from the *fifteenth* time slot through the *eighteenth* time slot of the quantum circuit in Fig. 4.9.

In the backend that is *simulator* of Open QASM with *thirty-two* quantum bits in **IBM**'s quantum computer, we use the command "run" to execute the program in Listing 4.1. The measured result appears in Fig. 4.10. From Fig. 4.10, we obtain the outcome 1000 (c[3] = 1 = q[0] = |1⟩, c[2] = 0 = q[1] = |0⟩, c[1] = 0 = q[2] = |0⟩ and c[0] = 0 = q[3] = |0⟩) with the probability 1 (100%). This is to say that with the possibility 1 (100%) we obtain that the value of quantum bit q[0] is |1⟩, the value of quantum bit q[1] is |0⟩, the value of quantum bit q[2] is |0⟩ and the value of quantum bit q[3] is |0⟩. The measured outcome 1000 (eight) with the probability 1 (100%)

Fig. 4.10 The signal frequency that is the number of the period of signals per quantum register with four quantum bits is 1000 with the probability 1 (100%) for a given oracular function O_f: $\{a_1 \ a_2 \ a_3 \ a_4 | \forall \ a_d \in \{0, 1\}$ for $1 \leq d \leq 4\} \rightarrow \{\frac{1}{\sqrt{2^4}} \times e^{\sqrt{-1} \times 2 \times \pi \times \frac{1}{2} \times a_4} |a_4 \in \{0, 1\}\}$

indicates that for a given oracular function O_f: $\{a_1\, a_2\, a_3\, a_4 \mid \forall\, a_d \in \{0, 1\}$ for $1 \le d \le 4\} \to \{\frac{1}{\sqrt{2^4}} \times e^{\sqrt{-1} \times 2 \times \pi \times \frac{1}{2} \times a_4} \mid a_4 \in \{0, 1\}\}$, its output rotates back to its starting value $\left(\frac{1}{\sqrt{2^4}}\right)$ *eight* times and the phase from its output rotates back to its starting value $0°$ *eight* times. This implies that the number of the period per sixteen outputs is eight and the frequency f of O_f is equal to eight. So, we obtain that the period r of O_f is $(16/8) = 2$ so that $O_f(a_1\, a_2\, a_3\, a_4) = O_f(a_1\, a_2\, a_3\, a_4 + 2)$ to any two inputs $a_1\, a_2\, a_3\, a_4$ and $a_1\, a_2\, a_3\, a_4 + 2$.

4.11.6 The Power of the Inverse Quantum Fourier Transform to Find the Period and the Frequency of Signals in a Given Oracular Function

After the five statements from line 5 through line 9 in Listing 4.1, sampling sixteen points in Fig. 4.7 encodes sixteen phases of outputs of a given oracular function that is O_f: $\{a_1\, a_2\, a_3\, a_4 \mid \forall\, a_d \in \{0, 1\}$ for $1 \le d \le 4\} \to \{\frac{1}{\sqrt{2^4}} \times e^{\sqrt{-1} \times 2 \times \pi \times \frac{1}{2} \times a_4} \mid a_4 \in \{0, 1\}\}$. Next, after the twelve statements from line 10 through line 21 in Listing 4.1, the quantum state vector with the four quantum bits q[0], q[1], q[2] and q[3] encodes the frequency of signals and the strength of the frequency of signals. Next, after the four statements from line 22 through line 25 in Listing 4.1, the measured outcome in Fig. 4.10 is 1000 with the possibility 100%. We use Fig. 4.11 to explain that the inverse Quantum Fourier transform has the same power as the inverse discrete Fourier transform to find the frequency f and the period r of the same oracular function O_f. We take the horizontal axis of Fig. 4.10 as the *new* horizontal axis of Fig. 4.11 and take the vertical axis of Fig. 4.10 as the *new* vertical axis of Fig. 4.11. Because the horizontal axis of Fig. 4.10 is to represent the measured outcome that is the value of the quantum register with four quantum bits and is the frequency of signals, the

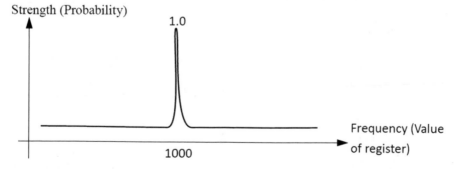

Fig. 4.11 The frequency of signal is the measured outcome of the quantum register with four quantum bits and its strength is the possibility of obtaining the outcome after the inverse quantum Fourier transform of sampling sixteen points in Fig. 4.7

new horizontal axis of Fig. 4.11 is to represent the frequency of signals. Similarly, since the vertical axis of Fig. 4.10 is to represent the possibility of obtaining each measured outcome that is the strength of the frequency of signals, the *new* vertical axis of Fig. 4.11 is to represent the strength of the frequency of signals. Because the measured outcome in Fig. 4.10 is 1000 (eight) with the possibility 100%, the frequency of signals in Fig. 4.11 is 1000 (eight) and the strength of the frequency of signals is 100%.

From Fig. 4.11, the frequency f of signals to the phase of each output of O_f is equal to 1000 (eight). This indicates that the signal rotates back to its starting value $0°$ *eight* times and there are eight periods of signals per sixteen time units. This gives that the period r of signals to the phase of each output of O_f is equal to $(16/8) = (2)$. Because the magnitude of each output to O_f is the same and is equal to $\left(\frac{1}{\sqrt{2^4}}\right)$ and the signal (the phase of each output to O_f) rotates back to its starting value $0°$ *eight* times, its output rotates back to its starting value $\left(\frac{1}{\sqrt{2^4}}\right)$ *eight* times. This is to say that the number of the period per sixteen outputs is 1000 (eight) and the frequency f of O_f is equal to eight. Therefore, this gives that the period r of O_f is $16/8 = 2$. This implies that the inverse quantum Fourier transform and the inverse discrete Fourier transform have the same power to find the frequency f and the period r of the same oracular function O_f.

4.12 Determine the Frequency and the Period of the Second Given Oracular Function

The second given oracular function is S_f: $\{a_1 \, a_2 \, a_3 \, a_4 \, |\forall \, a_d \in \{0, 1\}$ for $1 \leq d \leq 4\}$ $\rightarrow \{\frac{1}{\sqrt{2^4}} \times e^{\sqrt{-1} \times 2 \times \pi \times 0.a_3 a_4} \, |a_3$ and $a_4 \in \{0, 1\}\} = \{\frac{1}{\sqrt{2^4}} \times e^{\sqrt{-1} \times 2 \times \pi \times (\frac{1}{2} \times a_3 + \frac{1}{4} \times a_4)}$ $|a_3$ and $a_4 \in \{0, 1\}\}$. For the second given oracular function S_f, the value $(\frac{1}{\sqrt{2^4}} \times e^{\sqrt{-1} \times 2 \times \pi \times 0.a_3 a_4})$ is the amplitude (the weight) of each input $a_1 \, a_2 \, a_3 \, a_4$. The square of the absolute value of the amplitude (the weight) to an input $a_1 \, a_2 \, a_3 \, a_4$ is the possibility of getting the output of S_f that takes the input $a_1 \, a_2 \, a_3 \, a_4$ as its input value. The sum of the square of the absolute value to the amplitude (the weight) of each input $a_1 \, a_2 \, a_3 \, a_4$ is equal to one. Sixteen outputs of S_f that takes each input from $a_1^0 \, a_2^0 \, a_3^0 \, a_4^0$ through $a_1^1 \, a_2^1 \, a_3^1 \, a_4^1$ are respectively $\left(\frac{1}{\sqrt{2^4}}\right)$, $\left(\sqrt{-1} \times \frac{1}{\sqrt{2^4}}\right)$, $\left(-\frac{1}{\sqrt{2^4}}\right)$, $\left(-\sqrt{-1} \times \frac{1}{\sqrt{2^4}}\right)$, $\left(\frac{1}{\sqrt{2^4}}\right)$, $\left(\sqrt{-1} \times \frac{1}{\sqrt{2^4}}\right)$, $\left(-\frac{1}{\sqrt{2^4}}\right)$, $\left(-\sqrt{-1} \times \frac{1}{\sqrt{2^4}}\right)$, $\left(\frac{1}{\sqrt{2^4}}\right)$, $\left(\sqrt{-1} \times \frac{1}{\sqrt{2^4}}\right)$, $\left(-\frac{1}{\sqrt{2^4}}\right)$, $\left(-\sqrt{-1} \times \frac{1}{\sqrt{2^4}}\right)$, $\left(\frac{1}{\sqrt{2^4}}\right)$, $\left(\sqrt{-1} \times \frac{1}{\sqrt{2^4}}\right)$, $\left(-\frac{1}{\sqrt{2^4}}\right)$ and $\left(-\sqrt{-1} \times \frac{1}{\sqrt{2^4}}\right)$.

The period r of S_f is to satisfy $S_f(a_1 \, a_2 \, a_3 \, a_4) = S_f(a_1 \, a_2 \, a_3 \, a_4 + r)$ to any two inputs $a_1 \, a_2 \, a_3 \, a_4$ and $a_1 \, a_2 \, a_3 \, a_4 + r$. The number of the period per sixteen outputs is to the frequency f of S_f. This gives that the value of $r \times f$ is equal to sixteen (16). Hidden information and hidden patterns stored in the second given oracular function S_f are to that its output rotates back to its starting value $\left(\frac{1}{\sqrt{2^4}}\right)$ *four* times. This is to say that the number of the period per sixteen outputs is *four* and the frequency f of S_f is equal to four. Therefore, because the value of $r \times f$ is equal to sixteen (16), this gives that the period r of S_f is equal to $(16/4) = 4$. This implies that when S_f takes two inputs $a_1\,a_2\,a_3\,a_4$ and $a_1\,a_2\,a_3\,a_4 + 4$ as its input values, two outputs of S_f is the same to the two inputs. This indicates that $S_f(a_1\,a_2\,a_3\,a_4) = S_f(a_1\,a_2\,a_3\,a_4 + 4)$. For gaining the frequency f and the period r of S_f, it needs to complete at least sixteen exponential computations to $e^{\sqrt{-1} \times 2 \times \pi \times 0.a_3a_4}$ and multiplication of sixteen times.

4.12.1   Signals of Encoding the Period and the Frequency of Phases in the Second Given Oracular Function

For the second given oracular function S_f, the value $\left(\frac{1}{\sqrt{2^4}}\right)$ is the *magnitude* of the amplitude (the weight) to each input $a_1\,a_2\,a_3\,a_4$. The square of the absolute value for the *magnitude* of the amplitude (the weight) to an input $a_1\,a_2\,a_3\,a_4$ is the possibility of acquiring the output of S_f that takes the input $a_1\,a_2\,a_3\,a_4$ as its input value. The value $(2 \times \pi \times 0{\cdot}a_3a_4) = (2 \times \pi \times ((1/2) \times a_3 + (1/4) \times a_4))$ is the *phase* among the amplitudes of those sixteen inputs. The phase can take any value from 0 degree to $360°$. The phase of each input from $a_1^0\,a_2^0\,a_3^0\,a_4^0$ through $a_1^1\,a_2^1\,a_3^1\,a_4^1$ is respectively $0°, 90°, 180°, 270°, 0°, 90°, 180°, 270°, 0°, 90°, 180°, 270°, 0°, 90°, 180°$ and $270°$. Because the magnitude of each amplitude from sixteen outputs of S_f is all $\left(\frac{1}{\sqrt{2^4}}\right)$, we think of the input domain of S_f as the time domain and the phases from its sixteen outputs as signals. This implies that determining the frequency f and the period r of S_f is equivalent to figure out the frequency f and the period r of signals in the time domain (the input domain).

Since the phase of the output of each input from $a_1^0\,a_2^0\,a_3^0\,a_4^0$ through $a_1^1\,a_2^1\,a_3^1\,a_4^1$ is respectively $0°, 90°, 180°, 270°, 0°, 90°, 180°, 270°, 0°, 90°, 180°, 270°, 0°, 90°,$ $180°$ and $270°$, we take the sixteen input values as the corresponding sixteen time units and the sixteen phases as the sixteen samples of signals. Each sample encodes an angle. The angle can take 0 degree or $90°$ or $180°$ or $270°$. The sixteen input values from $a_1^0\,a_2^0\,a_3^0\,a_4^0$ through $a_1^1\,a_2^1\,a_3^1\,a_4^1$ corresponds to sixteen time units from zero through fifteen. We make use of Fig. 4.12 to show the reason of why figuring out the frequency f and the period r of S_f is equivalent to compute the frequency f and the period r of signals in the time domain (the input domain). In Fig. 4.12, the horizontal axis is to represent the time domain in which it includes the input domain of S_f and the vertical axis is to represent signals in which it includes the sixteen phases from its output. From Fig. 4.12, the signal rotates back to its starting value $0°$ *four* times. This implies that there are *four* periods of signals per sixteen time units and the frequency f of signals is equal to *four*. This gives that the period r of signals is equal to $(16/4) = (4)$. Since the magnitude of each output to S_f is the same and is equal to $\left(\frac{1}{\sqrt{2^4}}\right)$ and the signal (the phase of each output to S_f) rotates back to its

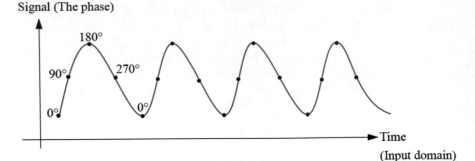

Fig. 4.12 Sampling sixteen points from each phase of sixteen outputs for the second given oracular function that is S_f: $\{a_1\ a_2\ a_3\ a_4\ |\forall\ a_d \in \{0, 1\}$ for $1 \le d \le 4\} \rightarrow \{\frac{1}{\sqrt{2^4}} \times e^{\sqrt{-1}\times 2\times\pi\times 0.a_3 a_4}\ |a_3$ and $a_4 \in \{0, 1\}\} = \{\frac{1}{\sqrt{2^4}} \times e^{\sqrt{-1}\times 2\times\pi\times\left(\frac{1}{2}\times a_3 + \frac{1}{4}\times a_4\right)}\ |a_3$ and $a_4 \in \{0, 1\}\}$

starting value $0°$ *four* times, its output rotates back to its starting value $\left(\frac{1}{\sqrt{2^4}}\right)$ *four* times. This is to say that the number of the period per sixteen outputs is *four* and the frequency f of S_f is equal to *four*. Therefore, this gives that the period r of S_f is $(16/4) = (4)$.

4.12.2 Circle Notation of Encoding the Period and the Frequency to Outputs of the Second Given Oracular Function

In Fig. 4.12 in subsection 4.12.1 sampling sixteen points only encodes sixteen phases from sixteen output of S_f, so we make use of circle notation to show how determining the frequency f and the period r of S_f is equivalent to deal with the frequency f and the period r of signals in the time domain (the input domain). In a circle, the filled *radius* encodes the *magnitude* of the amplitude to that S_f takes an input $a_1\ a_2\ a_3\ a_4$ as its input value. This is to say that the size (the shaded area) of the circle is directly proportional to the square of the absolute value of the magnitude of the amplitude to $S_f(a_1\ a_2\ a_3\ a_4)$. This gives that the size (the shaded area) of the circle is directly proportional to the possibility of obtaining the output of S_f that takes that input $a_1\ a_2$ $a_3\ a_4$ as its input value. A darker line drawn in the circle points out that the phase (a negative angle) rotates the circle *clockwise* or the phase (a positive angle) rotates the circle *counterclockwise*. A number $a_1\ a_2\ a_3\ a_4$ below the circle encodes an input to S_f. In Fig. 4.13, it includes sixteen outputs of the second given oracular function that is S_f: $\{\frac{1}{\sqrt{2^4}} \times e^{\sqrt{-1}\times 2\times\pi\times 0.a_3 a_4}\ |a_3$ and $a_4 \in \{0, 1\}\} = \{\frac{1}{\sqrt{2^4}} \times e^{\sqrt{-1}\times 2\times\pi\times\left(\frac{1}{2}\times a_3 + \frac{1}{4}\times a_4\right)}$ $|a_3$ and $a_4 \in \{0, 1\}\}$. In Fig. 4.13, sixteen circles encode sampling sixteen points

Fig. 4.13 Sixteen outputs of the second given oracular function that is $S_f: \{\frac{1}{\sqrt{2^4}} \times e^{\sqrt{-1} \times 2 \times \pi \times 0.a_3 a_4}$

$|a_3$ and $a_4 \in \{0, 1\}\} = \{\frac{1}{\sqrt{2^4}} \times e^{\sqrt{-1} \times 2 \times \pi \times \left(\frac{1}{2} \times a_3 + \frac{1}{4} \times a_4\right)} \mid a_3$ and $a_4 \in \{0, 1\}\}$

in Fig. 4.12 and sixteen numbers below each circle encode sixteen time units of the time domain (the input domain) of Fig. 4.12.

Since the shaded area of each circle in Fig. 4.13 is directly proportional to $\left(\frac{1}{\sqrt{2^4}}\right)^2$, the shaded area of each circle is the same. In Fig. 4.13, a number $a_1^0 \, a_2^0 \, a_3^0 \, a_4^0$ below the *left first* circle encodes a value zero that is an input to $S_f(a_1^0 \, a_2^0 \, a_3^0 \, a_4^0)$. The darker line drawn in the left first circle points out that the phase to output of $S_f(a_1^0 \, a_2^0 \, a_3^0 \, a_4^0)$ is a $0°$ and gives a $0°$ rotation to the left first circle. The radius of the left first circle is $\left(\frac{1}{\sqrt{2^4}}\right) = (1/4)$ that is the magnitude of output to $S_f(a_1^0 \, a_2^0 \, a_3^0 \, a_4^0)$. Similarly, in Fig. 4.13, a number $a_1^0 \, a_2^0 \, a_3^0 \, a_4^1$ below the *left second* circle encodes a value one that is an input to $S_f(a_1^0 \, a_2^0 \, a_3^0 \, a_4^1)$. The darker line drawn in the left second circle indicates that the phase to output of $S_f(a_1^0 \, a_2^0 \, a_3^0 \, a_4^1)$ is a $90°$ and gives a $90°$ rotation to the left second circle. The radius of the left second circle is $\left(\frac{1}{\sqrt{2^4}}\right) = (1/4)$ that is the magnitude of output to $S_f(a_1^0 \, a_2^0 \, a_3^0 \, a_4^1)$. Similarly, in Fig. 4.13, a number $a_1^0 \, a_2^0 \, a_3^1 \, a_4^0$ below the *left third* circle encodes a value two that is an input to $S_f(a_1^0 \, a_2^0 \, a_3^1 \, a_4^0)$. The darker line drawn in the left third circle points out that the phase to output of $S_f(a_1^0 \, a_2^0 \, a_3^1 \, a_4^0)$ is a $180°$ and gives a $180°$ rotation to the left third circle. The radius of the left third circle is $\left(\frac{1}{\sqrt{2^4}}\right) = (1/4)$ that is the magnitude of output to $S_f(a_1^0 \, a_2^0 \, a_3^1 \, a_4^0)$. Similarly, in Fig. 4.13, a number $a_1^0 \, a_2^0 \, a_3^1 \, a_4^1$ below the *left fourth* circle encodes a value three that is an input to $S_f(a_1^0 \, a_2^0 \, a_3^1 \, a_4^1)$. The darker line drawn in the left fourth circle indicates that the phase to output of $S_f(a_1^0 \, a_2^0 \, a_3^1 \, a_4^1)$ is a $270°$ and gives a $270°$ rotation to the left fourth circle. The radius of the left fourth circle is $\left(\frac{1}{\sqrt{2^4}}\right) = (1/4)$ that is the magnitude of output to $S_f(a_1^0 \, a_2^0 \, a_3^1 \, a_4^1)$.

Similarly, in Fig. 4.13, the radius of the left *fifth* circle through the last circle is all $\left(\frac{1}{\sqrt{2^4}}\right) = (1/4)$ that is the magnitude of output to $S_f(a_1^0 \, a_2^1 \, a_3^0 \, a_4^0)$ through $S_f(a_1^1 \, a_2^1 \, a_3^1 \, a_4^1)$. The darker line drawn in the left fifth circle through the last circle points out that the phase to output of $S_f(a_1^0 \, a_2^1 \, a_3^0 \, a_4^0)$ through $S_f(a_1^1 \, a_2^1 \, a_3^1 \, a_4^1)$ is respectively $0°, 90°, 180°, 270°, 0°, 90°, 180°, 270°, 0°, 90°, 180°$ and $270°$. They respectively give a $0°$ rotation, a $90°$ rotation, a $180°$ rotation and a $270°$ rotation three times to the left fifth circle through the last circle. Twelve numbers $a_1^0 \, a_2^1 \, a_3^0 \, a_4^0$ through $a_1^1 \, a_2^1 \, a_3^1 \, a_4^1$ below the left fifth circle trough the last circle encode twelve values four through fifteen that are subsequently twelve inputs to $S_f(a_1^0 \, a_2^1 \, a_3^0 \, a_4^0)$ through $O_f(a_1^1 \, a_2^1 \, a_3^1 \, a_4^1)$.

From Fig. 4.13, hidden patterns and information stored in signals of sixteen circles (sampling sixteen points) are that the signal (the phase) rotates back to its starting value 0° *four* times. This is to say that there are four periods of signals per sixteen inputs (per sixteen time units) and the frequency f of signals is equal to *four*. This means that the period r of signals is equal to $(16/4) = (4)$. Since the signal (the phase of each output to S_f) rotates back to its starting value 0° *four* times, its output rotates back to its starting value $\left(\frac{1}{\sqrt{2^4}}\right)$ *four* times. This indicates that the number of the period per sixteen outputs is *four* and the frequency f of S_f is equal to four. Thus, this gives that the period r of S_f is $16/4 = 4$. So, when the second given oracular function S_f takes two inputs $a_1\,a_2\,a_3\,a_4$ and $a_1\,a_2\,a_3\,a_4 + 4$ as its input values, it gives the same output to the two inputs. This is to say that $S_f(a_1\,a_2\,a_3\,a_4) = S_f(a_1\,a_2\,a_3\,a_4 + 4)$ to any two inputs $a_1\,a_2\,a_3\,a_4$ and $a_1\,a_2\,a_3\,a_4 + 4$. In this way, for gaining the frequency f and the period r of S_f, it just needs to implement four multiplications $(2 \times \pi \times 0.00)$ (0°), four multiplications $(2 \times \pi \times 0.01)$ (90°), four multiplications $(2 \times \pi \times 0.10)$ (180°), and four multiplications $(2 \times \pi \times 0.11)$ (270°). It very significantly enhances the performance of the first way to compute the frequency f and the period r of S_f.

4.12.3   Quantum Programs and Quantum Circuits of Implementing Sixteen Outputs of the Second Given Oracular Function

In Listing 4.2, the program is in the backend that is *simulator* of Open QASM with *thirty-two* quantum bits in **IBM**'s quantum computer. The program is to compute the frequency f and the period r of S_f so that $S_f(a_1\,a_2\,a_3\,a_4) = S_f(a_1\,a_2\,a_3\,a_4 + r)$ to any two inputs $a_1\,a_2\,a_3\,a_4$ and $a_1\,a_2\,a_3\,a_4 + r$. In Listing 4.2, we illustrate how to apply quantum gates with one quantum bit and the inverse quantum Fourier transform to determine the frequency f and the period r of S_f. Figure 4.14 is the quantum circuit of finding the frequency f and the period r of S_f. The statement "OPENQASM 2.0;"

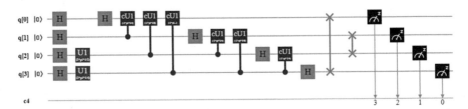

Fig. 4.14 The quantum circuit of figuring out the frequency f and the period r to the second given oracular function S_f: $\{a_1\,a_2\,a_3\,a_4 \mid \forall\, a_d \in \{0, 1\}$ for $1 \leq d \leq 4\} \rightarrow \{\frac{1}{\sqrt{2^4}} \times e^{\sqrt{-1} \times 2 \times \pi \times 0.a_3 a_4}\, |a_3$ and $a_4 \in \{0, 1\}\} = \{\frac{1}{\sqrt{2^4}} \times e^{\sqrt{-1} \times 2 \times \pi \times \left(\frac{1}{2} \times a_3 + \frac{1}{4} \times a_4\right)}\, |a_3$ and $a_4 \in \{0, 1\}\}$

on line one of Listing 4.2 is to indicate that the program is written with version 2.0 of Open QASM. Next, the statement "include "qelib1.inc";" on line two of Listing 4.2 is to continue parsing the file "qelib1.inc" as if the contents of the file were pasted at the location of the include statement, where the file "qelib1.inc" is **Quantum Experience (QE) Standard Header** and the path is specified relative to the current working directory.

Listing 4.2 The program of computing the frequency f and the period r to the second given oracular function S_f: $\{a_1\ a_2\ a_3\ a_4\ |\forall\ a_d \in \{0, 1\}$ for $1 \le d \le 4\} \to \{\frac{1}{\sqrt{2^4}} \times e^{\sqrt{-1} \times 2 \times \pi \times 0.a_3 a_4}\ |\ a_3$ and $a_4 \in \{0, 1\}\} = \{\frac{1}{\sqrt{2^4}} \times e^{\sqrt{-1} \times 2 \times \pi \times (\frac{1}{2} \times a_3 + \frac{1}{4} \times a_4)}|a_3$ and $a_4 \in \{0, 1\}\}$.

```
1.  OPENQASM 2.0;
2.  include "qelib1.inc";
3.  qreg q[4];
4.  creg c[4];
```

Next, the statement "qreg q[4];" on line three of Listing 4.2 is to declare that in the program there are four quantum bits. In the left top of Fig. 4.14, four quantum bits are subsequently q[0], q[1], q[2] and q[3]. The initial value of each quantum bit is set to state $|0\rangle$. We make use of four quantum bits q[0], q[1], q[2] and q[3] to encode the input domain $\{a_1\ a_2\ a_3\ a_4\ |\ \forall\ a_d \in \{0, 1\}$ for $1 \le d \le 4\}$ of S_f. This indicates that quantum bit q[0] encodes bit a_1, quantum bit q[1] encodes bit a_2, quantum bit q[2] encodes bit a_3 and quantum bit q[3] encodes bit a_4.

For the convenience of our explanation, $q[k]^0$ for $0 \le k \le 3$ is to encode the value 0 of q[k] and $q[k]^1$ for $0 \le k \le 3$ is to encode the value 1 of q[k]. Similarly, for the convenience of our explanation, an initial state vector of figuring out the frequency f and the period r of S_f so that $S_f(a_1\ a_2\ a_3\ a_4) = S_f(a_1\ a_2\ a_3\ a_4 + r)$ to any two inputs $a_1\ a_2\ a_3\ a_4$ and $a_1\ a_2\ a_3\ a_4 + r$ is
$$|B_0\rangle = |q[0]^0\rangle\ |q[1]^0\rangle\ |q[2]^0\rangle\ |q[3]^0\rangle = |0\rangle\ |0\rangle\ |0\rangle\ |0\rangle = |0000\rangle.$$
The corresponding decimal value of the initial state vector $|B_0\rangle$ is $2^3 \times q[0]^0 + 2^2 \times q[1]^0 + 2^1 \times q[2]^0 + 2^0 \times q[3]^0$. This is to say that quantum bit $q[0]^0$ is the most significant bit and quantum bit $q[3]^0$ is the least significant bit.

Next, the statement "creg c[4];" on line four of Listing 4.2 is to declare that there are four classical bits in the program. In the left bottom of Fig. 4.14, four classical bits are respectively c[0], c[1], c[2] and c[3]. The initial value of each classical bit is set to zero (0). For the convenience of our explanation, $c[k]^0$ for $0 \le k \le 3$ is to encode the value 0 of c[k] and $c[k]^1$ for $0 \le k \le 3$ is to encode the value 1 of c[k]. The corresponding decimal value of the four initial classical bits $c[3]^0\ c[2]^0\ c[1]^0\ c[0]^0$ is $2^3 \times c[3]^0 + 2^2 \times c[2]^0 + 2^1 \times c[1]^0 + 2^0 \times c[0]^0$. This implies that classical bit $c[3]^0$ is the most significant bit and classical bit $c[0]^0$ is the least significant bit.

Next, the four statements "h q[0];", "h q[1];" "h q[2];" and "h q[3];" on line five through eight of Listing 4.2 are to complete four Hadamard gates of the *first* time slot

Listing 4.2 continued...

// We use the following *six* statements to implement the second given oracular

// function S_f: $\{a_1\ a_2\ a_3\ a_4 \mid \forall\ a_d \in \{0, 1\}$ for $1 \le d \le 4\} \to \{\frac{1}{\sqrt{2^4}} \times$

$e^{\sqrt{-1}\times 2\times\pi\times 0.a_3a_4}$ |

// a_3 and $a_4 \in \{0, 1\}\} = \{\frac{1}{\sqrt{2^4}} \times e^{\sqrt{-1}\times 2\times\pi\times(\frac{1}{2}\times a_3+\frac{1}{4}\times a_4)} \mid a_3$ and $a_4 \in \{0,$

1}}.

  5. h q[0];

  6. h q[1];

  7. h q[2];

  8. h q[3];

  9. u1(2*pi*1/4) q[3];

 10. u1(2*pi*1/2) q[2];

of the quantum circuit in Fig. 4.14. The four statements "h q[0];", "h q[1];" "h q[2];" and "h q[3];" take the initial state vector $|B_0\rangle = |q[0]^0\rangle\ |q[1]^0\rangle\ |q[2]^0\rangle\ |q[3]^0\rangle$ as their input state vector. Since the initial state of each quantum bit is set to state $|0\rangle$, the four statements "h q[0];", "h q[1];" "h q[2];" and "h q[3];" actually complete

$$\begin{pmatrix} \frac{1}{\sqrt{2}} & \frac{1}{\sqrt{2}} \\ \frac{1}{\sqrt{2}} & -\frac{1}{\sqrt{2}} \end{pmatrix} \times \begin{pmatrix} 1 \\ 0 \end{pmatrix} = \begin{pmatrix} \frac{1}{\sqrt{2}} \\ \frac{1}{\sqrt{2}} \end{pmatrix} = \frac{1}{\sqrt{2}}\begin{pmatrix} 1 \\ 1 \end{pmatrix} = \frac{1}{\sqrt{2}}\left(\begin{pmatrix} 1 \\ 0 \end{pmatrix} + \begin{pmatrix} 0 \\ 1 \end{pmatrix}\right) = \frac{1}{\sqrt{2}}(|0\rangle + |1\rangle).$$

This indicates that the four statements convert four quantum bits q[0], q[1], q[2] and q[3] from one state $|0\rangle$ to another state $\frac{1}{\sqrt{2}}(|0\rangle + |1\rangle)$ (their superposition). Therefore, the superposition of the four quantum bits q[0], q[1], q[2] and q[3] is $(\frac{1}{\sqrt{2}}(|0\rangle + |1\rangle))$ $(\frac{1}{\sqrt{2}}(|0\rangle + |1\rangle))$ $(\frac{1}{\sqrt{2}}(|0\rangle + |1\rangle))$ $(\frac{1}{\sqrt{2}}(|0\rangle + |1\rangle))$. This is to say that we gain the following new state vector

$$|B_1\rangle = \left(\frac{1}{\sqrt{2}}(|q[0]^0\rangle + |q[0]^1\rangle)\right)\left(\frac{1}{\sqrt{2}}(|q[1]^0\rangle + |q[1]^1\rangle)\right)$$
$$\left(\frac{1}{\sqrt{2}}(|q[2]^0\rangle + |q[2]^1\rangle)\right)$$
$$\left(\frac{1}{\sqrt{2}}(|q[3]^0\rangle + |q[3]^1\rangle)\right).$$

Next, the two statements "u1(2*pi*1/4) q[3];" and "u1(2*pi*1/2) q[2];" on line 9 through line 10 in Listing 4.2 subsequently complete one rotation gate $\begin{pmatrix} 1 & 0 \\ 0 & e^{\sqrt{-1}\times 2\times\pi\times 1/4} \end{pmatrix}$ to quantum bit q[3] and another rotation gate

$\begin{pmatrix} 1 & 0 \\ 0 & e^{\sqrt{-1} \times 2 \times \pi \times 1/2} \end{pmatrix}$ to quantum bit q[2] in the *second* time slot of the quantum circuit in Fig. 4.14. The two statements "u1(2*pi*1/4) q[3];" and "u1(2*pi*1/2) q[2];" take the new state vector $(|B_1\rangle)$ as their input state vector. The state of quantum bit q[3] is $(\frac{1}{\sqrt{2}} (|q[3]^0\rangle + |q[3]^1\rangle))$ and the state of quantum bit q[2] is $(\frac{1}{\sqrt{2}} (|q[2]^0\rangle + |q[2]^1\rangle))$. Therefore, the statement "u1(2*pi*1/4) q[3];" actu-

ally implements $\begin{pmatrix} 1 & 0 \\ 0 & e^{\sqrt{-1} \times 2 \times \pi \times 1/4} \end{pmatrix} \times \begin{pmatrix} \frac{1}{\sqrt{2}} \\ \frac{1}{\sqrt{2}} \end{pmatrix} = \begin{pmatrix} \frac{1}{\sqrt{2}} \\ \frac{1}{\sqrt{2}} \times e^{\sqrt{-1} \times 2 \times \pi \times 1/4} \end{pmatrix} = \frac{1}{\sqrt{2}}$

$\begin{pmatrix} 1 \\ e^{\sqrt{-1} \times 2 \times \pi \times 1/4} \end{pmatrix} = \frac{1}{\sqrt{2}} (\begin{pmatrix} 1 \\ 0 \end{pmatrix} + \begin{pmatrix} 0 \\ e^{\sqrt{-1} \times 2 \times \pi \times 1/4} \end{pmatrix}) = \frac{1}{\sqrt{2}} (\begin{pmatrix} 1 \\ 0 \end{pmatrix} + e^{\sqrt{-1} \times 2 \times \pi \times 1/4}$

$\begin{pmatrix} 0 \\ 1 \end{pmatrix}) = \frac{1}{\sqrt{2}} (|0\rangle + e^{\sqrt{-1} \times 2 \times \pi \times 1/4} |1\rangle) = \frac{1}{\sqrt{2}} (|0\rangle + e^{\sqrt{-1} \times \pi/2} |1\rangle) = \frac{1}{\sqrt{2}} (|0\rangle + (\sqrt{-1})$

$|1\rangle) = \frac{1}{\sqrt{2}} (|q[3]^0\rangle + (\sqrt{-1}) |q[3]^1\rangle))$. This is to say that the statement "u1(2*pi*1/4) q[3];" adds the phase $e^{\sqrt{-1} \times 2 \times \pi \times 1/4} = e^{\sqrt{-1} \times \pi/2} = \cos(\pi/2) + \sqrt{-1} \times \sin(\pi/2)$ $= 0 + \sqrt{-1} \times 1 = \sqrt{-1}$ to the coefficient of the state $|1\rangle$ in the superposition of the quantum bit q[3] and does not change the coefficient of the state $|0\rangle$ in the superposition of the quantum bit q[3].

Similarly, the statement "u1(2*pi*1/2) q[2];" actually implements $\begin{pmatrix} 1 & 0 \\ 0 & e^{\sqrt{-1} \times 2 \times \pi \times 1/2} \end{pmatrix} \times \begin{pmatrix} \frac{1}{\sqrt{2}} \\ \frac{1}{\sqrt{2}} \end{pmatrix} = \begin{pmatrix} \frac{1}{\sqrt{2}} \\ \frac{1}{\sqrt{2}} \times e^{\sqrt{-1} \times 2 \times \pi \times 1/2} \end{pmatrix} = \frac{1}{\sqrt{2}} \begin{pmatrix} 1 \\ e^{\sqrt{-1} \times 2 \times \pi \times 1/2} \end{pmatrix}$

$= \frac{1}{\sqrt{2}} (\begin{pmatrix} 1 \\ 0 \end{pmatrix} + \begin{pmatrix} 0 \\ e^{\sqrt{-1} \times 2 \times \pi \times 1/2} \end{pmatrix}) = \frac{1}{\sqrt{2}} (\begin{pmatrix} 1 \\ 0 \end{pmatrix} + e^{\sqrt{-1} \times 2 \times \pi \times 1/2} \begin{pmatrix} 0 \\ 1 \end{pmatrix}) = \frac{1}{\sqrt{2}} (|0\rangle$

$+ e^{\sqrt{-1} \times 2 \times \pi \times 1/2} |1\rangle) = \frac{1}{\sqrt{2}} (|0\rangle + e^{\sqrt{-1} \times \pi} |1\rangle) = \frac{1}{\sqrt{2}} (|0\rangle + (-1) |1\rangle) = \frac{1}{\sqrt{2}} (|q[2]^0\rangle$ $+ (-1) |q[2]^1\rangle))$. This implies that the statement "u1(2*pi*1/2) q[2];" adds the phase $e^{\sqrt{-1} \times 2 \times \pi \times 1/2} = e^{\sqrt{-1} \times \pi} = \cos(\pi) + \sqrt{-1} \times \sin(\pi) = -1 + \sqrt{-1} \times 0 = -1$ to the coefficient of the state $|1\rangle$ in the superposition of the quantum bit q[2] and does not change the coefficient of the state $|0\rangle$ in the superposition of the quantum bit q[2]. There is no quantum gate to act on quantum bits q[0] through q[1] in the *second* time slot of the quantum circuit in Fig. 4.14, so their current states are not changed. This indicates that we get the following new state vector

$$|B_2\rangle = \left(\frac{1}{\sqrt{2}} (|q[0]^0\rangle + |q[0]^1\rangle) \right) \left(\frac{1}{\sqrt{2}} (|q[1]^0\rangle + |q[1]^1\rangle) \right)$$

$$\left(\frac{1}{\sqrt{2}} (|q[2]^0\rangle + e^{\sqrt{-1} \times 2 \times \pi \times 1/2} |q[2]^1\rangle) \right)$$

$$\left(\frac{1}{\sqrt{2}} (|q[3]^0\rangle + e^{\sqrt{-1} \times 2 \times \pi \times 1/4} |q[3]^1\rangle) \right)$$

$$= \left(\frac{1}{\sqrt{2}} (|q[0]^0\rangle + |q[0]^1\rangle) \right)$$

$$\left(\frac{1}{\sqrt{2}}\left(|q[1]^0\rangle + |q[1]^1\rangle\right)\right)$$

$$\left(\frac{1}{\sqrt{2}}\left(|q[2]^0\rangle + (-1)\ |q[2]^1\rangle\right)\right)$$

$$\left(\frac{1}{\sqrt{2}}\left(\big|q[3]^0\rangle + (\sqrt{-1})\big|q[3]^1\rangle\right)\right).$$

In the new state vector $(|B_2\rangle)$, the sixteen amplitudes from state $|q[0]^0\rangle\ |q[1]^0\rangle$ $|q[2]^0\rangle\ |q[3]^0\rangle$ through state $|q[0]^1\rangle\ |q[1]^1\rangle\ |q[2]^1\rangle\ |q[3]^1\rangle$ are subsequently $\left(\frac{1}{\sqrt{2^4}}\right)$, $\left(\sqrt{-1}\frac{1}{\sqrt{2^4}}\right)$, $\left(-\frac{1}{\sqrt{2^4}}\right)$, $\left(-\sqrt{-1}\frac{1}{\sqrt{2^4}}\right)$, $\left(\frac{1}{\sqrt{2^4}}\right)$, $\left(\sqrt{-1}\frac{1}{\sqrt{2^4}}\right)$, $\left(-\frac{1}{\sqrt{2^4}}\right)$, $\left(-\sqrt{-1}\frac{1}{\sqrt{2^4}}\right)$, $\left(\frac{1}{\sqrt{2^4}}\right)$, $\left(\sqrt{-1}\frac{1}{\sqrt{2^4}}\right)$, $\left(-\frac{1}{\sqrt{2^4}}\right)$, $\left(-\sqrt{-1}\frac{1}{\sqrt{2^4}}\right)$, $\left(\frac{1}{\sqrt{2^4}}\right)$, $\left(\sqrt{-1}\frac{1}{\sqrt{2^4}}\right)$, $\left(-\frac{1}{\sqrt{2^4}}\right)$, and $\left(-\sqrt{-1}\frac{1}{\sqrt{2^4}}\right)$. This means that in the new state vector $(|B_2\rangle)$, it makes use of the amplitude of each state to encode sixteen outputs to the second given oracular function S_f: $\{a_1\ a_2\ a_3\ a_4\ |\ \forall\ a_d \in \{0, 1\}$ for $1 \le d \le 4\} \to \{\frac{1}{\sqrt{2^4}} \times e^{\sqrt{-1} \times 2 \times \pi \times 0.a_3 a_4}\ |\ a_3$ and $a_4 \in \{0, 1\}\} = \{\frac{1}{\sqrt{2^4}} \times e^{\sqrt{-1} \times 2 \times \pi \times (\frac{1}{2} \times a_3 + \frac{1}{4} \times a_4)}\ |\ a_3$ and $a_4 \in \{0, 1\}\}$. Hidden information and hidden patterns stored in the new state vector $(|B_2\rangle)$ are to that the amplitude rotates back to its starting value $\left(\frac{1}{\sqrt{2^4}}\right)$ *four* times.

Similarly, in the new state vector $(|B_2\rangle)$, it uses the magnitude $\left(\frac{1}{\sqrt{2^4}}\right) = (1/4)$ of the amplitude of each state as the radius of each circle in Fig. 4.13. In Fig. 4.13, the shaded area of each circle is directly proportional to $(1/4)^2$, so the shaded area of each circle is the same. In the new state vector $(|B_2\rangle)$, the sixteen phases from state $|q[0]^0\rangle$ $|q[1]^0\rangle\ |q[2]^0\rangle\ |q[3]^0\rangle$ through state $|q[0]^1\rangle\ |q[1]^1\rangle\ |q[2]^1\rangle\ |q[3]^1\rangle$ are subsequently $0°$, $90°$, $180°$, $270°$, $0°$, $90°$, $180°$, $270°$, $0°$, $90°$, $180°$, $270°$, $0°$, $90°$, $180°$, and $270°$. They subsequently give a $0°$ rotation, a $90°$ rotation, a $180°$ rotation and a $270°$ rotation *four* times to the left first circle through the last circle in Fig. 4.13. In the new state vector $(|B_2\rangle)$, it applies the darker line drawn in the left first circle through the last circle in Fig. 4.13 to point out a $0°$ rotation, a $90°$ rotation, a $180°$ rotation and a $270°$ rotation *four* times to the sixteen circles in Fig. 4.13. This indicates that hidden information and hidden patterns stored in the new state vector $(|B_2\rangle)$ are to that the phase rotates back to its starting value $0°$ *four* times.

4.12.4   Use the Inverse Quantum Fourier Transform to Compute the Frequency and the Period of Outputs of the Second Given Oracular Function

Next, the *twelve* statements from line 11 through line 22 in Listing 4.2 implement the inverse quantum Fourier transform with four quantum bits. The statement "h q[0];"

Listing 4.2 continued...

// We use the following *twelve* statements to implement the inverse quantum
   // Fourier transform with four quantum bits.
11. h q[0];
12. cu1($-2*pi*1/4$) q[1],q[0];
13. cu1($-2*pi*1/8$) q[2],q[0];
14. cu1($-2*pi*1/16$) q[3],q[0];
15. h q[1];
16. cu1($-2*pi*1/4$) q[2],q[1];
17. cu1($-2*pi*1/8$) q[3],q[1];

on line 11 in Listing 4.2 completes one Hadamard gate in the third time slot of
Fig. 4.14. It takes the new state vector $|B_2\rangle$ as its input state vector. The current state
of quantum bit q[0] in $|B_2\rangle$ is $(\frac{1}{\sqrt{2}} (|q[0]^0\rangle + |q[0]^1\rangle))$, so the statement "h q[0];"
on line 11 in Listing 4.2 actually performs $\begin{pmatrix} \frac{1}{\sqrt{2}} & \frac{1}{\sqrt{2}} \\ \frac{1}{\sqrt{2}} & -\frac{1}{\sqrt{2}} \end{pmatrix} \times \begin{pmatrix} \frac{1}{\sqrt{2}} \\ \frac{1}{\sqrt{2}} \end{pmatrix} = \begin{pmatrix} 1 \\ 0 \end{pmatrix} = |0\rangle =$
$|q[0]^0\rangle$. This is to say that the statement "h q[0];" converts quantum bit q[0] from one
state $(\frac{1}{\sqrt{2}} (|q[0]^0\rangle + |q[0]^1\rangle))$ (its superposition) to another state $|q[0]^0\rangle$. Since there
is no quantum gate to act on quantum bits q[1] through q[3] in the *third* time slot
of the quantum circuit in Fig. 4.14, their current states are not changed. This means
that we acquire the following new state vector

$$| B_3\rangle = \left(|q[0]^0\rangle\right)\left(\frac{1}{\sqrt{2}}(|q[1]^0\rangle+|q[1]^1\rangle)\right)\left(\frac{1}{\sqrt{2}}(|q[2]^0\rangle + (-1)|q[2]^1\rangle)\right)$$
$$\left(\frac{1}{\sqrt{2}}(|q[3]^0\rangle + (\sqrt{-1})|q[3]^1\rangle)\right).$$

Next, the statement "cu1($-2*pi*1/4$) q[1],q[0];" on line 12 in Listing 4.2 is a
controlled rotation gate $\begin{bmatrix} 1 & 0 & 0 & 0 \\ 0 & 1 & 0 & 0 \\ 0 & 0 & 1 & 0 \\ 0 & 0 & 0 & e^{\sqrt{-1}\times-2\times\pi\times1/4} \end{bmatrix}$. The control bit is quantum bit
q[1] and the target bit is quantum bit q[0]. If the control bit is $|1\rangle$ and the target bit
is $|1\rangle$, then it adds the phase $e^{\sqrt{-1}\times-2\times\pi\times1/4}$ to the coefficient of the state $|1\rangle$ of the
target bit. Otherwise, it does not change the target bit. The statement "cu1($-2*pi*1/4$)
q[1],q[0];" on line 12 in Listing 4.2 takes the new state vector $|B_3\rangle$ as its input state
vector and finishes one controlled rotation gate in the fourth time slot of Fig. 4.14. The
state of the target bit q[0] is $(|q[0]^0\rangle)$, so the statement "cu1($-2*pi*1/4$) q[1],q[0];"
does not change the state $(|q[0]^0\rangle)$ of the target bit q[0]. Because there is no quantum
gate to act on quantum bits q[2] through q[3] in the *fourth* time slot of the quantum
circuit in Fig. 4.14, their current states are not changed. This means that we get the
following new state vector

$$|B_4\rangle = \left(|q[0]^0\rangle\right)\left(\frac{1}{\sqrt{2}}\left(|q[1]^0\rangle + |q[1]^1\rangle\right)\right)\left(\frac{1}{\sqrt{2}}\left(|q[2]^0\rangle + (-1)|q[2]^1\rangle\right)\right)$$
$$\left(\frac{1}{\sqrt{2}}\left(|q[3]^0\rangle + (\sqrt{-1})|q[3]^1\rangle\right)\right).$$

Next, the statement "cu1$(-2*pi*1/8)$ q[2],q[0];" on line 13 in Listing 4.2 is a

controlled rotation gate $\begin{bmatrix} 1 & 0 & 0 & 0 \\ 0 & 1 & 0 & 0 \\ 0 & 0 & 1 & 0 \\ 0 & 0 & 0 & e^{\sqrt{-1}\times -2\times\pi\times 1/8} \end{bmatrix}$. The control bit is quantum bit

q[2] and the target bit is quantum bit q[0]. If the control bit is $|1\rangle$ and the target bit is $|1\rangle$, then it adds the phase $e^{\sqrt{-1}\times -2\times\pi\times 1/8}$ to the coefficient of the state $|1\rangle$ of the target bit. Otherwise, it does not change the target bit. The statement "cu1$(-2*pi*1/8)$ q[2],q[0];" on line 13 in Listing 4.2 takes the new state vector $|B_4\rangle$ as its input state vector and implements one controlled rotation gate in the fifth time slot of Fig. 4.14. Because the state of the target bit q[0] is $(|q[0]^0\rangle)$, the statement "cu1$(-2*pi*1/8)$ q[2],q[0];" does not change the state $(|q[0]^0\rangle)$ of the target bit q[0]. In the *fifth* time slot of the quantum circuit in Fig. 4.14 there is no quantum gate to act on quantum bits q[1] and q[3], their current states are not changed. This implies that we obtain the following new state vector

$$|B_5\rangle = \left(|q[0]^0\rangle\right)\left(\frac{1}{\sqrt{2}}\left(|q[1]^0\rangle + |q[1]^1\rangle\right)\right)\left(\frac{1}{\sqrt{2}}\left(|q[2]^0\rangle + (-1)|q[2]^1\rangle\right)\right)$$
$$\left(\frac{1}{\sqrt{2}}\left(|q[3]^0\rangle + (\sqrt{-1})|q[3]^1\rangle\right)\right).$$

Next, the statement "cu1$(-2*pi*1/16)$ q[3],q[0];" on line 14 in Listing 4.2 is a

controlled rotation gate $\begin{bmatrix} 1 & 0 & 0 & 0 \\ 0 & 1 & 0 & 0 \\ 0 & 0 & 1 & 0 \\ 0 & 0 & 0 & e^{\sqrt{-1}\times -2\times\pi\times 1/16} \end{bmatrix}$. The control bit is quantum bit

q[3] and the target bit is quantum bit q[0]. If the control bit is $|1\rangle$ and the target bit is $|1\rangle$, then it adds the phase $e^{\sqrt{-1}\times -2\times\pi\times 1/16}$ to the coefficient of the state $|1\rangle$ of the target bit. Otherwise, it does not change the target bit. The statement "cu1$(-2*pi*1/16)$ q[3],q[0];" on line 14 in Listing 4.2 takes the new state vector $|B_5\rangle$ as its input state vector and completes one controlled rotation gate in the sixth time slot of Fig. 4.14. Since the state of the target bit q[0] is $(|q[0]^0\rangle)$, the statement "cu1$(-2*pi*1/16)$ q[3],q[0];" does not change the state $(|q[0]^0\rangle)$ of the target bit q[0]. In the *sixth* time slot of the quantum circuit in Fig. 4.14 there is no quantum gate to act on quantum bits q[1] and q[2], their current states are not changed. This indicates that we gain the following new state vector

$$|B_6\rangle = \left(|q[0]^0\rangle\right)\left(\frac{1}{\sqrt{2}}\left(|q[1]^0\rangle + |q[1]^1\rangle\right)\right)\left(\frac{1}{\sqrt{2}}\left(|q[2]^0\rangle + (-1)|q[2]^1\rangle\right)\right)$$

$$\left(\frac{1}{\sqrt{2}}\left(|q[3]^0\rangle + (\sqrt{-1})|q[3]^1\rangle\right)\right).$$

Next, the statement "h q[1];" on line 15 in Listing 4.2 implements one Hadamard gate in the seventh time slot of Fig. 4.14. It takes the new state vector $|B_6\rangle$ as its input state vector. The current state of quantum bit q[1] in $|B_6\rangle$ is ($\frac{1}{\sqrt{2}}$ ($|q[1]^0\rangle + |q[1]^1\rangle$))), so the statement "h q[1];" on line 15 in Listing 4.2 actually performs $\begin{pmatrix} \frac{1}{\sqrt{2}} & \frac{1}{\sqrt{2}} \\ \frac{1}{\sqrt{2}} & -\frac{1}{\sqrt{2}} \end{pmatrix}$

$\times \begin{pmatrix} \frac{1}{\sqrt{2}} \\ \frac{1}{\sqrt{2}} \end{pmatrix} = \begin{pmatrix} 1 \\ 0 \end{pmatrix} = |0\rangle = |q[1]^0\rangle$. This means that the statement "h q[1];" converts quantum bit q[1] from one state ($\frac{1}{\sqrt{2}}$ ($|q[1]^0\rangle + |q[1]^1\rangle$))) (its superposition) to another state $|q[1]^0\rangle$. Because there is no quantum gate to act on quantum bits q[0], q[2] and q[3] in the *seventh* time slot of the quantum circuit in Fig. 4.14, their current states are not changed. This is to say that we acquire the following new state vector

$$|B_7\rangle = \left(|q[0]^0\rangle\right)\left(|q[1]^0\rangle\right)\left(\frac{1}{\sqrt{2}}\left(|q[2]^0\rangle + (-1)|q[2]^1\rangle\right)\right)$$
$$\left(\frac{1}{\sqrt{2}}\left(|q[3]^0\rangle + (\sqrt{-1})\ |q[3]^1\rangle\right)\right).$$

Next, the statement "cu1(-2*pi*1/4) q[2],q[1];" on line 16 in Listing 4.2 is a controlled rotation gate $\begin{bmatrix} 1 & 0 & 0 & 0 \\ 0 & 1 & 0 & 0 \\ 0 & 0 & 1 & 0 \\ 0 & 0 & 0 & e^{\sqrt{-1}\times-2\times\pi\times1/4} \end{bmatrix}$. The control bit is quantum bit q[2] and the target bit is quantum bit q[1]. If the control bit is $|1\rangle$ and the target bit is $|1\rangle$, then it adds the phase $e^{\sqrt{-1}\times-2\times\pi\times1/4}$ to the coefficient of the state $|1\rangle$ of the target bit. Otherwise, it does not change the target bit. The statement "cu1(-2*pi*1/4) q[2],q[1];" on line 16 in Listing 4.2 takes the new state vector $|B_7\rangle$ as its input state vector and finishes one controlled rotation gate in the eighth time slot of Fig. 4.14. Because the state of the target bit q[1] is ($|q[1]^0\rangle$), the statement "cu1(-2*pi*1/4) q[2],q[1];" does not change the state ($|q[1]^0\rangle$) of the target bit q[1]. In the *eighth* time slot of the quantum circuit in Fig. 4.14 there is no quantum gate to act on quantum bits q[0] and q[3], their current states are not changed. Thus, we gain the following new state vector

$$|B_8\rangle = \left(|q[0]^0\rangle\right)\left(|q[1]^0\rangle\right)\left(\frac{1}{\sqrt{2}}\left(|q[2]^0\rangle + (-1)|q[2]^1\rangle\right)\right)$$
$$\left(\frac{1}{\sqrt{2}}\left(|q[3]^0\rangle + (\sqrt{-1})\ |q[3]^1\rangle\right)\right).$$

Next, the statement "cu1($-2*pi*1/8$) q[3],q[1];" on line 17 in Listing 4.2 is a

controlled rotation gate $\begin{bmatrix} 1 & 0 & 0 & 0 \\ 0 & 1 & 0 & 0 \\ 0 & 0 & 1 & 0 \\ 0 & 0 & 0 & e^{\sqrt{-1}\times -2\times \pi \times 1/8} \end{bmatrix}$. The control bit is quantum bit

q[3] and the target bit is quantum bit q[1]. If the control bit is $|1\rangle$ and the target bit is $|1\rangle$, then it adds the phase $e^{\sqrt{-1}\times -2\times \pi \times 1/8}$ to the coefficient of the state $|1\rangle$ of the target bit. Otherwise, it does not change the target bit. The statement "cu1($-2*pi*1/8$) q[3],q[1];" on line 17 in Listing 4.2 takes the new state vector $|B_8\rangle$ as its input state vector and implements one controlled rotation gate in the ninth time slot of Fig. 4.14. The state of the target bit q[1] is $(|q[1]^0\rangle)$, so the statement "cu1($-2*pi*1/8$) q[3],q[1];" does not change the state $(|q[1]^0\rangle)$ of the target bit q[1]. In the *ninth* time slot of the quantum circuit in Fig. 4.14 there is no quantum gate to act on quantum bits q[0] and q[2], their current states are not changed. This is to say that we obtain the following new state vector

$$|B_9\rangle = \left(|q[0]^0\rangle\right)\left(|q[1]^0\rangle\right)\left(\frac{1}{\sqrt{2}}\left(|q[2]^0\rangle + (-1)|q[2]^1\rangle\right)\right)$$
$$\left(\frac{1}{\sqrt{2}}\left(|q[3]^0\rangle + (\sqrt{-1})\ |q[3]^1\rangle\right)\right).$$

Next, the statement "h q[2];" on line 18 in Listing 4.2 completes one Hadamard gate in the tenth time slot of Fig. 4.14. It takes the new state vector $|B_9\rangle$ as its input state

Listing 4.2 continued…

```
18.  h q[2];
19.  cu1(-2*pi*1/4) q[3],q[2];
20.  h q[3];
21.  swap q[0],q[3];
22.  swap q[1],q[2];
```

vector. Since the current state of quantum bit q[2] in $|B_9\rangle$ is $(\frac{1}{\sqrt{2}}(|q[2]^0\rangle + (-1)|q[2]^1\rangle))$, the statement "h q[2];" on line 18 in Listing 4.2 actually implements $\begin{pmatrix} \frac{1}{\sqrt{2}} & \frac{1}{\sqrt{2}} \\ \frac{1}{\sqrt{2}} & -\frac{1}{\sqrt{2}} \end{pmatrix} \times \begin{pmatrix} \frac{1}{\sqrt{2}} \\ \frac{-1}{\sqrt{2}} \end{pmatrix} = \begin{pmatrix} 0 \\ 1 \end{pmatrix} = |1\rangle = |q[2]\rangle$. This means that the statement "h q[2];" converts quantum bit q[2] from one state $(\frac{1}{\sqrt{2}}(|q[2]^0\rangle + (-1)|q[2]^1\rangle))$ (its superposition) to another state $|q[2]^1\rangle$. Because there is no quantum gate to act on quantum bits q[0], q[1] and q[3] in the *tenth* time slot of the quantum circuit in Fig. 4.14, their current states are not changed. This implies that we get the following new state vector

$$|B_{10}\rangle = \left(|q[0]^0\rangle\right)\left(|q[1]^0\rangle\right)\left(|q[2]^1\rangle\right)\left(\frac{1}{\sqrt{2}}\left(|q[3]^0\rangle + (\sqrt{-1})|q[3]^1\rangle\right)\right)$$
$$= \left(|q[0]^0\rangle\right)\left(|q[1]^0\rangle\right)\left(\frac{1}{\sqrt{2}}\left(|q[2]^1\rangle|q[3]^0\rangle + (\sqrt{-1})|q[2]^1\rangle|q[3]^1\rangle\right)\right).$$

Next, the statement "cu1(-2*pi*1/4) q[3],q[2];" on line 19 in Listing 4.2 is a controlled rotation gate $\begin{bmatrix} 1 & 0 & 0 & 0 \\ 0 & 1 & 0 & 0 \\ 0 & 0 & 1 & 0 \\ 0 & 0 & 0 & e^{\sqrt{-1}\times -2\times\pi\times 1/4} \end{bmatrix}$. The control bit is quantum bit

q[3] and the target bit is quantum bit q[2]. If the control bit is $|1\rangle$ and the target bit is $|1\rangle$, then it adds the phase $e^{\sqrt{-1}\times -2\times\pi\times 1/4}$ to the coefficient of the state $|1\rangle$ of the target bit. Otherwise, it does not change the target bit. The statement "cu1$(-2*pi*1/4)$ q[3],q[2];" on line 19 in Listing 4.2 takes the new state vector $|B_{10}\rangle$ as its input state vector and completes one controlled rotation gate in the eleventh time slot of the quantum circuit in Fig. 4.14. Since the state of the target bit q[2] is $(|q[2]^1\rangle)$, the statement "cu1$(-2*pi*1/4)$ q[3],q[2];" adds the phase $e^{\sqrt{-1}\times -2\times\pi\times 1/4} = \cos(-\pi/2)$ $+(\sqrt{-1})\times\sin(-\pi/2) = \cos(\pi/2) + (-1)\times(\sqrt{-1})\times\sin(\pi/2) = 0 + (-1)\times(\sqrt{-1})$ $\times 1 = (-\sqrt{-1})$ to the coefficient of the state the state $(|q[2]^1\rangle)$ of the target bit q[2]. In the *eleventh* time slot of the quantum circuit in Fig. 4.14 there is no quantum gate to act on quantum bits q[0] and q[1], their current states are not changed. Hence, we acquire the following new state vector

$$|B_{11}\rangle = \left(|q[0]^0\rangle\right)\left(|q[1]^0\rangle\right)$$
$$\left(\frac{1}{\sqrt{2}}\left(|q[2]^1\rangle|q[3]^0\rangle + (-\sqrt{-1})(\sqrt{-1})|q[2]^1\rangle|q[3]^1\rangle\right)\right)$$
$$= \left(|q[0]^0\rangle\right)\left(|q[1]^0\rangle\right)\left(\frac{1}{\sqrt{2}}\left(|q[2]^1\rangle|q[3]^0\rangle + |q[2]^1\rangle|q[3]^1\rangle\right)\right)$$
$$= \left(|q[0]^0\rangle\right)\left(|q[1]^0\rangle\right)\left(|q[2]^1\rangle\right)\left(\frac{1}{\sqrt{2}}\left(|q[3]^0\rangle+|q[3]^1\rangle\right)\right).$$

Next, the statement "h q[3];" on line 20 in Listing 4.2 performs one Hadamard gate in the *twelfth* time slot of the quantum circuit in Fig. 4.14. It takes the new state vector $|B_{11}\rangle$ as its input state vector. The current state of quantum bit q[3] in $|B_{11}\rangle$ is $(\frac{1}{\sqrt{2}}(|q[3]^0\rangle + |q[3]^1\rangle))$, so the statement "h q[3];" on line 20 in Listing 4.2 actually finishes $\begin{pmatrix} \frac{1}{\sqrt{2}} & \frac{1}{\sqrt{2}} \\ \frac{1}{\sqrt{2}} & -\frac{1}{\sqrt{2}} \end{pmatrix} \times \begin{pmatrix} \frac{1}{\sqrt{2}} \\ \frac{1}{\sqrt{2}} \end{pmatrix} = \begin{pmatrix} 1 \\ 0 \end{pmatrix} = |0\rangle = |[q[3]^0\rangle$. This implies that the statement "h q[3];" converts quantum bit q[3] from one state $(\frac{1}{\sqrt{2}}(|q[3]^0\rangle + |q[3]^1\rangle))$ to another state $|q[3]^0\rangle$. Since there is no quantum gate to act on quantum bits q[0], q[1] and q[2] in the *twelfth* time slot of the quantum circuit in Fig. 4.14, their current states are not changed. This is to say that we gain the following new state vector

$$|B_{12}\rangle = \left(|q[0]^0\rangle\right)\left(|q[1]^0\rangle\right)\left(|q[2]^1\rangle\right)\left(|q[3]^0\rangle\right).$$

Next, the statement "swap q[0],q[3];" on line 21 in Listing 4.2 is a swap gate

$\begin{bmatrix} 1 & 0 & 0 & 0 \\ 0 & 0 & 1 & 0 \\ 0 & 1 & 0 & 0 \\ 0 & 0 & 0 & 1 \end{bmatrix}$ that is to exchange the information contained in the two quantum bits

q[0] and q[3]. It takes the new state vector $|B_{12}\rangle$ as its input state vector and implements one swap gate in the *thirteenth* time slot of the quantum circuit in Fig. 4.14. Because there is no quantum gate to act on quantum bits q[1] and q[2] in the *thirteenth* time slot of the quantum circuit in Fig. 4.14, their current states are not changed. This means that we obtain the following new state vector

$$|B_{13}\rangle = \left(|q[0]^0\rangle\right)\left(|q[1]^0\rangle\right)\left(|q[2]^1\rangle\right)\left(|q[3]^0\rangle\right).$$

Next, the statement "swap q[1],q[2];" on line 22 in Listing 4.2 is a swap gate

$\begin{bmatrix} 1 & 0 & 0 & 0 \\ 0 & 0 & 1 & 0 \\ 0 & 1 & 0 & 0 \\ 0 & 0 & 0 & 1 \end{bmatrix}$ that is to exchange the information contained in the two quantum bits

q[1] and q[2]. It takes the new state vector $|B_{13}\rangle$ as its input state vector and performs one swap gate in the *fourteenth* time slot of the quantum circuit in Fig. 4.14. Since there is no quantum gate to act on quantum bits q[0] and q[3] in the *fourteenth* time slot of the quantum circuit in Fig. 4.14, their current states are not changed. This is to say that we acquire the following new state vector

$$|B_{14}\rangle = \left(|q[0]^0\rangle\right)\left(|q[1]^1\rangle\right)\left(|q[2]^0\rangle\right)\left(|q[3]^0\rangle\right).$$

4.12.5 Reading Out the Frequency and the Period to Outputs of the Second Given Oracular Function by Means of Using Measurement in Quantum Programs

Quantum bit q[0] is the most significant bit and quantum bit q[3] is the least significant bit. Classical bit c[3] is the most significant bit and classical bit c[0] is the

Listing 4.2 continued...

```
23.  measure q[0] → c[3];
24.  measure q[1] → c[2];
25.  measure q[2] → c[1];
```

Fig. 4.15 The signal frequency that is the number of the period of signals per quantum register with four quantum bits is 0100 with the probability 1 (100%) for the second given oracular function S_f: $\{a_1\ a_2\ a_3\ a_4\ |\forall\ a_d \in \{0, 1\}\ \text{for}\ 1 \leq d \leq 4\} \rightarrow \{\frac{1}{\sqrt{2^4}} \times e^{\sqrt{-1} \times 2 \times \pi \times 0.a_3a_4}\ |a_3\ \text{and}\ a_4 \in \{0, 1\}\}$
$= \{\frac{1}{\sqrt{2^4}} \times e^{\sqrt{-1} \times 2 \times \pi \times (\frac{1}{2} \times a_3 + \frac{1}{4} \times a_4)}\ |a_3\ \text{and}\ a_4 \in \{0, 1\}\}$

> 26. measure q[3] → c[0];

least significant bit. Hence, the four statements "measure q[0] → c[3];", "measure q[1] → c[2];", "measure q[2] → c[1];" and "measure q[3] → c[0];" from line 23 through line 26 in Listing 4.2 are to read out the most significant quantum bit q[0] through the least significant quantum bit q[3]. They record the outcome of measurement by means of overwriting the most significant classical bit c[3] through the least significant classical bit c[0]. They implement the measurement from the *fifteenth* time slot through the *eighteenth* time slot of the quantum circuit in Fig. 4.14.

In the backend that is *simulator* of Open QASM with *thirty-two* quantum bits in **IBM**'s quantum computer, we make use of the command "run" to execute the program in Listing 4.2. The measured result appears in Fig. 4.15. From Fig. 4.15, we acquire the outcome 0100 (c[3] = 0 = q[0] = |0⟩, c[2] = 1 = q[1] = |1⟩, c[1] = 0 = q[2] = |0⟩ and c[0] = 0 = q[3] = |0⟩) with the probability 1 (100%). This indicates that with the possibility 1 (100%) we gain that the value of quantum bit q[0] is |0⟩, the value of quantum bit q[1] is |1⟩, the value of quantum bit q[2] is |0⟩ and the value of quantum bit q[3] is |0⟩. From the measured outcome 0100 (*four*) with the probability 1 (100%), output of the second given oracular function S_f rotates back to its starting value $\left(\frac{1}{\sqrt{2^4}}\right)$ *four* times and the phase from its output rotates back to its starting value 0° *four* times. This means that the number of the period per sixteen outputs is *four* and the frequency f of S_f is equal to *four*. Therefore, we get that the period r of S_f is (16/4) = 4 so that $S_f(a_1\ a_2\ a_3\ a_4) = S_f(a_1\ a_2\ a_3\ a_4 + 4)$ to any two inputs $a_1\ a_2\ a_3\ a_4$ and $a_1\ a_2\ a_3\ a_4 + 4$.

4.12.6   The Power of the Inverse Quantum Fourier Transform to Read Out the Frequency and the Period to Output of the Second Given Oracular Function

After the six statements from line 5 through line 10 in Listing 4.2, sampling sixteen points in Fig. 4.12 encodes sixteen phases of outputs of the second given oracular function that is S_f: $\{a_1 \ a_2 \ a_3 \ a_4 \mid \forall \ a_d \in \{0, 1\}$ for $1 \leq d \leq 4\} \rightarrow \{\frac{1}{\sqrt{2^4}} \times e^{\sqrt{-1} \times 2 \times \pi \times 0.a_3 a_4} \mid a_3$ and $a_4 \in \{0, 1\}\} = \{\frac{1}{\sqrt{2^4}} \times e^{\sqrt{-1} \times 2 \times \pi \times (\frac{1}{2} \times a_3 + \frac{1}{4} \times a_4)} \mid a_3$ and $a_4 \in \{0, 1\}\}$. Then, after the twelve statements from line 11 through line 22 in Listing 4.2, the quantum state vector with the four quantum bits q[0], q[1], q[2] and q[3] encodes the frequency of signals and the strength of the frequency of signals. Next, after the four statements from line 23 through line 26 in Listing 4.2, the measured outcome in Fig. 4.15 is 0100 with the possibility 100%. We make use of Fig. 4.16 to explain that the inverse Quantum Fourier transform has the same power as the inverse discrete Fourier transform to find the frequency f and the period r of the same oracular function S_f. We take the horizontal axis of Fig. 4.15 as the *new* horizontal axis of Fig. 4.16 and take the vertical axis of Fig. 4.15 as the *new* vertical axis of Fig. 4.16. The horizontal axis of Fig. 4.15 is to represent the measured outcome that is the value of the quantum register with four quantum bits and is the frequency of signals, so the *new* horizontal axis of Fig. 4.16 is to represent the frequency of signals. Similarly, because the vertical axis of Fig. 4.15 is to represent the possibility of gaining each measured outcome that is the strength of the frequency of signals, the *new* vertical axis of Fig. 4.16 is to represent the strength of the frequency of signals. Since the measured outcome in Fig. 4.15 is 0100 (four) with the possibility 100%, the frequency of signals in Fig. 4.16 is 0100 (four) and the strength of the frequency of signals is 100%.

From Fig. 4.16, the frequency f of signals to the phase of each output of S_f is equal to 0100 (four). This implies that the signal rotates back to its starting value $0°$

Fig. 4.16 After the inverse quantum Fourier transform of sampling sixteen points in Fig. 4.12, the frequency of signal is the measured outcome of the quantum register with four quantum bits and its strength is the possibility of reading out the outcome

four times and there are four periods of signals per sixteen time units. This means that the period r of signals to the phase of each output of S_f is equal to $(16/4) = (4)$. The magnitude of each output to S_f is the same and is equal to $\left(\frac{1}{\sqrt{2^4}}\right)$ and the signal (the phase of each output to S_f) rotates back to its starting value $0°$ *four* times, so its output rotates back to its starting value $\left(\frac{1}{\sqrt{2^4}}\right)$ *four* times. This indicates that the number of the period per sixteen outputs is 0100 (four) and the frequency f of S_f is equal to four. Thus, this gives that the period r of S_f is equal to $(16/4) = 4$. This means that the inverse quantum Fourier transform and the inverse discrete Fourier transform have the same power to find the frequency f and the period r of the same oracular function S_f.

4.13 Summary

In this chapter, we gave an introduction of complex roots of unity. Next, we described an Abelian group for n complex nth roots of unity together with a binary operation of multiplication. We also introduced properties of complex nth roots of unity. We then illustrated the discrete Fourier transform and the inverse discrete Fourier transform. We also introduced the quantum Fourier transform of implementing the discrete Fourier transform. Simultaneously, we described quantum circuits of implementing the quantum Fourier transform and gave assessment of time complexity for implementing the quantum Fourier transform. Next, we illustrated the inverse quantum Fourier transform of implementing the inverse discrete Fourier transform. Simultaneously, we introduced quantum circuits of implementing the inverse quantum Fourier transform and gave assessment of time complexity for implementing the inverse quantum Fourier transform. We then wrote two quantum programs to compute the frequency and the period to outputs of two given oracular functions.

4.14 Bibliographical Notes

In this chapter for more details about n complex nth roots of unity and their properties, the recommended book is (Cormen et al. 2009). For a more detailed description to the discrete Fourier transform and the inverse discrete Fourier transform, the recommended books are (Cormen et al. 2009; Nielsen and Chuang 2000; Imre and Balazs 2007; Lipton and Regan 2014; Silva 2018; Johnston et al. 2019). The two famous articles (Coppersmith 1964; Shor 1994) gave the original version of the Quantum Fourier transform and the inverse quantum Fourier transform. A good illustration for the product state decomposition of the quantum Fourier transform and the inverse quantum Fourier transform is the two famous articles in [Griffiths and Niu 1996; Cleve et al. 1998]. A good introduction to the instructions of Open QASM is the famous article in (Cross 2017).

4.15   Exercises

4.1   Please write a quantum program to determine the frequency and the period for outputs of the third given oracular function that is T_f: $\{a_1\ a_2\ a_3\ a_4 \mid \forall\ a_d \in \{0, 1\}$ for $1 \leq d \leq 4\} \rightarrow \{\frac{1}{\sqrt{2^4}} \times e^{\sqrt{-1} \times 2 \times \pi \times 0.a_2a_3a_4} \mid a_2, a_3$ and $a_4 \in \{0, 1\}\}$.

4.2   Please write a quantum program to compute the frequency and the period for outputs of the fourth given oracular function that is F_f: $\{a_1\ a_2\ a_3 \mid \forall\ a_d \in \{0, 1\}$ for $1 \leq d \leq 3\} \rightarrow \{\frac{1}{\sqrt{2^3}} \times e^{\sqrt{-1} \times 2 \times \pi \times 0.a_3} \mid a_3 \in \{0, 1\}\}$.

4.3   Please write a quantum program to figure out the frequency and the period for outputs of the fifth given oracular function that is I_f: $\{a_1\ a_2\ a_3 \mid \forall\ a_d \in \{0, 1\}$ for $1 \leq d \leq 3\} \rightarrow \{\frac{1}{\sqrt{2^3}} \times e^{\sqrt{-1} \times 2 \times \pi \times 0.a_2a_3} \mid a_2$ and $a_3 \in \{0, 1\}\}$.

References

Cleve, R., Ekert, A., Macciavello, C., Mosca, M.: Quantum algorithms revisited. In: Proceedings Royal Society London, Ser: A, vol. 454, pp. 339–354 (1998). E-print quant-ph/9708016

Coppersmith, D.: An approximate Fourier transform useful in quantum factoring. IBM research report. Technical Report RC 19642, December 1994. E-print quant-ph/0201067

Cormen, T.H., Leiserson, C.E., Rivest, R.L., Stein, C.: Introduction to Algorithms, 3rd edn. The MIT Press. ISBN-13: 978-0262033848, ISBN-10: 9780262033848 (2009)

Cross, A.W., Bishop, L.S., Smolin, J.A., Gambetta, J.M.: Open Quantum Assembly Language (2017). https://arxiv.org/abs/1707.03429

Griffiths, R.B., Niu, C.S.: Semiclassical Fourier transform for quantum computation. Phys. Rev. Lett. 76(17), 3228–3231 (1996). E-print quant-ph/9511007

Imre, S., Balazs, F.: Quantum Computation and Communications: An Engineering Approach. John Wiley & Sons Ltd., UK, 2007, ISBN-10: 047086902X and ISBN-13: 978-0470869024 (2005)

Johnston, E.R., Harrigan, N., Gimeno-Segovia, M.: Programming Quantum Computers: Essential Algorithms and Code Samples. O'Reilly Media, Inc., ISBN-13: 978-1492039686, ISBN-10: 1492039683 (2019)

Lipton, R.J., Regan, K.W.: Quantum Algorithms via Linear Algebra: A Primer. The MIT Press. ISBN 978-0-262-02839-4 (2014)

Nielsen, Chuang, I.L.: Quantum Computation and Quantum Information. Cambridge University Press, New York, NY. ISBN-10: 9781107002173 and ISBN-13: 978-1107002173 (2000)

Shor, P.: Algorithms for quantum computation: discrete logarithms and factoring. In: Proceedings 35th Annual Symposium on Foundations of Computer Science, Santa Fe, pp. 124–134, November 20–22 1994

Silva, V.: Practical Quantum Computing for Developers: Programming Quantum Rigs in the Cloud using Python, Quantum Assembly Language and IBM Q Experience. Apress, December 13. ISBN-10: 1484242173 and ISBN-13: 978-1484242179 (2018)

Chapter 5
Order-Finding and Factoring

The inverse quantum Fourier transform and the quantum Fourier transform are the quantum circuits of implementing the Fourier transform and they can be applied to solve a variety of interesting questions. In this chapter, we now introduce two of the most interesting of those questions that are respectively the *order-finding problem* and the *factoring problem*. Miller in 1976 proved that solving the order-finding problem is equivalent to solve the factoring problem. For the RSA public-key cryptosystem, People have currently installed more than 400,000,000 copies of its algorithms and it is the primary cryptosystem used for security on the Internet and World Wide Web. The security of the RSA public-key cryptosystem is dependent on that the problem of factoring a big nature number into the production of two large prime numbers is intractable on a classical computer.

Shor's order-finding algorithm can solve the problems of order-finding and factoring exponential faster than any conventional computer. By means of using Shor's algorithm to factor a big nature number with 1024 bits into the production of two prime numbers with 512 bits each, Imre and Ferenc in (Imre and Ferenc 2005) indicate that the execution time is approximately 0.01 s. This is to say that Shor's algorithm will make the RSA public-key cryptosystem obsolete once its reliable physical implementation becomes available on the market. In this chapter, we first introduce a little background in number theory. Next, we explain how the order-finding problem implies the ability to factor as well. We also explain how shor's algorithm solves the order-finding problem. Next, we describe how to write quantum algorithms to implement Shor's algorithm for solving the simplest case in the problems of order-finding and factoring.

© The Author(s), under exclusive license to Springer Nature Switzerland AG 2021
W.-L. Chang and A. V. Vasilakos, *Fundamentals of Quantum Programming in IBM's Quantum Computers*, Studies in Big Data 81,
https://doi.org/10.1007/978-3-030-63583-1_5

5.1   Introduction to Fundamental of Number Theory

We denote the set **Z** of *integers* is $\{\ldots, -3, -2, -1, 0, 1, 2, 3, \ldots\}$. We may often refer to the set of *non-negative integers* to be $\{0, 1, 2, 3, \ldots\}$ and the set of *positive integers* to be $\{1, 2, 3, \ldots\}$. This is to say that 0 (zero) is one element in the set of *non-negative integers* and is not one element in the set of *positive integers*. We may occasionally say natural numbers that mean *positive integers* and the set of natural numbers is to $\{1, 2, 3, \ldots\}$.

More formally, given any positive integers w and n, we represent uniquely w in the following form

$$w = q \times n + r, \tag{5.1}$$

where q is a non-negative integer that is the quotient (result) of dividing w by n and the *remainder r* lies in the range 0 to $(n - 1)$. If the value of the remainder r is equal to zero, then we say that in this case n is a *factor* or a *divisor* of w. Otherwise, n is not a factor of w. Notice that 1 and w are always factors of w. Modular arithmetic is simply ordinary arithmetic in which we just pay attention to remainders. We make use of the notation (mod N) to point out that we are working in modular arithmetic, with respect to the positive integer N. For example, because 1, 3, 5, 7, 9 and 11 all have the same remainder (1) when divided by 2, we write $1 = 3 = 5 = 7 = 9 = 11$ (mod 2).

The *greatest common divisor* or *factor* of integers, c and d, is the largest integer which is a divisor or a factor of both c and d. We write the number as $\gcd(c, d)$. For instance, the greatest common divisor or factor of 14 and 10 is 2. An easy method of obtaining this number is to enumerate the positive divisors of 14 (1, 2, 7, 14) and 10 (1, 2, 5, 10), and then pick out the largest common element in the two lists that is equal to two (2). Integers, c and d, are said to be *co-prime* if their greatest common divisor is 1. Because the greatest common divisor of 3 and 5 is 1, we say that 3 and 5 are co-prime. A *prime number* is an integer greater than 1, which has only itself and 1 as factors. If a number is an integer greater than 1 and it is not a prime number, then we call it as a composite number. The first few prime numbers are 2, 3, 5, 7, 11, 13, 17, 19, 23, ... The most important single fact about the positive integers perhaps is that we may represent uniquely them as a product of factors, which are prime numbers. Let b be any integer greater than 1. Then b has a *prime factorization* of the following form

$$b = p_1^{b_1} \times p_2^{b_2} \times \cdots \times p_n^{b_n}, \tag{5.2}$$

where $p_1, p_2, \ldots, p_n$ are distinct prime numbers, and $b_1, b_2, \ldots, b_n$ are positive integers. For small numbers, finding the prime factorization by trial and error is very easy. For example, for a small number 10, its prime factorization is $10 = 2^1 \times 5^1$. Though huge effort aimed at finding one method that can efficiently determine the prime factorization of large numbers, there is no efficient method known in a digital computer to complete the task.

5.2   Description to Euclid's Algorithm

Euclid's algorithm is a much more efficient method of computing the greatest common divisor. Figure 5.1 is the flowchart of Euclid's algorithm. We use an example to explain how Euclid's algorithm works out the greatest common divisor. The example is to determine the greatest common divisor of two positive integers c $= 15$ and $d = 12$. From the first execution of Statement S_1 in Fig. 5.1, it obtains the quotient $q = 15/12 = 1$. Next, from the first execution of Statement S_2 in Fig. 5.1, it gets the remainder $r = 15 \pmod{12} = 3$. Because the value of r is not equal to zero, on the first execution of Statement S_3 in Fig. 5.1, it returns to a *false*. Therefore, next, from the first execution of Statement S_6 in Fig. 5.1, it gains the new value of the dividend $c = 12$. Similarly, from the first execution of Statement S_7 in Fig. 5.1, it acquires the new value of the divisor $d = 3$.

Next, from the second execution of Statement S_1 in Fig. 5.1, it obtains the quotient $q = 12/3 = 4$. From the second execution of Statement S_2 in Fig. 5.1, it gets the

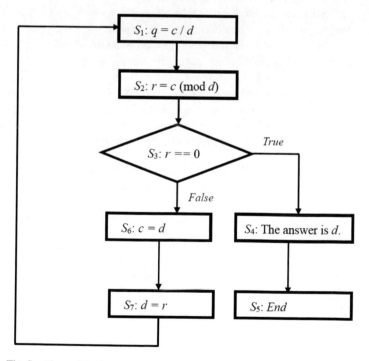

Fig. 5.1 The flowchart of Euclid's algorithm

remainder $r = 12 \pmod 3 = 0$. Since the value of r is equal to zero, on the second execution of Statement S_3 in Fig. 5.1, it returns to a *true*. Next, from the first execution of Statement S_4 in Fig. 5.1, it gives that the answer is 3. This is to say that the greatest common divisor of 15 and 12 is 3. Next, from the first execution of Statement S_5 in Fig. 5.1, it terminates the execution of Euclid's algorithm.

What resources does Euclid's algorithm consume? We assume that two positive integers, c and d, may be represented as bit strings of at most L bits each. This implies that none of the quotient q and the remainder r can be more than L bits long. Therefore, we may suppose that using L bit arithmetic does each computation. From Fig. 5.1, the divide-and-remainder operation is the heart of Euclid's algorithm. At most using the divide-and-remainder operation of $O(L)$ times completes Euclid's algorithm. Because each divide-and-remainder operation requires $O(L^2)$ operations, the total cost of Euclid's algorithm is $O(L^3)$.

5.3   Illustration to Quadratic Congruence

We assume that N is a composite number with n bits. 1 and N are two trivial factors of N itself. We also suppose that there is a function β that is $\{X|0 \le X \le N\} \rightarrow \{X^2 \pmod N\}$. The domain of the function β is $\{X|0 \le X \le N\}$ and its range is $\{X^2 \pmod N\}$. If there is an integer $0 \le X \le N$ such that $\beta(X) = X^2 = C \pmod N$, i.e., the congruence has a solution, then C is said to be a quadratic congruence $\pmod N$. Quadratic congruence $\pmod N$ is a NP-complete problem in (Manders and Adleman 1978). If the value of C is equal to one, then four integer solutions for $X^2 = 1 \pmod N$ are, respectively, $b, N - b, 1$ and $N - 1$, where $1 < b < (N/2)$ and $(N/2) < N - b < N - 1$. 1 and $N - 1$ are trivial solutions and b and $N - b$ are non-trivial solutions. This is a special case of quadratic congruence $\pmod N$ and it is still a NP-complete problem. Lemma 5.1 is used to show that we can determine a factor of N if we can find a non-trivial solution $X \ne \pm 1 \pmod N$ to the equation $X^2 = 1 \pmod N$.

Lemma 5.1 *We assume that N is a composite number with n bits, and X is a non-trivial solution to the equation $X^2 = 1 \pmod N$ in the range $0 \le X \le N$, that is, neither $X = 1 \pmod N$ nor $X = N - 1 = -1 \pmod N$. Then at least one of $gcd(N, X - 1)$ and $gcd(N, X + 1)$ is a non-trivial factor of N that can be determined using Euclid's algorithm with $O(n^3)$ operations.*

Proof Because $X^2 = 1 \pmod N$, it must be that N divides $X^2 - 1 = (X + 1) \times (X - 1)$. Since $X \ne 1$ and $X \ne N - 1$, it must be that N does not divide $(X + 1)$ and does not divide $(X - 1)$. This is to say that N must have a common factor with one or the other of $(X + 1)$ and $(X - 1)$ and $1 < X < N - 1$. Therefore, we obtain $X - 1 < X + 1 < N$. From the condition $X - 1 < X + 1 < N$, we know that the common factor cannot be N itself. Applying Euclid's algorithm with $O(n^3)$ operations we may figure out $gcd(N, X - 1)$ and $gcd(N, X + 1)$ and therefore gain a non-trivial factor of N.   ∎

We consider one example in which N is equal to 15 and any given function β that is $\{X|0 \leq X \leq 15\} \to \{X^2 \pmod{15}\}$. The domain of the given function β is $\{X|0 \leq X \leq 15\}$ and its range is $\{X^2 \pmod{15}\}$. Sixteen outputs of $\beta(X)$ from the first input zero through the last input fifteen are subsequently 0, 1, 4, 9, 1, 10, 6, 4, 4, 6, 10, 1, 9, 4, 1 and 0. Four inputs 1, 4, 11 and 14 to satisfy $X^2 = 1 \pmod{15}$. Therefore, four integer solutions for $X^2 = 1 \pmod{15}$ are, respectively, 4, 11, 1 and 14. 1 and 14 are trivial solutions. 4 and 11 are non-trivial solutions. Because $4^2 = 1 \pmod{15}$ and $11^2 = 1 \pmod{15}$, it must be that 15 divides $4^2 - 1 = (4+1) \times (4-1)$ and $11^2 - 1 = (11+1) \times (11-1)$. Hence, 15 must have a common factor with one or the other of $(4+1)$ and $(4-1)$ and 15 must have a common factor with one or the other of $(11+1)$ and $(11-1)$. This is to say that using Euclid's algorithm we may figure out $\gcd(15, 5) = 5$ and $\gcd(15, 3) = 3$ or $\gcd(15, 12) = 3$ and $\gcd(15, 10) = 5$. This means that 15 has a prime factorization that is to $15 = 5 \times 3$.

5.4   Introduction to Continued Fractions

Between the continuum of real numbers and integers, there are many valuable connections. The theory of *continued fractions* is one such beautiful connection. If c and d are integers, then we call (c/d) as the *rational fraction* or *rational number*. A *finite simple continued fraction* is denoted by a finite collection $q[1], q[2], q[3], \ldots, q[i]$ of positive integers,

$$(q[1], q[2], q[3], \ldots, q[i]) = q[1] + \cfrac{1}{q[2] + \cfrac{1}{q[3] + \cdots + \frac{1}{q[i]}}}. \tag{5.3}$$

We denote the kth convergent $(1 \leq k \leq i)$ to this continued fraction to be

$$(q[1], q[2], q[3], \ldots, q[k]) = q[1] + \cfrac{1}{q[2] + \cfrac{1}{q[3] + \cdots + \frac{1}{q[k]}}}. \tag{5.4}$$

Figure 5.2 is the flowchart of the continued fractional algorithm. We can use the continued fractional algorithm to determine a finite collection $q[1], q[2], q[3], \ldots, q[i]$ of positive integers in (5.3) for representing continued fraction of a *rational fraction*, (c/d). Therefore, applying the right-hand side equivalence in (5.3) we can describe c/d as

$$\frac{c}{d} = q[1] + \cfrac{1}{q[2] + \cfrac{1}{q[3] + \cdots + \frac{1}{q[i]}}}. \tag{5.5}$$

We make use of one example to explain how the continued fractional algorithm in Fig. 5.2 determines the continued fractional representation of (c/d) if $c = 31$ and $d = 13$ and the corresponding convergent. From the first execution of statement S_0

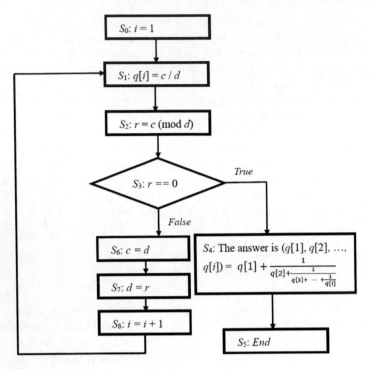

Fig. 5.2 The flowchart of the continued fractional algorithm

through statement S_2, it obtains $i = 1$, $q[1] = c/d = 31/13 = 2$ and $r = 31/(\text{mod } 13) = 5$. This is to split $(31/13)$ into its integer and fractional part and to invert its fractional part,

$$\frac{31}{13} = 2 + \frac{5}{13} = 2 + \frac{1}{\frac{13}{5}}. \tag{5.6}$$

Because the value of r is equal to 5, from the first execution of statement S_3, it returns to a *false*. Thus, next, from the first execution of statement S_6 through statement S_8, it gets that the new value of the *numerator c* is equal to 13, the new value of the *denominator d* is equal to 5 and the value of index variable i is equal to 2.

Next, from the second execution of statement S_1 through statement S_2, it obtains $q[2] = c/d = 13/5 = 2$ and $r = 13/(\text{mod } 5) = 3$. These steps—split then invert—are now used to $(13/5)$, giving

$$\frac{31}{13} = 2 + \frac{1}{2 + \frac{3}{5}} = 2 + \frac{1}{2 + \frac{1}{\frac{5}{3}}}. \tag{5.7}$$

Since the value of r is equal to 3, from the second execution of statement S_3, it returns to a *false*. Hence, next, from the second execution of statement S_6 through statement S_8, it gets that the new value of the numerator c is equal to 5, the new value of the denominator d is equal to 3 and the value of index variable i is equal to 3.

Next, from the third execution of statement S_1 through statement S_2, it gains $q[3]$ $= c/d = 5/3 = 1$ and $r = 5/(\text{mod } 3) = 2$. These steps—split then invert—are now used to (5/3), giving

$$\frac{31}{13} = 2 + \cfrac{1}{2 + \cfrac{1}{1 + \frac{2}{3}}} = 2 + \cfrac{1}{2 + \cfrac{1}{1 + \frac{1}{\frac{3}{2}}}}. \tag{5.8}$$

Because the value of r is equal to 2, from the third execution of statement S_3, it returns to a *false*. Hence, next, from the third execution of statement S_6 through statement S_8, it obtains that the new value of the numerator c is equal to 3, the new value of the denominator d is equal to 2 and the value of index variable i is equal to 4.

Next, from the fourth execution of statement S_1 through statement S_2, it acquires $q[4] = c/d = 3/2 = 1$ and $r = 3/(\text{mod } 2) = 1$. These steps—split then invert—are now used to (3/2), giving

$$\frac{31}{13} = 2 + \cfrac{1}{2 + \cfrac{1}{1 + \cfrac{1}{1 + \frac{1}{2}}}} = 2 + \cfrac{1}{2 + \cfrac{1}{1 + \cfrac{1}{1 + \frac{1}{\frac{2}{1}}}}}. \tag{5.9}$$

Since the value of r is equal to 1, from the fourth execution of statement S_3, it returns to a *false*. Thus, next, from the fourth execution of statement S_6 through statement S_8, it obtains that the new value of the numerator c is equal to 2, the new value of the denominator d is equal to 1 and the value of index variable i is equal to 5.

Next, from the fifth execution of statement S_1 through statement S_2, it acquires $q[5] = c/d = 2/1 = 2$ and $r = 2/(\text{mod } 1) = 0$. Because the value of r is equal to zero, this is to split (2/1) into its integer and fractional part and *not* to invert its fractional part. This means that $(2/1) = 2 + \frac{0}{1} = 2$. Therefore, these steps—split then *no* invert—are now used to (2/1), giving

$$\frac{31}{13} = 2 + \cfrac{1}{2 + \cfrac{1}{1 + \cfrac{1}{1 + \frac{1}{2 + \frac{0}{1}}}}} = 2 + \cfrac{1}{2 + \cfrac{1}{1 + \cfrac{1}{1 + \frac{1}{2}}}}. \tag{5.10}$$

Because the value of r is equal to 0, from the fifth execution of statement S_3, it returns to a *true*. Therefore, next, from the first execution of statement S_4, the answer is to the continued fractional representation of (31/13)

$$\frac{31}{13} = (q[1] = 2, q[2] = 2, q[3] = 1, q[4] = 1, q[5] = 2) = 2 + \cfrac{1}{2 + \cfrac{1}{1 + \cfrac{1}{1 + \frac{1}{2}}}}.$$

$$(5.11)$$

Next, from the first execution of Statement S_5, it terminates the execution of the continued fractional algorithm. For a rational number (31/13), the first convergent through the fifth convergent are subsequently $(q[1]) = 2$, $(q[1], q[2]) = 2 + \frac{1}{2} = \frac{5}{2}$, $(q[1], q[2], q[3]) = 2 + \frac{1}{2 + \frac{1}{1}} = \frac{7}{3}$, $(q[1], q[2], q[3], q[4]) = 2 + \frac{1}{2 + \frac{1}{1 + \frac{1}{1}}} = \frac{12}{5}$ and $(q[1], q[2], q[3], q[4], q[5]) = 2 + \cfrac{1}{2 + \cfrac{1}{1 + \cfrac{1}{1 + \frac{1}{2}}}} = \frac{31}{13}$.

What resources does the continued fractional algorithm in Fig. 5.2 consume for obtaining a continued fractional expansion to a rational number $c/d > 1$, where c and d are integers of L bits? This is to say that how many values of $q[k]$ for $1 \leq k \leq i$ in (5.3) must be determined from the continued fractional algorithm in Fig. 5.2. From statement S_1 and statement S_2 in the continued fractional algorithm of Fig. 5.2, each quotient $q[k]$ for $1 \leq k \leq i$ is at most L bits long, and the remainder r is at most L bits long. Thus, we may assume that making use of L bit arithmetic does each computation. From Fig. 5.2, the divide-and-remainder operation is the heart of the continued fractional algorithm. At most applying the divide-and-remainder operation of $O(L)$ times completes the continued fractional algorithm. Because each divide-and-remainder operation requires $O(L^2)$ operations, the total cost of the continued fractional algorithm is $O(L^3)$.

5.5   Order-Finding and Factoring

We assume that N is a positive integer, the greatest common divisor of X and N is one for $1 \leq X \leq N$ and X is co-prime to N. The *order* of X modulo N is to the *least* positive integer r such that $X^r = 1 \pmod{N}$. The *ordering-finding problem* is to compute r, given X and N. There is no efficient algorithm on a classical computer to solve the ordering-finding problem and the problem of factoring numbers. However, solving the problem of factoring numbers is equivalent to solve the ordering-finding problem. This is to say that if there is one efficient algorithm to solve the ordering-finding problem on a quantum computer, then it can solve the problem of factoring numbers quickly. We use Lemma 5.2 to demonstrate that we can determine a factor of N if we can find the *order* r of X modulo N to satisfy $X^r = 1 \pmod{N}$ and to be even such that a non-trivial solution $X^{\frac{r}{2}} \neq \pm 1 \pmod{N}$ to the equation $X^r = \left(X^{\frac{r}{2}}\right)^2 = 1 \pmod{N}$. Simultaneously, we use Lemma 5.3 to show representation theorem for the greatest common divisor of two integers c and d.

Lemma 5.2 *We suppose that N is a composite number with n bits, and the order r of X modulo N satisfies $X^r = 1 \pmod{N}$ and is even such that a non-trivial solution $X^{\frac{r}{2}} \neq \pm 1 \pmod{N}$ to the equation $X^r = \left(X^{\frac{r}{2}}\right)^2 = 1 \pmod{N}$ in the range $0 \leq X^{\frac{r}{2}}$*

$\leq N$. *That is that neither* $X^{\frac{r}{2}} = 1$ *(mod N) nor* $X^{\frac{r}{2}} = N - 1 = -1$ *(mod N). Then at least one of gcd(N, $X^{\frac{r}{2}} - 1$) and gcd(N, $X^{\frac{r}{2}} + 1$) is a non-trivial factor of N that can be determined using Euclid's algorithm with $O(n^3)$ operations.*

Proof Since $X^r = \left(X^{\frac{r}{2}}\right)^2 = 1$ (mod N), it must be that N divides $\left(X^{\frac{r}{2}}\right)^2 - 1 = \left(X^{\frac{r}{2}} + 1\right) \times \left(X^{\frac{r}{2}} - 1\right)$. Because $X^{\frac{r}{2}} \neq 1$ and $X^{\frac{r}{2}} \neq N - 1$, it must be that N does not divide $\left(X^{\frac{r}{2}} + 1\right)$ and does not divide $\left(X^{\frac{r}{2}} - 1\right)$. This means that N must have a common factor with one or the other of $\left(X^{\frac{r}{2}} + 1\right)$ and $\left(X^{\frac{r}{2}} - 1\right)$ and $1 < X^{\frac{r}{2}} < N - 1$. Thus, we get $X^{\frac{r}{2}} - 1 < X^{\frac{r}{2}} + 1 < N$. From the condition $X^{\frac{r}{2}} - 1 < X^{\frac{r}{2}} + 1 < N$, we see that the common factor cannot be N itself. Using Euclid's algorithm with $O(n^3)$ operations we may compute gcd(N, $X^{\frac{r}{2}} - 1$) and gcd(N, $X^{\frac{r}{2}} + 1$) and hence obtain a non-trivial factor of N. ∎

Lemma 5.3 *The greatest common divisor of two integers c and d is the least positive integer that can be written in the form $c \times u + d \times v$, where u and v are integers.*

Proof Let $t = c \times u + d \times v$ be the smallest positive integer written in this form. Let w is the greatest common divisor of c and d. Therefore, w is a divisor to both c and d and it is a divisor of t. This means that $w \leq t$. For completing the proof, we show $t \leq w$ by demonstrating that t is a divisor of both c and d. The proof is by contradiction. We assume that t is not a divisor of c. Then $c = q \times t + r$, where the remainder r is in the range 1 to $t - 1$. Rearranging this equation $c = q \times t + r$ and using $t = c \times u + d \times v$, we obtain $r = c \times (1 - q \times u) + d \times (-q \times v)$ that is a positive integer that is a linear combination of c and d. Because r is smaller than t, this contradicts the definition of t as the smallest positive integer written in a linear combination of c and d. Therefore, we infer that t must divide c. Similarly, by symmetry t must also be a divisor of d. This means that $t \leq w$ and $w \leq t$. Therefore, we complete the proof. ∎

Lemma 5.4 *We assume that integer a divides both integer c and integer d. Then a divides the greatest common divisor of both c and d.*

Proof From Lemma 5.3, the greatest common divisor of c and d is $c \times u + d \times v$, where u and v are integers. Because a divides both c and d, it must also divide $c \times u + d \times v$. Therefore, we at once infer that a divides the greatest common divisor of both c and d. ∎

When does a number, c, have a multiplicative inverse in modular arithmetic? This is to ask, given c and N, when does there exist a d such that $c \times d = 1$ (mod N)? We consider one example in which $3 \times 4 = 1$ (mod 11). This gives that the number 3 has a multiplicative inverse 4 in arithmetic modulo 11. On the other hand, trial and error explains that 3 has no multiplicative inverse modulo 6. Determining multiplicative inverse in modular arithmetic is actually related to the greatest common divisor by the notion *co-primality*: integer c and integer d are *co-prime* if their greatest common divisor is 1 (one). For example, 3 and 11 are co-prime, because the positive divisors of 3 is 1 and 3 and the positive divisors of 11 is 1 and 11. We use Lemma 5.5 to characterize the existence of multiplicative inverse in modular arithmetic applying co-primality. Simultaneously, we use Lemma 5.6 to show that the order r of X modulo N satisfies $r \leq N$.

Lemma 5.5 *Let N be an integer that is greater than 1. An integer X has a multiplicative inverse modulo N if and only if the greatest common divisor of X and N is 1.*

Proof We use $\gcd(X, N)$ to represent the greatest common divisor of X and N. We assume that X has a multiplicative inverse, which we define X^{-1}, modulo N. Then $X \times X^{-1} = 1 \pmod{N}$. This gives that $X \times X^{-1} = u \times N + 1$ for some integer u, and hence $X \times X^{-1} + (-u) \times N = 1$. From Lemma 5.3, we obtain $\gcd(X, N) = X \times X^{-1} + (-u) \times N = 1$. Therefore, we at once conclude that $\gcd(X, N) = 1$.

Conversely, if $\gcd(X, N) = 1$, then from Lemma 5.3 there must exist integer y and integer z such that $X \times y + z \times N = 1$. After applying the modular operation to both sides we obtain $X \times y \pmod{N} + z \times N \pmod{N} = 1 \pmod{N}$. As $z \times N \pmod{N} = 0$, we have then that $X \times y \pmod{N} = 1 \pmod{N}$. This means that the remainde r of $X \times y$ modulo N is equal to one. Therefore, we obtain $X \times y = 1 \pmod{N}$. So, $y = X^{-1}$ is the multiplicative inverse of X.   ∎

Lemma 5.6 *Let N be an integer greater than 1 and $1 \leq X \leq N$. X and N are co-prime and r is the least positive integer such that $X^r = 1 \pmod{N}$. Then $r \leq N$.*

Proof Consider a sequence of different order values for X: $X^0 \pmod{N}$, $X^1 \pmod{N}$, $X^2 \pmod{N}$, ..., $X^{N-1} \pmod{N}$, $X^N \pmod{N}$. Under the modular operation there can only be N unique values $0, 1, 2, ..., N - 1$ for $X^i \pmod{N}$ in the above sequence. If there are more items in the above sequence than N, some $X^i \pmod{N}$ will have the same value when the modular operation is applied (there are $N + 1$ items in the above sequence). For example, let $N = 5$ and $X = 2$. Then $X^1 = 2 \pmod 5$ and $X^5 = 2 \pmod 5$.

Hence, among the different items in the above sequence, there are two items that are equivalent under the modular operation, $X^n = X^m \pmod{N}$, where we can assume, without loss of generality, that $n > m$ and $n, m \leq N$. Since from Lemma 5.5 $\gcd(X, N) = 1$, we know that there exists a multiplicative inverse X^{-1} of X such that $X \times X^{-1} = 1 \pmod{N}$. Because the greatest common divisor of X and N is one, the greatest common divisor of X^m and N is equal to one. From Lemma 5.5, $\gcd(X^m, N) = 1$, we know that there exists a multiplicative inverse X^{-m} of X^m such that $X^m \times X^{-m} = 1 \pmod{N}$. Next, multiply both sides of the modular operation, $X^n = X^m \pmod{N}$, by X^{-m} to obtain $X^n \times X^{-m} = X^m \times X^{-m} \pmod{N}$ and $X^{n-m} = X^{m-m} \pmod{N} = X^0 \pmod{N} = 1 \pmod{N}$. From the statements above we have that $r = n - m$. Furthermore, as $n, m \leq N$ and $n > m$, it follows that $r = n - m \leq N$.   ∎

5.6 Compute the Order of 2 Modulo 15 and the Prime Factorization for 15

We would like to find the prime factorization for $N = 15$. We need to search for the nontrivial factor for $N = 15$. From Lemmas 5.1 and 5.6, we select a number $X = 2$ so that the greatest common divisor of $X = 2$ and $N = 15$ is 1 (one). This is to say that $X = 2$ is co-prime to $N = 15$. From Lemma 5.6, the order r of 2 modulo 15 satisfies $r \leq 15$. Because the number of bit representing $N = 15$ is four bits long, we also only need to use four bits that represent the value of r.

Determining the order r of 2 modulo 15 is equivalent to determine the period r of a given oracular function P_f: $\{r_1 r_2 r_3 r_4 \mid \forall r_d \in \{0, 1\}$ for $1 \leq d \leq 4\} \rightarrow \{2^{r_1 r_2 r_3 r_4}$ (mod 15) $\mid \forall r_d \in \{0, 1\}$ for $1 \leq d \leq 4\}$. The period r of P_f is to satisfy $P_f(r_1 r_2 r_3 r_4) = P_f(r_1 r_2 r_3 r_4 + r)$ to any two inputs $(r_1 r_2 r_3 r_4)$ and $(r_1 r_2 r_3 r_4 + r)$. Sixteen outputs of P_f that takes each input from $r_1^0 r_2^0 r_3^0 r_4^0$ through $r_1^1 r_2^1 r_3^1 r_4^1$ are subsequently 1, 2, 4, 8, 1, 2, 4, 8, 1, 2, 4, 8, 1, 2, 4 and 8. The frequency f of P_f is equal to the number of the period per sixteen outputs and is equal to four. The period r of P_f is the *reciprocal* of the frequency f of P_f. Thus, we obtain $r = \frac{1}{\frac{f}{16}} = \frac{16}{f} = \frac{16}{4} = 4$ and $r \times f = 4 \times 4 = 16$.

On the other hand, we think of the input domain of P_f as the time domain and its output as signals. Computing the order r of 2 modulo 15 is equivalent to determine the period r and the frequency f of signals in the time domain (the input domain). Because the output of each input from $r_1^0 r_2^0 r_3^0 r_4^0$ through $r_1^1 r_2^1 r_3^1 r_4^1$ is subsequently 1, 2, 4, 8, 1, 2, 4, 8, 1, 2, 4, 8, 1, 2, 4 and 8, we take the sixteen input values as the corresponding sixteen time units and the sixteen outputs as the sixteen samples of signals. Each sample encodes an output of P_f. The output can take 1, 2, 4 or 8. The sixteen input values from $r_1^0 r_2^0 r_3^0 r_4^0$ through $r_1^1 r_2^1 r_3^1 r_4^1$ corresponds to sixteen time units from zero through fifteen.

We use Fig. 5.3 to explain the reason of why computing the order r of 2 modulo 15 is equivalent to determine the period r and the frequency f of signals in the time domain (the input domain). In Fig. 5.3, the horizontal axis is to represent the time domain in which it contains the input domain of P_f and the vertical axis is to represent signals in which it consists of the sixteen outputs of P_f. For convenience of presentation, we use variable k to represent the decimal value of each binary input and make use of 2^k mod 15 to represent $2^{r_1 r_2 r_3 r_4}$ (mod 15).

From Fig. 5.3, hidden patterns and information stored in a given oracular function P_f are that the signal rotates back to its *first* signal (output with 1) *four* times. Its signal rotates back to its *second* signal (output with 2) *four* times. Its signal rotates back to its *third* signal (output with 4) *four* times and its signal rotates back to its *fourth* signal (output with 8) *four* times. This indicates that there are four periods of signals per sixteen time units and the frequency f of signals is equal to four.

Because in Fig. 5.3 the period r of signals is the *reciprocal* of the frequency f of signals, the period r of signals is $\frac{1}{\frac{4}{16}} = 16/4 = 4$. The period $r = 4$ of signals in Fig. 5.3 satisfies $P_f(r_1^0 r_2^1 r_3^0 r_4^0) = 2^{r_1^0 r_2^1 r_3^0 r_4^0}$ (mod 15) $= 2^4$ (mod 15) $= 16$ (mod 15) $=$

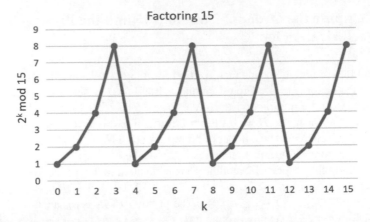

Fig. 5.3 Sampling sixteen points from sixteen outputs of a given oracular function that is P_f: $\{r_1$ $r_2 \, r_3 \, r_4 \mid \forall \, r_d \in \{0, 1\}$ for $1 \leq d \leq 4\} \rightarrow \{2^{r_1 r_2 r_3 r_4} \pmod{15} \mid \forall \, r_d \in \{0, 1\}$ for $1 \leq d \leq 4\}$

1 (mod 15), so the period $r = 4$ of signals in P_f is equivalent to the order $r = 4$ of 2 modulo 15. The cost to find the order $r = 4$ of 2 modulo 15 is to implement sixteen (2^4) operations of modular exponentiation, $2^{r_1 r_2 r_3 r_4} \pmod{15}$. Since $r = 4$ is even and is less than 15, from Lemma 5.2, we use Euclid's algorithm to compute $\gcd(15, 2^{\frac{4}{2}} + 1)$ and $\gcd(15, 2^{\frac{4}{2}} - 1)$. This is to say that two nontrivial factors for $N = 15$ are respectively 5 and 3. Therefore, the prime factorization for $N = 15$ is $N = 5 \times 3$.

Because $\left(\frac{1}{\frac{4}{16}} = 16/4 \right)$ is a rational number and is an integer, we make use of the continued fractional algorithm in Fig. 5.2 to determine the continued fractional representation of (c/d) if $c = 16$ and $d = 4$ and the corresponding convergent for explaining how the continued fractional algorithm works out in real applications. From the first execution of statement S_0 through statement S_2, it gets $i = 1$, $q[1] = c/d = 16/4 = 4$ and $r = 16 \pmod{4} = 0$. This is to split (16 / 4) into its integer and fractional part and not to invert its fractional part,

$$\frac{16}{4} = 4 + \frac{0}{4} = 4. \tag{5.12}$$

Because the value of r is equal to 0, from the first execution of statement S_3, it returns to a *true*. Thus, next, from the first execution of statement S_4, the answer is to the continued fractional representation of (16/4)

$$\frac{16}{4} = (q[1] = 4) = 4. \tag{5.13}$$

Next, from the first execution of Statement S_5, it terminates the execution of the continued fractional algorithm. For a rational number (16/4), the first convergent is $(q[1]) = 4 = \frac{4}{1}$ that is the closest to $\left(\frac{1}{\frac{4}{16}} = \frac{16}{4} \right)$ and is actually equal to $\frac{16}{4}$. This means that the first convergent $(q[1]) = 4 = \frac{4}{1}$ is equal to the period $r = \frac{r}{1}$. Hence, we obtain that the period r is equal to the numerator 4 of the first convergent. Because the numerator $r = 4$ of the first convergent is less than $N = 15$, the numerator $r = 4$ is equivalent to that the order $r = 4$ of 2 modulo 15 satisfies $2^4 = 1 \pmod{15}$.

5.7 Determine the Order of 2 Modulo 21 and the Prime Factorization for 21

We want to search for the prime factorization for $N = 21$. We need to find the nontrivial factor for $N = 21$. From Lemmas 5.1 and 5.6, we select a number $X = 2$ so that the greatest common divisor of $X = 2$ and $N = 21$ is 1 (one). This indicates that $X = 2$ is co-prime to $N = 21$. From Lemma 5.6, the order r of 2 modulo 21 satisfies $r \leq 21$. The number of bit representing $N = 21$ is five bits long, so we only need to make use of five bits that encode the value of r.

Computing the order r of 2 modulo 21 is equivalent to figure out the period r of a given oracular function A_f: $\{r_1 r_2 r_3 r_4 r_5 \mid \forall r_d \in \{0, 1\}$ for $1 \leq d \leq 5\} \rightarrow \{2^{r_1 r_2 r_3 r_4 r_5}$ $\pmod{21} \mid \forall r_d \in \{0, 1\}$ for $1 \leq d \leq 5\}$. The period r of A_f is to satisfy $A_f(r_1 r_2 r_3 r_4$ $r_5) = A_f(r_1 r_2 r_3 r_4 r_5 + r)$ to any two inputs $(r_1 r_2 r_3 r_4 r_5)$ and $(r_1 r_2 r_3 r_4 r_5 + r)$. Thirty-two outputs of A_f that takes each input from $r_1^0 r_2^0 r_3^0 r_4^0 r_5^0$ through $r_1^1 r_2^1 r_3^1 r_4^1 r_5^1$ are subsequently 1, 2, 4, 8, 16, 11, 1, 2, 4, 8, 16, 11, 1, 2, 4, 8, 16, 11, 1, 2, 4, 8, 16, 11, 1, 2, 4, 8, 16, 11, 1 and 2. The frequency f of A_f is equal to the number of the period per thirty-two outputs. The period r of A_f is the *reciprocal* of the frequency f of A_f. Hence, we obtain $r = \frac{r}{1} = \frac{1}{\frac{f}{32}} = \frac{32}{f}$ and $r \times f = 32 \times 1 = 32$.

On the other hand, we think of the input domain of A_f as the time domain and its output as signals. Figuring out the order r of 2 modulo 21 is equivalent to compute the period r and the frequency f of signals in the time domain (the input domain). The output of each input from $r_1^0 r_2^0 r_3^0 r_4^0 r_5^0$ through $r_1^1 r_2^1 r_3^1 r_4^1 r_5^1$ is subsequently 1, 2, 4, 8, 16, 11, 1, 2, 4, 8, 16, 11, 1, 2, 4, 8, 16, 11, 1, 2, 4, 8, 16, 11, 1, 2, 4, 8, 16, 11, 1 and 2. Therefore, we take the thirty-two input values as the corresponding thirty-two time units and the thirty-two outputs as the thirty-two samples of signals. Each sample encodes an output of A_f. The output can take 1, 2, 4, 8, 16 or 11. The thirty-two input values from $r_1^0 r_2^0 r_3^0 r_4^0 r_5^0$ through $r_1^1 r_2^1 r_3^1 r_4^1 r_5^1$ corresponds to thirty-two time units from zero through thirty-one.

We apply Fig. 5.4 to explain the reason of why determining the order r of 2 modulo 21 is equivalent to determine the period r and the frequency f of signals in the time domain (the input domain). In Fig. 5.4, the horizontal axis is to represent the time domain in which it contains the input domain of A_f and the vertical axis is to represent signals in which it consists of the thirty-two outputs of A_f. For convenience of presentation, we make use of variable k to represent the decimal value of each binary input and use $2^k \bmod 21$ to represent $2^{r_1 r_2 r_3 r_4 r_5} \pmod{21}$.

From Fig. 5.4, hidden patterns and information stored in a given oracular function A_f are that the signal rotates back to its *first* signal (output with 1) *six* times. Its signal

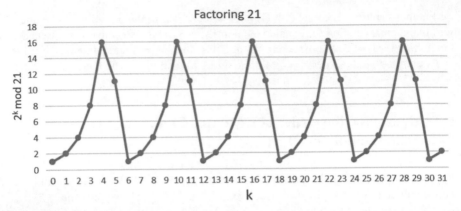

Fig. 5.4 Sampling thirty-two points from thirty-two outputs of a given oracular function that is A_f: $\{r_1 r_2 r_3 r_4 r_5 \mid \forall r_d \in \{0, 1\}$ for $1 \le d \le 5\} \rightarrow \{2^{r_1 r_2 r_3 r_4 r_5} \pmod{21} \mid \forall r_d \in \{0, 1\}$ for $1 \le d \le 5\}$

rotates back to its *second* signal (output with 2) *six* times. Its signal rotates back to its *third* signal (output with 4) *five* times and its signal rotates back to its *fourth* signal (output with 8) *five* times. Its signal rotates back to its *fifth* signal (output with 16) *five* times and its signal rotates back to its *sixth* signal (output with 11) *five* times. This is to say that there are $\left(5\frac{2}{6}\right)$ periods of signals per thirty-two time units and the frequency f of signals is equal to $\left(5\frac{2}{6}\right)$.

Since in Fig. 5.4 the period r of signals is the *reciprocal* of the frequency f of signals, the period r of signals is $\frac{1}{5\frac{2}{6}} = \frac{1}{\frac{32}{32}} = \frac{32}{6} = 6/1 = 6$. The period $r = 6$ of

signals in Fig. 5.4 satisfies $A_f(r_1^0 \ r_2^0 \ r_3^1 \ r_4^1 \ r_5^0) = 2^{r_1^0 r_2^0 r_3^1 r_4^1 r_5^0} \pmod{21} = 2^6 \pmod{21}$ $= 64 \pmod{21} = 1 \pmod{21}$, so the period $r = 6$ of signals in A_f is equivalent to the order $r = 6$ of 2 modulo 21. The cost to find the order $r = 6$ of 2 modulo 21 is to implement thirty-two (2^5) operations of modular exponentiation, $2^{r_1 r_2 r_3 r_4 r_5} \pmod{21}$. Because $r = 6$ is even and is less than 21, from Lemma 5.2, we use Euclid's algorithm to compute $\gcd(21, 2^{\frac{6}{2}} + 1)$ and $\gcd(21, 2^{\frac{6}{2}} - 1)$. This implies that two nontrivial factors for $N = 21$ are respectively 3 and 7. Thus, the prime factorization for $N = 21$ is $N = 3 \times 7$.

Because $\left(\frac{1}{5\frac{2}{6}} = \frac{1}{\frac{32}{32}} = \frac{32}{6} = 6/1\right)$ is a rational number and is an integer, we apply

the continued fractional algorithm in Fig. 5.2 to determine the continued fractional representation of (c/d) if $c = 6$ and $d = 1$ and the corresponding convergent for explaining how the continued fractional algorithm works out in real applications. From the first execution of statement S_0 through statement S_2, it gets $i = 1$, $q[1]$ $= c/d = 6/1 = 6$ and $r = 6 \pmod 1 = 0$. This is to split $(6/1)$ into its integer and fractional part and not to invert its fractional part,

$$\frac{6}{1} = 6 + \frac{0}{1} = 6. \tag{5.14}$$

Because the value of r is equal to 0, from the first execution of statement S_3, it returns to a *true*. Hence, next, from the first execution of statement S_4, the answer is to the continued fractional representation of $(6/1)$

$$\frac{6}{1} = (q[1] = 6) = 6. \tag{5.15}$$

Next, from the first execution of Statement S_5, it terminates the execution of the continued fractional algorithm. For a rational number $(6/1)$, the first convergent is $(q[1]) = 6 = \frac{6}{1}$ that is the closest to $\left(\frac{1}{\frac{5\frac{2}{6}}{32}} = \frac{1}{\frac{\frac{6}{32}}{32}} = \frac{32}{32} = \frac{6}{1} \right)$ and is actually equal to $\frac{6}{1}$. This indicates that the first convergent $(q[1]) = 6 = \frac{6}{1}$ is equal to the period $r = \frac{r}{1}$. Thus, we obtain that the period r is equal to the numerator 6 of the first convergent. Since the numerator $r = 6$ of the first convergent is less than $N = 21$, the numerator $r = 6$ is equivalent to that the order $r = 6$ of 2 modulo 21 satisfies $2^6 = 1 \pmod{21}$.

5.8 Calculate the Order of 2 Modulo 35 and the Prime Factorization for 35

We would like to find the prime factorization for $N = 35$. We need to search for the nontrivial factor for $N = 35$. From Lemmas 5.1 and 5.6, we select a number $X = 2$ so that the greatest common divisor of $X = 2$ and $N = 35$ is 1 (one). This implies that $X = 2$ is co-prime to $N = 35$. Because from Lemma 5.6, the order r of 2 modulo 35 satisfies $r \leq 35$. The number of bit representing $N = 35$ is six bits long, we only need to use six bits that encode the value of r.

Calculating the order r of 2 modulo 35 is equivalent to compute the period r of a given oracular function B_f: $\{r_1\, r_2\, r_3\, r_4\, r_5\, r_6 \mid \forall\, r_d \in \{0, 1\}$ for $1 \leq d \leq 6\} \rightarrow$ $\{2^{r_1 r_2 r_3 r_4 r_5 r_6} \pmod{35} \mid \forall\, r_d \in \{0, 1\}$ for $1 \leq d \leq 6\}$. The period r of B_f is to satisfy $B_f(r_1\, r_2\, r_3\, r_4\, r_5\, r_6) = B_f(r_1\, r_2\, r_3\, r_4\, r_5\, r_6 + r)$ to any two inputs $(r_1\, r_2\, r_3\, r_4\, r_5\, r_6)$ and $(r_1\, r_2\, r_3\, r_4\, r_5\, r_6 + r)$. The front *twenty-four* outputs of B_f that takes each input from $r_1^0 r_2^0 r_3^0 r_4^0 r_5^0 r_6^0$ through $r_1^1 r_2^1 r_3^1 r_4^1 r_5^1 r_6^1$ are subsequently 1, 2, 4, 8, 16, 32, 29, 23, 11, 22, 9, 18, 1, 2, 4, 8, 16, 32, 29, 23, 11, 22, 9 and 18. The middle *twenty-four* outputs of B_f are respectively 1, 2, 4, 8, 16, 32, 29, 23, 11, 22, 9, 18, 1, 2, 4, 8, 16, 32, 29, 23, 11, 22, 9 and 18. The last *sixteen* outputs of B_f are subsequently 1, 2, 4, 8, 16, 32, 29, 23, 11, 22, 9, 18, 1, 2, 4 and 8. The frequency f of B_f is equal to the number of the period per sixty-four outputs. The period r of B_f is the *reciprocal* of the frequency f of B_f. Therefore, we obtain $r = \frac{r}{1} = \frac{1}{\frac{1}{64}} = \frac{64}{f}$ and $r \times f = 64 \times 1 = 64$.

On the other hand, we think of the input domain of B_f as the time domain and its output as signals. Determining the order r of 2 modulo 35 is equivalent to figure out the period r and the frequency f of signals in the time domain (the input domain). The front *twenty-four* outputs of each input from $r_1^0 r_2^0 r_3^0 r_4^0 r_5^0 r_6^0$ through $r_1^1 r_2^1 r_3^1 r_4^1 r_5^1 r_6^1$ are

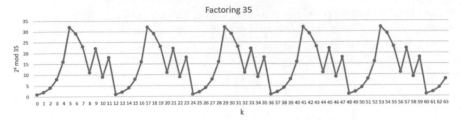

Fig. 5.5 Sampling sixty-four points from sixty-four outputs of a given oracular function that is B_f: $\{r_1\,r_2\,r_3\,r_4\,r_5\,r_6 \mid \forall\,r_d \in \{0,1\}$ for $1 \le d \le 6\} \rightarrow \{2^{r_1 r_2 r_3 r_4 r_5 r_6}$ (mod 35) $\mid \forall\,r_d \in \{0,1\}$ for $1 \le d \le 6\}$

subsequently 1, 2, 4, 8, 16, 32, 29, 23, 11, 22, 9, 18, 1, 2, 4, 8, 16, 32, 29, 23, 11, 22, 9 and 18. The middle *twenty-four* outputs are respectively 1, 2, 4, 8, 16, 32, 29, 23, 11, 22, 9, 18, 1, 2, 4, 8, 16, 32, 29, 23, 11, 22, 9 and 18. The last *sixteen* outputs are subsequently 1, 2, 4, 8, 16, 32, 29, 23, 11, 22, 9, 18, 1, 2, 4 and 8. Hence, we take the sixty-four input values as the corresponding sixty-four time units and the sixty-four outputs as the sixty-four samples of signals. Each sample encodes an output of B_f. The output can take 1, 2, 4, 8, 16, 32, 29, 23, 11, 22, 9 or 18. The sixty-four input values from $r_1^0 r_2^0 r_3^0 r_4^0 r_5^0 r_6^0$ through $r_1^1 r_2^1 r_3^1 r_4^1 r_5^1 r_6^1$ corresponds to sixty-four time units from zero through sixty-three.

We apply Fig. 5.5 to show the reason of why computing the order r of 2 modulo 35 is equivalent to determine the period r and the frequency f of signals in the time domain (the input domain). In Fig. 5.5, the horizontal axis is to represent the time domain in which it consists of the input domain of B_f and the vertical axis is to represent signals in which it includes the sixty-four outputs of B_f. For convenience of presentation, we use variable k to represent the decimal value of each binary input and use 2^k mod 35 to represent $2^{r_1 r_2 r_3 r_4 r_5 r_6}$ (mod 35).

From Fig. 5.5, hidden patterns and information stored in a given oracular function B_f are that the signal rotates back to its *first* signal (output with 1) *six* times. Its signal rotates back to its *second* signal (output with 2) *six* times. Its signal rotates back to its *third* signal (output with 4) *six* times and its signal rotates back to its *fourth* signal (output with 8) *six* times. Its signal rotates back to its *fifth* signal (output with 16) *five* times and its signal rotates back to its *sixth* signal (output with 32) *five* times. Its signal rotates back to its *seventh* signal (output with 29) *five* times and its signal rotates back to its *eighth* signal (output with 23) *five* times. Its signal rotates back to its *ninth* signal (output with 11) *five* times and its signal rotates back to its *tenth* signal (output with 22) *five* times. Its signal rotates back to its *eleventh* signal (output with 9) *five* times and its signal rotates back to its *twelfth* signal (output with 18) *five* times. This indicates that there are $\left(5\frac{4}{12}\right)$ periods of signals per sixty-four time units and the frequency f of signals is equal to $\left(5\frac{4}{12}\right)$.

Since in Fig. 5.5 the period r of signals is the *reciprocal* of the frequency f of signals, the period r of signals is $\frac{1}{5\frac{4}{12}} = \frac{1}{\frac{64}{12}} = \frac{64}{\frac{64}{12}} = 12/1 = 12$. The period $r = 12$ of

signals in Fig. 5.5 satisfies $B_f(r_1^0 r_2^0 r_3^0 r_4^0 r_5^0 r_6^0) = 2^{r_1^0 r_2^0 r_3^1 r_4^1 r_5^0 r_6^0}$ (mod 35) $= 2^{12}$ (mod 35)

$= 4096 \pmod{35} = 1 \pmod{35}$, so the period $r = 12$ of signals in B_f is equivalent to the order $r = 12$ of 2 modulo 35. The cost to find the order $r = 12$ of 2 modulo 35 is to implement sixty-four (2^6) operations of modular exponentiation, $2^{r_1 r_2 r_3 r_4 r_5 r_6} \pmod{35}$. Because $r = 12$ is even and is less than 35, from Lemma 5.2, we use Euclid's algorithm to compute $\gcd(35, 2^{\frac{12}{2}} + 1)$ and $\gcd(35, 2^{\frac{12}{2}} - 1)$. This implies that two nontrivial factors for $N = 35$ are respectively 5 and 7. Thus, the prime factorization for $N = 35$ is $N = 5 \times 7$.

Because $\left(\frac{1}{5\frac{4}{12}} = \frac{1}{\frac{64}{12}} = \frac{64}{64} = 12/1 \right)$ is a rational number and is an integer, we make use of the continued fractional algorithm in Fig. 5.2 to determine the continued fractional representation of (c/d) if $c = 12$ and $d = 1$ and the corresponding convergent for explaining how the continued fractional algorithm works out in real applications. From the first execution of statement S_0 through statement S_2, it gets $i = 1$, $q[1] = c/d = 12/1 = 12$ and $r = 12 \pmod{1} = 0$. This is to split $(12/1)$ into its integer and fractional part and not to invert its fractional part,

$$\frac{12}{1} = 12 + \frac{0}{1} = 12. \tag{5.16}$$

Since the value of r is equal to 0, from the first execution of statement S_3, it returns to a *true*. Thus, next, from the first execution of statement S_4, the answer is to the continued fractional representation of $(12/1)$

$$\frac{12}{1} = (q[1] = 12) = 12. \tag{5.17}$$

Next, from the first execution of Statement S_5, it terminates the execution of the continued fractional algorithm. For a rational number $(12/1)$, the first convergent is $(q[1]) = 12 = \frac{12}{1}$ that is the closest to $\left(\frac{1}{5\frac{4}{12}} = \frac{1}{\frac{64}{12}} = \frac{64}{64} = \frac{12}{1} \right)$ and is actually equal to $\frac{12}{1}$. This is to say that the first convergent $(q[1]) = 12 = \frac{12}{1}$ is equal to the period $r = \frac{r}{1}$. Therefore, we obtain that the period r is equal to the numerator 12 of the first convergent. The numerator $r = 12$ of the first convergent is less than $N = 35$, so the numerator $r = 12$ is equivalent to that the order $r = 12$ of 2 modulo 35 satisfies $2^{12} = 1 \pmod{35}$.

5.9   Determine the Order of 5 Modulo 33 and the Prime Factorization for 33

We would like to search for the prime factorization for $N = 33$. We need to find the nontrivial factor for $N = 33$. From Lemmas 5.1 and 5.6, we select a number $X = 5$ so that the greatest common divisor of $X = 5$ and $N = 33$ is 1 (one). This is to say that $X = 5$ is co-prime to $N = 33$. Since from Lemma 5.6, the order r of 5 modulo

33 satisfies $r \leq 33$. The number of bit representing $N = 33$ is six bits long, we only need to use six bits that encode the value of r.

Computing the order r of 5 modulo 33 is equivalent to calculate the period r of a given oracular function C_f: $\{r_1\ r_2\ r_3\ r_4\ r_5\ r_6 \mid \forall\ r_d \in \{0, 1\}$ for $1 \leq d \leq 6\} \rightarrow \{5^{r_1 r_2 r_3 r_4 r_5 r_6} \pmod{33} \mid \forall\ r_d \in \{0, 1\}$ for $1 \leq d \leq 6\}$. The period r of C_f is to satisfy $C_f(r_1\ r_2\ r_3\ r_4\ r_5\ r_6) = C_f(r_1\ r_2\ r_3\ r_4\ r_5\ r_6 + r)$ to any two inputs $(r_1\ r_2\ r_3\ r_4\ r_5\ r_6)$ and $(r_1\ r_2\ r_3\ r_4\ r_5\ r_6 + r)$. The front *twenty* outputs of C_f that takes each input from $r_1^0 r_2^0 r_3^0 r_4^0 r_5^0 r_6^0$ through $r_1^1 r_2^1 r_3^1 r_4^1 r_5^1 r_6^1$ are subsequently 1, 5, 25, 26, 31, 23, 16, 14, 4, 20, 1, 5, 25, 26, 31, 23, 16, 14, 4 and 20. The middle *twenty* outputs of C_f are respectively 1, 5, 25, 26, 31, 23, 16, 14, 4, 20, 1, 5, 25, 26, 31, 23, 16, 14, 4 and 20. The last *twenty-four* outputs of C_f are subsequently 1, 5, 25, 26, 31, 23, 16, 14, 4, 20, 1, 5, 25, 26, 31, 23, 16, 14, 4, 20, 1, 5, 25 and 26. The frequency f of C_f is equal to the number of the period per sixty-four outputs. The period r of C_f is the *reciprocal* of the frequency f of C_f. Hence, we get $r = \frac{r}{1} = \frac{1}{\frac{1}{64}} = \frac{64}{f}$ and $r \times f = 64 \times 1 = 64$.

On the other hand, we think of the input domain of C_f as the time domain and its output as signals. Calculating the order r of 5 modulo 33 is equivalent to determine the period r and the frequency f of signals in the time domain (the input domain). The front *twenty* outputs of each input from $r_1^0 r_2^0 r_3^0 r_4^0 r_5^0 r_6^0$ through $r_1^1 r_2^1 r_3^1 r_4^1 r_5^1 r_6^1$ are subsequently 1, 5, 25, 26, 31, 23, 16, 14, 4, 20, 1, 5, 25, 26, 31, 23, 16, 14, 4 and 20. The middle *twenty* outputs are respectively 1, 5, 25, 26, 31, 23, 16, 14, 4, 20, 1, 5, 25, 26, 31, 23, 16, 14, 4 and 20. The last *twenty-four* outputs are subsequently 1, 5, 25, 26, 31, 23, 16, 14, 4, 20, 1, 5, 25, 26, 31, 23, 16, 14, 4, 20, 1, 5, 25 and 26. Therefore, we take the sixty-four input values as the corresponding sixty-four time units and the sixty-four outputs as the sixty-four samples of signals. Each sample encodes an output of C_f. The output can take 1, 5, 25, 26, 31, 23, 16, 14, 4 or 20. The sixty-four input values from $r_1^0 r_2^0 r_3^0 r_4^0 r_5^0 r_6^0$ through $r_1^1 r_2^1 r_3^1 r_4^1 r_5^1 r_6^1$ corresponds to sixty-four time units from zero through sixty-three.

We use Fig. 5.6 to explain the reason of why figuring out the order r of 5 modulo 33 is equivalent to compute the period r and the frequency f of signals in the time domain (the input domain). In Fig. 5.6, the horizontal axis is to represent the time domain that is the input domain of C_f. The vertical axis is to represent signals that encode the sixty-four outputs of C_f. For convenience of presentation, we make use

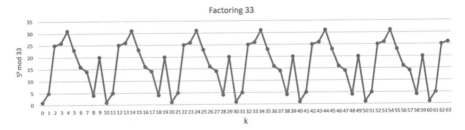

Fig. 5.6 Sampling sixty-four points from sixty-four outputs of a given oracular function that is C_f: $\{r_1\ r_2\ r_3\ r_4\ r_5\ r_6 \mid \forall\ r_d \in \{0, 1\}$ for $1 \leq d \leq 6\} \rightarrow \{5^{r_1 r_2 r_3 r_4 r_5 r_6} \pmod{33} \mid \forall\ r_d \in \{0, 1\}$ for $1 \leq d \leq 6\}$

of variable k to represent the decimal value of each binary input and apply 5^k mod 33 to represent $5^{r_1 r_2 r_3 r_4 r_5 r_6}$ (mod 33).

From Fig. 5.6, hidden patterns and information stored in a given oracular function C_f are that the signal rotates back to its *first* signal (output with 1) *seven* times. Its signal rotates back to its *second* signal (output with 5) *seven* times. Its signal rotates back to its *third* signal (output with 25) *seven* times and its signal rotates back to its *fourth* signal (output with 26) *seven* times. Its signal rotates back to its *fifth* signal (output with 31) *six* times and its signal rotates back to its *sixth* signal (output with 23) *six* times. Its signal rotates back to its *seventh* signal (output with 16) *six* times and its signal rotates back to its *eighth* signal (output with 14) *six* times. Its signal rotates back to its *ninth* signal (output with 4) *six* times and its signal rotates back to its *tenth* signal (output with 20) *six* times. This implies that there are $\left(6\frac{4}{10}\right)$ periods of signals per sixty-four time units and the frequency f of signals is equal to $\left(6\frac{4}{10}\right)$.

Because in Fig. 5.6 the period r of signals is the *reciprocal* of the frequency f of signals, the period r of signals is $\frac{1}{\frac{6\frac{4}{10}}{64}} = \frac{1}{\frac{64}{10}} = \frac{64}{\frac{64}{10}} = 10/1 = 10$. The period $r = 10$ of

signals in Fig. 5.6 satisfies $C_f(r_1^0\ r_2^0\ r_3^1\ r_4^0\ r_5^1\ r_6^0) = 5^{r_1^0 r_2^0 r_3^1 r_4^0 r_5^1 r_6^0}$ (mod 33) $= 5^{10}$ (mod 33) $= 9765625$ (mod 33) $= 1$ (mod 33), so the period $r = 10$ of signals in C_f is equivalent to the order $r = 10$ of 5 modulo 33. The cost to find the order $r = 10$ of 5 modulo 33 is to implement sixty-four (2^6) operations of modular exponentiation, $5^{r_1 r_2 r_3 r_4 r_5 r_6}$ (mod 33). Since $r = 10$ is even and is less than 33, from Lemma 5.2, we use Euclid's algorithm to compute $\gcd(33, 5^{\frac{10}{2}} + 1)$ and $\gcd(33, 5^{\frac{10}{2}} - 1)$. This indicates that two nontrivial factors for $N = 33$ are respectively 3 and 11. Therefore, the prime factorization for $N = 33$ is $N = 3 \times 11$.

Since $\left(\frac{1}{\frac{6\frac{4}{10}}{64}} = \frac{1}{\frac{64}{10}} = \frac{64}{\frac{64}{10}} = 10/1\right)$ is a rational number and is an integer, we use

the continued fractional algorithm in Fig. 5.2 to determine the continued fractional representation of (c/d) if $c = 10$ and $d = 1$ and the corresponding convergent for explaining how the continued fractional algorithm works out in real applications. From the first execution of statement S_0 through statement S_2, it obtains $i = 1$, $q[1] = c/d = 10/1 = 10$ and $r = 10$ (mod 1) $= 0$. This is to split (10/1) into its integer and fractional part and not to invert its fractional part,

$$\frac{10}{1} = 10 + \frac{0}{1} = 10. \tag{5.18}$$

Because the value of r is equal to 0, from the first execution of statement S_3, it returns to a *true*. Therefore, next, from the first execution of statement S_4, the answer is to the continued fractional representation of (10/1)

$$\frac{10}{1} = (q[1] = 10) = 10. \tag{5.19}$$

Next, from the first execution of Statement S_5, it terminates the execution of the continued fractional algorithm. For a rational number (10/1), the first convergent is

$(q[1]) = 10 = \frac{10}{1}$ that is the closest to $\left(\frac{1}{6\frac{4}{10}} = \frac{1}{\frac{64}{10}} = \frac{64}{64} = \frac{10}{1} \right)$ and is actually equal

to $\frac{10}{1}$. This means that the first convergent $(q[1]) = 10 = \frac{10}{1}$ is equal to the period $r = \frac{r}{1}$. Hence, we get that the period r is equal to the numerator 10 of the first convergent. Because the numerator $r = 10$ of the first convergent is less than $N = 33$, the numerator $r = 10$ is equivalent to that the order $r = 10$ of 5 modulo 33 satisfies $5^{10} = 1$ (mod 33).

5.10 The Possibility of Finding the Even Order of X Modulo N

We assume that the set $\mathbf{Z} = \{..., -3, -2, -1, 0, 1, 2, 3, ...\}$ of integers and the set $Y = \{0, 1, 2, 3, ...\}$ of natural numbers. The notion $d \mid a$ (read "d *divides* a") means that $a = q \times d$ for some integer q and a is a *multiple* of d and d is a divisor of a. For example, 15 is a multiple of 1, 3, 5, and 15 and 1, 3, 5, and 15 are the divisors of 15. Every integer a is divisible by the *trivial divisor* 1 and a. Nontrivial divisor of integer a are also called *factors* of a. For example, the factors (the nontrivial divisors) of 15 are 3 and 5.

An integer $a > 1$ whose only divisors are the trivial divisor a and 1 is said to be a *prime* number (or, more simply, a *prime*). The first six prime, in order, are 2, 3, 5, 7, 11 and 13. An integer $a > 1$ that is not a prime is said to be a *composite* number (or, more simply, a *composite*). For example, 15 is a composite because it has the factors 3 and 5. The integer 1 is said to be a *unit* and is neither prime nor composite. Similarly, the integer 0 and all negative integers are neither prime nor composite. From (5.1), for given any positive integers w and N, there are unique integers q and r such that $0 \leq r < N$ and $w = q \times N + r$. Integer q is the *quotient* (result) of dividing w by N. Integer r is the *remainder* of dividing w by N and we write $r = w$ (mod N). Given any integer N, we can partition the integers into those that are multiples of N and those that are not multiples of N. By classifying the multiples and the non-multiples of N in light of their remainders when divided by N, we can obtain the refinement of this partition.

According to their remainders modulo N, the integer can be divided into N equivalent classes. The *equivalent class modulo N* including an integer w is

$$[w]_N = \{w + q \times N : q \in \mathbf{Z}\}. \tag{5.20}$$

For example, $[2]_5 = \{..., 2, 7, 12, 17, ...\}$. If the remainder of w modulo N is the same as that of a modulo N, then we can say that writing $a \in [w]_N$ is the same as writing $a = w$ (mod N). The set of all such equivalent classes is

$$\mathbf{Z}_N = \{[w]_N : 0 \leq w \leq N - 1\} = \{0, 1, 2, 3, ... N - 1\}. \tag{5.21}$$

In (5.21), we use 0 to represent $[0]_N$, we apply 1 to represent $[1]_N$, we make use of 2 to represent $[2]_N$ and so on with that we use apply $N-1$ to represent $[N-1]_N$. We use its least nonnegative element to represent each class.

Because the equivalent class of two integers uniquely determines the equivalent class of their sum, product or difference, we can easily define addition, multiplication and subtraction operations for $\mathbf{Z}_N$. This is to say that if $c = c^1 \pmod{N}$ and $d = d^1 \pmod{N}$, then

$$c + d = c^1 + d^1 \pmod{N}, c \times d = c^1 \times d^1 \pmod{N} \text{ and } c - d = c^1 - d^1 \pmod{N} \tag{5.22}$$

Therefore, we denote addition, multiplication and subtraction modulo N, defined $+_N$, $\times_N$ and $-_N$, as follows:

$$[c]_N +_N [d]_N = [c+d]_N, [c]_N \times_N [d]_N = [c \times d]_N \text{ and } [c]_N -_N [d]_N = [c-d]_N. \tag{5.23}$$

Applying this definition of addition modulo N in (5.23), we define the **additive group modulo** N as $(\mathbf{Z}_N, +_N)$. We use Lemma 5.7 to show that the system $(\mathbf{Z}_N, +_N)$ is a finite abelian group.

Lemma 5.7 *The system $(\mathbf{Z}_N, +_N)$ is a finite abelian group.*

Proof For any two elements $[c]_N$ and $[d]_N$ in $\mathbf{Z}_N$, from (5.20) through (5.23), we obtain that $0 \le c \le N-1$, $0 \le d \le N-1$ and $[c]_N +_N [d]_N = [c+d]_N$. If $0 \le c + d \le N-1$, then $[c]_N +_N [d]_N = [c+d]_N$ is an element in $\mathbf{Z}_N$. If $N \le c + d \le 2 \times N - 2$, then $[c]_N +_N [d]_N = [c+d]_N = [c+d-N]_N$. Because $0 \le c + d - N \le N-2$, it is one element in $\mathbf{Z}_N$. This means that the system $(\mathbf{Z}_N, +_N)$ is closed.

For any three elements $[c]_N$, $[d]_N$ and $[e]_N$ in $\mathbf{Z}_N$, from (5.23), we obtain $([c]_N +_N [d]_N) +_N [e]_N = ([c+d]_N) +_N [e]_N = [(c+d)+e]_N = [c+(d+e)]_N = [c]_N +_N ([d+e]_N) = [c]_N +_N ([d]_N +_N [e]_N)$. This indicates that the system $(\mathbf{Z}_N, +_N)$ satisfies the associativity of $+_N$.

For any two elements $[c]_N$ and $[d]_N$ in $\mathbf{Z}_N$, from (5.23), we obtain $[c]_N +_N [d]_N = [c+d]_N = [d+c]_N = [d]_N +_N [c]_N$. This implies that the system $(\mathbf{Z}_N, +_N)$ satisfies the commutativity of $+_N$.

The identity element of the system $(\mathbf{Z}_N, +_N)$ is $[0]_N$ because for any element $[c]_N$ in $\mathbf{Z}_N$, from (5.23), we have $[c]_N +_N [0]_N = [c+0]_N = [c]_N = [0+c]_N = [0]_N +_N [c]_N$. The additive inverse of any element $[c]_N$ in $\mathbf{Z}_N$ is $[N-c]_N$ because $[c]_N +_N [N-c]_N = [c+N-c]_N = [N]_N = [0]_N$. The number of elements in the system $(\mathbf{Z}_N, +_N)$ is N, so it is finite. Therefore, from the statements above, we at once infer that the system $(\mathbf{Z}_N, +_N)$ is a finite abelian group. ∎

The set $\mathbf{Z}_N^*$ is the set of elements in $\mathbf{Z}_N$ that are relatively prime to N and is

$$\mathbf{Z}_N^* = \{[w]_N \in \mathbf{Z}_N : \gcd(w, N) = 1\}. \tag{5.24}$$

Because $[w]_N = \{w + q \times N : q \in \mathbf{Z}\}$ and $\gcd(w, N) = 1$, we have $\gcd(w + q \times N, N) = 1$. For example, $\mathbf{Z}_{15}^* = \{[1]_{15}, [2]_{15}, [4]_{15}, [7]_{15}, [8]_{15}, [11]_{15}, [13]_{15}, [14]_{15}\}$. Using the definition of multiplication modulo N in (5.23), we denote the multiplicative group modulo N as $(\mathbf{Z}_N^*, \times_N)$. We make use of Lemma 5.8 to demonstrate that the system $(\mathbf{Z}_N^*, \times_N)$ is a finite abelian group.

Lemma 5.8 *The system $(\mathbf{Z}_N^*, \times_N)$ is a finite abelian group.*

Proof For any two elements $[c]_N$ and $[d]_N$ in $\mathbf{Z}_N^*$, from (5.20) through (5.23), we get that $0 \leq c \leq N - 1, 0 \leq d \leq N - 1, \gcd(c, N) = 1, \gcd(d, N) = 1$ and $[c]_N \times_N [d]_N = [c \times d]_N$. Because $\gcd(c, N) = 1$ and $\gcd(d, N) = 1$, we have $\gcd(c \times d, N) = 1$. This means that $[c]_N \times_N [d]_N = [c \times d]_N$ is an element in $\mathbf{Z}_N^*$. Therefore, the system $(\mathbf{Z}_N^*, \times_N)$ is closed.

For any three elements $[c]_N, [d]_N$ and $[e]_N$ in $\mathbf{Z}_N^*$, from (5.23), we obtain $([c]_N \times_N [d]_N) \times_N [e]_N = ([c \times d]_N) \times_N [e]_N = [(c \times d) \times e]_N = [c \times (d \times e)]_N = [c]_N \times_N ([d + e]_N) = [c]_N \times_N ([d]_N) \times_N [e]_N)$. This is to say that the system $(\mathbf{Z}_N^*, \times_N)$ satisfies the associativity of $\times_N$.

For any two elements $[c]_N$ and $[d]_N$ in $\mathbf{Z}_N^*$, from (5.23), we get $[c]_N \times_N [d]_N = [c \times d]_N = [d \times c]_N = [d]_N \times_N [c]_N$. This indicates that the system $(\mathbf{Z}_N^*, \times_N)$ satisfies the commutativity of $\times_N$.

For any element $[c]_N$ in $\mathbf{Z}_N^*$, from (5.23), we have $[c]_N \times_N [1]_N = [c \times 1]_N = [c]_N = [1 \times c]_N = [1]_N \times_N [c]_N$. This indicates that the identity element of the system $(\mathbf{Z}_N^*, \times_N)$ is $[1]_N$.

Any element $[c]_N$ in $\mathbf{Z}_N^*$ satisfies $\gcd(c, N) = 1$. Therefore, from Lemma 5.5, there exists a unique multiplicative inverse $[c^{-1}]_N$ of $[c]_N$, modulo N, such that $[c]_N \times_N [c^{-1}]_N = [c \times c^{-1}]_N = [1]_N = [c^{-1} \times c]_N = [c^{-1}]_N \times_N [c]_N$. The number of elements in the system $(\mathbf{Z}_N^*, \times_N)$ is less than N, so it is finite. Hence, from the statements above, we at once derive that the system $(\mathbf{Z}_N^*, \times_N)$ is a finite abelian group. ∎

The size of $\mathbf{Z}_N^*$ is known as *Euler's phi function* $\phi(N)$ satisfies the following equation

$$\phi(N) = N \times \left(\prod_{p|N} \left(1 - \frac{1}{p} \right) \right), \tag{5.25}$$

where p runs over all the primes dividing n (including N itself, if N is a prime). For example, because the prime divisors of 15 are 3 and 5, we obtain $\phi(15) = 15 \times (1 - (1/3)) \times (1 - (1/5)) = 15 \times (2/3) \times (4/5) = 8$. This is to say that the size of $\mathbf{Z}_{15}^*$ is eight (8) and $\mathbf{Z}_{15}^*$ is equal to $\{[1]_{15}, [2]_{15}, [4]_{15}, [7]_{15}, [8]_{15}, [11]_{15}, [13]_{15}, [14]_{15}\}$. If p is a prime, then p itself is the only prime divisor. Therefore, from (5.25), we obtain $\phi(p) = p \times (1 - (1/p)) = p - 1$. The only integers that are less than p^a and are not co-prime to p^a are the multiples of p: $p, 2 \times p, \ldots, (p^{a-1-1}) \times p$, from which we infer

$$\phi(p^a) = (p^a - 1) - (p^{a-1} - 1) = p^a - p^{a-1} = p^{a-1} \times (p - 1). \tag{5.26}$$

Furthermore, if c and d are co-prime, then $\phi(c \times d)$ satisfies the following equation

$$\phi(c \times d) = \phi(c) \times \phi(d). \tag{5.27}$$

On the other hand, when N is a power of an odd prime p, $N = p^a$. It turns out that $\mathbf{Z}_N^* = \mathbf{Z}_p^{a*}$ is a *cyclic* group, that is, there is an element h in $\mathbf{Z}_p^{a*}$ which generates $\mathbf{Z}_p^{a*}$ in the sense that any other element y may be written $y = h^m \pmod{N} = h^m \pmod{p^a}$ for some non-negative integer m. We use Lemmas 5.9 and 5.10 to explain the possibility of finding the even order that are not equal to $N - 1$ of X modulo N.

Lemma 5.9 *We assume that p is an odd prime and 2^b is the largest power of 2 dividing $\phi(p^a)$. Then with probability exactly one-half 2^b divides the order modulo p^a of a randomly chosen element of $\mathbf{Z}_p^{a*}$.*

Proof Because p is an odd prime, from (5.26) we obtain that $\phi(p^a) = p^{a-1} \times (p - 1)$ is even. Because $\phi(p^a)$ is even and 2^b divides $\phi(p^a)$, we obtain $b \geq 1$. Since $\mathbf{Z}_p^{a*}$ is a *cyclic* group, there exists an element h in $\mathbf{Z}_p^{a*}$ which generates $\mathbf{Z}_p^{a*}$ in the sense that any other element X may be written $X = h^m \pmod{p^a}$ for some m in the range 1 through $\phi(p^a)$ that is the size of $\mathbf{Z}_p^{a*}$. Let r the order of h^m modulo p^a and consider two cases. The first case is when m is odd. Since h is co-prime to (p^a) and $\phi(p^a)$ is the size of $\mathbf{Z}_p^{a*}$, $\phi(p^a)$ is the least value such that $h^{\phi(p^a)} = 1 \pmod{p^a}$. Because $(h^m)^r = h^{m \times r} = 1 \pmod{p^a}$, we infer that $\phi(p^a)$ divides $(m \times r)$. Since m is odd, $\phi(p^a)$ is even, 2^b divides $\phi(p^a)$ and $\phi(p^a)$ divides $(m \times r)$ and 2^b divides $(m \times r)$, we infer that 2^b divides r. The second case is when m is even. Because h is co-prime to p^a and m is even, we infer that $h^{m/2}$ modulo p^a is co-prime to p^a. Therefore, we have $\left(h^{m \times \phi(p^a)/2}\right) = \left(h^{\phi(p^a)}\right)^{m/2} = (1)^{m/2} = 1 \pmod{p^a}$. Because r is the order of h^m modulo p^a that is the least value such that $(h^m)^r = h^m \times {}^r = 1 \pmod{p^a}$, we infer that r divides $(\phi(p^a)/2)$ and r is less than 2^b that is the largest power of 2 dividing $\phi(p^a)$. Thus, we infer that 2^b does not divide r.

Because the value of m is in the range 1 through $\phi(p^a)$ that is even and is the size of $\mathbf{Z}_p^{a*}$, we may partition $\mathbf{Z}_p^{a*}$ into two sets of equal size. The first set of equal size is those that may be written $h^m \pmod{p^a}$ with that m is odd, for which 2^b divides r that is the order of h^m modulo p^a. The second set of equal size is those that may be written $h^m \pmod{p^a}$ with that m is even, for which 2^b does not divide r that is the order of h^m modulo p^a. Therefore, with probability $(1/2)$ the integer 2^b divides the order r of a randomly chosen element $\mathbf{Z}_p^{a*}$, and with probability $(1/2)$ it does not. ■

Lemma 5.10 *We assume that $N = p_1^{a_1} \times \cdots \times p_m^{a_m}$ is the prime factorization of an odd composite positive integer. Let X be chosen uniformly at random from $\mathbf{Z}_N^*$ and let r be the order of X modulo N. Then $P(r$ is even and $X^{r/2} \neq -1 \pmod{N}) \geq 1 - (1/2^m)$.*

Proof We show that $P(r$ is odd or $X^{r/2} = -1 \pmod{N}) \leq 1/2^m$. According to the Chinese remainder theorem, selecting X uniformly at random from $\mathbf{Z}_N^*$ is equivalent to selecting X_k independently and uniformly at random from $\mathbf{Z}_{p_k^{a_k}}^*$, and satisfying that $X = X_k \pmod{p_k^{a_k}}$ for $1 \leq k \leq m$. Let r_k be the order of X_k modulo $\left(p_k^{a_k}\right)$. Let 2^{b_k} be the largest power of 2 dividing r_k and 2^b is the largest power of 2 dividing r. Because X is co-prime to N $(p_1^{a_1} \times \cdots \times p_m^{a_m})$, $\phi(N) = \phi(p_1^{a_1} \times \cdots \times p_m^{a_m})$ is the size of $\mathbf{Z}_N^*$ that is the least value such that $X^{\phi(N)} = X^{\phi\left(p_1^{a_1} \times \cdots \times p_m^{a_m}\right)} = 1 \pmod{N}$. Since r is the order of X modulo N that is the least value such that $X^r = 1 \pmod{N}$, we have $r = \phi(p_1^{a_1} \times \cdots \times p_m^{a_m}) = \phi(p_1^{a_1}) \times \cdots \times \phi(p_m^{a_m})$. Because X_k is co-prime to $(p_k^{a_k})$ for $1 \leq k \leq m$, $\phi(p_k^{a_k})$ is the size of $\mathbf{Z}_{p_k^{a_k}}^*$ that is the least value such that $X_k^{\phi\left(p_k^{a_k}\right)} = 1 \pmod{(p_k^{a_k})}$. Since r_k is the order of X_k modulo $(p_k^{a_k})$ that is the least value such that $X_k^r = 1 \pmod{(p_k^{a_k})}$, we have $r_k = \phi(p_k^{a_k})$ for $1 \leq k \leq m$. Because $r = \phi(p_1^{a_1}) \times \ldots \times \phi(p_m^{a_m})$ and $r_k = \phi(p_k^{a_k})$ for $1 \leq k \leq m$, we infer that r_k divides r for $1 \leq k \leq m$. We will show that to have r odd or $X^{r/2} = -1 \pmod{N}$ it is necessary that b_k takes the same value for $1 \leq k \leq m$. The result then follows, as from Lemma 5.9 the probability of this occurring is at most $(1/2) \times (1/2) \times \cdots \times (1/2) = 1/2^m$.

We consider the first case is when r is odd. Because r_k divides r for $1 \leq k \leq m$, we infer r_k is odd. Because 2^{b_k} divides r_k for $1 \leq k \leq m$, we obtain $b_k = 0$ for $1 \leq k \leq m$. The second case is when r is even and $X^{r/2} = -1 \pmod{N}$. This is to say that $X^{r/2} = N - 1 = p_1^{a_1} \times \cdots \times p_m^{a_m} - 1$. Therefore, we have $X^{r/2} = N - 1 = p_1^{a_1} \times \cdots \times p_m^{a_m} - 1 = -1 \pmod{(p_k^{a_k})}$. So we obtain that r_k does not divide $(r/2)$. Because r_k divides r for $1 \leq k \leq m$, we must have $b_k = b$ for $1 \leq k \leq m$. Since $P(r$ is even and $X^{r/2} \neq -1 \pmod{N}) + P(r$ is odd or $X^{r/2} = -1 \pmod{N}) = 1$ and $P(r$ is odd or $X^{r/2} = -1 \pmod{N}) \leq 1/2^m$, we have $P(r$ is even and $X^{r/2} \neq -1 \pmod{N}) \geq 1 - (1/2^m)$. ∎

5.11 Public Key Cryptography and the RSA Cryptosystem

A consumer wants to buy something on the internet. He would like to transmit his credit card number over the internet in such a way that only the company offering the products that he is buying can receive the number. A *cryptographic protocol* or a *cryptosystem* on the internet can achieve such private communication. Effective cryptosystems make it easy for two parties who want to communicate each other, but make it very difficult for the eavesdropper to eavesdrop on the content of the conversation.

A particularly important class of cryptosystems are the *public key cryptosystems*. In a public key cryptosystem, Mary wants to send messages to her friends and to receive messages sent by her friends. She must first generate two *cryptographic keys*. One is a *public key*, P, and the other is a secret key, S. After Mary has generated her keys, she announces or publishes the public key so that anybody can gain access to the public key.

John is Mary's good friend. He would like to send a private message to Mary. Therefore, John first gets a copy of Mary's public key P. Then, he encrypts the private message he wants to send Mary, making use of Mary's public key P to complete the encryption. Because the public key and the encoded message is the only information available to an eavesdropper, it will be impossible for the eavesdropper to recover the message. However, Mary has the secret key S that is not available to an eavesdropper. She uses the secret key S to decrypt the encrypted message and obtains the *original* message. This transformation known as decryption is inverse to encryption, allowing Mary to recover John's private message.

The most widely used of public key cryptosystems is the **RSA** cryptosystem, named **RSA** for the initials of its creators, Rivest, Shamir, and Adleman. The presumed security of the **RSA** cryptosystem is based on the apparent difficulty of factoring on a digital computer. Now Mary wishes to generate public and secret keys for use with the **RSA** cryptosystem. She makes use of the following procedure to generate them:

(1)   Choose two large prime numbers, p and q.
(2)   Calculate the product $N = p \times q$.
(3)   Choose at random a small odd integer, e, which is relatively prime to $\phi(N) = (p-1) \times (q-1)$.
(4)   Calculate d, the multiplicative inverse of e, modulo $\phi(N)$.
(5)   The public key is the pair $P = (e, N)$.
(6)   The secret key is the pair $S = (d, N)$.

Now John uses the public key (e, N) to encrypt a message M to send to Mary. We assume that the message M has only $\log_2^N$ bits, as longer messages may be encrypted by means of breaking M up into blocks of at most $\log_2^N$ bits and then encrypting the blocks separately. The encryption procedure for a single block is to calculate:

$$E(M) = M^e (\mathrm{mod}\ N). \tag{5.28}$$

$E(M)$ is the encrypted version of the message M, which John sends to Mary. Mary is able to decrypt quickly the message applying her secret key $S = (d, N)$, simply by raising the encrypted message to the dth power:

$$D(E(M)) = M^{e \times d} (\mathrm{mod}\ N) = M(\mathrm{mod}\ N). \tag{5.29}$$

How can the **RSA** cryptosystem be broken? The answer is to that if we can efficiently factor a big composite number N into the production of two big prime numbers p and q. Then, we can extract p and q. This means that we can efficiently calculate $\phi(N) = (p-1) \times (q-1)$. Next, we can efficiently compute d, the multiplicative inverse of e, modulo $\phi(N)$. Therefore, we can completely determine the secret key (d, N). So, if factoring large numbers were easy then breaking the RSA cryptosystem would be easy.

5.12 Implementing The Controlled-Swap Gate of Three Quantum Bits

A (8×8) matrix $CSWAP$ and its conjugate transpose $\overline{CSWAP}$ are respectively

$$
\begin{pmatrix}
1 & 0 & 0 & 0 & 0 & 0 & 0 & 0 \\
0 & 1 & 0 & 0 & 0 & 0 & 0 & 0 \\
0 & 0 & 1 & 0 & 0 & 0 & 0 & 0 \\
0 & 0 & 0 & 1 & 0 & 0 & 0 & 0 \\
0 & 0 & 0 & 0 & 1 & 0 & 0 & 0 \\
0 & 0 & 0 & 0 & 0 & 0 & 1 & 0 \\
0 & 0 & 0 & 0 & 0 & 1 & 0 & 0 \\
0 & 0 & 0 & 0 & 0 & 0 & 0 & 1
\end{pmatrix}
\text{ and }
\begin{pmatrix}
1 & 0 & 0 & 0 & 0 & 0 & 0 & 0 \\
0 & 1 & 0 & 0 & 0 & 0 & 0 & 0 \\
0 & 0 & 1 & 0 & 0 & 0 & 0 & 0 \\
0 & 0 & 0 & 1 & 0 & 0 & 0 & 0 \\
0 & 0 & 0 & 0 & 1 & 0 & 0 & 0 \\
0 & 0 & 0 & 0 & 0 & 0 & 1 & 0 \\
0 & 0 & 0 & 0 & 0 & 1 & 0 & 0 \\
0 & 0 & 0 & 0 & 0 & 0 & 0 & 1
\end{pmatrix}.
\tag{5.30}
$$

Because $CSWAP \times \overline{CSWAP}$ is equal to a (8×8) identify matrix and $\overline{CSWAP} \times CSWAP$ is equal to a (8×8) identify matrix, matrix $CSWAP$ and matrix $\overline{CSWAP}$ are both a unitary matrix (a unitary operator). Matrix $CSWAP$ is to matrix representation of a $CSWAP$ (*controlled-SWAP*) gate of three quantum bits. The left picture in Fig. 5.7 is the first graphical circuit representation of a $CSWAP$ gate with three quantum bits. Quantum bit $|C_1\rangle$ at the bottom in the left picture in Fig. 5.7 is the controlled bit, and quantum bit $|S_1\rangle$ at the top and quantum bit $|S_2\rangle$ at the middle in the left picture in Fig. 5.7 are both the target bits. The functionality of the $CSWAP$ gate is to that if the controlled bit $|C_1\rangle$ is equal to $|1\rangle$, then it exchanges the information contained in the two target bits $|S_1\rangle$ and $|S_2\rangle$. Otherwise, it does not exchange the information contained in the two target bits $|S_1\rangle$ and $|S_2\rangle$. The middle picture in Fig. 5.7 is the second graphical circuit representation of a $CSWAP$ gate with three quantum bits. The right picture in Fig. 5.7 is to the graphical circuit representation of implementing the $CSWAP$ gate by means of using three $CCNOT$ gates. In the right picture in Fig. 5.7, if the controlled bit $|C_1\rangle$ is equal to $|1\rangle$, then using three $CNOT$ gates implements one $SWAP$ gate to exchange the information contained in the two target bits $|S_1\rangle$ and $|S_2\rangle$. Otherwise, it does not implement three $CNOT$ gates to complete one $SWAP$ gate and to exchange the information contained in the two target bits $|S_1\rangle$ and $|S_2\rangle$.

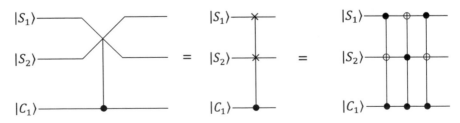

Fig. 5.7 Circuit Representation of a **CSWAP** gate with three quantum bits

5.12.1 Quantum Programs to Implement
The Controlled-Swap Gate of Three Quantum Bits

In **IBM Q** Experience, it does not provide one quantum instruction (operation) of implementing the **CCNOT** gate (the Toffoli gate) with three quantum bits. We decompose **CCNOT** gate into *six* **CNOT** gates and *nine* gates of one quantum bits that are shown in Fig. 5.8. In Fig. 5.8, H is the Hadamard gate, $T = \begin{bmatrix} 1 & 0 \\ 0 & e^{\sqrt{-1} \times \frac{\pi}{4}} \end{bmatrix}$ and $T^+ = \begin{bmatrix} 1 & 0 \\ 0 & e^{-1 \times \sqrt{-1} \times \frac{\pi}{4}} \end{bmatrix}$. In the backend *simulator* with thirty-two quantum bits, there is no limit for connectivity of a **CNOT** gate among thirty-two quantum bits.

In Listing 5.1, the program in the backend *simulator* with thirty-two quantum bits in **IBM**'s quantum computer is the *first* example of the *fifth* chapter in which we illustrate how to write a quantum program to implement a **CSWAP** gate with three quantum bits. Figure 5.9 is the corresponding quantum circuit of the program in Listing 5.1. For the convenience of our presentation, there are four instructions in the same line in Listing 5.1. We use "instruction number" or "line number" to indicate the order of the execution to each instruction in Listing 5.1.

The statement "OPENQASM 2.0;" on instruction number one in the first line of Listing 5.1 is to point out that the program is written with version 2.0 of Open QASM. Next, the statement "include "qelib1.inc";" on instruction number two in the first line of Listing 5.1 is to continue parsing the file "qelib1.inc" as if the contents of the file were pasted at the location of the include statement, where the file "qelib1.inc" is **Quantum Experience (QE) Standard Header** and the path is specified relative to the current working directory.

Next, the statement "qreg q[3];" on instruction number three in the first line of Listing 5.1 is to declare that in the program there are three quantum bits. In the left top of Fig. 5.9, three quantum bits are subsequently q[0], q[1] and q[2]. The initial value of each quantum bit is set to |0⟩. We use three quantum bits q[0], q[1] and

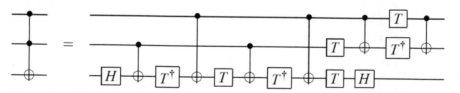

Fig. 5.8 Decomposing **CCNOT** gate into six **CNOT** gates and nine gates of one bit

Fig. 5.9 The corresponding quantum circuit of the program in Listing 5.1

q[2] to subsequently encode the first target bit $|S_1\rangle$, the second target bit $|S_2\rangle$ and the controlled bit $|C_1\rangle$.

Listing 5.1 The Program of Implementing a *CSWAP* Gate of Three Quantum Bits.

```
1. OPENQASM 2.0;   2. include "qelib1.inc";   3. qreg q[3];   4. creg c[3];
5. x q[0];              6. x q[2];
// Implement the first CCNOT gate in the right picture of Fig. 5.7 with two
// controlled bits q[2] and q[0] and target bit q[1].

    7. barrier q[0], q[1], q[2];   8. h q[1];        9. cx q[0], q[1];      10. tdg
    q[1];
    11. cx q[2], q[1];             12. t q[1];      13. cx q[0], q[1];     14. tdg
    q[1];
    15. cx q[2], q[1];             16. t q[0];     17. t q[1];            18. h q[1];
    19. cx q[2], q[0];             20 tdg q[0];    21 t q[2];             22. cx
    q[2], q[0];

// Implement the second CCNOT gate in the right picture of Fig. 5.7 with
two
// controlled bits q[2] and q[1] and target bit q[0].

    23. barrier q[0], q[1], q[2];   24 h q[0];      25. cx q[1], q[0];      26. tdg
    q[0];
    27. cx q[2], q[0];             28. t q[0];      29. cx q[1], q[0];      30. tdg
    q[0];
    31. cx q[2], q[0];             32. t q[1];     33. t q[0];            34. h q[0];
    35. cx q[2], q[1];             36. tdg q[1];   37. t q[2];            38. cx
    q[2], q[1];

// Implement the third CCNOT gate in the right picture of Fig. 5.7 with two
// controlled bits q[2] and q[0] and target bit q[1].

    39. barrier q[0], q[1], q[2];   40. h q[1];      41. cx q[0], q[1];      42. tdg
    q[1];
    43. cx q[2], q[1];             44. t q[1];      45. cx q[0], q[1];      46. tdg
    q[1];
    47. cx q[2], q[1];             48. t q[0];     49. t q[1];            50. h q[1];
    51. cx q[2], q[0];             52. tdg q[0];   53. t q[2];            54. cx
    q[2], q[0];

    55. measure q[0] -> c[0];
    56. measure q[1] -> c[1];
    57. measure q[2] -> c[2];
```

For the convenience of our explanation, q[k]^0 for $0 \leq k \leq 2$ is to represent the value 0 of q[k] and q[k]^1 for $0 \leq k \leq 2$ is to represent the value 1 of q[k]. Similarly, for the convenience of our explanation, an initial state vector of implementing a **CSWAP** gate is as follows:

$$|\Phi_0\rangle = \left|q[2]^0\right\rangle\left|q[1]^0\right\rangle\left|q[0]^0\right\rangle = |0\rangle|0\rangle|0\rangle = |000\rangle.$$

Then, the statement "creg c[3];" on instruction number four in the first line of Listing 5.1 is to declare that there are three classical bits in the program. In the left bottom of Fig. 5.9, three classical bits are respectively c[0], c[1] and c[2]. The initial value of each classical bit is set to 0. Classical bit c[2] is the most significant bit and classical bit c[0] is the least significant bit.

Next, the two statements "x q[0];" and "x q[2];" on line number *five* through line number *six* in the second line of Listing 5.1 implement two **NOT** gates to quantum bit q[0] and quantum bit q[2] in the *first* time slot of the quantum circuit in Fig. 5.9. They both actually complete $\begin{pmatrix} 0 & 1 \\ 1 & 0 \end{pmatrix} \times \begin{pmatrix} 1 \\ 0 \end{pmatrix} = \begin{pmatrix} 0 \\ 1 \end{pmatrix} = |1\rangle$. This indicates that converting q[0] from one state $|0\rangle$ to another state $|1\rangle$ and converting q[2] from one state $|0\rangle$ to another state $|1\rangle$ are completed. Since in the *first* time slot of the quantum circuit in Fig. 5.9 there is no quantum gate to act on quantum bit q[1], its state $|0\rangle$ is not changed. Therefore, we have the following new state vector

$$|\Phi_1\rangle = \left|q[2]^1\right\rangle\left|q[1]^0\right\rangle\left|q[0]^1\right\rangle.$$

In the state vector $|\Phi_1\rangle = |q[2]^1\rangle \, |q[1]^0\rangle \, |q[0]^1\rangle$, quantum bit $|q[2]^1\rangle$ is the input state $|1\rangle$ of the controlled bit $|C_1\rangle$ in the **CSWAP** gate in Fig. 5.7. Quantum bit $|q[1]^0\rangle$ is the input state $|0\rangle$ of the target bit $|S_2\rangle$ in the **CSWAP** gate in Fig. 5.7. Quantum bit $|q[0]^1\rangle$ is the input state $|1\rangle$ of the target bit $|S_1\rangle$ in the **CSWAP** gate in Fig. 5.7. Next, the statement "barrier q[0], q[1], q[2];" on line number seven in the *sixth* line of Listing 5.1 implements one barrier instruction to prevent optimization from reordering gates across its source line in the *second* time slot of the quantum circuit in Fig. 5.9. Next, from instruction number *eight* through instruction number *twenty-two* in Listing 5.1, the fifteen statements are "h q[1];" "cx q[0], q[1];", "tdg q[1];", "cx q[2], q[1];", "t q[1];", "cx q[0], q[1];", "tdg q[1];", "cx q[2], q[1];", "t q[0];", "t q[1];", "h q[1];", "cx q[2], q[0];", "tdg q[0];", "t q[2];" and "cx q[2], q[0];". They take the state vector $|\Phi_1\rangle = |q[2]^1\rangle \, |q[1]^0\rangle \, |q[0]^1\rangle$ as their input and implement the first **CCNOT** gate with two controlled bits q[2] and q[0] and one target bit q[1] from the *third* time slot through the *fifteen* time slot of Fig. 5.9. Because the two controlled bits q[2] and q[0] are both state $|1\rangle$, the state $|0\rangle$ of the target bit q[1] is converted into state $|1\rangle$. Therefore, we have the following new state vector

$$|\Phi_{15}\rangle = \left|q[2]^1\right\rangle\left|q[1]^1\right\rangle\left|q[0]^1\right\rangle.$$

Next, the statement "barrier q[0], q[1], q[2];" on instruction number twenty-three in Listing 5.1 implements one barrier instruction to prevent optimization from reordering gates across its source line in the *sixteenth* time slot of the quantum circuit in Fig. 5.9. Next, from instruction number *twenty-four* through instruction number *thirty-eight* in Listing 5.1, the fifteen statements are "h q[0];", "cx q[1], q[0];", "tdg q[0];", "cx q[2], q[0];", "t q[0];", "cx q[1], q[0];", "tdg q[0];", "cx q[2], q[0];", "t q[1];", "t q[0];", "h q[0];", "cx q[2], q[1];", "tdg q[1];", "t q[2];" and "cx q[2], q[1];". They take the state vector $|\Phi_{15}\rangle = |q[2]^1\rangle \, |q[1]^1\rangle \, |q[0]^1\rangle$ as their input and implement the second **CCNOT** gate with two controlled bits q[2] and q[1] and one target bit q[0] from the *seventeenth* time slot through the *twenty-eighth* time slot of Fig. 5.9. Since the two controlled bits q[2] and q[1] are both state $|1\rangle$, the state $|1\rangle$ of the target bit q[0] is converted into state $|0\rangle$. Thus, we obtain the following new state vector

$$|\Phi_{28}\rangle = \left|q[2]^1\right\rangle\left|q[1]^1\right\rangle\left|q[0]^0\right\rangle.$$

Next, the statement "barrier q[0], q[1], q[2];" on instruction number *thirty-nine* in Listing 5.1 implements one barrier instruction to prevent optimization from reordering gates across its source line in the *thirty-ninth* time slot of the quantum circuit in Fig. 5.9. Next, from instruction number *forty* through instruction number *fifty-four* in Listing 5.1, the fifteen statements are "h q[1];", "cx q[0], q[1];", "tdg q[1];", "cx q[2], q[1];", "t q[1];", "cx q[0], q[1];", "tdg q[1];", "cx q[2], q[1];", "t q[0];", "t q[1];", "h q[1];", "cx q[2], q[0];", "tdg q[0];", "t q[2];" and "cx q[2], q[0];". They take the state vector $|\Phi_{28}\rangle = |q[2]^1\rangle \, |q[1]^1\rangle \, |q[0]^0\rangle$ as their input and implement the third **CCNOT** gate with two controlled bits q[2] and q[0] and one target bit q[1] from the *thirtieth* time slot through the *forty-second* time slot of Fig. 5.9. Because the first controlled bit q[2] is state $|1\rangle$ and the second controlled bit q[0] is state $|0\rangle$, the state $|1\rangle$ of the target bit q[1] is not changed. Hence, we get the following new state vector

$$|\Phi_{42}\rangle = \left|q[2]^1\right\rangle\left|q[1]^1\right\rangle\left|q[0]^0\right\rangle.$$

Next, the three statements "measure q[0] -> c[0];", "measure q[1] -> c[1];" and "measure q[2] -> c[2];" from instruction number fifty-five through instruction number in Listing 5.1 is to measure the first quantum bit q[0], the second quantum bit q[1] and the third quantum bit q[2]. They record the measurement outcome by overwriting the first classical bit c[0], the second classical bit c[1] and the third classical bit c[2]. In the backend *simulator* with thirty-two quantum bits in **IBM**'s quantum computers, we use the command "run" to execute the program in Listing 5.1. The measured result appears in Fig. 5.10. From Fig. 5.10, we obtain the answer 110 ($c[2] = 1 = q[2] = |1\rangle$, $c[1] = 1 = q[1] = |1\rangle$ and $c[0] = 0 = q[0] = |0\rangle$)) with the probability 100%. Because the input state of the target bit $|S_1\rangle$ is state $|1\rangle$, the input state of the target bit $|S_2\rangle$ is state $|0\rangle$ and the input state of the controlled bit $|C_1\rangle$ is $|1\rangle$ in the **CSWAP** gate in Fig. 5.7, the information contained in the two target bits $|S_1\rangle$ and $|S_2\rangle$ are exchanged. Therefore, we have the final state of the target bit $|S_1\rangle$ encoded

Fig. 5.10 After the measurement to the program in Listing 5.1 is completed, we obtain the answer 110 with the probability 100%

by $|q[0]^0\rangle$ is state $|0\rangle$ and the final state of the target bit $|S_2\rangle$ encoded by $|q[1]^1\rangle$ is state $|1\rangle$ with the probability 100%.

5.13 Shor's Order-Finding Algorithm

For positive integers X and N with that the value of X is less than the value of N and the greatest common factor for them is one, the order (the period) of X modulo N is to the least positive integer r such that $X^r = 1 \pmod{N}$. The order-finding problem is to compute the order for some given X and N. On a digital computer, no algorithm *known* solves the problem with the number of the bit of specifying N that is L to be greater than or equal to $\log_2(N)$, by means of using resources polynomial in the $O(L)$ bits needed to specify the problem. In this section, we explain how Shor's order-finding algorithm is an efficient quantum algorithm to order finding.

Quantum circuit to Shor's order-finding algorithm is schematically depicted in Fig. 5.11. The first quantum register of n quantum bits is $\left(\otimes_{k=1}^{n}|p_k^0\rangle\right)$ and the initial state of each quantum bit is the $|0\rangle$ state. Quantum bit $|p_1^0\rangle$ is the most significant bit and quantum bit $|p_n^0\rangle$ is the least significant bit. Because the order of X modulo N is less than or equal to N, n is greater than or equal to $\log_2(N)$. Its decimal value is equal to $p_1 \times 2^{n-1} + p_2 \times 2^{n-2} + p_3 \times 2^{n-3} + \cdots + p_n \times 2^{n-n}$. The second quantum register of L quantum bits is $\left(\left(\otimes_{y=1}^{L-1}|w_y^0\rangle\right) \otimes \left(|w_L^1\rangle\right)\right)$. The initial state of each quantum bit in the front $(L-1)$ quantum bits is the $|0\rangle$ state. The initial state of the *least significant* quantum bit $\left(|w_L^1\rangle\right)$ is the $|1\rangle$ state. Quantum bit $|w_1^0\rangle$ is the most significant bit and quantum bit $|w_L^1\rangle$ is the least significant bit. Its decimal value is equal to $w_1 \times 2^{L-1} + w_2 \times 2^{L-2} + w_3 \times 2^{L-3} + \cdots + p_L \times 2^{L-L}$. From Fig. 5.11, the initial state vector is

$$|\varphi_0\rangle = \left(\otimes_{k=1}^{n}|p_k^0\rangle\right) \otimes \left(\left(\otimes_{y=1}^{L-1}|w_y^0\rangle\right) \otimes \left(|w_L^1\rangle\right)\right). \tag{5.31}$$

From Fig. 5.11, the initial state vector $|\varphi_0\rangle$ in (5.31) is followed by n Hadamard gates on the first (upper) quantum register. This gives that the new state vector is

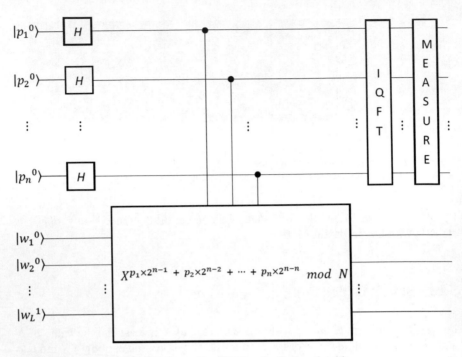

Fig. 5.11 Quantum circuit of implementing Shor's order-finding algorithm

$$|\varphi_1\rangle = \frac{1}{\sqrt{2^n}}\left(\left(\otimes_{k=1}^{n}|p_k^0\rangle + |p_k^1\rangle\right) \otimes \left(\left(\otimes_{y=1}^{L-1}|w_y^0\rangle\right) \otimes \left(|w_L^1\rangle\right)\right)\right)$$

$$= \frac{1}{\sqrt{2^n}}\left(\sum_{p=0}^{2^n-1}|P\rangle \otimes \left(\left(\otimes_{y=1}^{L-1}|w_y^0\rangle\right) \otimes \left(|w_L^1\rangle\right)\right)\right). \qquad (5.32)$$

Next, from Fig. 5.11, the new state vector $|\varphi_1\rangle$ in (5.32) is followed by a quantum gate $|X^P \bmod N\rangle = |X^{p_1 \times 2^{n-1} + p_2 \times 2^{n-2} + \cdots + p_n \times 2^{n-n}} \bmod N\rangle$ operating on both quantum registers. This gives that the new state vector is

$$|\varphi_2\rangle = \frac{1}{\sqrt{2^n}}\left(\sum_{p=0}^{2^n-1}|P\rangle|X^p \bmod N\rangle\right). \qquad (5.33)$$

Because the order of X modulo N is r, terms of $|\varphi_2\rangle$ in (5.33) can be regrouped as r equivalent classes according to the computational basis states with the same remainder of $|X^P \bmod N\rangle$. The first equivalent class is $\{r \times y + 0 \mid 0 \leq y \leq \lfloor(2^n - 0)/r\rfloor\}$, where $\lfloor(2^n - 0)/r\rfloor$ is to obtain the greatest integer that is less than or equal to $((2^n - 0)/r)$. The second equivalent class is $\{r \times y + 1 \mid 0 \leq y \leq \lfloor(2^n - 1)/r\rfloor\}$. The third equivalent class is $\{r \times y + 2 \mid 0 \leq y \leq \lfloor(2^n - 2)/r\rfloor\}$. The fourth equivalent class is $\{r \times y + 3 \mid 0 \leq y \leq \lfloor(2^n - 3)/r\rfloor\}$. The rth equivalent class is $\{r \times y + (r - 1) \mid 0 \leq y \leq \lfloor (2^n - (r - 1))/r \rfloor \}$. Not all the equivalent classes have the same number of elements. However, if r divides 2^n, then the number of elements in each equivalent class is the same. We assume that for $0 \leq P \leq (r - 1)$, $Y_P = \lfloor (2^n - P)/r \rfloor$. In light of the statements above, we rewrite the new state vector $|\varphi_2\rangle$ in (5.33) as follows

$$|\varphi_2\rangle = \sum_{P=0}^{r-1} \left(\frac{1}{\sqrt{2^n}} \sum_{y=0}^{Y_P} |r \times y + P\rangle \right) |X^P \bmod N\rangle. \tag{5.34}$$

For the convenience of our presentation, we assume that $|\varphi_{2P}\rangle = \left(\frac{1}{\sqrt{2^n}} \sum_{y=0}^{Y_P} |r \times y + P\rangle \right)$ for $0 \leq P \leq (r - 1)$.

As in Fig. 5.11, the last step before measurement is to complete the inverse quantum Fourier transform (an **IQFT**) on the first (upper) quantum register. The superposition principle allows the unitary operator to act one by one on each $|\varphi_{2P}\rangle$. Therefore, we obtain the following new state vector

$$|\varphi_3\rangle = \frac{1}{\sqrt{2^n}} \sum_{P=0}^{r-1} \sum_{y=0}^{Y_P} \frac{1}{\sqrt{2^n}} \sum_{i=0}^{2^n-1} e^{-\sqrt{-1} \times \frac{2\times\pi}{2^n} \times i \times (y \times r + P)} |i\rangle > |X^P \bmod N\rangle.$$

$$= \sum_{i=0}^{2^n-1} \sum_{P=0}^{r-1} \sum_{y=0}^{Y_P} \frac{e^{-\sqrt{-1} \times \frac{2\times\pi}{2^n} \times i \times (y \times r + P)}}{2^n} |i\rangle |X^P \bmod N\rangle. \tag{5.35}$$

For the convenience of our presentation, we assume that $\varphi_{iP} = \left(\sum_{y=0}^{Y_P} \frac{e^{-\sqrt{-1} \times \frac{2\times\pi}{2^n} \times i \times (y \times r + P)}}{2^n} \right)$ for $0 \leq P \leq (r - 1)$ and $0 \leq i \leq (2^n - 1)$. Coefficients φ_{iP} and $|\varphi_{iP}|^2$ subsequently represent the amplitude and the probability of measuring $|i\rangle |X^P \bmod N >$ at the output of the circuit in Fig. 5.11. The probability amplitudes may cancel each other while increasing the probability of measuring a suitable state.

For the convenience of our presentation, we assume that $P(i)$ represents the probability of measuring $|i\rangle |X^P \bmod N\rangle$ at the output of the circuit in Fig. 5.11. Basic probability theory guarantees that

$$P(i) = \sum_{P=0}^{r-1} |\varphi_{iP}|^2 = \sum_{P=0}^{r-1} \left| \sum_{y=0}^{Y_P} \frac{e^{-\sqrt{-1} \times \frac{2\times\pi}{2^n} \times i \times (y \times r + P)}}{2^n} \right|^2$$

$$= \sum_{P=0}^{r-1} \left| e^{-\sqrt{-1} \times \frac{2\times\pi}{2^n} \times i \times P} \right|^2 \times \left| \sum_{y=0}^{Y_P} \frac{\left(e^{-\sqrt{-1} \times \frac{2\times\pi}{2^n} \times i \times r} \right)^y}{2^n} \right|^2 \tag{5.36}$$

Since basic probability theory ensures that $\left| e^{-\sqrt{-1} \times \frac{2 \times \pi}{2^n} \times i \times P} \right|^2 =$
$\left(e^{-\sqrt{-1} \times \frac{2 \times \pi}{2^n} \times i \times P} \right) \times \left(e^{\sqrt{-1} \times \frac{2 \times \pi}{2^n} \times i \times P} \right) = 1$, we can rewrite $P(i)$ in (5.36) as
follows

$$P(i) = \sum_{P=0}^{r-1} 1 \times \left| \sum_{y=0}^{Y_P} \frac{\left(e^{-\sqrt{-1} \times \frac{2 \times \pi}{2^n} \times i \times r} \right)^y}{2^n} \right|^2$$

$$= \sum_{P=0}^{r-1} \left| \frac{1}{2^n} \sum_{y=0}^{Y_P} \left(e^{-\sqrt{-1} \times \frac{2 \times \pi}{2^n} \times i \times r} \right)^y \right|^2$$

$$= \frac{1}{2^{2 \times n}} \times \left(\sum_{P=0}^{r-1} \left| \sum_{y=0}^{Y_P} \left(e^{-\sqrt{-1} \times \frac{2 \times \pi}{2^n} \times i \times r} \right)^y \right|^2 \right). \tag{5.37}$$

For realizing a sum of geometrical sequence, we discuss the *ideal* case and the
practice case. If the argument of the absolute value operator is $e^{-\sqrt{-1} \times \frac{2 \times \pi}{2^n} \times i \times r} = 1$,
then $(i \times r / 2^n)$ is an integer and we can rewrite $P(i)$ in (5.37) as follows

$$P(i) = \frac{1}{2^{2 \times n}} \times \left(\sum_{P=0}^{r-1} \left| \sum_{y=0}^{Y_P} 1^y \right|^2 \right).$$

$$= \frac{1}{2^{2 \times n}} \times \left(\sum_{P=0}^{r-1} (Y_P + 1)^2 \right). \tag{5.38}$$

We call (5.38) as the *ideal* case. If the argument of the absolute value operator
is $e^{-\sqrt{-1} \times \frac{2 \times \pi}{2^n} \times i \times r} \neq 1$, then $(i \times r / 2^n)$ is not an integer and we can rewrite $P(i)$ in
(5.37) as follows

$$P(i) = \frac{1}{2^{2 \times n}} \times \left(\sum_{P=0}^{r-1} \left| \frac{1 - e^{-\sqrt{-1} \times \frac{2 \times \pi}{2^n} \times i \times r \times (1 + Y_P)}}{1 - e^{-\sqrt{-1} \times \frac{2 \times \pi}{2^n} \times i \times r}} \right|^2 \right). \tag{5.39}$$

We call (5.39) as the *practice* case. Because $\left| 1 - e^{\sqrt{-1} \times \theta} \right|^2 = 4 \times \sin^2(\theta / 2)$ and
$\sin(-\theta / 2) = -\sin(\theta / 2)$ and $\sin^2(-\theta / 2) = \sin(-\theta / 2) \times \sin(-\theta / 2) = (-\sin(\theta / 2)) \times$
$(-\sin(\theta / 2)) = \sin^2(\theta / 2)$, we can rewrite $P(i)$ in (5.39) as follows

$$P(i) = \frac{1}{2^{2 \times n}} \times \left(\sum_{P=0}^{r-1} \frac{4 \times \sin^2\left(\frac{-2 \times \pi \times i \times r \times (Y_P + 1)}{2^n} \times \frac{1}{2} \right)}{4 \times \sin^2\left(\frac{-2 \times \pi \times i \times r}{2^n} \times \frac{1}{2} \right)} \right)$$

$$= \frac{1}{2^{2 \times n}} \times \left(\sum_{P=0}^{r-1} \frac{\sin^2\left(\frac{\pi \times i \times r \times (Y_P + 1)}{2^n}\right)}{\sin^2\left(\frac{\pi \times i \times r}{2^n}\right)} \right). \tag{5.40}$$

5.14 Quantum Circuits of Factoring 15

We want to complete the prime factorization for $N = 15$. We need to find the nontrivial factor for $N = 15$. From Lemmas 5.1 and 5.6, we select a number $X = 2$ so that the greatest common divisor of $X = 2$ and $N = 15$ is 1 (one). This indicates that $X = 2$ is co-prime to $N = 15$. From Lemma 5.6, the order r of 2 modulo 15 satisfies $r \leq 15$. Since the number of bit representing $N = 15$ is four bits long, we also only need to make use of four bits that represent the value of r. If the value of r is an even, then the first nontrivial factor for $N = 15$ is equal to $\gcd(2^{r/2} + 1, N)$ and the second nontrivial factor for $N = 15$ is equal to $\gcd(2^{r/2} - 1, N)$.

Computing the order r of 2 modulo 15 is equivalent to calculate the period r of a given oracular function O_f: $\{p_1 p_2 p_3 p_4 \mid \forall p_d \in \{0, 1\}$ for $1 \leq d \leq 4\} \rightarrow \{2^{p_1 p_2 p_3 p_4}$ (mod 15) $\mid \forall p_d \in \{0, 1\}$ for $1 \leq d \leq 4\}$. An input variable $p_1 p_2 p_3 p_4$ is four bits long. Bit p_1 is the *most* significant bit and bit p_4 is the *least* significant bit. The corresponding decimal value is equal to $p_1 \times 2^{4-1} + p_2 \times 2^{4-2} + p_3 \times 2^{4-3} + p_4 \times 2^{4-4} = P$. The period r of O_f is to satisfy $O_f(p_1 p_2 p_3 p_4) = O_f(p_1 p_2 p_3 p_4 + r)$ to any two inputs $(p_1 p_2 p_3 p_4)$ and $(p_1 p_2 p_3 p_4 + r)$.

For implementing the operation of modular exponentiation, $2^{p_1 p_2 p_3 p_4}$ (mod 15), we assume that an auxiliary variable $w_1 w_2 w_3 w_4$ is four bits long. Bit w_1 is the *most* significant bit and bit w_4 is the *least* significant bit. The corresponding decimal value is equal to $w_1 \times 2^{4-1} + w_2 \times 2^{4-2} + w_3 \times 2^{4-3} + w_4 \times 2^{4-4} = W$. The initial value of each bit in the front *three* bits is zero (0). The initial value of the *least significant* bit w_4 is one (1).

5.14.1 Flowchart of Computing the Order R of X Modulo N

Figure 5.12 is to flowchart of computing the order r of X modulo N. In Fig. 5.12, in statement S_1, it sets the value of an auxiliary variable W to be one. It sets the value of X to be two and sets the value of N to be fifteen. Because the number of bits to represent N is four, it sets the value of an auxiliary variable n to be four. It sets the index variable j of the first loop to one. Next, in statement S_2, it executes the conditional judgement of the first loop. If the value of j is less than 2^n, then *next executed* instruction is statement S_3. Otherwise, in statement S_8, it sets the value of the order r to zero. This is to say that we cannot find the order r of X modulo N. Next, in statement S_9, it executes an *End* instruction to terminate the task that is to find the order r of X modulo N.

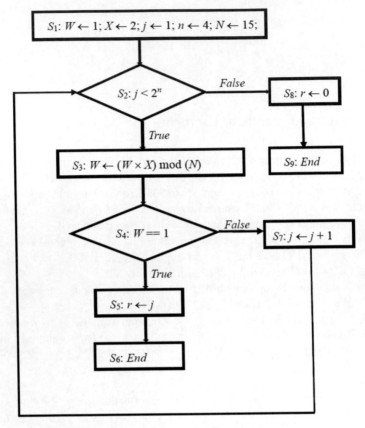

Fig. 5.12   Flowchart of computing the order r of X modulo N

In statement S_3, it completes one multiplication instruction and one modular instruction and stores the result into an auxiliary variable W. On the jth execution of statement S_3, it actually completes $X^j \pmod{N}$ and stores the result into an auxiliary variable W. Because the range of the value to the index variable j is from one through $2^n - 1$, it actually at most completes $X^1 \pmod{N}$ through $X^{2^n-1} \pmod{N}$. As $X^0 \pmod{N} = 1 \pmod{N}$, in statement S_1 it sets an auxiliary variable W to one that is to complete $X^0 \pmod{N}$.

Next, in statement S_4, it executes the conditional judgement to decide whether the value of W is equal to one. If the value of W is equal to one, then next executed instruction is statement S_5. In statement S_5, it sets the value of the order r to be the value of j. This indicates that we have found the order r of X modulo N. Next, in statement S_6, it executes an *End* instruction to terminate the task that is to find the order r of X modulo N.

However, if the value of W in statement S_4 is not equal to one, then next executed instruction is statement S_7. Next, in statement S_7, it increases the value of the index variable j. Repeat to execute statement S_2 through statement S_7 until in statement S_2, the conditional judgement becomes a *false* value or in statement S_6, it executes an *End* instruction to terminate the task. From Fig. 5.12, the total number of multiplication instruction and modular instruction is at most $(2^n - 1)$ multiplication instructions and $(2^n - 1)$ modular instructions. This is to say that the cost of finding the order r of X modulo N is at most to complete $(2^n - 1)$ multiplication instructions and $(2^n - 1)$ modular instructions.

5.14.2   Implementing Modular Exponentiation X^P *(Mod N)*

In Fig. 5.12, in sequent model, it computes each modular exponentiation X^P (mod N) to $0 \le P \le 2^n - 1$. However, in Fig. 5.11, in parallel model, it simultaneously computes each modular exponentiation X^P (mod N) to $0 \le P \le 2^n - 1$. In Fig. 5.11, it uses a quantum gate $|X^P \bmod N> = | X^{p_1 \times 2^{n-1} + p_2 \times 2^{n-2} + \cdots + p_n \times 2^{n-n}} \bmod N>$ operating on both quantum registers. The method for figuring out the modular exponentiation X^P (mod N) has two stages. The first stage makes use of modular multiplication to compute X^2 (mod N), by squaring X (mod N). Next, it computes X^4 (mod N) by squaring X^2 (mod N). Then, it computes X^8 (mod N) by squaring X^4 (mod N). In this way, it continues to compute X^{2^k} (mod N) for all k up to $(n - 1)$. We make use of $n = O(L)$, so a total of $(n - 1) = O(L)$ squaring operations is completed at a cost of $O(L^2)$ each (this cost assumes that the circuit used to do the squaring operation implements a kind of multiplication). Therefore, a total cost of the first stage is $O(L^3)$.

The second stage of the method is to complete the following observation

$$X^P(\bmod N) = X^{p_1 \times 2^{n-1} + p_2 \times 2^{n-2} + \cdots + p_n \times 2^{n-n}} (\bmod N)$$

$$= \left(X^{p_1 \times 2^{n-1}} (\bmod N) \right) \times_N \left(X^{p_2 \times 2^{n-2}} (\bmod N) \right)$$

$$\times_N \cdots \times_N \left(X^{p_n \times 2^{n-n}} (\bmod N) \right). \tag{5.41}$$

Completing $(n - 1) = O(L)$ modular multiplications with a cost $O(L^2)$ each, we see that using $O(L^3)$ gates can compute this product in (5.41). This is sufficiently efficient for finding the order r of X modulo N. Of course, methods that are more efficient are possible if there are the circuits of the better multiplication.

5.14.3   Computing the Order R of $(X = 2)$ Modulo $(N = 15)$

The *order* of $(X = 2)$ modulo $(N = 15)$ is to the *least* positive integer r such that $2^r = 1$ (mod 15). Because $r \le (N = 15)$ and the number of bits representing $(N = 15)$ is four bits long, the number of bits representing r is four bits long. Therefore, an input variable $p_1\, p_2\, p_3\, p_4$ is four bits long. Bit p_1 is the *most* significant bit and bit p_4 is the *least* significant bit. The corresponding decimal value is equal to $p_1 \times 2^{4-1}$

$+ p_2 \times 2^{4-2} + p_3 \times 2^{4-3} + p_4 \times 2^{4-4} = P$. This is to say that the range of the value to P is from zero through fifteen. If P is to the *least* positive integer such that $2^P = 1 \pmod{15}$, then the value of r is equal to the value of P. We use p_d^0 to represent the value of p_d to be zero for $1 \leq d \leq 4$ and apply p_d^1 to represent the value of p_d to be one for $1 \leq d \leq 4$.

Because the remainder for $2^P \pmod{15}$ to $0 \leq P \leq 15$ is from zero through fourteen, we assume that an auxiliary variable $w_1\, w_2\, w_3\, w_4$ is four bits long and we use it to store the result of computing $2^P \pmod{15}$ to $0 \leq P \leq 15$. We use w_d^0 to represent the value of w_d to be zero for $1 \leq d \leq 4$ and apply w_d^1 to represent the value of w_d to be one for $1 \leq d \leq 4$. Bit w_1 is the *most* significant bit and bit w_4 is the *least* significant bit. The corresponding decimal value is equal to $w_1 \times 2^{4-1} + w_2 \times 2^{4-2} + w_3 \times 2^{4-3} + w_4 \times 2^{4-4} = W$. The initial value of each bit in the front *three* bits is zero. The initial value of the *least significant* bit w_4 is one. This is to say that $W = w_1^0 \times 2^{4-1} + w_2^0 \times 2^{4-2} + w_3^0 \times 2^{4-3} + w_4^1 \times 2^{4-4} = 1$.

Computing the order r of 2 modulo 15 is equivalent to calculate the period r of a given oracular function O_f: $\{p_1\, p_2\, p_3\, p_4 \mid \forall\, p_d \in \{0, 1\}$ for $1 \leq d \leq 4\} \rightarrow \{2^{p_1 p_2 p_3 p_4} \pmod{15} \mid \forall\, p_d \in \{0, 1\}$ for $1 \leq d \leq 4\}$. The period r of O_f is to satisfy $O_f(p_1\, p_2\, p_3\, p_4) = O_f(p_1\, p_2\, p_3\, p_4 + r)$ to any two inputs $(p_1\, p_2\, p_3\, p_4)$ and $(p_1\, p_2\, p_3\, p_4 + r)$. The first stage of computing $O_f(p_1\, p_2\, p_3\, p_4) = 2^{p_1 p_2 p_3 p_4} \pmod{15}$ is to use modular multiplication to compute $2^2 \pmod{15}$, by squaring $2 \pmod{15}$. We get the following result

$$2^2 \pmod{15} = (2 \pmod{15})^2 = (2 \pmod{15}) \times_{15} (2 \pmod{15}) = 4 \pmod{15} = 4. \tag{5.42}$$

Next, it computes $2^4 \pmod{15}$ by squaring $2^2 \pmod{15}$ and we get the following result

$$2^4 \pmod{15} = \left(2^2 \pmod{15}\right)^2 = \left(2^2 \pmod{15}\right)$$
$$\times_{15} \left(2^2 \pmod{15}\right) = 4^2 \pmod{15} = 1. \tag{5.43}$$

Then, it computes $2^8 \pmod{15}$ by squaring $2^4 \pmod{15}$ and we get the following result

$$2^8 \pmod{15} = \left(2^4 \pmod{15}\right)^2 = \left(2^4 \pmod{15}\right)$$
$$\times_{15} \left(2^4 \pmod{15}\right) = 1^2 \pmod{15} = 1. \tag{5.44}$$

Next, the second stage of computing $O_f(p_1\, p_2\, p_3\, p_4) = 2^{p_1 p_2 p_3 p_4} \pmod{15}$ is to complete the following observation

$$2^P \pmod{15} = 2^{p_1 \times 2^3 + p_2 \times 2^2 + p_3 \times 2^1 + p_4 \times 2^0} \pmod{15} = \left(2^{p_1 \times 2^3} \pmod{15}\right) \times_{15}$$
$$\left(2^{p_2 \times 2^2} \pmod{15}\right) \times_{15} \left(2^{p_3 \times 2^1} \pmod{15}\right) \times_{15} \left(2^{p_4 \times 2^0} \pmod{15}\right). \tag{5.45}$$

If the value of bit p_1 is equal to one (1), then $(2^{p_1^1 \times 2^3} \pmod{15}) = (2^8 \pmod{15})$ $= 1$. Otherwise, $(2^{p_1^0 \times 2^3} \pmod{15}) = (2^0 \pmod{15}) = 1$. This is to say that $(2^{p_1 \times 2^3} \pmod{15}) = 1$. Next, if the value of bit p_2 is equal to one (1), then $(2^{p_2^1 \times 2^2} \pmod{15}) = (2^4 \pmod{15}) = 1$. Otherwise, $(2^{p_2^0 \times 2^2} \pmod{15}) = (2^0 \pmod{15}) = 1$. This indicates that $(2^{p_2 \times 2^2} \pmod{15}) = 1$. Therefore, we can rewrite the equation in (5.45) as follows

$$2^P \pmod{15} = \left(2^{p_3 \times 2^1} \pmod{15}\right) \times_{15} \left(2^{p_4 \times 2^0} \pmod{15}\right). \qquad (5.46)$$

According to the equation in (5.46), sixteen outputs of O_f that takes each input from $p_1^0 p_2^0 p_3^0 p_4^0$ through $p_1^1 p_2^1 p_3^1 p_4^1$ are subsequently 1, 2, 4, 8, 1, 2, 4, 8, 1, 2, 4, 8, 1, 2, 4 and 8. The binary values to one (1), two (2), four (4) and eight (8) are subsequently $w_1^0 w_2^0 w_3^0 w_4^1$, $w_1^0 w_2^0 w_3^1 w_4^0$, $w_1^0 w_2^1 w_3^0 w_4^0$ and $w_1^1 w_2^0 w_3^0 w_4^0$. The frequency f of O_f is equal to the number of the period per sixteen outputs. This gives that $r \times f = 16$. Hidden patterns and information stored in a given oracular function O_f are to that its output rotates back to its starting value (1) *four* times. This implies that the number of the period per sixteen outputs is *four* and the frequency f of O_f is equal to four. The period r of O_f is the *reciprocal* of the frequency f of O_f. Thus, we obtain $r = \frac{1}{f} = \frac{16}{f} = \frac{16}{4} = 4$ and $r \times f = 4 \times 4 = 16$. Therefore, the order r of $(X = 2)$ modulo $(N = 15)$ is equal to four.

5.14.4 Reduction of Implementing Modular Exponentiation 2^P (Mod 15)

In (5.46), the oracular function is $O_f(p_1 \ p_2 \ p_3 \ p_4) = 2^{p_1 p_2 p_3 p_4} \pmod{15} = (2^{p_3 \times 2^1} \pmod{15}) \times_{15} (2^{p_4 \times 2^0} \pmod{15})$. If the value of bit p_3 is equal to one (1), then $(2^{p_3^1 \times 2^1} \pmod{15}) = (2^2 \pmod{15}) = 4$. Otherwise, $(2^{p_3^0 \times 2^1} \pmod{15}) = (2^0 \pmod{15}) = 1$. This is to say that if the value of bit p_3 is equal to one (1), then the instruction "$(2^{p_3^1 \times 2^1} \pmod{15}) = (2^2 \pmod{15}) = 4$" can be implemented by means of multiply any auxiliary variable $w_1^0 w_2^0 w_3^0 w_4^1$ by 2^2. Otherwise, the instruction "$(2^{p_3^0 \times 2^1} \pmod{15}) = (2^0 \pmod{15}) = 1$" are not implemented.

Similarly, if the value of bit p_4 is equal to one (1), then $(2^{p_4^1 \times 2^0} \pmod{15}) = (2^1 \pmod{15}) = 2$. Otherwise, $(2^{p_4^0 \times 2^0} \pmod{15}) = (2^0 \pmod{15}) = 1$. This indicates that if the value of bit p_4 is equal to one (1), then the instruction "$(2^{p_4^1 \times 2^0} \pmod{15}) = (2^1 \pmod{15}) = 2$" can be implemented by means of multiply the auxiliary variable $w_1^0 w_2^0 w_3^0 w_4^1$ by 2^1. Otherwise, the instruction "$(2^{p_4^0 \times 2^0} \pmod{15}) = (2^0 \pmod{15}) = 1$" are not executed.

Of course, using a simple bit shift can achieve multiplication by 2 (or indeed any power of 2) on any binary auxiliary variable. Computing 2^P requires P multiplications by 2 on any binary auxiliary variable. Completing left bit shift of P times can implement P multiplications by 2 on any binary auxiliary variable. This implies that computing 2^P requires completing left bit shift of P times on any binary auxiliary variable.

For example, in our example we use bits p_3 and p_4 as the controlled bits. If the value of the controlled bit p_4 is equal to one (1), then we use left bit shift of one time on the binary auxiliary variable $w_1^0 w_2^0 w_3^0 w_4^1$ to implement the instruction "$(2^{p_4^1 \times 2^0}$ (mod 15)) $= (2^1$ (mod 15)) $= 2$". Left bit shift of one time is to that it exchanges each bit w_k for $1 \leq k \leq 4$ with the next highest weighted position. On the execution of the first time, it exchanges bit w_1^0 with bit w_2^0 and the result is $w_2^0 w_1^0 w_3^0 w_4^1$. Next, on the execution of the second time, it exchanges bit w_1^0 with bit w_3^0 and the result is $w_2^0 w_1^0 w_3^0 w_4^1$. Next, on the execution of the third time, it exchanges bit w_1^0 with bit w_4^1 and the result is $w_2^0 w_3^0 w_4^1 w_1^0$.

The corresponding decimal value of $w_2^0 w_3^0 w_4^1 w_1^0$ is two (2). This means that it implement the instruction "$(2^{p_4^1 \times 2^0}$ (mod 15)) $= (2^1$ (mod 15)) $= 2$". We can use three **CSWAP** gates to implement them. Similarly, if the value of the controlled bit p_3 is equal to one (1), then we complete a shift by two bits to implement the instruction "$(2^{p_3^1 \times 2^1}$ (mod 15)) $= (2^2$ (mod 15)) $= 4$". A shift by two bits is to that it exchanges the bit at the weighted position (2^3) with another bit at the weighted position (2^1) and exchanges at the weighted position (2^2) with another bit at the weighted position (2^0). We can make use of two **CSWAP** gates to implement them. This reduction makes us not to implement multiplication circuits and modular circuits.

5.14.5   Initialize Quantum Registers of Quantum Circuits to Find the Order **R** of (**X** = 2) Modulo (**N** = 15)

We use the quantum circuit in Fig. 5.13 to find the order r of $(X = 2)$ modulo $(N = 15)$. The first (upper) quantum register has four quantum bits. Bit $|p_1\rangle$ is the *most* significant bit and bit $|p_4\rangle$ is the *least* significant bit. The corresponding decimal value is equal to $p_1 \times 2^{4-1} + p_2 \times 2^{4-2} + p_3 \times 2^{4-3} + p_4 \times 2^{4-4} = P$. The initial state of each quantum bit $|p_d\rangle$ for $1 \leq d \leq 4$ is set to state $|0\rangle$. The second (lower) quantum register has four quantum bits. Bit $|w_1\rangle$ is the *most* significant bit and bit $|w_4\rangle$ is the *least* significant bit. The corresponding decimal value is equal to $w_1 \times 2^{4-1} + w_2 \times 2^{4-2} + w_3 \times 2^{4-3} + w_4 \times 2^{4-4} = W$. The initial state of the front three quantum bits $|w_d\rangle$ for $1 \leq d \leq 3$ is set to state $|0\rangle$. The initial state of the least significant quantum bits $|w_d\rangle$ for $1 \leq d \leq 3$ is set to state $|1\rangle$.

In Listing 5.2, the program is in the backend that is *simulator* of Open QASM with *thirty-two* quantum bits in **IBM**'s quantum computer. The program is to find the order r of $(X = 2)$ modulo $(N = 15)$. Figure 5.14 is the corresponding quantum circuit of the program in Listing 5.2 and is to implement the quantum circuit of finding the order r of $(X = 2)$ modulo $(N = 15)$ in Fig. 5.13.

The statement "OPENQASM 2.0;" on line one of Listing 5.2 is to indicate that the program is written with version 2.0 of Open QASM. Then, the statement "include "qelib1.inc";" on line two of Listing 5.2 is to continue parsing the file "qelib1.inc" as

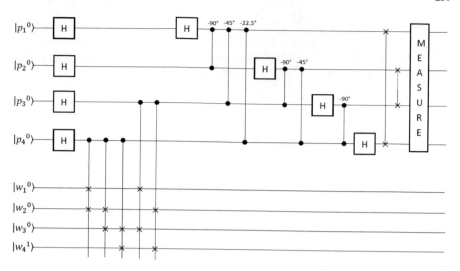

Fig. 5.13 Quantum circuits of finding the order r of $(X = 2)$ modulo $(N = 15)$

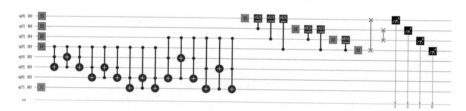

Fig. 5.14 Implementing quantum circuits of finding the order r of $(X = 2)$ modulo $(N = 15)$ in Fig. 5.13

if the contents of the file were pasted at the location of the include statement, where the file "qelib1.inc" is **Quantum Experience (QE) Standard Header** and the path is specified relative to the current working directory.

Listing 5.2 The Program of Finding the Order r of $(X = 2)$ Modulo $(N = 15)$.

```
1.  OPENQASM 2.0;
2.  include "qelib1.inc";
3.  qreg q[8];
4.  creg c[4];
5.  x q[7];
```

Next, the statement "qreg q[8];" on line three of Listing 5.2 is to declare that in the program there are *eight* quantum bits. In the left top of Fig. 5.14, eight quantum

bits are subsequently q[0], q[1], q[2], q[3], q[4], q[5], q[6] and q[7]. The initial value
of each quantum bit is set to state $|0\rangle$. We use four quantum bits q[0], q[1], q[2] and
q[3] to respectively encode four quantum bits $|p_1\rangle$, $|p_2\rangle$, $|p_3\rangle$ and $|p_4\rangle$ in Fig. 5.13.
We apply four quantum bits q[4], q[5], q[6] and q[7] to respectively encode four
quantum bits $|w_1\rangle$, $|w_2\rangle$, $|w_3\rangle$ and $|w_4\rangle$ in Fig. 5.13.

For the convenience of our explanation, q[k]^0 for $0 \le k \le 7$ is to represent the
value 0 of q[k] and q[k]^1 for $0 \le k \le 7$ is to represent the value 1 of q[k]. Next,
the statement "creg c[4];" on line four of Listing 5.2 is to declare that there are four
classical bits in the program. In the left bottom of Fig. 5.14, four classical bits are
subsequently c[0], c[1], c[2] and c[3]. The initial value of each classical bit is set to
zero (0). For the convenience of our explanation, c[k]^0 for $0 \le k \le 3$ is to represent
the value 0 of c[k] and c[k]^1 for $0 \le k \le 3$ is to represent the value 1 of c[k]. The
corresponding decimal value of the four initial classical bits c[3]^0 c[2]^0 c[1]^0 c[0]^0
is $2^3 \times$ c[3]$^0 + 2^2 \times$ c[2]$^0 + 2^1 \times$ c[1]$^0 + 2^0 \times$ c[0]^0. This indicates that classical
bit c[3]^0 is the most significant bit and classical bit c[0]^0 is the least significant bit.
Next, the statement "x q[7];" on line five of Listing 5.2 is to convert the state $|0\rangle$ of
quantum bit $|q[7]\rangle$ into the state $|1\rangle$. For the convenience of our explanation, an initial
state vector of finding the order r of $(X = 2)$ modulo $(N = 15)$ is

$$|\Omega_0\rangle = |q[0]^0\rangle|q[1]^0\rangle|q[2]^0\rangle|q[3]^0\rangle|q[4]^0\rangle|q[5]^0\rangle|q[6]^0\rangle|q[7]^1\rangle. \qquad (5.47)$$

5.14.6   Quantum Superposition to Compute 2^P (Mod 15)

In the initial state vector $|\Omega_0\rangle$ in (5.47), quantum bits $|q[0]^0\rangle$ $|q[1]^0\rangle$ $|q[2]^0\rangle$ $|q[3]^0\rangle$
encode four quantum bits $|p_1^0\rangle$, $|p_2^0\rangle$, $|p_3^0\rangle$ and $|p_4^0\rangle$ of the first register in Fig. 5.13
and are the *precision* register. We use the precision register to represent the values of
P that we shall pass to modular exponentiation 2^P (mod 15). We shall make use of
quantum superposition to evaluate modular exponentiation 2^P (mod 15) for multiple
values of P in parallel, so we use four statements "h q[0];", "h q[1];", "h q[2];" and
"h q[3];" from line *six* through line *nine* in Listing 5.2 to place the *precision* register
into

Listing 5.2 Continued…

```
6.  h q[0];
7.  h q[1];
8.  h q[2];
9.  h q[3];
```

a superposition of all possible values. Therefore, we have the following new state vector

$$|\Omega_1\rangle = \left(\frac{1}{\sqrt{2}}(|q[0]^0\rangle + |q[0]^1\rangle)\right)\left(\frac{1}{\sqrt{2}}(|q[1]^0\rangle + |q[1]^1\rangle)\right)\left(\frac{1}{\sqrt{2}}(|q[2]^0\rangle + |q[2]^1\rangle)\right)$$

$$\left(\frac{1}{\sqrt{2}}(|q[3]^0\rangle + |q[3]^1\rangle)\right)(|q[4]^0\rangle|q[5]^0\rangle|q[6]^0\rangle|q[7]^1\rangle)$$

$$= \frac{1}{\sqrt{2^4}}\left(\sum_{p=0}^{2^4-1}|P\rangle\right)(|q[4]^0\rangle|q[5]^0\rangle|q[6]^0\rangle|q[7]^1\rangle). \qquad (5.48)$$

This way makes each state $(|P\rangle\ |q[4]^0\rangle\ |q[5]^0\rangle\ |q[6]^0\rangle\ |q[7]^1\rangle)$ in the new state vector $(|\Omega_1\rangle)$ in (5.48) to be ready to be treated as a separate input to a parallel computation.

5.14.7   Implementing Conditional Multiply-by-2 for Computing 2^P (Mod 15)

In the initial state vector $|\Omega_0\rangle$ in (5.47), quantum bits $|q[4]^0\rangle\ |q[5]^0\rangle\ |q[6]^0\rangle\ |q[7]^1\rangle$ encode four quantum bits $|w_1^0\rangle$, $|w_2^0\rangle$, $|w_3^0\rangle$ and $|w_4^1\rangle$ of the second register in Fig. 5.13 and are the *auxiliary* register. We now would like to complete modular exponentiation 2^P (mod 15) $= (2^{p_3 \times 2^1}$ (mod 15)) $\times_{15} (2^{p_4 \times 2^0}$ (mod 15)) in (5.46) on the superposition of inputs we have within the precision register, and we shall apply the auxiliary register to hold and to store the results. We use three **CSWAP** gates to implement the instruction $(2^{p_4 \times 2^0}$ (mod 15)) that is conditional multiply-by-2. Three statements "ccx q[3], q[4], q[5];", "ccx q[3], q[5], q[4];" and "ccx q[3], q[4], q[5];" from line ten through line twelve in Listing 5.2 are three **CCNOT** gates to implement the *first* **CSWAP** gate in Fig. 5.13. In the first **CSWAP** gate, quantum bit q[3] is its controlled bit and quantum bits q[4] and q[5] are its target bits. In the **CCNOT** instruction, the first operand and the second operand are its two controlled bits and the third operand is its target bit.

Listing 5.2 Continued...

//Implement the first **CSWAP** gate.

```
10.  ccx q[3], q[4], q[5];
11.  ccx q[3], q[5], q[4];
12.  ccx q[3], q[4], q[5];
```

//Implement the second **CSWAP** gate.

13. ccx q[3], q[5], q[6];
14. ccx q[3], q[6], q[5];
15. ccx q[3], q[5], q[6];

//Implement the third **CSWAP** gate.

16. ccx q[3], q[6], q[7];
17. ccx q[3], q[7], q[6];
18. ccx q[3], q[6], q[7];

The three **CCNOT** gates exchange the quantum bit at the weighted position (2^3) of the work register with the quantum bit at the weighted position (2^2) of the work register. This means that the new state vector is

$$|\Omega_2\rangle = \frac{1}{\sqrt{2^4}}(|q[0]^0\rangle + |q[0]^1\rangle)(|q[1]^0\rangle + |q[1]^1\rangle)(|q[2]^0\rangle + |q[2]^1\rangle)(|q[3]^0\rangle|q[4]^0\rangle$$

$$|q[5]^0\rangle|q[6]^0\rangle|q[7]^1\rangle + |q[3]^1\rangle|q[5]^0\rangle|q[4]^0\rangle|q[6]^0\rangle|q[7]^1\rangle). \tag{5.49}$$

Next, in the second **CSWAP** gate in Fig. 5.13, quantum bit q[3] is its controlled bit and quantum bits q[5] and q[6] are its target bits. Three **CCNOT** gates "ccx q[3], q[5], q[6];", "ccx q[3], q[6], q[5];" and "ccx q[3], q[5], q[6];" from line thirteen through line fifteen in Listing 5.2 are to implement the *second* **CSWAP** gate in Fig. 5.13. This indicates that the new state vector is

$$|\Omega_3\rangle = \frac{1}{\sqrt{2^4}}(|q[0]^0\rangle + |q[0]^1\rangle)(|q[1]^0\rangle + |q[1]^1\rangle)(|q[2]^0\rangle + |q[2]^1\rangle)(|q[3]^0\rangle|q[4]^0\rangle$$

$$|q[5]^0\rangle|q[6]^0\rangle|q[7]^1\rangle + |q[3]^1\rangle|q[5]^0\rangle|q[6]^0\rangle|q[4]^0\rangle|q[7]^1\rangle). \tag{5.50}$$

Then, in the third **CSWAP** gate in Fig. 5.13, quantum bit q[3] is its controlled bit and quantum bits q[6] and q[7] are its target bits. Three **CCNOT** gates "ccx q[3], q[6], q[7];", "ccx q[3], q[7], q[6];" and "ccx q[3], q[6], q[7];" from line sixteen through line eighteen in Listing 5.2 are to implement the *third* **CSWAP** gate in Fig. 5.13. This implies that the new state vector is

$$|\Omega_4\rangle = \frac{1}{\sqrt{2^4}}(|q[0]^0\rangle + |q[0]^1\rangle)(|q[1]^0\rangle + |q[1]^1\rangle)(|q[2]^0\rangle + |q[2]^1\rangle)(|q[3]^0\rangle|q[4]^0\rangle$$

$$|q[5]^0\rangle|q[6]^0\rangle|q[7]^1\rangle + |q[3]^1\rangle|q[5]^0\rangle|q[6]^0\rangle|q[7]^1\rangle|q[4]^0\rangle). \tag{5.51}$$

In the new state vector $|\Omega_4\rangle$ in (5.51), if the least significant quantum bit q[3] in the precision register is the $|1\rangle$ state, then the value of its work register is two (2). Otherwise, the value of its work register is not changed and is still one (1).

5.14.8   *Implementing Conditional Multiply-by-4 for Computing 2^P (Mod 15)*

Because the work register in the new state vector $|\Omega_4\rangle$ in (5.51) holds and stores the result of computing $(2^{p_4 \times 2^0} \pmod{15})$, next we want to complete the instruction $(2^{p_3 \times 2^1} \pmod{15})$. If the value of quantum bit $|q[2]\rangle$ that encodes quantum bit $|p_3\rangle$ in the precision register is one (1), then that indicates to complete the instruction $(2^{p_3 \times 2^1} \pmod{15})$ to require another two multiplications by two (2) on the work register. We use two **CSWAP** gates to implement the instruction $(2^{p_3 \times 2^1} \pmod{15})$ that is conditional multiply-by-4. In Fig. 5.13, the fourth **CSWAP** gate is to exchange the quantum bit at the weighted position (2^3) of the work register with another quantum bit at the weighted position (2^1) of the work register if the value of the quantum bit at the weighted position (2^1) of the precision register is state $|1\rangle$. Next, in Fig. 5.13, the fifth **CSWAP** gate is to exchange the quantum bit at the weighted position (2^2) of the work register with another quantum bit at the weighted position (2^0) of the work register if the value of the quantum bit at the weighted position (2^1) of the precision register is state $|1\rangle$.

Three statements "ccx q[2], q[4], q[6];", "ccx q[2], q[6], q[4];" and "ccx q[2], q[4], q[6];" from line nineteen through line twenty-one in Listing 5.2 are three **CCNOT** gates to implement the *fourth* **CSWAP** gate in Fig. 5.13. In the fourth **CSWAP** gate, quantum bit q[2] is its controlled bit and quantum bits q[4] and q[6] are its target bits. In the **CCNOT** instruction, the first operand and the second operand are its two controlled bits and the third operand is its target bit.

Listing 5.2 Continued…

*//Implement the fourth **CSWAP** gate.*

19. ccx q[2], q[4], q[6];
20. ccx q[2], q[6], q[4];
21. ccx q[2], q[4], q[6];

*//Implement the fifth **CSWAP** gate.*

22. ccx q[2], q[5], q[7];
23. ccx q[2], q[7], q[5];
24. ccx q[2], q[5], q[7];

The three **CCNOT** gates exchange the quantum bit at the weighted position (2^3) of the work register with the quantum bit at the weighted position (2^1) of the work register. This means that the new state vector is

$$|\Omega_5\rangle = \frac{1}{\sqrt{2^4}}\left(\left|q[0]^0\right\rangle + \left|q[0]^1\right\rangle\right)\left(\left|q[1]^0\right\rangle + \left|q[1]^1\right\rangle\right)\left(\left|q[2]^0\right\rangle\left|q[3]^0\right\rangle\left|q[4]^0\right\rangle\left|q[5]^0\right\rangle\left|q[6]^0\right\rangle\right.$$

$$\left|q[7]^1\right\rangle + \left|q[2]^0\right\rangle\left|q[3]^1\right\rangle\left|q[5]^0\right\rangle\left|q[6]^0\right\rangle\left|q[7]^1\right\rangle\left|q[4]^0\right\rangle + \left(\left|q[2]^1\right\rangle\left|q[3]^0\right\rangle\left|q[6]^0\right\rangle\left|q[5]^0\right\rangle\right.$$

$$\left.\left|q[4]^0\right\rangle\left|q[7]^1\right\rangle + \left|q[2]^1\right\rangle\left|q[3]^1\right\rangle\left|q[7]^1\right\rangle\left|q[6]^0\right\rangle\left|q[5]^0\right\rangle\left|q[4]^0\right\rangle\right). \tag{5.52}$$

Next, in the fifth **CSWAP** gate in Fig. 5.13, quantum bit q[2] is its controlled bit and quantum bits q[5] and q[7] are its target bits. Three **CCNOT** gates "ccx q[2],q[5],q[7];", "ccx q[2],q[7],q[5];" and "ccx q[2],q[5],q[7];" from line twenty-two through line twenty-four in Listing 5.2 are to implement the *fifth* **CSWAP** gate in Fig. 5.13. This means that the new state vector is

$$|\Omega_6\rangle = \frac{1}{\sqrt{2^4}}\left(\left(\left|q[0]^0\right\rangle + \left|q[0]^1\right\rangle\right)\left(\left|q[1]^0\right\rangle + \left|q[1]^1\right\rangle\right)\left(\left|q[2]^0\right\rangle\left|q[3]^0\right\rangle\left|q[4]^0\right\rangle\left|q[5]^0\right\rangle\left|q[6]^0\right\rangle\right.\right.$$

$$\left|q[7]^1\right\rangle + \left|q[2]^0\right\rangle\left|q[3]^1\right\rangle\left|q[5]^0\right\rangle\left|q[6]^0\right\rangle\left|q[7]^1\right\rangle\left|q[4]^0\right\rangle + \left(\left|q[2]^1\right\rangle\left|q[3]^0\right\rangle\left|q[6]^0\right\rangle\left|q[7]^1\right\rangle\right.$$

$$\left.\left|q[4]^0\right\rangle\left|q[5]^0\right\rangle + \left|q[2]^1\right\rangle\left|q[3]^1\right\rangle\left|q[7]^1\right\rangle\left|q[4]^0\right\rangle\left|q[5]^0\right\rangle\left|q[6]^0\right\rangle\right)\right). \tag{5.53}$$

In the new state vector $|\Omega_6\rangle$ in (5.53), it shows how we have now managed to compute $2^P \pmod{15} = (2^{P_3 \times 2^1} \pmod{15}) \times_{15} (2^{P_4 \times 2^0} \pmod{15})$ in (5.46) on every value of P from the precision register in superposition.

5.14.9 *Implementing Inverse Quantum Fourier Transform of Four Quantum Bits*

In Fig. 5.13, by completing an inverse quantum Fourier transform on the precision register, it effectively transform the precision register state into a superposition of the *periodic* signal's component frequencies. Twelve statements from

Listing 5.2 Continued...

//Implement an inverse quantum Fourier transform.

```
25.  h q[0];
26.  cu1(-2*pi*1/4) q[1],q[0];
27.  cu1(-2*pi*1/8) q[2],q[0];
28.  cu1(-2*pi*1/16) q[3],q[0];

29.  h q[1];
30.  cu1(-2*pi*1/4) q[2],q[1];
31.  cu1(-2*pi*1/8) q[3],q[1];

32.  h q[2];
33.  cu1(-2*pi*1/4) q[3],q[2];
```

```
34.  h q[3];
35.  swap q[0],q[3];
36.  swap q[1],q[2];
```

line twenty-five through line thirty-six in Listing 5.2 implement an inverse quantum Fourier transform on the precision register. They take the new state vector $|\Omega_6\rangle$ in (5.53) as their input state vector. They produce the following new state vector

$$
|\Omega_7\rangle = \left(\frac{1}{\sqrt{2^2}} \left(|q[0]^0\rangle|q[1]^0\rangle|q[2]^0\rangle|q[3]^0\rangle \right) + \frac{1}{\sqrt{2^2}} \left(|q[0]^0\rangle|q[1]^1\rangle|q[2]^0\rangle|q[3]^0\rangle \right) \right.
$$
$$
+ \frac{1}{\sqrt{2^2}} \left(|q[0]^1\rangle|q[1]^0\rangle|q[2]^0\rangle|q[3]^0\rangle \right) + \frac{1}{\sqrt{2^2}} \left. \left(|q[0]^1\rangle|q[1]^1\rangle|q[2]^0\rangle|q[3]^0\rangle \right) \right)
$$
$$
\left(|q[4]^0\rangle|q[5]^0\rangle|q[6]^0\rangle|q[7]^1\rangle + |q[5]^0\rangle|q[6]^0\rangle|q[7]^1\rangle|q[4]^0\rangle + |q[6]^0\rangle|q[7]^1\rangle \right)
$$
$$
|q[4]^0\rangle|q[5]^0\rangle + |q[7]^1\rangle|q[4]^0\rangle|q[5]^0\rangle|q[6]^0\rangle). \tag{5.54}
$$

5.14.10 Read the Quantum Result

Finally, four statements "measure q[0] -> c[3];", "measure q[1] -> c[2];", "measure q[2] -> c[1];" and "measure q[3] -> c[0];" from line thirty-seven through line forty in Listing 5.2 implement the measurement on the precision register in Fig. 5.13. They measure four quantum bits q[0], q[1], q[2] and q[3] of the precision register. They record the measurement outcome by overwriting four classical bits c[3], c[2], c[1] and c[0].

Listing 5.2 Continued...

//Implement one measurement on the precision register

```
37.  measure q[0] -> c[3];
38.  measure q[1] -> c[2];
39.  measure q[2] -> c[1];
40.  measure q[3] -> c[0];
```

In the backend *simulator* with thirty-two quantum bits in **IBM**'s quantum computers, we use the command "run" to execute the program in Listing 5.2. Figure 5.15 shows the measured result. From Fig. 5.15, we obtain that a computational basis state 0000 (c[3] = 0 = q[0] = |0⟩, c[2] = 0 = q[1] = |0⟩, c[1] = 0 = q[2] = |0⟩ and c[0] = 0 = q[3] = |0 ⟩) has the probability 24.512% (0.24512). On the other hand, we get that a computational basis state 0100 (c[3] = 0 = q[0]

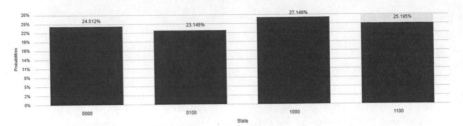

Fig. 5.15 A computational basis state 0000 has the probability 24.512% (0.24512), a computational basis state 0100 has the probability 23.145% (0.23145), a computational basis state 1000 has the probability 27.148% (0.27148) and a computational basis state 1100 has the probability 25.195% (0.25195)

$= |0\rangle$, $c[2] = 1 = q[1] = |1\rangle$, $c[1] = 0 = q[2] = |0\rangle$ and $c[0] = 0 = q[3] = |0\rangle$) has the probability 23.145% (0.23145). Alternatively, we gain that a computational basis state 1000 ($c[3] = 1 = q[0] = |1\rangle$, $c[2] = 0 = q[1] = |0\rangle$, $c[1] = 0 = q[2] = |0\rangle$ and $c[0] = 0 = q[3] = |0\rangle$) has the probability 27.148% (0.27148). On the other hand, we obtain that a computational basis state 1100 ($c[3] = 1 = q[0] = |1\rangle$, $c[2] = 1 = q[1] = |1\rangle$, $c[1] = 0 = q[2] = |0\rangle$ and $c[0] = 0 = q[3] = |0\rangle$) has the probability 25.195% (0.25195).

We select the computational basis state 0100 ($c[3] = 0 = q[0] = |0\rangle$, $c[2] = 1 = q[1] = |1\rangle$, $c[1] = 0 = q[2] = |0\rangle$ and $c[0] = 0 = q[3] = |0\rangle$) with the probability 23.145% (0.23145) as the measured result. Because the decimal value of the computational basis state 1000 is four (4) and ($2^4 / 4$) is a rational number, we use the continued fractional algorithm in Fig. 5.2 to determine the continued fractional representation of (c / d) if $c = 2^4 = 16$ and $d = 4$ and the corresponding convergent. From the first execution of statement S_0 through statement S_2, it gets $i = 1$, $q[1] = c / d = 16 / 4 = 4$ and $r = 16 \pmod 4 = 0$. This is to split (16 /4) into its integer and fractional part and not to invert its fractional part,

$$\frac{16}{4} = 4 + \frac{0}{4} = 4. \tag{5.55}$$

Since the value of r is equal to 0, from the first execution of statement S_3, it returns to a *true*. Thus, next, from the first execution of statement S_4, the answer is to the continued fractional representation of (16 /4)

$$\frac{16}{4} = (q[1] = 4) = 4. \tag{5.56}$$

Next, from the first execution of Statement S_5, it terminates the execution of the continued fractional algorithm. For a rational number (16 /4), the first convergent is ($q[1]$) $= 4 = \frac{4}{1}$ and is the closest one to ($\frac{16}{4}$) with numerator less than 15. Therefore, we check $2^4 \pmod{15}$ which equals one (1) and we find the order r to be four (4). Because the order r is even, from **Lemma** 5.2, we use Euclid's algorithm to compute

$\gcd(15, 2^{\frac{4}{2}} + 1)$ and $\gcd(15, 2^{\frac{4}{2}} - 1)$. This implies that two nontrivial factors for N = 15 are respectively 5 and 3. Therefore, the prime factorization for $N = 15$ is $N = 5 \times 3$.

5.15   Assessment to Complexity of Shor's Order-Finding Algorithm

In Fig. 5.11, the precision register (the first register or the upper register) with n quantum bits represents the order r of X modulo N. Because the value of r is less than or equal to the value of N, the value of n is the smallest integer greater than or equal to $\log_2^N$ and we can write it as $n = \lceil \log_2^N \rceil$. In Fig. 5.11, the work register (the second register or the lower register) with L quantum bits is to store the result of computing X^P (mod N) for $0 \le p \le 2^n - 1$. Because the result of computing X^P (mod N) for $0 \le p \le 2^n - 1$ is less than the value of N, the value of L is the smallest integer greater than or equal to $\log_2^N$ and we can write it as $L = \lceil \log_2^N \rceil$. The value of X is less than the value of N, so we can use L bits to represent the value of X and the value of N.

In Fig. 5.11, in parallel model, it simultaneously computes each modular exponentiation X^P (mod N) to $0 \le P \le 2^n - 1$. In Fig. 5.11, it applies a quantum gate $|X^P$ mod $N\rangle = | X^{p_1 \times 2^{n-1} + p_2 \times 2^{n-2} + \cdots + p_n \times 2^{n-n}}$ mod $N >$ operating on the precision register and the work register. The method to calculate the modular exponentiation X^P (mod N) contains two stages. The first stage uses modular multiplication to compute X^2 (mod N), by squaring X (mod N). Then, it figures out X^4 (mod N) by squaring X^2 (mod N). Then, it calculates X^8 (mod N) by squaring X^4 (mod N). In this way, it continues to figure out X^{2^k} (mod N) for all k up to $(n - 1)$.

Since $n = \lceil \log_2^N \rceil$ and $L = \lceil \log_2^N \rceil$, we apply $n = O(L)$. Therefore, in the first stage it completes a total of $(n - 1) = O(L)$ squaring operations. A squaring operation consists of a multiplication instruction and a modular instruction. Multiplicand and multiplier in the multiplication instruction are both X and they are L bits long. Product in the multiplication instruction is $(2 \times L)$ bits long and it is dividend in the modular instruction. Divisor in the modular instruction is N and it is $n = O(L)$ bits long. The number of the auxiliary carry bit in the multiplication instruction is $(2 \times L + 1)$ bits and the number of the auxiliary borrow bit in the modular instruction is $(2 \times L + 1)$ bits.

Because the cost of the circuit to implement one multiplication instruction is $O(L^2)$ digital logic gates and the cost of the circuit to implement one modular instruction is $O(L^2)$ digital logic gates, the cost of the quantum circuit to implement one squaring operation is $O(L^2)$ quantum gates. Therefore, the cost to complete $(n - 1) = O(L)$ squaring operations is $O(L^3)$ quantum gates and a total cost of implementing the first stage is $O(L^3)$ quantum gates.

Next, the second stage of the method is to complete X^P (mod N) = $(X^{p_1 \times 2^{n-1}}$ (mod N)) $\times_N (X^{p_2 \times 2^{n-2}}$ (mod N)) $\times_N \cdots \times_N (X^{p_n \times 2^{n-n}}$ (mod N)). It implements (n

$- 1) = O(L)$ modular multiplications. Each modular multiplication consists of a multiplication instruction and a modular instruction. Multiplicand and multiplier in the multiplication instruction are L bits long. Product in the multiplication instruction is $(2 \times L)$ bits long and it is dividend in the modular instruction. Divisor in the modular instruction is N and it is $n = O(L)$ bits long. The number of the auxiliary carry bit in the multiplication instruction is $(2 \times L + 1)$ bits and the number of the auxiliary borrow bit in the modular instruction is $(2 \times L + 1)$ bits.

Since the cost of the circuit to implement one multiplication instruction is $O(L^2)$ digital logic gates and the cost of the circuit to implement one modular instruction is $O(L^2)$ digital logic gates, the cost of the quantum circuit to implement one modular multiplication is $O(L^2)$ quantum gates. Hence, the cost to complete $(n - 1) = O(L)$ modular multiplication is $O(L^3)$ quantum gates and a total cost of implementing the second stage is $O(L^3)$ quantum gates. This means that the cost of the quantum circuit to implement $|X^P \bmod N \rangle = |X^{p_1 \times 2^{n-1} + p_2 \times 2^{n-2} + \cdots + p_n \times 2^{n-n}} \bmod N \rangle$ operating on the precision register and the work register is $O(L^3)$ quantum gates.

Next, in Fig. 5.11, it completes one inverse quantum Fourier transform. The cost of implementing one inverse quantum Fourier transform is $O(L^2)$ quantum gates. Finally, in Fig. 5.11, it completes a measurement on the precision register. Thus, in Shor's order-finding algorithm, the cost to compute the order r of X modulo N is $O(L)$ quantum bits and $O(L^3)$ quantum gates.

5.16 Summary

In this chapter, we gave an introduction of fundamental number theory. Next, we described **Euclid's** algorithm. We also introduced quadratic congruence. We then illustrated continued fractions. We also introduced the two problems of order finding and factoring. Next, we described how to compute the order of 2 modulo 15 and the prime factorization for 15. We also illustrated how to calculate the order of 2 modulo 21 and the prime factorization for 21. Next, we introduced how to calculate the order of 2 modulo 35 and the prime factorization for 35. We also introduced how to figure out the order of 5 modulo 33 and the prime factorization for 33. We then described the possibility of finding the even order of X modulo N. We also illustrated public key cryptography and the RSA cryptosystem. Next, we introduced how to implement the **controlled-swap** gate of three quantum bits. We also described Shor's order-finding algorithm. We then illustrated how to design quantum circuits of factoring 15. We also gave assessment of complexity of Shor's order-finding algorithm.

5.17 Bibliographical Notes

In this chapter for more details about an introduction of fundamental and advanced knowledge of number theory, the recommended books are (Hardy and Wright 1979; Nielsen and Chuang 2000; Imre and Balazs 2005; Lipton and Regan 2014; Silva 2018; Johnston et al. 2019). For a more detailed description to Shor's order-finding algorithm, the recommended article and books are (Shor 1994; Nielsen and Chuang 2000; Imre and Balazs 2005; Lipton and Regan 2014; Silva 2018; Johnston et al. 2019). A good introduction to the instructions of Open QASM is the famous article in (Cross et al. 2017).

5.18 Exercises

5.1 Let c and d be integer, and let r be the remainder when c is divided by d. Then provided $r \neq 0$, please prove the equation $\gcd(c, d) = \gcd(d, r)$.

5.2 (**Chinese remainder theorem**) we assume that $y_1, \ldots, y_n$ are positive integers such that any pair y_i and y_j ($i \neq j$) are co-prime. Then the system of equations

$$z = c_1 (\mathrm{mod}\ y_1)$$
$$z = c_2 (\mathrm{mod}\ y_2)$$
$$\cdots$$
$$z = c_n (\mathrm{mod}\ y_n)$$

has a solution. Moreover, any two solutions to this system of equations are equal modulo $Y = y_1\, y_2\, \ldots\, y_n$. Please prove **Chinese remainder theorem**.

5.3 We assume that p and k are integers and p is a prime in the range 1 to $p - 1$. Then prime p divides $\binom{p}{k}$. Please prove it.

5.4 (**Fermat's little theorem**) we assume that p is a prime, and a is an integer. Then $a^p = a \ (\mathrm{mod}\ p)$. If integer a is not divisible by prime p then $a^{p-1} = 1 \ (\mathrm{mod}\ p)$. Please prove them.

5.5 The *Euler* function $\varphi(n)$ is defined to be the number of positive integers which are less than n and are co-prime to n. We assume that a is co-prime to n. Then $a^{\varphi(n)} = 1 \ (\mathrm{mod}\ n)$.

5.6 If $(i\,/2^n)$ is a rational fraction and z and r are positive integers that satisfy $|(z\,/r) - (i\,/2^n)| \leq (1\,/(2 \times r^2))$, then $(z\,/r)$ is a convergent of the continued fraction of $(i\,/2^n)$.

5.7 Prove that $\left|1 - e^{\sqrt{-1} \times \theta}\right|^2 = 4 \times \sin^2(\theta\,/2)$.

References

Cross, A.W., Bishop, L.S., Smolin, J.A., Gambetta, J.M.: Open quantum assembly language. 2017. https://arxiv.org/abs/1707.03429

Hardy, G.H., Wright, E.M.: An Introduction to the Theory of Numbers, 5th edn. Oxford University Press, New York (1979). ISBN: 0198531702

Imre, S., Balazs, F.: Quantum Computation and Communications: An Engineering Approach. Wiley, UK (2007). ISBN-10: 047086902X and ISBN-13: 978-0470869024, 2005

Johnston, E.R., Harrigan, N., Gimeno-Segovia, M.: Programming Quantum Computers: essential algorithms and code samples. O'Reilly Media, Inc., **ISBN**-13: 978-1492039686, **ISBN**-10: 1492039683, 2019

Lipton, R.J., Regan, K.W.: Quantum Algorithms via Linear Algebra: A Primer. The MIT Press (2014). ISBN 978-0-262-02839-4

Manders, K.L., Adleman, L.M.: NP-complete decision problems for binary quadratics. J. Comput. Syst. Sci. **16**(2), 168–184 (1978)

Nielsen, M.A., Chuang, I.L.: Quantum Computation and Quantum Information. Cambridge University Press, New York, NY, 2000, ISBN-10: 9781107002173 and ISBN-13: 978-1107002173

Shor, P.: Algorithms for quantum computation: discrete logarithms and factoring. In: Proceedings 35th Annual Symposium on Foundations of Computer Science, Santa Fe, pp. 124–134, November 20-22 1994

Silva, V.: Practical Quantum Computing for Developers: Programming Quantum Rigs in the Cloud using Python, Quantum Assembly Language and IBM Q Experience. Apress, December 13, 2018, ISBN-10: 1484242173 and ISBN-13: 978-1484242179

Chapter 6
Phase Estimation and Its Applications

A *decision* problem is a problem in which it has only two possible outputs (yes or no) on any input of n bits. An output "yes" in the decision problem is to the number of solutions not to be zero and another output "no" in the decision problem is to the number of solutions to be zero. An example of a decision problem is deciding whether a given Boolean formula, $F(x_1, x_2) = x_1 \wedge x_2$, has solutions that satisfy $F(x_1, x_2)$ to have a *true* value or not, where the value of two Boolean variables x_1 and x_2 is either true (1) or false (0) and "$\wedge$" is the **AND** operation of two operands. For the convenience of the presentation, Boolean variable $x_1{}^0$ is to represent the value 0 (zero) of Boolean variable x_1 and Boolean variable $x_1{}^1$ is to represent the value 1 (one) of Boolean variable x_1. Boolean variable $x_2{}^0$ is to represent the value 0 (zero) of Boolean variable x_2 and Boolean variable $x_2{}^1$ is to represent the value 1 (one) of Boolean variable x_2.

A decision procedure is in the form of an algorithm to solve a decision problem. A decision procedure for the decision problem "a given Boolean formula, $F(x_1, x_2)$ $= x_1 \wedge x_2$, does it have solutions that satisfy $F(x_1, x_2)$ to have a *true* value?" would implement $x_1 \wedge x_2$ (the **AND** operation of two operands) of four times according to four different inputs $x_1{}^0 x_2{}^0, x_1{}^0 x_2{}^1, x_1{}^1 x_2{}^0$ and $x_1{}^1 x_2{}^1$. After it executes each **AND** operation, it finds the fourth input $x_1{}^1 x_2{}^1$ that satisfies $F(x_1, x_2)$ to have a *true* value. Finally, it gives an output "yes" to the decision problem. This implies that the number of solutions is not equal to zero. If time complexity of a decision procedure to solve a decision problem with the input of n bits is $O(2^n)$, then the decision problem is a NP-Complete problem.

We assume that a $(2^n \times 2^n)$ unitary matrix (operator) U has a $(2^n \times 1)$ eigenvector $|u\rangle$ with eigenvalue $e^{\sqrt{-1} \times 2 \times \pi \times \theta}$ such that $U \times |u\rangle = e^{\sqrt{-1} \times 2 \times \pi \times \theta} \times |u\rangle$, where the value of θ is *unknown* and is real. The purpose of the phase estimate algorithm is to estimate the value of θ. Deciding whether there exist solutions for a problem with the input of n bits is equivalent to estimate the value of θ. In this chapter, we first describe how the phase estimate algorithm works on quantum computers and various kinds of real applications. We illustrate how to write quantum programs to

© The Author(s), under exclusive license to Springer Nature Switzerland AG 2021
W.-L. Chang and A. V. Vasilakos, *Fundamentals of Quantum Programming in IBM's Quantum Computers*, Studies in Big Data 81,
https://doi.org/10.1007/978-3-030-63583-1_6

compute and estimate the value of θ to that any given a $(2^n \times 2^n)$ unitary matrix (operator) U has a $(2^n \times 1)$ eigenvector $|u\rangle$ with eigenvalue $\left(e^{\sqrt{-1} \times 2 \times \pi \times \theta}\right)$. Next, we explain the reason of why deciding whether there exist solutions for a problem with the input of n bits is equivalent to estimate the value of θ. We also explain how the quantum-counting algorithm determines the number of solutions for a decision problem with the input of n bits. Next, we introduce how to write quantum algorithms to implement the quantum-counting algorithm that is a real application of the phase estimate algorithm for computing the number of solutions for various kinds of real applications with the input of n bits.

6.1 Phase Estimation

We use the quantum circuit shown in Fig. 6.1 to implement the phase estimation algorithm. It uses two quantum registers. At the left top in Fig. 6.1, the first register $\left(\otimes_{k=t}^{1} |y_k^0\rangle\right)$ contains t quantum bits initially in the state $|0\rangle$. Quantum bit $|y_t^0\rangle$ is the most significant bit. Quantum bit $|y_1^0\rangle$ is the least significant bit. The corresponding decimal value of the first register is $(|y_t^0\rangle \times 2^{t-1}) + \cdots + (|y_2^0\rangle \times 2^{2-1}) + (|y_1^0\rangle \times 2^{1-1})$. How we select t that is dependent on two things. The first thing is to that the number of bits of accuracy we wish to have in our estimation for the value of θ. The second thing is to that with what probability we wish the phase estimation algorithm

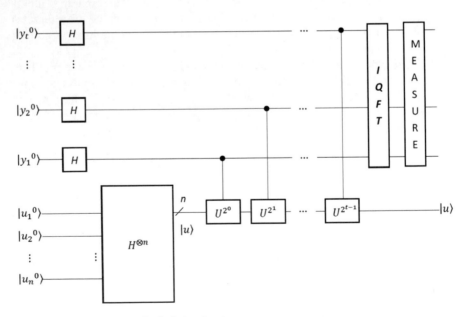

Fig. 6.1 Quantum circuit of calculating the phase

to be successful. The dependence of t on these quantities appear naturally from the following analysis.

At the left bottom in Fig. 6.1, the second register $\left(\otimes_{j=1}^{n} \left| u_j^0 \right\rangle \right)$ contains n quantum bits initially in the state $|0\rangle$. Quantum bit $|u_1^0\rangle$ is the most significant bit. Quantum bit $|u_n^0\rangle$ is the least significant bit. The corresponding decimal value of the second register is $(|u_1^0\rangle \times 2^{n-1}) + (|u_2^0\rangle \times 2^{n-2}) + \cdots + (|y_n^0\rangle \times 2^{n-n})$. How we select n that is dependent on a thing. The thing is to the size of the input for various kinds of real applications. This means that we select n that actually is to the number of bits of input for a problem. For the convenience of the presentation, the following initial state vector is

$$|\varphi_0\rangle = \left(\otimes_{k=t}^{1} \left| y_k^0 \right\rangle \right) \otimes \left(\otimes_{j=1}^{n} \left| u_j^0 \right\rangle \right). \tag{6.1}$$

6.1.1 Initialization of Phase Estimation

In Fig. 6.1, the circuit begins by means of using a Hadamard transform on the *first* register $\left(\otimes_{k=t}^{1} \left| y_k^0 \right\rangle \right)$ and another Hadamard transform on the *second* register $\left(\otimes_{j=1}^{n} \left| u_j^0 \right\rangle \right)$. A superposition of the first register is $\left(\frac{1}{\sqrt{2}} \left(\otimes_{k=t}^{1} \left(\left| y_k^0 \right\rangle + \left| y_k^1 \right\rangle \right) \right) \right)$. Another superposition of the second register is $\left(|u\rangle = \frac{1}{\sqrt{2^n}} \left(\otimes_{j=1}^{n} \left(\left| u_j^0 \right\rangle + \left| u_j^1 \right\rangle \right) \right) \right)$. This is to say that the superposition of the second register begins in the new state vector $\left(|u\rangle = \frac{1}{\sqrt{2^n}} \left(\otimes_{j=1}^{n} \left(\left| u_j^0 \right\rangle + \left| u_j^1 \right\rangle \right) \right) \right)$, and consists of n quantum bits as is necessary to store $(|u\rangle)$. The new state vector $(|u\rangle)$ is an eigenstate (eigenvector) of U. Therefore, this gives that the following new state vector is

$$
\begin{aligned}
|\varphi_1\rangle &= \left(\frac{1}{\sqrt{2^t}} \left(\otimes_{k=t}^{1} \left(\left| y_k^0 \right\rangle + \left| y_k^1 \right\rangle \right) \right) \right) \otimes \left(\frac{1}{\sqrt{2^n}} \left(\otimes_{j=1}^{n} \left(\left| u_j^0 \right\rangle + \left| u_j^1 \right\rangle \right) \right) \right) \\
&= \left(\frac{1}{\sqrt{2^t}} \left(\otimes_{k=t}^{1} \left(\left| y_k^0 \right\rangle + \left| y_k^1 \right\rangle \right) \right) \right) \otimes (|u\rangle). \tag{6.2}
\end{aligned}
$$

6.1.2 Controlled-U Operations on the Superposition of the Second Register to Phase Estimation

Next, in Fig. 6.1, the circuit implements application of controlled-U operations on the superposition of the second register that is the state $(|u\rangle)$, with U raised to successive powers of two. Because the effect of one application of unitary operator U on its eigenvector (eigenstate) $(|u\rangle)$ is $\left(U \times |u\rangle = e^{\sqrt{-1} \times 2 \times \pi \times \theta} \times |u\rangle \right)$, the effect of repeated application of unitary operator U on its eigenvector (eigenstate) ($|u\rangle$) is

$$
\begin{aligned}
U^a |u\rangle = U^{a-1} U |u\rangle &= U^{a-1} \left(e^{\sqrt{-1} \times 2 \times \pi \times \theta} \times |u\rangle \right) \\
&= e^{\sqrt{-1} \times 2 \times \pi \times \theta} \times \left(U^{a-1} |u\rangle \right) = e^{\sqrt{-1} \times 2 \times \pi \times \theta} \times e^{\sqrt{-1} \times 2 \times \pi \times \theta} \\
&\times \cdots \times e^{\sqrt{-1} \times 2 \times \pi \times \theta} |u\rangle = e^{\sqrt{-1} \times 2 \times \pi \times \theta \times a} \times |u\rangle.
\end{aligned}
\tag{6.3}
$$

Implementing one controlled-U operation that has its eigenvector (eigenstate) ($|u\rangle$) and its eigenvalue $e^{\sqrt{-1} \times 2 \times \pi \times \theta}$ is to that if the controlled quantum bit is the state $|1\rangle$, then it completes one application of unitary operator U, $\left(U \times |u\rangle = e^{\sqrt{-1} \times 2 \times \pi \times \theta} \times |u\rangle \right)$. Otherwise, it does not complete one application of unitary operator U.

Similarly, implementing repeated application of one controlled-U operation that has its eigenvector (eigenstate) ($|u\rangle$) and its eigenvalue $e^{\sqrt{-1} \times 2 \times \pi \times \theta}$ is to that if the controlled quantum bit is the state $|1\rangle$, then it completes repeated application of unitary operator U, $\left(U^a \times |u\rangle = e^{\sqrt{-1} \times 2 \times \pi \times \theta \times a} \times |u\rangle \right)$. Otherwise, it does not complete repeated application of unitary operator U.

In the new state vector $|\varphi_1\rangle$ in (6.2), each quantum bit in the first register is currently in its superposition. A superposition $\left(\frac{1}{\sqrt{2}} (|y_1^0\rangle + |y_1^1\rangle) \right)$ at the weighted position 2^0 is the controlled quantum bit of implementing controlled-U^{2^0} operations on the superposition of the second register that is the state ($|u\rangle$). This gives that the following new state vector is

$$
\begin{aligned}
|\varphi_2\rangle &= \left(\frac{1}{\sqrt{2^t}} (\otimes_{k=t}^{2} (|y_k^0\rangle + |y_k^1\rangle)) \right) \otimes \left(|y_1^0\rangle |u\rangle + e^{\sqrt{-1} \times 2 \times \pi \times \theta \times 2^0} |y_1^1\rangle |u\rangle \right) \\
&= \left(\frac{1}{\sqrt{2^t}} (\otimes_{k=t}^{2} (|y_k^0\rangle + |y_k^1\rangle)) \right) \otimes \left(|y_1^0\rangle + e^{\sqrt{-1} \times 2 \times \pi \times \theta \times 2^0} |y_1^1\rangle \right) \otimes (|u\rangle).
\end{aligned}
\tag{6.4}
$$

Altering the phase of the state $|y_1^1\rangle$ is from one (1) to become $\left(e^{\sqrt{-1} \times 2 \times \pi \times \theta \times 2^0} \right)$. We call it as *phase kickback*.

Next, in the new state vector $|\varphi_2\rangle$ in (6.4), a superposition $\left(\frac{1}{\sqrt{2}} (|y_2^0\rangle + |y_2^1\rangle) \right)$ at the weighted position 2^1 is the controlled quantum bit of implementing controlled-U^{2^1} operations on the superposition of the second register that is the state ($|u\rangle$). This means that the following new state vector is

$$
\begin{aligned}
|\varphi_3\rangle &= \left(\frac{1}{\sqrt{2^t}} (\otimes_{k=t}^{3} (|y_k^0\rangle + |y_k^1\rangle)) \right) \otimes \left(|y_2^0\rangle + e^{\sqrt{-1} \times 2 \times \pi \times \theta \times 2^1} |y_2^1\rangle \right) \\
&\otimes \left(|y_1^0\rangle + e^{\sqrt{-1} \times 2 \times \pi \times \theta \times 2^0} |y_1^1\rangle \right) \otimes (|u\rangle).
\end{aligned}
\tag{6.5}
$$

Because of *phase kickback*, the phase of the state $|y_2^1\rangle$ is from one (1) to become $\left(e^{\sqrt{-1}\times 2\times \pi \times \theta \times 2^1}\right)$.

Next, in the new state vector $|\varphi_3\rangle$ in (6.5), a superposition $\left(\frac{1}{\sqrt{2}}(|y_3^0\rangle + |y_3^1\rangle)\right)$ at the weighted position 2^2 through a superposition $\left(\frac{1}{\sqrt{2}}(|y_t^0\rangle + |y_t^1\rangle)\right)$ at the weighted position 2^{t-1} are the controlled quantum bits of implementing controlled-U^{2^2} operations through controlled-$U^{2^{t-1}}$ operations on the superposition of the second register that is the state $(|u\rangle)$. This gives that the following new state vector is

$$|\varphi_4\rangle = \left(\frac{1}{\sqrt{2^t}}\left(|y_t^0\rangle + e^{\sqrt{-1}\times 2\times \pi \times \theta \times 2^{t-1}}|y_t^1\rangle\right)\right) \otimes \left(|y_{t-1}^0\rangle + e^{\sqrt{-1}\times 2\times \pi \times \theta \times 2^{t-2}}|y_{t-1}^1\rangle\right)$$

$$\otimes \cdots \otimes \left(|y_2^0\rangle + e^{\sqrt{-1}\times 2\times \pi \times \theta \times 2^1}|y_2^1\rangle\right) \otimes \left(|y_1^0\rangle + e^{\sqrt{-1}\times 2\times \pi \times \theta \times 2^0}|y_1^1\rangle\right)\right) \otimes (|u\rangle)$$

$$= \left(\frac{1}{\sqrt{2^t}}\left(\sum_{Y=0}^{2^t-1} e^{\sqrt{-1}\times 2\times \pi \times \theta \times Y}|Y\rangle\right)\right) \otimes (|u\rangle). \tag{6.6}$$

Because of *phase kickback*, the phase of the state $|Y\rangle$ for $0 \le Y \le 2^t - 1$ is from one (1) to become $\left(e^{\sqrt{-1}\times 2\times \pi \times \theta \times Y}\right)$. From this description above, the second quantum register stays in the state $(|u\rangle)$ through the computation.

6.1.3 Inverse Quantum Fourier Transform on the Superposition of the First Register to Phase Estimation

Next, in Fig. 6.1, the circuit implements the **inverse Quantum Fourier transform** on the superposition of the first register. It takes the new state vector $(|\varphi_4\rangle)$ in (6.6) as its input state vector. The output state of the **inverse Quantum Fourier transform** on the superposition of the first register is

$$|\varphi_5\rangle = \left(\sum_{Y=0}^{2^t-1} \frac{1}{\sqrt{2^t}} e^{\sqrt{-1}\times 2\times \pi \times \theta \times Y} \frac{1}{\sqrt{2^t}} \sum_{i=0}^{2^t-1} e^{-\sqrt{-1}\times 2\times \pi \times \frac{i}{2^t} \times Y}|i\rangle\right) \otimes (|u\rangle)$$

$$= \left(\frac{1}{2^t}\left(\sum_{Y=0}^{2^t-1}\sum_{i=0}^{2^t-1} e^{\sqrt{-1}\times 2\times \pi \times Y \times (\theta - \frac{i}{2^t})}|i\rangle\right)\right) \otimes (|u\rangle)$$

$$= \left(\sum_{i=0}^{2^t-1}\sum_{Y=0}^{2^t-1} \frac{1}{2^t}\left(e^{\sqrt{-1}\times 2\times \pi \times (\theta - \frac{i}{2^t})}\right)^Y |i\rangle\right) \otimes (|u\rangle). \tag{6.7}$$

From this description above, the second quantum register still stays in the state $(|u\rangle)$ through the computation. From the new state vector $(|\varphi_5\rangle)$ in (6.7), the probability amplitude of $|i\rangle$ is

$$\phi_i = \frac{1}{2^t} \times \left(\sum_{Y=0}^{2^t-1} \left(e^{\sqrt{-1} \times 2 \times \pi \times \left(\theta - \frac{i}{2^t}\right)} \right)^Y \right). \tag{6.8}$$

6.1.4   Idealistic Phase Estimation

The probability amplitude of $|i\rangle$ is simply the sum of a geometrical sequence with quotient $q = \left(e^{\sqrt{-1} \times 2 \times \pi \times \left(\theta - \frac{i}{2^t}\right)} \right)$. On one hand if the value of θ may be expressed in t bits in the first quantum register, as $\theta = 0.y_t \, y_{t-1} \dots y_2 \, y_1 = (y_t \, y_{t-1} \dots y_2 \, y_1/2^t)$. Then the value of θ actually is equal to $(i/2^t)$ for $0 \leq i \leq 2^t - 1$ and is an integer multiple of $(1/2^t)$. This gives that the quotient q is $e^{\sqrt{-1} \times 2 \times \pi \times \left(\frac{i}{2^t} - \frac{i}{2^t}\right)} = e^{\sqrt{-1} \times 2 \times \pi \times 0} = 1$, the probability amplitude of $|i\rangle$ is $\frac{1}{2^t} \times \left(\sum_{Y=0}^{2^t-1} 1^Y \right) = \frac{1}{2^t} \times \left(\sum_{Y=0}^{2^t-1} 1 \right) = \frac{1}{2^t} \times 2^t = 1$ and any other probability amplitudes disappear. This is the *ideal case* of phase estimation. Finally, in Fig. 6.1, after a measurement on the output state of the inverse quantum Fourier transform to the superposition of the first register is completed, we obtain the computational basis state $|i\rangle$ with the successful probability 1 (100%). This indicates that the value of θ is equal to $(i/2^t)$ with the successful probability 1 (100%). Therefore, we obtain the eigenvalue $\left(e^{\sqrt{-1} \times 2 \times \pi \times \frac{i}{2^t}} \right)$ with the successful probability 1 (100%).

6.1.5   Phase Estimation in Practical Cases

On the other hand if the value of θ may not be expressed in t bits in the first quantum register. This is to say that $\theta \neq 0.y_t \, y_{t-1} \dots y_2 \, y_1 \neq (y_t \, y_{t-1} \dots y_2 \, y_1/2^t)$. Then the quotient q is $e^{\sqrt{-1} \times 2 \times \pi \times \left(\theta - \frac{i}{2^t}\right)} \neq 1$ and we can rewrite the probability amplitude of $|i\rangle$ in (6.8) as follows

$$\phi_i = \frac{1}{2^t} \times \frac{1 - q^{2^t}}{1 - q} = \frac{1}{2^t} \times \frac{1 - \left(e^{\sqrt{-1} \times 2 \times \pi \times \left(\theta - \frac{i}{2^t}\right)} \right)^{2^t}}{1 - e^{\sqrt{-1} \times 2 \times \pi \times \left(\theta - \frac{i}{2^t}\right)}}$$

$$= \frac{1}{2^t} \times \frac{1 - e^{\sqrt{-1} \times 2 \times \pi \times \left(2^t \times \theta - i\right)}}{1 - e^{\sqrt{-1} \times 2 \times \pi \times \left(\theta - \frac{i}{2^t}\right)}}. \tag{6.9}$$

This gives another good explanation of uncertainty and thus appearing inaccuracy when measuring the output of the **inverse quantum Fourier transform** in Fig. 6.1. The probability of measuring a suitable state $|i\rangle$ on the first register in Fig. 6.1 is

$$|\phi_i|^2 = \frac{1}{2^{2\times t}} \frac{\left|1 - e^{\sqrt{-1}\times 2\times\pi\times\left(2^t\times\theta-i\right)}\right|^2}{\left|1 - e^{\sqrt{-1}\times 2\times\pi\times\left(\theta-\frac{i}{2^t}\right)}\right|^2}. \tag{6.10}$$

Because $\left|1 - e^{\sqrt{-1}\times\gamma}\right|^2 = 4\sin^2(\gamma/2)$, we can rewrite $|\phi_i|^2$ in (6.10) as follows

$$|\phi_i|^2 = \frac{1}{2^{2\times t}} \times \frac{4\times\sin^2\left(\frac{2\times\pi\times\left(2^t\times\theta-i\right)}{2}\right)}{4\times\sin^2\left(\frac{2\times\pi\times\left(\theta-\frac{i}{2^t}\right)}{2}\right)}$$

$$= \frac{1}{2^{2\times t}} \times \frac{\sin^2\left(\frac{2\times\pi\times\left(2^t\times\theta-i\right)}{2}\right)}{\sin^2\left(\frac{2\times\pi\times\left(\theta-\frac{i}{2^t}\right)}{2}\right)}. \tag{6.11}$$

This is the *practical case* of phase estimation. Finally, in Fig. 6.1, after a measurement on the output state of the inverse quantum Fourier transform to the superposition of the first register is completed, we obtain the computational basis state $|i\rangle$ with the probability $\left(\frac{1}{2^{2\times t}} \times \frac{\sin^2\left(\frac{2\times\pi\times\left(2^t\times\theta-i\right)}{2}\right)}{\sin^2\left(\frac{2\times\pi\times\left(\theta-\frac{i}{2^t}\right)}{2}\right)}\right)$. Because $(i/2^t) = (y_t\, y_{t-1} \ldots y_2\, y_1/2^t)$ $= 0.y_t\, y_{t-1} \ldots y_2\, y_1$, $(i/2^t)$ is an estimated value to the value of θ with the probability $\left(\frac{1}{2^{2\times t}} \times \frac{\sin^2\left(\frac{2\times\pi\times\left(2^t\times\theta-i\right)}{2}\right)}{\sin^2\left(\frac{2\times\pi\times\left(\theta-\frac{i}{2^t}\right)}{2}\right)}\right)$. Hence, we only obtain an *estimated* eigenvalue $\left(e^{\sqrt{-1}\times 2\times\pi\times\frac{i}{2^t}}\right)$ with the probability $\left(\frac{1}{2^{2\times t}} \times \frac{\sin^2\left(\frac{2\times\pi\times\left(2^t\times\theta-i\right)}{2}\right)}{\sin^2\left(\frac{2\times\pi\times\left(\theta-\frac{i}{2^t}\right)}{2}\right)}\right)$.

This is to say that if more than one $|\phi_i|^2$ differs from zero then there is a nonzero probability of receiving different estimated phases (eigenvalues) after the measurement when repeating to execute the circuit of phase estimation in Fig. 6.1.

6.1.6 *Performance and Requirement to Phase Estimation*

The phase estimation algorithm allows one to estimate the value of the phase θ to an eigenvalue $\left(e^{\sqrt{-1}\times 2\times\pi\times\theta}\right)$ of a unitary operator U with its eigenvector ($|u\rangle$). From the analysis in Sect. 6.1.4, if the value of the phase θ is to $\theta = 0.y_t\, y_{t-1} \cdots y_2\, y_1 = (y_t\, y_{t-1} \cdots y_2\, y_1/2^t)$ that is to a t bit binary expansion of the first quantum register, then in the circuit of Fig. 6.1 the outcome of the final measurement is $|i\rangle$ with the probability 100%. Because $|i\rangle$ is a t bit binary expansion of the first quantum register, we obtain that the value of the phase θ is equal to $(i/2^t)$ with the probability 100%. This is the *ideal* case.

On the other hand, from the analysis in Sect. 6.1.5, if the value of the phase θ is not a t bit binary expansion of the first quantum register, then the outcome of the

final measurement is $|i\rangle$ with the probability $\left(\dfrac{1}{2^{2\times t}} \times \dfrac{\sin^2\left(\frac{2\times\pi\times(2^t\times\theta - i)}{2} \right)}{\sin^2\left(\frac{2\times\pi\times\left(\theta - \frac{i}{2^t}\right)}{2} \right)} \right)$. Let Y be the

integer in the range 0 to $2^t - 1$ so that $(Y/2^t) = (y_t\, y_{t-1} \cdots y_2\, y_1/2^t) = (0.y_t\, y_{t-1} \cdots y_2\, y_1)$ is the best t bit approximation to the value of the phase θ and $(Y/2^t)$ is less than the value of the phase θ. This indicates that the difference $\delta = \theta - (Y/2^t)$ between θ and $(Y/2^t)$ satisfies $0 \le \delta \le (1/2^t)$. We assume that the outcome of the final measurement in the circuit of Fig. 6.1 is $|i\rangle$. We aim to bound the probability of obtaining a value of i such that $|i - Y| > \varepsilon$, where ε is a positive integer characterizing our desired tolerance to error. The probability of measuring such a state $|i\rangle$ is

$$P(|i - Y| > \varepsilon) \le \frac{1}{2 \times (\varepsilon - 1)}. \tag{6.12}$$

We assume that we would like to approximate the value of the phase θ to an accuracy 2^{-t}, that is, we select $\varepsilon = 2^{t-n} - 1$. By means of using $t = n + q$ quantum bits in the circuit of Fig. 6.1, we see from (6.12) that the probability of obtaining an approximation correct to this accuracy is at least

$$P(|i - Y| \le \varepsilon) = 1 - P(|i - Y| > \varepsilon) = 1 - \frac{1}{2 \times (\varepsilon - 1)}$$

$$= 1 - \frac{1}{2 \times (2^{t-n} - 1 - 1)} = 1 - \frac{1}{2 \times (2^{t-n} - 2)}. \tag{6.13}$$

Therefore to successfully obtain the value of the phase θ accurate to t bits with probability of success at least $1 - \alpha = 1 - \frac{1}{2\times(2^{t-n}-2)}$, we select

$$t = n + \lceil \log_2(2 + (1/(2 \times \alpha))) \rceil. \tag{6.14}$$

Because $\alpha = \frac{1}{2\times(2^{t-n}-2)}$, we obtain $\alpha \times (2 \times (2^{t-n} - 2)) = 1$. This is to say that $2^{t-n} - 2 = (1/(2 \times \alpha))$ and $2^{t-n} = (1/(2 \times \alpha)) + 2$ and $\log_2(2^{t-n}) = \log_2(2 + (1/(2 \times \alpha)))$ and $t = n + \lceil \log_2(2 + (1/(2 \times \alpha))) \rceil$. This is the result in (6.14).

6.1.7 Assessment to Complexity of Phase Estimation

In the circuit of Fig. 6.1, the number of quantum bits to the *first* register $\left(\otimes_{k=t}^{1}\left|y_k^0\right\rangle\right)$ is t quantum bits and the number of quantum bits to the *second* register $\left(\otimes_{j=1}^{n}\left|u_j^0\right\rangle\right)$ is n quantum bits. Therefore, space complexity of phase estimation is $O(t+n)$ quantum bits. The *first* stage in the circuit of Fig. 6.1 is to implement $(t+n)$ Hadamard gates.

Next, the *second* stage in the circuit of Fig. 6.1 is to implement application of controlled-U operations on the superposition of the second register that is the state $(|u\rangle)$, with U raised to successive powers of two. The $U1(\lambda)$ gate is $U1(\lambda) = U1(\text{lambda}) = \begin{pmatrix} 1 & 0 \\ 0 & e^{\sqrt{-1}\times\lambda} \end{pmatrix}$ for that λ (lambda) is a real value. If the value of λ is equal to $(2 \times \pi \times \theta \times 2^{k-1})$ to $1 \le k \le t$, then it can implement a controlled-$U^{2^{k-1}}$ operation to $1 \le k \le t$. This is to say that a total cost of completing the second stage is to implement t $U1(\lambda)$ gates.

Next, the *third* stage in the circuit of Fig. 6.1 is to implement the inverse quantum Fourier transform on the superposition of the first register. A total cost of completing the inverse quantum Fourier transform is to implement $O(t^2)$ quantum gates. Finally, reading out the output state of the inverse quantum Fourier transform on the super-position of the first register is to implement one measurement. Because from the statements above a total cost of completing phase estimation is $O(t^2 + n)$ quantum gates, time complexity of phase estimation is to $O(t^2 + n)$ quantum gates.

6.2 Computing Eigenvalue of a $(2^2 \times 2^2)$ Unitary Matrix U with a $(2^2 \times 1)$ Eigenvector $|u\rangle$ in Phase Estimation

We use the circuit in Fig. 6.2 to compute eigenvalue of a $(2^2 \times 2^2)$ unitary matrix U with a $(2^2 \times 1)$ eigenvector $|u\rangle$. It makes use of two quantum registers. At the left

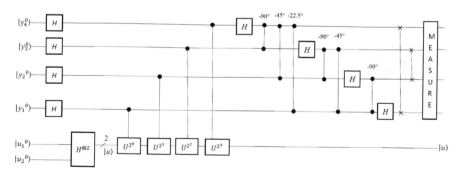

Fig. 6.2 Quantum circuit for calculating eigenvalue of a $(2^2 \times 2^2)$ unitary matrix U with a $(2^2 \times 1)$ eigenvector $|u\rangle$

top in Fig. 6.2, the first register $\left(\otimes_{k=4}^{1}\left|y_{k}^{0}\right\rangle\right)$ contains *four* quantum bits initially in the state $|0\rangle$. Quantum bit $|y_4{}^0\rangle$ is the most significant bit. Quantum bit $|y_1{}^0\rangle$ is the least significant bit. The corresponding decimal value of the first register is $(|y_4{}^0\rangle \times 2^{4-1}) + (|y_3{}^0\rangle \times 2^{3-1}) + (|y_2{}^0\rangle \times 2^{2-1}) + (|y_1{}^0\rangle \times 2^{1-1})$. At the left bottom in Fig. 6.2, the second register $\left(\otimes_{j=1}^{2}\left|u_{j}^{0}\right\rangle\right)$ contains *two* quantum bits initially in the state $|0\rangle$. Quantum bit $|u_1{}^0\rangle$ is the most significant bit. Quantum bit $|u_2{}^0\rangle$ is the least significant bit. The corresponding decimal value of the second register is $(|u_1{}^0\rangle \times 2^{2-1}) + (|u_2{}^0\rangle \times 2^{2-2})$. For the convenience of the presentation, the following initial state vector is

$$|\varphi_0\rangle = \left(\otimes_{k=4}^{1}\left|y_{k}^{0}\right\rangle\right) \otimes \left(\otimes_{j=1}^{2}\left|u_{j}^{0}\right\rangle\right). \tag{6.15}$$

6.2.1 Initialize Quantum Registers to Calculate Eigenvalue of a $(2^2 \times 2^2)$ Unitary Matrix U with a $(2^2 \times 1)$ Eigenvector $|u\rangle$ in Phase Estimation

In Listing 6.1, the program is in the backend that is *simulator* of Open QASM with *thirty-two* quantum bits in **IBM**'s quantum computer. The program is to compute eigenvalue of a $(2^2 \times 2^2)$ unitary matrix U with a $(2^2 \times 1)$ eigenvector $|u\rangle$ in phase estimation. Figure 6.3 is the corresponding quantum circuit of the program in Listing 6.1 and is to implement the quantum circuit of Fig. 6.2 to compute eigenvalue of a $(2^2 \times 2^2)$ unitary matrix U with a $(2^2 \times 1)$ eigenvector $|u\rangle$ in phase estimation.

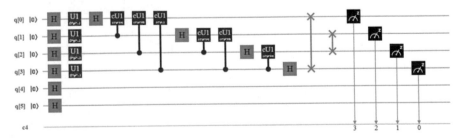

Fig. 6.3 Implementing quantum circuits of Fig. 6.2 to compute eigenvalue of a $(2^2 \times 2^2)$ unitary matrix U with a $(2^2 \times 1)$ eigenvector $|u\rangle$ in phase estimation

Listing 6.1 The program of computing eigenvalue of a $(2^2 \times 2^2)$ unitary matrix U with a $(2^2 \times 1)$ eigenvector $|u\rangle$ in phase estimation.

```
1.  OPENQASM 2.0;
2.  include "qelib1.inc";
3.  qreg q[6];
4.  creg c[4];
```

The statement "OPENQASM 2.0;" on line one of Listing 6.1 is to point out that the program is written with version 2.0 of Open QASM. Next, the statement "include "qelib1.inc";" on line two of Listing 6.1 is to continue parsing the file "qelib1.inc" as if the contents of the file were pasted at the location of the include statement, where the file "qelib1.inc" is **Quantum Experience (QE) Standard Header** and the path is specified relative to the current working directory.

Then, the statement "qreg q[6];" on line three of Listing 6.1 is to declare that in the program there are *six* quantum bits. In the left top of Fig. 6.3, six quantum bits are subsequently q[0], q[1], q[2], q[3], q[4] and q[5]. The initial value of each quantum bit is set to state $|0\rangle$. We make use of four quantum bits q[0], q[1], q[2] and q[3] to respectively encode four quantum bits $|y_4\rangle$, $|y_3\rangle$, $|y_2\rangle$ and $|y_1\rangle$ in Fig. 6.2. We use two quantum bits q[4] and q[5] to respectively encode two quantum bits $|u_1\rangle$ and $|u_2\rangle$ in Fig. 6.2. For the convenience of our explanation, $q[k]^0$ for $0 \le k \le 5$ is to represent the value 0 of q[k] and $q[k]^1$ for $0 \le k \le 5$ is to represent the value 1 of q[k]. Because quantum bit $|y_4{}^0\rangle$ is the most significant bit and quantum bit $|y_1{}^0\rangle$ is the least significant bit, quantum bit $|q[0]^0\rangle$ is the most significant bit and quantum bit $|q[3]^0\rangle$ is the least significant bit. The corresponding decimal value of the first register in Fig. 6.3 is $(|q[0]^0\rangle \times 2^{4-1}) + (|q[1]^0\rangle \times 2^{3-1}) + (|q[2]^0\rangle \times 2^{2-1}) + (|q[3]^0\rangle \times 2^{1-1})$.

Next, the statement "creg c[4];" on line four of Listing 6.1 is to declare that there are four classical bits in the program. In the left bottom of Fig. 6.3, four classical bits are subsequently c[0], c[1], c[2] and c[3]. The initial value of each classical bit is set to zero (0). For the convenience of our explanation, $c[k]^0$ for $0 \le k \le 3$ is to represent the value 0 of c[k] and $c[k]^1$ for $0 \le k \le 3$ is to represent the value 1 of c[k]. The corresponding decimal value of the four initial classical bits $c[3]^0$ $c[2]^0$ $c[1]^0$ $c[0]^0$ is $2^3 \times c[3]^0 + 2^2 \times c[2]^0 + 2^1 \times c[1]^0 + 2^0 \times c[0]^0$. This indicates that classical bit $c[3]^0$ is the most significant bit and classical bit $c[0]^0$ is the least significant bit. For the convenience of our explanation, we can rewrite the initial state vector $|\varphi_0\rangle = \left(\otimes_{k=4}^1 |y_k^0\rangle\right) \otimes \left(\otimes_{j=1}^2 |u_j^0\rangle\right)$ in (6.15) in Fig. 6.2 as follows

$$|\varphi_0\rangle = |q[0]^0\rangle |q[1]^0\rangle |q[2]^0\rangle |q[3]^0\rangle |q[4]^0\rangle |q[5]^0\rangle. \qquad (6.16)$$

6.2.2 Superposition of Quantum Registers to Calculate Eigenvalue of a $(2^2 \times 2^2)$ Unitary Matrix U with a $(2^2 \times 1)$ Eigenvector $|u\rangle$ in Phase Estimation

In Fig. 6.2, the first stage of the circuit is to implement a Hadamard transform with four Hadamard gates on the *first* register $\left(\otimes_{k=4}^{1}|y_k^0\rangle\right)$ and another Hadamard transform with two Hadamard gates on the *second* register $\left(\otimes_{j=1}^{2}|u_j^0\rangle\right)$. The *six* statements "h q[0];", "h q[1];", "h q[2];", "h q[3];", "h q[4];" and "h q[5];" on line *five* of Listing 6.1 through line *ten* of Listing 6.1 is to implement *six* Hadamard gates on the first register and the second register. They complete each Hadamard gate in the first time slot of Fig. 6.3 and perform the first stage of the circuit in Fig. 6.2.

Listing 6.1 continued…

//Implement a Hadamard transform on two registers

  5.   h q[0];
  6.   h q[1];
  7.   h q[2];
  8.   h q[3];
  9.   h q[4];
 10. h q[5];

A superposition of the first register is $\left(\frac{1}{\sqrt{2^4}}(\otimes_{k=4}^{1}(|y_k^0\rangle + |y_k^1\rangle))\right)=$ $\left(\frac{1}{\sqrt{2^4}}(\otimes_{a=0}^{3}(|q[a]^0\rangle + |q[a]^1\rangle))\right)$. Another superposition of the second register is $\left(|u\rangle = \frac{1}{\sqrt{2^2}}\left(\otimes_{j=1}^{2}\left(|u_j^0\rangle + |u_j^1\rangle\right)\right)\right)= \frac{1}{\sqrt{2^2}}(\otimes_{b=4}^{5}(|q[b]^0\rangle + |q[b]^1\rangle))$. This is to say that the superposition of the second register begins in the new state vector $\left(|u\rangle = \frac{1}{\sqrt{2^2}}\left(\otimes_{j=1}^{2}\left(|u_j^0\rangle + |u_j^1\rangle\right)\right)\right)= \frac{1}{\sqrt{2^2}}(\otimes_{b=4}^{5}(|q[b]^0\rangle + |q[b]^1\rangle))$ and contains *two* quantum bits as is necessary to store $(|u\rangle)$. The new state vector $(|u\rangle)$ is an eigenstate (eigenvector) of U. Therefore, this gives that the following new state vector is

$$|\varphi_1\rangle = \left(\frac{1}{\sqrt{2^4}}(\otimes_{k=4}^{1}(|y_k^0\rangle + |y_k^1\rangle))\right) \otimes \left(\frac{1}{\sqrt{2^2}}(\otimes_{j=1}^{2}(|u_j^0\rangle + |u_j^1\rangle))\right)$$

$$= \left(\frac{1}{\sqrt{2^4}}(\otimes_{k=4}^{1}(|y_k^0\rangle + |y_k^1\rangle))\right) \otimes (|u\rangle)$$

$$= \left(\frac{1}{\sqrt{2^4}}(\otimes_{a=0}^{3}(|q[a]^0\rangle + |q[a]^1\rangle))\right) \otimes \left(\frac{1}{\sqrt{2^2}}(\otimes_{b=4}^{5}(|q[b]^0\rangle + |q[b]^1\rangle))\right)$$

$$= \left(\frac{1}{\sqrt{2^4}}(\otimes_{a=0}^{3}(|q[a]^0\rangle + |q[a]^1\rangle))\right) \otimes (|u\rangle). \tag{6.17}$$

6.2.3 Controlled-U Operations on the Superposition of the Second Register to Determine Eigenvalue of a $(2^2 \times 2^2)$ Unitary Matrix U with a $(2^2 \times 1)$ Eigenvector $|u\rangle$ in Phase Estimation

In the new state vector $|\varphi_1\rangle$ in (6.17), each quantum bit in the first register is currently in its superposition. The value of the first register is from state $(\otimes_{k=4}^{1}|y_k^0\rangle)$ (zero) encoded by state $(\otimes_{a=0}^{3}|q[a]^0\rangle)$ through state $(\otimes_{k=4}^{1}|y_k^1\rangle)$ (fifteen) encoded by state $(\otimes_{a=0}^{3}|q[a]^1\rangle)$. The circuit of Fig. 6.2 can precisely estimate sixteen phases. This is to say that the first register with four quantum bits can precisely represent sixteen phases. Sixteen phases are subsequently $(0/2^4)$, $(1/2^4)$, $(2/2^4)$, $(3/2^4)$, $(4/2^4)$, $(5/2^4)$, $(6/2^4)$, $(7/2^4)$, $(8/2^4)$, $(9/2^4)$, $(10/2^4)$, $(11/2^4)$, $(12/2^4)$, $(13/2^4)$, $(14/2^4)$ and $(15/2^4)$. The corresponding sixteen phase angles are subsequently $(2 \times \pi \times 0/2^4)$, $(2 \times \pi \times 1/2^4)$, $(2 \times \pi \times 2/2^4)$, $(2 \times \pi \times 3/2^4)$, $(2 \times \pi \times 4/2^4)$, $(2 \times \pi \times 5/2^4)$, $(2 \times \pi \times 6/2^4)$, $(2 \times \pi \times 7/2^4)$, $(2 \times \pi \times 8/2^4)$, $(2 \times \pi \times 9/2^4)$, $(2 \times \pi \times 10/2^4)$, $(2 \times \pi \times 11/2^4)$, $(2 \times \pi \times 12/2^4)$, $(2 \times \pi \times 13/2^4)$, $(2 \times \pi \times 14/2^4)$ and $(2 \times \pi \times 15/2^4)$.

Say that we are trying to determine an eigenvalue of $90°$. This is to say that the effect of one application of unitary operator U on its eigenvector (eigenstate) $(|u\rangle)$ is $\left(U \times |u\rangle e^{\sqrt{-1} \times 2 \times \pi \times \theta} \times |u\rangle = e^{\sqrt{-1} \times 2 \times \pi \times \frac{4}{2^4}} \times |u\rangle\right)$. So, the effect of repeated application of unitary operator U on its eigenvector (eigenstate) $(|u\rangle)$ is

$$U^a|u\rangle = e^{\sqrt{-1} \times 2 \times \pi \times \theta \times a}|u\rangle = e^{\sqrt{-1} \times 2 \times \pi \times \frac{4}{2^4} \times a} \times |u\rangle. \tag{6.18}$$

A superposition $\left(\frac{1}{\sqrt{2}}(|y_1^0\rangle + |y_1^1\rangle)\right)$ that is encoded by $\left(\frac{1}{\sqrt{2}}(|q[3]^0\rangle + |q[3]^1\rangle)\right)$ at the weighted position 2^0 is the controlled quantum bit of implementing controlled-U^{2^0} operations on the superposition of the second register that is the state $(|u\rangle)$. Similarly, a superposition $\left(\frac{1}{\sqrt{2}}(|y_2^0\rangle + |y_2^1\rangle)\right)$ that is encoded by $\left(\frac{1}{\sqrt{2}}(|q[2]^0\rangle + |q[2]^1\rangle)\right)$ at the weighted position 2^1 is the controlled quantum bit of implementing controlled-U^{2^1} operations on the superposition of the second register that is the state $(|u\rangle)$. Next, a superposition $\left(\frac{1}{\sqrt{2}}(|y_3^0\rangle + |y_3^1\rangle)\right)$ that is encoded by $\left(\frac{1}{\sqrt{2}}(|q[1]^0\rangle + |q[1]^1\rangle)\right)$ at the weighted position 2^2 is the controlled quantum bit of implementing controlled-U^{2^2} operations on the superposition of the second register that is the state $(|u\rangle)$. Next, a superposition $\left(\frac{1}{\sqrt{2}}(|y_4^0\rangle + |y_4^1\rangle)\right)$ that is encoded by $\left(\frac{1}{\sqrt{2}}(|q[0]^0\rangle + |q[0]^1\rangle)\right)$ at the weighted position 2^3 is the controlled quantum bit of implementing controlled-U^{2^3} operations on the superposition of the second register that is the state $(|u\rangle)$.

The *four* statements from line *eleven* through line *fourteen* in Listing 6.1 are "u1(2 * pi * 4/16 * 1) q[3];", "u1(2 * pi * 4/16 * 2) q[2];", "u1(2 * pi * 4/16 * 4) q[1];" and "u1(2 * pi * 4/16 * 8) q[0];". They take the new state vector $(|\varphi_1\rangle)$ in (6.17) as their input state vector and implement each controlled-U operation on the superposition of the second register in the *second* time slot of Fig. 6.3 and in the *second* stage of Fig. 6.2. They alert the phase of the state $|y_1^1\rangle$ $(|q[3]^1\rangle)$ is from one (1) to become $\left(e^{\sqrt{-1}\times 2\times\pi\times\frac{4}{16}\times 2^0}\right) = \left(e^{\sqrt{-1}\times 2\times\pi\times\frac{4}{16}\times 1}\right)$. They alert the

phase of the state $|y_2^1\rangle$ $(|q[2]^1\rangle)$ is from one (1) to become $\left(e^{\sqrt{-1}\times 2\times\pi\times\frac{4}{16}\times 2^1}\right) =$

$\left(e^{\sqrt{-1}\times 2\times\pi\times\frac{4}{16}\times 2}\right)$. They alert the phase of the state $|y_3^1\rangle$ $(|q[1]^1\rangle)$ is from one (1)

to become $\left(e^{\sqrt{-1}\times 2\times\pi\times\frac{4}{16}\times 2^2}\right) = \left(e^{\sqrt{-1}\times 2\times\pi\times\frac{4}{16}\times 4}\right)$ and alert the phase of the state

$|y_4^1\rangle$ $(|q[0]^1\rangle)$ is from one (1) to become $\left(e^{\sqrt{-1}\times 2\times\pi\times\frac{4}{16}\times 2^3}\right) = \left(e^{\sqrt{-1}\times 2\times\pi\times\frac{4}{16}\times 8}\right)$.
This gives that the following new state vector is

$$|\varphi_2\rangle = \left(\frac{1}{\sqrt{2^4}}\left(|y_4^0\rangle + e^{\sqrt{-1}\times 2\times\pi\times\frac{4}{16}\times 2^3}|y_4^1\rangle\right) \otimes \left(|y_3^0\rangle + e^{\sqrt{-1}\times 2\times\pi\times\frac{4}{16}\times 2^2}|y_3^1\rangle\right)\right.$$

$$\otimes\left(|y_2^0\rangle + e^{\sqrt{-1}\times 2\times\pi\times\frac{4}{16}\times 2^1}|y_2^1\rangle\right) \otimes \left.\left(|y_1^0\rangle + e^{\sqrt{-1}\times 2\times\pi\times\frac{4}{16}\times 2^0}|y_1^1\rangle\right)\right) \otimes (|u\rangle)$$

$$= \left(\frac{1}{\sqrt{2^4}}\left(|y_4^0\rangle + e^{\sqrt{-1}\times 2\times\pi\times\frac{4}{16}\times 8}|y_4^1\rangle\right) \otimes \left(|y_3^0\rangle + e^{\sqrt{-1}\times 2\times\pi\times\frac{4}{16}\times 4}|y_3^1\rangle\right)\right.$$

$$\otimes\left(|y_2^0\rangle + e^{\sqrt{-1}\times 2\times\pi\times\frac{4}{16}\times 2}|y_2^1\rangle\right) \otimes \left.\left(|y_1^0\rangle + e^{\sqrt{-1}\times 2\times\pi\times\frac{4}{16}\times 1}|y_1^1\rangle\right)\right) \otimes (|u\rangle)$$

$$= \left(\frac{1}{\sqrt{2^4}}\left(|q[0]^0\rangle + e^{\sqrt{-1}\times 2\times\pi\times\frac{4}{16}\times 8}|q[0]^1\rangle\right)\right.$$

$$\otimes\left(|q[0]^1\rangle + e^{\sqrt{-1}\times 2\times\pi\times\frac{4}{16}\times 4}|q[0]^1\rangle\right)$$

$$\otimes\left(|q[2]^0\rangle + e^{\sqrt{-1}\times 2\times\pi\times\frac{4}{16}\times 2}|q[2]^1\rangle\right)$$

$$\otimes\left.\left(|q[3]^0\rangle + e^{\sqrt{-1}\times 2\times\pi\times\frac{4}{16}\times 1}|q[3]^1\rangle\right)\right) \otimes (|u\rangle)$$

$$= \left(\frac{1}{\sqrt{2^4}}\left(\sum_{Y=0}^{2^4-1} e^{\sqrt{-1}\times 2\times\pi\times\frac{4}{16}\times Y}|Y\rangle\right)\right) \otimes (|u\rangle). \tag{6.19}$$

Listing 6.1 continued...
//Implement controlled-U operations on the superposition of the second register
11. u1(2 * pi * 4/16 * 1) q[3];
12. u1(2 * pi * 4/16 * 2) q[2];
13. u1(2 * pi * 4/16 * 4) q[1];
14. u1(2 * pi * 4/16 * 8) q[0];

From this description above, the second quantum register stays in the state $(|u\rangle)$
through the computation. Because of *phase kickback*, the phase of the state $|Y\rangle$ for 0
$\leq Y \leq 2^4 - 1$ is from one (1) to become $\left(e^{\sqrt{-1}\times 2\times\pi\times\frac{4}{16}\times Y}\right)$. In the state vector $(|\varphi_2\rangle)$
in (6.19), it contains sixteen phase angles from state $|0\rangle$ through state $|15\rangle$. The front

eight phase angles are $(90° \times 0 = 0°)$, $(90° \times 1 = 90°)$, $(90° \times 2 = 180°)$, $(90° \times 3 = 270°)$, $(90° \times 4 = 360° = 0°)$, $(90° \times 5 = 450° = 90°)$, $(90° \times 6 = 540° = 180°)$ and $(90° \times 7 = 630° = 270°)$. The last eight phase angles are $(90° \times 8 = 720° = 0°)$, $(90° \times 9 = 810° = 90°)$, $(90° \times 10 = 900° = 180°)$, $(90° \times 11 = 990° = 270°)$, $(90° \times 12 = 1080° = 0°)$, $(90° \times 13 = 1170° = 90°)$, $(90° \times 14 = 1260° = 180°)$ and $(90° \times 15 = 1350° = 270°)$. The phase angle rotates back to its starting value $0°$ *four* times.

6.2.4   The Inverse Quantum Fourier Transform on the Superposition of the First Register to Compute Eigenvalue of a $(2^2 \times 2^2)$ Unitary Matrix U with a $(2^2 \times 1)$ Eigenvector |u⟩ in Phase Estimation

Listing 6.1 continued…

//Implement one inverse quantum Fourier transform on the superposition of the first

//Register

```
15.  h q[0];
16.  cu1(−2 * pi * 1/4) q[1], q[0];
17.  cu1(−2 * pi * 1/8) q[2], q[0];
18.  cu1(−2 * pi * 1/16) q[3], q[0];

19.  h q[1];
20.  cu1(−2 * pi * 1/4) q[2], q[1];
21.  cu1(−2 * pi * 1/8) q[3], q[1];

22.  h q[2];
23.  cu1(−2 * pi * 1/4) q[3], q[2];

24.  h q[3];

25.  swap q[0], q[3];
26.  swap q[1], q[2];
```

Hidden patterns and information stored in the state vector $(|\varphi_2\rangle)$ in (6.19) are to that its phase angle rotates back to its starting value $0°$ *four* times. This implies that the number of the period per sixteen phase angles is *four* and the frequency is equal to *four* (16/4). The twelve statements from line *fifteen* through line *twenty-six* in Listing 6.1 implement each quantum operation from the *third* time slot through the *fourteenth* time slot in Fig. 6.3. They actually implement each quantum operation of completing an **inverse quantum Fourier transform** on the superposition of the first register in Fig. 6.2. They take the state vector $(|\varphi_2\rangle)$ in (6.19) as their input state vector. Because the **inverse quantum Fourier transform** effectively transforms the state of the first register into a superposition of the *periodic* signal's component frequencies, they produce the following state vector

$$|\varphi_3\rangle = \left(\sum_{Y=0}^{2^4-1} \frac{1}{\sqrt{2^4}} e^{\sqrt{-1} \times 2 \times \pi \times \frac{4}{2^4} \times Y} \frac{1}{\sqrt{2^4}} \sum_{i=0}^{2^4-1} e^{-\sqrt{-1} \times 2 \times \pi \times \frac{i}{2^4} \times Y} |i\rangle \right) \otimes (|u\rangle)$$

$$= \left(\frac{1}{2^4} \left(\sum_{Y=0}^{2^4-1} \sum_{i=0}^{2^4-1} e^{\sqrt{-1} \times 2 \times \pi \times Y \times \left(\frac{4}{2^4} - \frac{i}{2^4} \right)} |i\rangle \right) \right) \otimes (|u\rangle)$$

$$= \left(\sum_{i=0}^{2^4-1} \sum_{Y=0}^{2^4-1} \frac{1}{2^4} \left(e^{\sqrt{-1} \times 2 \times \pi \times \left(\frac{4}{2^4} - \frac{i}{2^4} \right)} \right)^Y |i\rangle \right) \otimes (|u\rangle). \qquad (6.20)$$

6.2.5 Read the Quantum Result to Figure Out Eigenvalue of a $(2^2 \times 2^2)$ Unitary Matrix U with a $(2^2 \times 1)$ Eigenvector |u⟩ in Phase Estimation

Finally, the four statements "measure q[0] → c[3];", "measure q[1] → c[2];", "measure q[2] → c[1];" and "measure q[3] → c[0];" from line *twenty-seven* through line *thirty* in Listing 6.1 implement a measurement. They measure the output state of the inverse quantum Fourier transform to the superposition of the first register in Fig. 6.3 and in Fig. 6.2. This is to say that they measure four quantum bits q[0], q[1], q[2] and q[3] of the first register and record the measurement outcome by overwriting four classical bits c[3], c[2], c[1] and c[0].

Listing 6.1 continued...

//Complete a measurement on the first register
27. measure q[0] → c[3];
28. measure q[1] → c[2];
29. measure q[2] → c[1];
30. measure q[3] → c[0];

In the backend *simulator* with thirty-two quantum bits in **IBM**'s quantum computers, we use the command "run" to execute the program in Listing 6.1. Figure 6.4 shows the measured result. From Fig. 6.4, we obtain that a computational basis state 0100 (c[3] = 0 = q[0] = |0⟩, c[2] = 1 = q[1] = |1⟩, c[1] = 0 = q[2] = |0⟩ and c[0] = 0 = q[3] = |0⟩)) has the probability 100%. This is to say that the value of θ is equal to (4/16). Therefore, we obtain that eigenvalue of a $(2^2 \times 2^2)$

Fig. 6.4 A computational basis state 0100 has the probability 100%

unitary matrix U with a $(2^2 \times 1)$ eigenvector $|u\rangle$ is equal to $\left(e^{\sqrt{-1} \times 2 \times \pi \times \frac{4}{2^4}}\right)$ with the probability 100%.

6.3   Quantum Counting to a Decision Problem with Any Input of n Bits in Real Applications of Phase Estimation

A *decision* problem is a problem in which it has only two possible outputs (yes or no) on any input of n bits. An output "yes" in the decision problem is to the number of solutions not to be zero and another output "no" in the decision problem is to the number of solutions to be zero. Solving a decision problem with any input of n bits is equivalent to solve one interesting problem with any input of n bits that is to from an unsorted database including 2^n items with each item has n bits how many items satisfy any given condition and we would like to find the number of solutions. If the number of solutions is not equal to zero, then there is an output "yes" in the decision problem with any input of n bits. Otherwise, there is an output "no" in the decision problem with any input of n bits.

A common formulation of a decision problem with any input of n bits is as follows. For any given oracular function $O_f: \{u_1 \, u_2 \, \ldots \, u_{n-1} \, u_n \mid \forall \, u_j \in \{0, 1\} \text{ for } 1 \le j \le n\} \rightarrow \{0, 1\}$, its domain is $\{u_1 \, u_2 \, \ldots \, u_{n-1} \, u_n \mid \forall \, u_j \in \{0, 1\} \text{ for } 1 \le j \le n\}$ and its range is $\{0, 1\}$. The decision problem with any input of n bits is asking to how many elements from its domain satisfy the condition $O_f(u_1 \, u_2 \, \ldots \, u_{n-1} \, u_n) = 1$. If the number of elements from its domain that satisfy $O_f(u_1 \, u_2 \, \ldots \, u_{n-1} \, u_n)$ to have a true value (1) is not equal to zero, then an output is "yes" to the decision problem with any input of n bits. Otherwise, an output is "no" for the decision problem with any input of n bits.

6.3.1 Binary Search Trees for Representing the Domain of a Decision Problem with Any Input of **n** Bits

A *tree* is a finite set of one or more nodes such that there is a specially designated node called the *root* and the remaining nodes are partitioned into $v \geq 0$ disjoint sets $T_1, \ldots, T_v$, where each of these sets is a tree. $T_1, \ldots, T_v$ are called the subtrees of the root. A *binary tree* is a finite set of nodes that is either empty or contains a root and two disjoint binary trees called the *left* subtree and the *right* subtree.

For any given oracular function O_f: $\{u_1 \, u_2 \, \ldots \, u_{n-1} \, u_n \mid \forall \, u_j \in \{0, 1\}$ for $1 \leq j \leq n\} \to \{0, 1\}$, its domain is $\{u_1 \, u_2 \, \ldots \, u_{n-1} \, u_n \mid \forall \, u_j \in \{0, 1\}$ for $1 \leq j \leq n\}$ and its range is $\{0, 1\}$. A decision problem with any input of n bits is asking to how many elements from its domain satisfy the condition $O_f(u_1 \, u_2 \, \ldots \, u_{n-1} \, u_n)$ to have a true value (1). We make use of a binary tree in Fig. 6.5 to represent the *structure* of the domain that is $\{u_1 \, u_2 \, \ldots \, u_{n-1} \, u_n \mid \forall \, u_j \in \{0, 1\}$ for $1 \leq j \leq n\}$. In the binary tree in Fig. 6.5, a node stands for a bit of one element in $\{u_1 \, u_2 \, \ldots \, u_{n-1} \, u_n \mid \forall \, u_j \in \{0, 1\}$ for $1 \leq j \leq n\}$. The root of the binary tree in Fig. 6.5 is u_1. The value of the *left* branch of each node represents that the value of the corresponding bit is equal to zero (0) and the value of the *right* branch of each node stands for that the value of the corresponding bit is equal to one (1). Since the value of the left branch of each node is less than the value of the right branch of each node, we regard the binary tree in Fig. 6.5 as a binary search tree.

The binary search tree in Fig. 6.5 includes 2^n subtrees and each subtree encodes one element in $\{u_1 \, u_2 \, \ldots \, u_{n-1} \, u_n \mid \forall \, u_j \in \{0, 1\}$ for $1 \leq j \leq n\}$. For example, the first subtree $(u_1)\text{--}^0\text{--}(u_2)\text{--}^0\text{--} \, \ldots \, (u_{n-1})\text{--}^0\text{--}(u_n)\text{--}^0\text{--}$ encodes the first element $\{u_1{}^0 \, u_2{}^0$

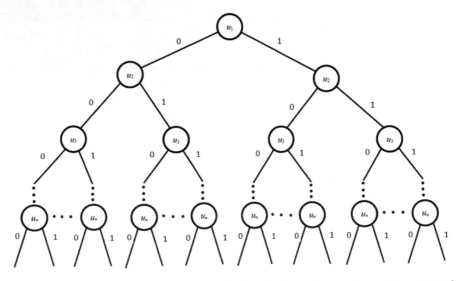

Fig. 6.5 A binary search tree for representing the domain of a decision problem with any input of n bits

... $u_{n-1}{}^0 u_n{}^0$}. The second subtree (u_1)--^{0}--(u_2)--^{0}-- ... (u_{n-1})--^{0}--(u_n)--^{1}-- encodes the second element {$u_1{}^0 u_2{}^0$ ... $u_{n-1}{}^0 u_n{}^1$}. The last subtree (u_1)--^{1}--(u_2)--^{1}-- ... (u_{n-1})--^{1}--(u_n)--^{1}-- encodes the last element {$u_1{}^1 u_2{}^1$ ... $u_{n-1}{}^1 u_n{}^1$}.

6.3.2 Flowchart of Solving a Decision Problem with Any Input of n Bits

Figure 6.6 is flowchart of solving a decision problem with any input of n bits. On the execution of the first statement, S_1, it sets the initial value of $u_1 u_2 \ldots u_{n-1} u_n$ to zero (0). Next, on the execution of the *second* statement, S_2, it judges whether $O_f(u_1 u_2 \ldots u_{n-1} u_n)$ has a true value (1) or not. If it returns a true value, then on the execution of the *third* statement, S_3, it generates that an output is "yes". Next, on the execution of the *fourth* statement, S_4, it executes one "End" instruction to terminate

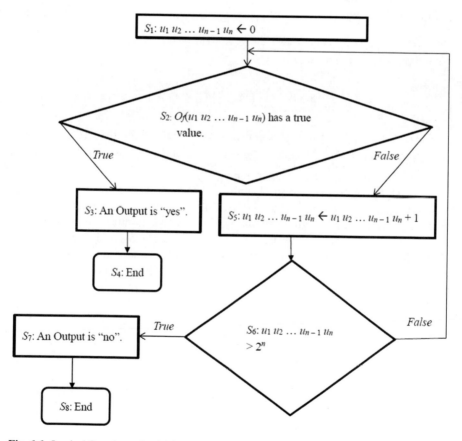

Fig. 6.6 Logical flowchart of solving a decision problem with any input of n bits

the processing of solving a decision problem with any input of n bits. Otherwise, on the execution the *fifth* statement, S_5, it increases the value of $u_1 u_2 \ldots u_{n-1} u_n$. Next, on the execution of the *sixth* statement, S_6, it judges whether the value of $u_1 u_2 \ldots u_{n-1} u_n$ is greater than 2^n or not. If it returns a true value, then on the execution of the *seventh* statement, S_7, it produces that an output is "no". Next, on the execution of the *eighth* statement, S_8, it executes one "End" instruction to terminate the processing of solving a decision problem with any input of n bits. Otherwise, it goes to statement S_2 and continues to execute statement S_2.

6.3.3 *Geometrical Interpretation to Solve a Decision Problem with Any Input of n Bits*

Binary search trees in Fig. 6.5 encode $\{u_1 u_2 \ldots u_{n-1} u_n \mid \forall\, u_j \in \{0, 1\}$ for $1 \leq j \leq n\}$ that is the domain of a decision problem with any input of n bits. We assume that an initial state vector ($|\phi_0\rangle$) is $\left(\otimes_{j=1}^{n} \left| u_j^0 \right\rangle\right)$. We begin to make use of a Hadamard transform $\left(\otimes_{j=1}^{n} H\right)$ on the initial state vector ($|\phi_0\rangle$) that is the register $\left(\otimes_{j=1}^{n} \left| u_j^0 \right\rangle\right)$. A superposition of the register is

$$|\phi_1\rangle = \frac{1}{\sqrt{2^n}}\left(\otimes_{j=1}^{n}\left(\left| u_j^0\right\rangle + \left| u_j^1\right\rangle\right)\right). \tag{6.21}$$

The new state vector ($|\phi_1\rangle$) encodes each subtree in Fig. 6.5 with that the amplitude of each subtree is $\left(\frac{1}{\sqrt{2^n}}\right)$. This is to say that it encodes each element of the domain to a decision problem with any input of n bits.

In the state vector ($|\phi_1\rangle$) in (6.21), subtrees (elements) that satisfy $O_f(u_1 u_2 \ldots u_{n-1} u_n)$ to have a true value (1) are referred as *marked* states and ones that do not result in a solution are referred as *unmarked* states. We assume that N is equal to 2^n. We also assume that in the state vector ($|\phi_1\rangle$) in (6.21), S stands for the number of solution(s) and ($N - S$) stands for the number of non-solution(s) to a decision problem with any input of n bits. We build two superpositions comprising uniformly distributed computational basis states

$$|\varphi\rangle = \frac{1}{\sqrt{N - S}}\left(\sum_{O_f(u_1 u_2 \cdots u_n)=0} |u_1 u_2 \cdots u_n\rangle\right), \tag{6.22}$$

$$|\lambda\rangle = \frac{1}{\sqrt{S}}\left(\sum_{O_f(u_1 u_2 \cdots u_n)=1} |u_1 u_2 \cdots u_n\rangle\right). \tag{6.23}$$

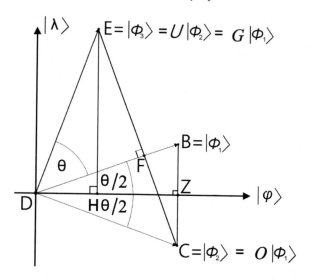

Fig. 6.7 Geometrical interpretation of solving a decision problem with any input of n bits in a two-dimensional Hilbert space spanned by $(|\varphi\rangle)$ and $(|\lambda\rangle)$

Because the inner product of $|\varphi\rangle$ and $|\lambda\rangle$ is equal to zero and the length of $|\varphi\rangle$ and $|\lambda\rangle$ is equal to one, $|\varphi\rangle$ and $|\lambda\rangle$ form an orthonormal basis of a two-dimensional Hilbert space which is depicted in Fig. 6.7. In Fig. 6.7, Point D is the *original* point of the two-dimensional Hilbert space and its coordinate is $(0, 0)$.

The state vector $(|\phi_1\rangle)$ in (6.21) can be expressed as a linear combination of $(|\varphi\rangle)$ and $(|\lambda\rangle)$ in a two-dimensional Hilbert space of Fig. 6.7 in the following way

$$|\phi_1\rangle = \frac{1}{\sqrt{N}} \left(\sum_{O_f(u_1 u_2 \cdots u_n)=0} |u_1 u_2 \cdots u_n\rangle \right.$$
$$\left. + \sum_{O_f(u_1 u_2 \cdots u_n)=1} |u_1 u_2 \cdots u_n\rangle \right)$$
$$= \left(\frac{\sqrt{N-S}}{\sqrt{N}} |\varphi\rangle + \frac{\sqrt{S}}{\sqrt{N}} |\lambda\rangle \right). \tag{6.24}$$

From (6.24), coordinate of $(|\phi_1\rangle)$ in a two-dimensional Hilbert space of Fig. 6.7 is $\left(\frac{\sqrt{N-S}}{\sqrt{N}}, \frac{\sqrt{S}}{\sqrt{N}} \right)$ and is strictly related to the angle between $(|\phi_1\rangle)$ and $(|\varphi\rangle)$ denoted by $\left(\frac{\theta}{2}\right)$ which is depicted in Fig. 6.7. Point B is coordinate point of $(|\phi_1\rangle)$.

In the quantum search algorithm introduced in the third Chapter, the Oracle O multiplies the probability amplitude of the answer(s) by -1 and leaves any other amplitude unchanged. We use the Oracle O to operate on the state vector $(|\phi_1\rangle)$ in (6.21) and obtain the new state vector $|\phi_2\rangle = O(|\phi_1\rangle)$ that can be expressed as a linear combination of ($|\varphi\rangle$) and ($|\lambda\rangle$) in a two-dimensional Hilbert space of Fig. 6.7 in the following way

$$|\phi_2\rangle = \left(\frac{\sqrt{N-S}}{\sqrt{N}} |\varphi\rangle + \left(-\frac{\sqrt{S}}{\sqrt{N}} |\lambda\rangle \right) \right). \tag{6.25}$$

From (6.25), coordinate of ($|\phi_2\rangle$) in a two-dimensional Hilbert space of Fig. 6.7 is $\left(\frac{\sqrt{N-S}}{\sqrt{N}}, -\frac{\sqrt{S}}{\sqrt{N}} \right)$ and is depicted in Fig. 6.7. Point C is coordinate point of ($|\phi_2\rangle$). The angle between ($|\phi_2\rangle$) and ($|\varphi\rangle$) is actually equal to $\left(\frac{\theta}{2} \right)$ that is depicted in Fig. 6.7. The Oracle O is equivalent to a reflection about axis $|\varphi\rangle$ in the two-dimensional geometrical interpretation of Fig. 6.7. Because in Fig. 6.7 point Z is the intersection of line $\overline{BC}$ and axis $|\varphi\rangle$ in which they are vertical each other, we obtain its coordinate to be $\left(\frac{\sqrt{N-S}}{\sqrt{N}}, 0 \right)$.

In the quantum search algorithm introduced in the third Chapter, the unitary operator U is the inversion about the average. The Grover operator G consists of two transformations on the index register that are U and O. We apply the unitary operator U to operate on the state vector ($|\phi_2\rangle$) in (6.25) and get the new state vector $|\phi_3\rangle = U(|\phi_2\rangle) = (U)(O)|\phi_1\rangle = G(|\phi_1\rangle)$. The new state vector ($|\phi_3\rangle$) can be expressed as a linear combination of ($|\varphi\rangle$) and ($|\lambda\rangle$) in a two-dimensional Hilbert space of Fig. 6.7 in the following way

$$|\phi_3\rangle = \left(\frac{\sqrt{N-S}}{\sqrt{N}} \times \left(\frac{N-4\times S}{N} \right) |\varphi\rangle + \frac{\sqrt{S}}{\sqrt{N}} \times \left(\frac{3\times N - 4\times S}{N} \right) |\lambda\rangle \right). \tag{6.26}$$

From (6.26), coordinate of ($|\phi_3\rangle$) in a two-dimensional Hilbert space of Fig. 6.7 is $\left(\frac{\sqrt{N-S}}{\sqrt{N}} \times \left(\frac{N-4\times S}{N} \right), \frac{\sqrt{S}}{\sqrt{N}} \times \left(\frac{3\times N-4\times S}{N} \right) \right)$ and is depicted in Fig. 6.7. Point E is coordinate point of ($|\phi_3\rangle$). The angle between ($|\phi_3\rangle$) and ($|\phi_1\rangle$) is actually equal to (θ) that is depicted in Fig. 6.7. The unitary operator U (the inversion about the average) in Fig. 6.7 reflects its input state ($|\phi_2\rangle$) over ($|\phi_1\rangle$) to ($|\phi_3\rangle$) in the two-dimensional geometrical interpretation of Fig. 6.7. In Fig. 6.7, point F is the intersection of line $\overline{EC}$ and line $\overline{DB}$ in which they are vertical each other and point H is the intersection of line $\overline{EH}$ and axis $|\varphi\rangle$ in which they are vertical each other.

6.3.4   Determine the Matrix of the Grover Operator in Geometrical Interpretation to Solve a Decision Problem with Any Input of **n** Bits

From Fig. 6.7, point B is $\left(\frac{\sqrt{N-S}}{\sqrt{N}}, \frac{\sqrt{S}}{\sqrt{N}}\right)$, point D is $(0, 0)$ and point Z is $\left(\frac{\sqrt{N-S}}{\sqrt{N}}, 0\right)$. The length of line $\overline{DB}$ is one (1), the length of line $\overline{DZ}$ is $\left(\sqrt{\frac{N-S}{N}}\right)$ and the length of line $\overline{BZ}$ is $\left(\sqrt{\frac{S}{N}}\right)$. Therefore, we obtain that $\sin(\theta/2) = \left(\sqrt{\frac{S}{N}}/1\right) = \left(\sqrt{\frac{S}{N}}\right)$ and $\cos(\theta/2) = \left(\sqrt{\frac{N-S}{N}}/1\right) = \left(\sqrt{\frac{N-S}{N}}\right)$. Because coordinate of $(|\phi_1\rangle)$ in Fig. 6.7 is $\left(\frac{\sqrt{N-S}}{\sqrt{N}}, \frac{\sqrt{S}}{\sqrt{N}}\right)$, its coordinate is also equal to $(\cos(\theta/2), \sin(\theta/2))$ in the basis of $(|\varphi\rangle)$ and $(|\lambda\rangle)$. From Fig. 6.7, $\sin(\theta + (\theta/2)) = \left(\frac{\sqrt{S}}{\sqrt{N}} \times \left(\frac{3 \times N - 4 \times S}{N}\right)\right)$ and $\cos(\theta + (\theta/2)) = \left(\frac{\sqrt{N-S}}{\sqrt{N}} \times \left(\frac{N - 4 \times S}{N}\right)\right)$ are obtained. Since coordinate of $|\phi_3\rangle$ in Fig. 6.7 is $\left(\frac{\sqrt{N-S}}{\sqrt{N}} \times \left(\frac{N - 4 \times S}{N}\right), \frac{\sqrt{S}}{\sqrt{N}} \times \left(\frac{3 \times N - 4 \times S}{N}\right)\right)$, its coordinate is also equal to $(\cos(\theta + (\theta/2)), \sin(\theta + (\theta/2)))$ in the basis of $(|\varphi\rangle)$ and $(|\lambda\rangle)$. From Fig. 6.7, the matrix of the Grover operator G in the basis of $(|\varphi\rangle)$ and $(|\lambda\rangle)$ is

$$G = \begin{bmatrix} \cos(\theta) & -\sin(\theta) \\ \sin(\theta) & \cos(\theta) \end{bmatrix}_{2\times2}. \tag{6.27}$$

The matrix of the Grover operator G in the basis of $(|\varphi\rangle)$ and $(|\lambda\rangle)$ is a unitary matrix (a unitary operator) because of $\left(\begin{bmatrix} \cos(\theta) & -\sin(\theta) \\ \sin(\theta) & \cos(\theta) \end{bmatrix}_{2\times2} \times \right.$

$$\begin{bmatrix} \cos(\theta) & \sin(\theta) \\ -\sin(\theta) & \cos(\theta) \end{bmatrix}_{2\times2} = \begin{bmatrix} \cos(\theta) & \sin(\theta) \\ -\sin(\theta) & \cos(\theta) \end{bmatrix}_{2\times2} \times \left. \begin{bmatrix} \cos(\theta) & -\sin(\theta) \\ \sin(\theta) & \cos(\theta) \end{bmatrix}_{2\times2} = \begin{bmatrix} 1 & 0 \\ 0 & 1 \end{bmatrix}_{2\times2}\right).$$

The eigenvalues of the Grover operator G in the basis of $(|\varphi\rangle)$ and $(|\lambda\rangle)$ are

$$\left(e^{\sqrt{-1}\times\theta}\right) \text{ and } \left(e^{-\sqrt{-1}\times\theta}\right). \tag{6.28}$$

The value of θ is a real. The corresponding eigenvectors of the Grover operator G in the basis of $(|\varphi\rangle)$ and $(|\lambda\rangle)$ are

$$|V_1\rangle = \frac{e^{\sqrt{-1}\times\gamma}}{\sqrt{2}} \begin{bmatrix} \sqrt{-1} \\ 1 \end{bmatrix}_{2\times1} \text{ and } |V_2\rangle = \frac{e^{\sqrt{-1}\times\gamma}}{\sqrt{2}} \begin{bmatrix} -\sqrt{-1} \\ 1 \end{bmatrix}_{2\times1}. \tag{6.29}$$

The value of γ is a real.

6.3.5   Quantum Counting Circuit to Solve a Decision Problem with Any Input of **n** Bits

From Fig. 6.7, we can figure out the projection of $|\phi_1\rangle$ onto axis $|\varphi\rangle$ that is $\sin(\theta/2) = \left(\sqrt{\frac{S}{N}}/1\right) = \left(\frac{S}{N}\right)$. The value of S is to the number of solutions that is how many elements in the domain $\{u_1 \ u_2 \ \dots \ u_{n-1} \ u_n \ | \ \forall \ u_j \in \{0, 1\}$ for $1 \leq j \leq n\}$ satisfy $O_f(u_1 \ u_2 \ \dots \ u_{n-1} \ u_n)$ to have a true value. Because $S = (\sin(\theta/2))^2 \times N$ and the value of N is known, if we can determine the value of θ, then we can compute the value of S that is the number of solutions. If the value of S is not equal to zero, then an output is "yes" to a decision problem with any input of n bits. Otherwise, an output is "no" to the decision problem with any input of n bits.

Figure 6.8 is quantum-counting circuits that are a real application of phase estimation. In Fig. 6.8, if an eigenvalue generated from controlled Grover operations are $\left(e^{\sqrt{-1}\times\theta}\right)$, then we use controlled Grover operations followed by **inverse quantum Fourier transform** to find the best approximation of t bits to the value of θ. Otherwise, we use controlled Grover operations followed by **quantum Fourier transform** to find the best approximation of t bits to the value of θ. In Fig. 6.8, a superposition of the second register is the state vector $|u\rangle$. The state vector $|u\rangle$ is a superposition of $(|\varphi\rangle)$ in (6.22) and $(|\lambda\rangle)$ in (6.23). Because $|V_1\rangle$ and $|V_2\rangle$ in (6.29) form an orthonormal basis of the space spanned by $(|\varphi\rangle)$ in (6.22) and $(|\lambda\rangle)$ in (6.23), the

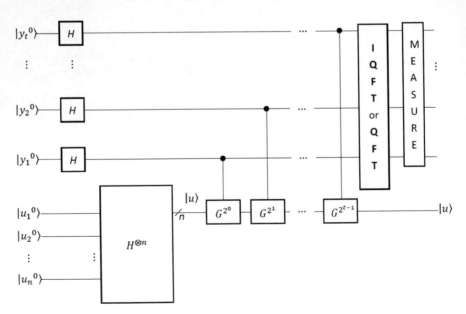

Fig. 6.8 Quantum counting circuits to calculate the number of solutions to a decision problem with the input of n bits

state vector $|u\rangle$ in Fig. 6.8 can be expressed as a linear combination of $|V_1\rangle$ and $|V_2\rangle$ in (6.29).

6.4   Determine the Number of Solutions to the Independent-Set Problem in a Graph with Two Vertices and One Edge in Phase Estimation

We assume that graph G has is a set V of vertices and a set E of edges. We also suppose that V is $\{v_1, \ldots, v_n\}$ in which each element v_j for $1 \leq j \leq n$ is a vertex in graph G. We assume that E is $\{(v_a, v_b)|\ v_a \in V$ and $v_b \in V\}$. We use $G = (V, E)$ to represent it. We assume that $|V|$ is the number of vertices in V and $|E|$ is the number of edges in E. We also suppose that $|V|$ is equal to n and $|E|$ is equal to m. The value of m is at most equal to $((n \times (n - 1))/2)$. For graph $G = (V, E)$, its *complementary* graph is $\overline{G} = (V, \overline{E})$ in which each edge in $\overline{E}$ is out of E. This is to say that $\overline{E}$ is $\{(v_c, v_d)|\ v_c \in V$ and $v_d \in V$ and $(v_c, v_d) \notin E\}$. We assume that $|\overline{E}|$ is the number of edges in $\overline{E}$. The number of edges in $\overline{E}$ is $(((n \times (n - 1))/2) - m)$. An *independent-set* of graph G with n vertices and m edges is a subset $V^1 \subseteq V$ of vertices such that for all $v_c, v_d \in V^1$, the edge (v_c, v_d) is *not* in E. The independent-set problem of graph G with n vertices and m edges is to find a *maximum-sized* independent set in G.

   Consider that in Fig. 6.9, a graph G^1 contains two vertices $\{v_1, v_2\}$ and one edge $\{(v_1, v_2)\}$ and its complementary graph $\overline{G^1}$ includes the same vertices and zero edge. This is an example of a decision problem that is deciding whether a graph G^1 in Fig. 6.9 has a *maximum-sized* independent set or not. All of the subsets of vertex are $\{\}$ that is an empty set, $\{v_1\}$, $\{v_2\}$ and $\{v_1, v_2\}$. Because in $\{v_1, v_2\}$, the edge (v_1, v_2) is one edge of graph G^1, $\{v_1, v_2\}$ does not satisfy definition of an independent set. For other three subsets of vertex that are $\{\}$ that is an empty set, $\{v_1\}$ and $\{v_2\}$, there is no edge in them to connect to other *distinct* vertex. Therefore, they satisfy definition of an independent set. So, all of the independent sets in graph G^1 are $\{\}$ that is an empty set, $\{v_1\}$ and $\{v_2\}$. Since the number of vertex in them are subsequently zero, one and one, the *maximum-sized* independent set for graph G^1 is $\{v_1\}$ and $\{v_2\}$. Finally, for the decision problem "a graph G^1 in Fig. 6.9, does it have a *maximum-sized* independent set?" it gives an output "yes".

   For any graph G with n vertices and m edges, all possible independent sets are 2^n possible choices consisting of legal and illegal independent sets in G. Each possible choice corresponds to a subset of vertices in G. Hence, we assume that Y is a set of 2^n

Fig. 6.9 A graph G^1 has two vertices and one edge

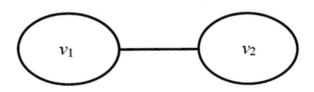

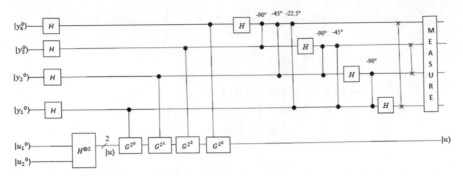

Fig. 6.10 Quantum circuit for deciding whether a graph G^1 with two vertices and one edge in Fig. 6.9 has a *maximum-sized* independent set or not

possible choices and Y is equal to $\{u_1 u_2 \ldots u_{n-1} u_n \mid \forall u_j \in \{0, 1\}$ for $1 \leq j \leq n\}$. This indicates that the length of each element in Y is n bits and each element represents one of 2^n possible choices. For the sake of presentation, we suppose that $u_j^{\,0}$ is that the value of u_j is zero and $u_j^{\,1}$ is that the value of u_j is one. If an element $u_1 u_2 \ldots u_{n-1} u_n$ in Y is a legal independent set and the value of u_j for $1 \leq j \leq n$ is one, then $u_j^{\,1}$ represents that the jth vertex is within the legal independent set. If an element $u_1 u_2 \ldots u_{n-1} u_n$ in Y is a legal independent set and the value of u_j for $1 \leq j \leq n$ is zero, then $u_j^{\,0}$ represents that the jth vertex is not within the legal independent set. We use superposition of a register with n quantum bits $\left(\frac{1}{\sqrt{2^n}} \left(\otimes_{j=1}^n \left(\left| u_j^0 \right\rangle + \left| u_j^1 \right\rangle \right) \right) \right)$ to encode a set of 2^n possible choices, $Y = \{u_1 u_2 \ldots u_{n-1} u_n \mid \forall u_j \in \{0, 1\}$ for $1 \leq j \leq n\}$.

Deciding whether a graph G^1 with two vertices and one edge in Fig. 6.9 has a *maximum-sized* independent set or not is equivalent to compute the number of solution to the same problem. Therefore, we make use of the circuit in Fig. 6.10 to determine the number of solution to the independent set problem in a graph G^1 with two vertices and one edge in Fig. 6.9. It uses two quantum registers. At the left top in Fig. 6.10, the first register $\left(\otimes_{k=4}^1 \left| y_k^0 \right\rangle \right)$ includes *four* quantum bits initially in the state $|0\rangle$. Quantum bit $|y_4^0\rangle$ is the most significant bit. Quantum bit $|y_1^0\rangle$ is the least significant bit. The corresponding decimal value of the first register is $(|y_4^0\rangle \times 2^{4-1}) + (|y_3^0\rangle \times 2^{3-1}) + (|y_2^0\rangle \times 2^{2-1}) + (|y_1^0\rangle \times 2^{1-1})$. At the left bottom in Fig. 6.10, the second register $\left(\otimes_{j=1}^2 \left| u_j^0 \right\rangle \right)$ contains *two* quantum bits initially in the state $|0\rangle$. Quantum bit $|u_1\rangle$ encodes the *first* vertex v_1 in graph G^1 in Fig. 6.9 and is the most significant bit. Quantum bit $|u_2\rangle$ encodes the *second* vertex v_2 in graph G^1 in Fig. 6.9 and is the least significant bit. Quantum bits $|u_1^1\rangle |u_2^1\rangle$ encodes $\{v_1, v_2\}$ that is a subset of two vertices. Quantum bits $|u_1^1\rangle |u_2^0\rangle$ encodes $\{v_1\}$ that is a subset of one vertex. Quantum bits $|u_1^0\rangle |u_2^1\rangle$ encodes $\{v_2\}$ that is a subset of one vertex. Quantum bits $|u_1^0\rangle |u_2^0\rangle$ encodes $\{\}$ that is an empty subset without any vertex. Of course, the corresponding decimal value of the second register is $(|u_1^0\rangle \times 2^{2-1}) + (|u_2^0\rangle \times 2^{2-2})$. For the convenience of the presentation, the following initial state vector is

$$|\varphi_0\rangle = \left(\otimes_{k=4}^{1} |y_k^0\rangle\right) \otimes \left(\otimes_{j=1}^{2} |u_j^0\rangle\right). \tag{6.30}$$

6.4.1 Initialize Quantum Registers to Calculate the Number of Solutions to the Independent-Set Problem in a Graph with Two Vertices and One Edge in Phase Estimation

In Listing 6.2, the program is in the backend that is *simulator* of Open QASM with *thirty-two* quantum bits in **IBM**'s quantum computer. The program is to calculate the number of solutions to the independent-set problem in graph G^1 with two vertices and one edge in Fig. 6.9. Figure 6.11 is the corresponding quantum circuit of the program in Listing 6.2 and is to implement the quantum circuit of Fig. 6.10 to calculate the number of solutions to the independent-set problem in graph G^1 with two vertices and one edge in Fig. 6.9.

Listing 6.2 The program of computing the number of solutions to the independent-set problem in graph G^1 with two vertices and one edge in Fig. 6.9.

```
1.  OPENQASM 2.0;
2.  include "qelib1.inc";

3.  qreg q[6];
4.  creg c[4];
```

The statement "OPENQASM 2.0;" on line one of Listing 6.2 is to indicate that the program is written with version 2.0 of Open QASM. Then, the statement "include "qelib1.inc";" on line two of Listing 6.2 is to continue parsing the file "qelib1.inc" as if the contents of the file were pasted at the location of the include statement, where the file "qelib1.inc" is **Quantum Experience (QE) Standard Header** and the path is specified relative to the current working directory.

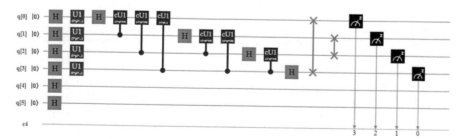

Fig. 6.11 Implementing quantum circuits of Fig. 6.10 to compute the number of solutions to the independent-set problem in graph G^1 with two vertices and one edge in Fig. 6.9

Next, the statement "qreg q[6];" on line three of Listing 6.2 is to declare that in the program there are *six* quantum bits. In the left top of Fig. 6.11, six quantum bits are respectively q[0], q[1], q[2], q[3], q[4] and q[5]. The initial value of each quantum bit is set to state |0⟩. We use four quantum bits q[0], q[1], q[2] and q[3] to subsequently encode four quantum bits $|y_4\rangle$, $|y_3\rangle$, $|y_2\rangle$ and $|y_1\rangle$ in Fig. 6.10. We apply two quantum bits q[4] and q[5] to respectively encode two quantum bits $|u_1\rangle$ and $|u_2\rangle$ in Fig. 6.10. For the convenience of our explanation, $q[k]^0$ for $0 \leq k \leq 5$ is to represent the value 0 of q[k] and $q[k]^1$ for $0 \leq k \leq 5$ is to represent the value 1 of q[k]. Since quantum bit $|y_4^0\rangle$ is the most significant bit and quantum bit $|y_1^0\rangle$ is the least significant bit, quantum bit $|q[0]^0\rangle$ is the most significant bit and quantum bit $|q[3]^0\rangle$ is the least significant bit. The corresponding decimal value of the first register in Fig. 6.11 is $(|q[0]^0\rangle \times 2^{4-1}) + (|q[1]^0\rangle \times 2^{3-1}) + (|q[2]^0\rangle \times 2^{2-1}) + (|q[3]^0\rangle \times 2^{1-1})$.

Then, the statement "creg c[4];" on line four of Listing 6.2 is to declare that there are four classical bits in the program. In the left bottom of Fig. 6.11, four classical bits are respectively c[0], c[1], c[2] and c[3]. The initial value of each classical bit is set to zero (0). For the convenience of our explanation, $c[k]^0$ for $0 \leq k \leq 3$ is to represent the value 0 of c[k] and $c[k]^1$ for $0 \leq k \leq 3$ is to represent the value 1 of c[k]. The corresponding decimal value of the four initial classical bits $c[3]^0$ $c[2]^0$ $c[1]^0$ $c[0]^0$ is $2^3 \times c[3]^0 + 2^2 \times c[2]^0 + 2^1 \times c[1]^0 + 2^0 \times c[0]^0$. This is to say that classical bit $c[3]^0$ is the most significant bit and classical bit $c[0]^0$ is the least significant bit. For the convenience of our explanation, we can rewrite the initial state vector $|\varphi_0\rangle = \left(\otimes_{k=4}^1 |y_k^0\rangle\right) \otimes \left(\otimes_{j=1}^2 |u_j^0\rangle\right)$ in (6.30) in Fig. 6.10 as follows

$$|\varphi_0\rangle = \left(\otimes_{k=4}^1 |y_k^0\rangle\right) \otimes \left(\otimes_{j=1}^2 |u_j^0\rangle\right) = |q[0]^0\rangle|q[1]^0\rangle|q[2]^0\rangle|q[3]^0\rangle|q[4]^0\rangle|q[5]^0\rangle. \tag{6.31}$$

6.4.2 Superposition of Quantum Registers to Compute the Number of Solutions to the Independent-Set Problem in a Graph with Two Vertices and One Edge in Phase Estimation

In Fig. 6.10, the first stage of the circuit is to implement a Hadamard transform with four Hadamard gates on the *first* register $\left(\otimes_{k=4}^1 |y_k^0\rangle\right)$ and another Hadamard transform with two Hadamard gates on the *second* register $\left(\otimes_{j=1}^2 |u_j^0\rangle\right)$. The *six* statements "h q[0];", "h q[1];", "h q[2];", "h q[3];", "h q[4];" and "h q[5];" on line *five* of Listing 6.2 through line *ten* of Listing 6.2 is to implement *six* Hadamard gates on the first register and the second register. They perform each Hadamard gate in the first time slot of Fig. 6.11 and complete the first stage of the circuit in Fig. 6.10.

Listing 6.2 continued...

//Implement a Hadamard transform on two registers
```
 5.  h q[0];
 6.  h q[1];
 7.  h q[2];
 8.  h q[3];
 9.  h q[4];
10.  h q[5];
```

A superposition of the first register is $\left(\frac{1}{\sqrt{2^4}}\left(\otimes_{k=4}^{1}\left(\left|y_k^0\right\rangle + \left|y_k^1\right\rangle\right)\right)\right)$ = $\left(\frac{1}{\sqrt{2^4}}\left(\otimes_{a=0}^{3}\left(\left|q[a]^0\right\rangle + \left|q[a]^1\right\rangle\right)\right)\right)$. Another superposition of the second register is $\left(|u\rangle = \frac{1}{\sqrt{2^2}}\left(\otimes_{j=1}^{2}\left(\left|u_j^0\right\rangle + \left|u_j^1\right\rangle\right)\right)\right)$ $= \frac{1}{\sqrt{2^2}}\left(\otimes_{b=4}^{5}\left(\left|q[b]^0\right\rangle + \left|q[b]^1\right\rangle\right)\right)$. This implies that the superposition of the second register begins in the new state vector $\left(|u\rangle = \frac{1}{\sqrt{2^2}}\left(\otimes_{j=1}^{2}\left(\left|u_j^0\right\rangle + \left|u_j^1\right\rangle\right)\right)\right)$ $= \frac{1}{\sqrt{2^2}}\left(\otimes_{b=4}^{5}\left(\left|q[b]^0\right\rangle + \left|q[b]^1\right\rangle\right)\right)$ and contains *two* quantum bits as is necessary to store $(|u\rangle)$. In superposition of the second register $(|u\rangle)$, state $\left(\left|u_1^1\right\rangle\left|u_2^1\right\rangle\right)$ that is encoded by state $\left(\left|q[4]^1\right\rangle\left|q[5]^1\right\rangle\right)$ with the amplitude $(1/2)$ encodes $\{v_1, v_2\}$ that is a subset of two vertices. State $\left(\left|u_1^1\right\rangle\left|u_2^0\right\rangle\right)$ that is encoded by state $\left(\left|q[4]^1\right\rangle\left|q[5]^0\right\rangle\right)$ with the amplitude $(1/2)$ encodes $\{v_1\}$ that is a subset of one vertex. State $\left(\left|u_1^0\right\rangle\left|u_2^1\right\rangle\right)$ that is encoded by state $\left(\left|q[4]^0\right\rangle\left|q[5]^1\right\rangle\right)$ with the amplitude $(1/2)$ encodes $\{v_2\}$ that is a subset of one vertex. State $\left(\left|u_1^0\right\rangle\left|u_2^0\right\rangle\right)$ that is encoded by state $\left(\left|q[4]^0\right\rangle\left|q[5]^0\right\rangle\right)$ with the amplitude $(1/2)$ encodes $\{\ \}$ that is an empty subset without vertex. The new state vector $(|u\rangle)$ is an eigenstate (eigenvector) of G that is the Grover operator and is a unitary operator. Thus, this gives that the following new state vector is

$$
\begin{aligned}
|\varphi_1\rangle &= \left(\frac{1}{\sqrt{2^4}}\left(\otimes_{k=4}^{1}\left(\left|y_k^0\right\rangle + \left|y_k^1\right\rangle\right)\right)\right) \otimes \left(\frac{1}{\sqrt{2^2}}\left(\otimes_{j=1}^{2}\left(\left|u_j^0\right\rangle + \left|u_j^1\right\rangle\right)\right)\right) \\
&= \left(\frac{1}{\sqrt{2^4}}\left(\otimes_{k=4}^{1}\left(\left|y_k^0\right\rangle + \left|y_k^1\right\rangle\right)\right)\right) \otimes (|u\rangle) \\
&= \left(\frac{1}{\sqrt{2^4}}\left(\otimes_{a=0}^{3}\left(\left|q[a]^0\right\rangle + \left|q[a]^1\right\rangle\right)\right)\right) \otimes \left(\frac{1}{\sqrt{2^2}}\left(\otimes_{b=4}^{5}\left(\left|q[b]^0\right\rangle + \left|q[b]^1\right\rangle\right)\right)\right) \\
&= \left(\frac{1}{\sqrt{2^4}}\left(\otimes_{a=0}^{3}\left(\left|q[a]^0\right\rangle + \left|q[a]^1\right\rangle\right)\right)\right) \otimes (|u\rangle). \quad\quad (6.32)
\end{aligned}
$$

6.4.3 Controlled-G Operations on the Superposition of the Second Register to Determine the Number of Solutions to the Independent-Set Problem in a Graph with Two Vertices and One Edge in Phase Estimation

In the new state vector $|\varphi_1\rangle$ in (6.32), each quantum bit in the first register is currently in its superposition. The value of the first register is from state $\left(\otimes_{k=4}^1 |y_k^0\rangle\right)$ (zero) encoded by state $\left(\otimes_{a=0}^3 |q[a]^0\rangle\right)$ through state $\left(\otimes_{k=4}^1 |y_k^1\rangle\right)$ (fifteen) encoded by state $\left(\otimes_{a=0}^3 |q[a]^1\rangle\right)$ with that the amplitude of each state is (1/4). The circuit of Fig. 6.10 can precisely estimate sixteen phases. This indicates that the first register with four quantum bits can precisely represent sixteen phases. Sixteen phases are respectively $(0/2^4), (1/2^4), (2/2^4), (3/2^4), (4/2^4), (5/2^4), (6/2^4), (7/2^4), (8/2^4), (9/2^4), (10/2^4), (11/2^4), (12/2^4), (13/2^4), (14/2^4)$ and $(15/2^4)$. The corresponding sixteen phase angles are respectively $(2 \times \pi \times 0/2^4), (2 \times \pi \times 1/2^4), (2 \times \pi \times 2/2^4), (2 \times \pi \times 3/2^4), (2 \times \pi \times 4/2^4), (2 \times \pi \times 5/2^4), (2 \times \pi \times 6/2^4), (2 \times \pi \times 7/2^4), (2 \times \pi \times 8/2^4), (2 \times \pi \times 9/2^4), (2 \times \pi \times 10/2^4), (2 \times \pi \times 11/2^4), (2 \times \pi \times 12/2^4), (2 \times \pi \times 13/2^4), (2 \times \pi \times 14/2^4)$ and $(2 \times \pi \times 15/2^4)$.

Say that we are trying to compute an eigenvalue of $90°$. The number of solutions for the independent-set problem in a graph G^1 with two vertices and one edge in Fig. 6.9 is $S = N \times (\sin(\theta/2))^2 = 4 \times (\sin(90°/2))^2 = 4 \times (1/2) = 2$. This gives that the answer is two for determining the number of solutions for the independent-set problem in a graph G^1 with two vertices and one edge in Fig. 6.9. Therefore, the effect of one application of the Grover operator G on its eigenvector (eigenstate) $(|u\rangle)$ is $(G \times |u\rangle = e^{\pm\sqrt{-1}\times 2\times\pi\times\theta} \times |u\rangle = e^{\pm\sqrt{-1}\times 2\times\pi\times\frac{4}{2^4}} \times |u\rangle)$. So, the effect of repeated application of the Grover operator G on its eigenvector (eigenstate) $(|u\rangle)$ is

$$G^a |u\rangle = e^{\pm\sqrt{-1}\times 2\times\pi\times\theta\times a} |u\rangle = e^{\pm\sqrt{-1}\times 2\times\pi\times\frac{4}{2^4}\times a} \times |u\rangle. \qquad (6.33)$$

A superposition $\left(\frac{1}{\sqrt{2}}(|y_1^0\rangle + |y_1^1\rangle)\right)$ that is encoded by $\left(\frac{1}{\sqrt{2}}(|q[3]^0\rangle + |q[3]^1\rangle)\right)$ at the weighted position 2^0 is the controlled quantum bit of implementing controlled-G^{2^0} operations on the superposition of the second register that is the state $(|u\rangle)$. Similarly, a superposition $\left(\frac{1}{\sqrt{2}}(|y_2^0\rangle + |y_2^1\rangle)\right)$ that is encoded by $\left(\frac{1}{\sqrt{2}}(|q[2]^0\rangle + |q[2]^1\rangle)\right)$ nat the weighted position 2^1 is the controlled quantum bit of implementing controlled-G^{2^1} operations on the superposition of the second register that is the state $(|u\rangle)$. Then, a superposition $\left(\frac{1}{\sqrt{2}}(|y_3^0\rangle + |y_3^1\rangle)\right)$ that is encoded by $\left(\frac{1}{\sqrt{2}}(|q[1]^0\rangle + |q[1]^1\rangle)\right)$ at the weighted position 2^2 is the controlled quantum bit of implementing controlled-G^{2^2} noperations on the superposition of the second register that is the state $(|u\rangle)$. Next, a superposition $\left(\frac{1}{\sqrt{2}}(|y_4^0\rangle + |y_4^1\rangle)\right)$ that is encoded by

$\left(\frac{1}{\sqrt{2}}\left(|q[0]^0\rangle + |q[0]^1\rangle\right)\right)$ at the weighted position 2^3 is the controlled quantum bit of implementing controlled-G^{2^3} operations on the superposition of the second register that is the state $(|u\rangle)$.

The Grover operator G has two eigenvalues $\left(e^{\sqrt{-1}\times 2\times\pi\times\theta}\right)$ and $\left(e^{-\sqrt{-1}\times 2\times\pi\times\theta}\right)$. We assume that it generates the eigenvalue $\left(e^{\sqrt{-1}\times 2\times\pi\times\theta}\right) = \left(e^{\sqrt{-1}\times 2\times\pi\times\frac{4}{2^4}}\right)$. The *four* statements from line *eleven* through line *fourteen* in Listing 6.2 are "u1(2 * pi * 4/16 * 1) q[3];", "u1(2 * pi * 4/16 * 2) q[2];", "u1(2 * pi * 4/16 * 4) q[1];" and "u1(2 * pi * 4/16 * 8) q[0];".

Listing 6.2 continued...

//Implement controlled-G operations on the superposition of the second register
```
11.  u1(2 * pi * 4/16 * 1) q[3];
12.  u1(2 * pi * 4/16 * 2) q[2];
13.  u1(2 * pi * 4/16 * 4) q[1];
14.  u1(2 * pi * 4/16 * 8) q[0];
```

They take the new state vector $(|\varphi_1\rangle)$ in (6.32) as their input state vector and implement each controlled-G operation on the superposition of the second register in the *second* time slot of Fig. 6.11 and in the *second* stage of Fig. 6.10. They alert the phase of the state $|y_1{}^1\rangle$ $(|q[3]^1\rangle)$ is from one (1) to become $\left(e^{\sqrt{-1}\times 2\times\pi\times\frac{4}{16}\times 2^0}\right) = \left(e^{\sqrt{-1}\times 2\times\pi\times\frac{4}{16}\times 1}\right)$. They alert the phase of the state $|y_2{}^1\rangle$ $(|q[2]^1\rangle)$ is from one (1) to become $\left(e^{\sqrt{-1}\times 2\times\pi\times\frac{4}{16}\times 2^1}\right) = \left(e^{\sqrt{-1}\times 2\times\pi\times\frac{4}{16}\times 2}\right)$. They alert the phase of the state $|y_3{}^1\rangle$ $(|q[1]^1\rangle)$ is from one (1) to become $\left(e^{\sqrt{-1}\times 2\times\pi\times\frac{4}{16}\times 2^2}\right) = \left(e^{\sqrt{-1}\times 2\times\pi\times\frac{4}{16}\times 4}\right)$ nnand alert the phase of the state $|y_4{}^1\rangle$ $(|q[0]^1\rangle)$ is from one (1) to become $\left(e^{\sqrt{-1}\times 2\times\pi\times\frac{4}{16}\times 2^3}\right) = \left(e^{\sqrt{-1}\times 2\times\pi\times\frac{4}{16}\times 8}\right)$. This gives that the following new state vector is

$$|\varphi_2\rangle = \left(\frac{1}{\sqrt{2^4}}\left(|y_4^0\rangle + e^{\sqrt{-1}\times 2\times\pi\times\frac{4}{16}\times 2^3}|y_4^1\rangle\right) \otimes \left(|y_3^0\rangle + e^{\sqrt{-1}\times 2\times\pi\times\frac{4}{16}\times 2^2}|y_3^1\rangle\right)\right.$$
$$\left. \otimes \left(|y_2^0\rangle + e^{\sqrt{-1}\times 2\times\pi\times\frac{4}{16}\times 2^1}|y_2^1\rangle\right) \otimes \left(|y_1^0\rangle + e^{\sqrt{-1}\times 2\times\pi\times\frac{4}{16}\times 2^0}|y_1^1\rangle\right)\right) \otimes (|u\rangle)$$
$$= \left(\frac{1}{\sqrt{2^4}}\left(|y_4^0\rangle + e^{\sqrt{-1}\times 2\times\pi\times\frac{4}{16}\times 8}|y_4^1\rangle\right) \otimes \left(|y_3^0\rangle + e^{\sqrt{-1}\times 2\times\pi\times\frac{4}{16}\times 4}|y_3^1\rangle\right)\right.$$
$$\left. \otimes \left(|y_2^0\rangle + e^{\sqrt{-1}\times 2\times\pi\times\frac{4}{16}\times 2}|y_2^1\rangle\right) \otimes \left(|y_1^0\rangle + e^{\sqrt{-1}\times 2\times\pi\times\frac{4}{16}\times 1}|y_1^1\rangle\right)\right) \otimes (|u\rangle)$$
$$= \left(\frac{1}{\sqrt{2^4}}\left(|q[0]^0\rangle + e^{\sqrt{-1}\times 2\times\pi\times\frac{4}{16}\times 8}|q[0]^1\rangle\right)\right.$$

$$\otimes \left(|q[1]^0\rangle + e^{\sqrt{-1}\times 2\times\pi\times\frac{4}{16}\times 4}|q[1]^1\rangle \right)$$

$$\otimes \left(|q[2]^0\rangle + e^{\sqrt{-1}\times 2\times\pi\times\frac{4}{16}\times 2}|q[2]^1\rangle \right)$$

$$\otimes \left(|q[3]^0\rangle + e^{\sqrt{-1}\times 2\times\pi\times\frac{4}{16}\times 1}|q[3]^1\rangle \right) \right) \otimes (|u\rangle)$$

$$= \left(\frac{1}{\sqrt{2^4}} \left(\sum_{Y=0}^{2^4-1} e^{\sqrt{-1}\times 2\times\pi\times\frac{4}{16}\times Y}|Y\rangle \right) \right) \otimes (|u\rangle). \tag{6.34}$$

From this description above, the second quantum register stays in the state $(|u\rangle)$ through the computation. Because of *phase kickback*, the phase of the state $|Y\rangle$ for $0 \le Y \le 2^4 - 1$ is from one (1) to become $\left(e^{\sqrt{-1}\times 2\times\pi\times\frac{4}{16}\times Y} \right)$. In the state vector $(|\varphi_2\rangle)$ in (6.34), it includes sixteen phase angles from state $|0\rangle$ through state $|15\rangle$. The front eight phase angles are $(90° \times 0 = 0°)$, $(90° \times 1 = 90°)$, $(90° \times 2 = 180°)$, $(90° \times 3 = 270°)$, $(90° \times 4 = 360° = 0°)$, $(90° \times 5 = 450° = 90°)$, $(90° \times 6 = 540° = 180°)$ and $(90° \times 7 = 630° = 270°)$. The last eight phase angles are $(90° \times 8 = 720° = 0°)$, $(90° \times 9 = 810° = 90°)$, $(90° \times 10 = 900° = 180°)$, $(90° \times 11 = 990° = 270°)$, $(90° \times 12 = 1080° = 0°)$, $(90° \times 13 = 1170° = 90°)$, $(90° \times 14 = 1260° = 180°)$ and $(90° \times 15 = 1350° = 270°)$. The phase angle rotates back to its starting value $0°$ *four* times.

6.4.4   The Inverse Quantum Fourier Transform on the Superposition of the First Register to Compute the Number of Solutions to the Independent-Set Problem in a Graph with Two Vertices and One Edge in Phase Estimation

Listing 6.2 continued…

//Implement one inverse quantum Fourier transform on the superposition of the first

//Register
```
15.  h q[0];
16.  cu1(−2 * pi * 1/4) q[1], q[0];
17.  cu1(−2 * pi * 1/8) q[2], q[0];
18.  cu1(−2 * pi * 1/16) q[3], q[0];

19.  h q[1];
20.  cu1(−2 * pi * 1/4) q[2], q[1];
21.  cu1(−2 * pi * 1/8) q[3], q[1];
```

22. h q[2];
23. cu1(−2 * pi * 1/4) q[3], q[2];

24. h q[3];

25. swap q[0], q[3];
26. swap q[1], q[2];

Hidden patterns and information stored in the state vector $(|\varphi_2\rangle)$ in (6.34) are to that its phase angle rotates back to its starting value $0°$ *four* times. This is to say that the number of the period per sixteen phase angles is *four* and the frequency is equal to *four* (16/4). The twelve statements from line *fifteen* through line *twenty-six* in Listing 6.2 complete each quantum operation from the *third* time slot through the *fourteenth* time slot in Fig. 6.11. They actually implement each quantum operation of performing an **inverse quantum Fourier transform** on the superposition of the first register in Fig. 6.10. They take the state vector $(|\varphi_2\rangle)$ in (6.34) as their input state vector. Since the **inverse quantum Fourier transform** effectively transforms the state of the first register into a superposition of the *periodic* signal's component frequencies, they generate the following state vector

$$
\begin{aligned}
|\varphi_3\rangle &= \left(\sum_{Y=0}^{2^4-1} \frac{1}{\sqrt{2^4}} e^{\sqrt{-1}\times 2\times \pi \times \frac{4}{2^4} \times Y} \frac{1}{\sqrt{2^4}} \sum_{i=0}^{2^4-1} e^{-\sqrt{-1}\times 2\times \pi \times \frac{i}{2^4} \times Y} |i\rangle \right) \otimes (|u\rangle) \\
&= \left(\frac{1}{\sqrt{2^4}} \left(\sum_{Y=0}^{2^4-1} \sum_{i=0}^{2^4-1} e^{\sqrt{-1}\times 2\times \pi \times Y \times \left(\frac{4}{2^4}-\frac{i}{2^4}\right)} |i\rangle \right) \right) \otimes (|u\rangle) \\
&= \left(\sum_{i=0}^{2^4-1} \sum_{Y=0}^{2^4-1} \frac{1}{\sqrt{2^4}} \left(e^{\sqrt{-1}\times 2\times \pi \times \left(\frac{4}{2^4}-\frac{i}{2^4}\right)} \right)^Y |i\rangle \right) \otimes (|u\rangle).
\end{aligned}
\tag{6.35}
$$

6.4.5 Read the Quantum Result to Figure out the Number of Solutions to the Independent-Set Problem in a Graph with Two Vertices and One Edge in Phase Estimation

Finally, the four statements "measure q[0] → c[3];", "measure q[1] → c[2];", "measure q[2] → c[1];" and "measure q[3] → c[0];" from line *twenty-seven* through line *thirty* in Listing 6.2 implement a measurement. They measure the output state of the inverse quantum Fourier transform to the superposition of the first register in Fig. 6.11 and in Fig. 6.10. This is to say that they measure four quantum bits q[0], q[1], q[2]

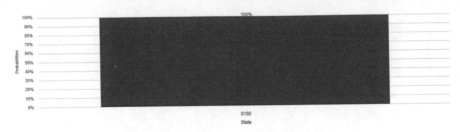

Fig. 6.12 A computational basis state 0100 has the probability 100%

and q[3] of the first register and record the measurement outcome by overwriting four classical bits c[3], c[2], c[1] and c[0].

> **Listing 6.2 continued...**
>
> //Complete a measurement on the first register
> 27. measure q[0] → c[3];
> 28. measure q[1] → c[2];
> 29. measure q[2] → c[1];
> 30. measure q[3] → c[0];

In the backend *simulator* with thirty-two quantum bits in **IBM**'s quantum computers, we use the command "run" to execute the program in Listing 6.2. Figure 6.12 shows the measured result. From Fig. 6.12, we get that a computational basis state 0100 ($c[3] = 0 = q[0] = |0\rangle$, $c[2] = 1 = q[1] = |1\rangle$, $c[1] = 0 = q[2] = |0\rangle$ and $c[0] = 0 = q[3] = |0\rangle$) has the probability 100%. This indicates that the phase angle is $\theta = 2 \times \pi \times (4/16) = 90°$ with the probability 100%. Hence, The number of solutions for the independent-set problem in a graph G^1 with two vertices and one edge in Fig. 6.9 is $S = N \times (\sin(\theta/2))^2 = 4 \times (\sin(90°/2))^2 = 4 \times (1/2) = 2$. This is to say that the answer with the probability 100% is two for computing the number of solutions for the independent-set problem in a graph G^1 with two vertices and one edge in Fig. 6.9. Therefore, an output is "yes" to a decision problem that is deciding whether a graph G^1 in Fig. 6.9 has a *maximum-sized* independent set or not.

6.5 Summary

In this chapter, we illustrated that a *decision* problem is a problem in which it has only two possible outputs (yes or no) on any input of n bits. An output "yes" in the decision problem on any input of n bits is to the number of solutions not to be zero and another output "no" in the decision problem on any input of n bits is to the number of solutions to be zero. Next, we described that a $(2^n \times 2^n)$ unitary matrix (operator) U has a $(2^n \times 1)$ eigenvector $|u\rangle$ with eigenvalue $e^{\sqrt{-1} \times 2 \times \pi \times \theta}$ such that $U \times |u\rangle = e^{\sqrt{-1} \times 2 \times \pi \times \theta} \times |u\rangle$, where the value of θ is *unknown* and is real. We then illustrated how the phase estimate algorithm with the what possibility estimates the value of θ. We also described time complexity, space complexity and performance of the phase estimate algorithm. Next, we introduced how to design quantum circuits and write quantum programs for computing eigenvalue of a $(2^2 \times 2^2)$ unitary matrix U with a $(2^2 \times 1)$ eigenvector $|u\rangle$. Next, we described how the quantum-counting algorithm determines the number of solutions for a decision problem with the input of n bits. We also illustrated time complexity, space complexity and performance of the quantum-counting algorithm. We then introduced how to design quantum circuits and write quantum programs to determine the number of solution to the independent set problem in a graph G^1 with two vertices and one edge.

6.6 Bibliographical Notes

In this chapter for more details about an introduction of the phase estimation algorithm, the recommended books are Nielsen and Chuang (2000), Imre and Balazs (2005), Lipton and Regan (2014), Silva (2018), Johnston et al. (2019). For a more detailed description to binary search trees, the recommended book is Horowitz et al. (2003). For a more detailed introduction to the discrete Fourier transform and the inverse discrete Fourier transform, the recommended books are Cormen et al. (2009), Nielsen and Chuang (2000), Imre and Balazs (2005), Lipton and Regan (2014), Silva (2018), Johnston et al. (2019). The two famous articles (Coppersmith 1994; Shor 1994) gave the original version of the Quantum Fourier transform and the inverse quantum Fourier transform. A good illustration for the product state decomposition of the quantum Fourier transform and the inverse quantum Fourier transform is the two famous articles in Griffiths and Niu (1996), Cleve et al. (1998). For a more detailed description to the quantum-counting algorithm, the recommended article and books are Brassard et al. (1998), Nielsen and Chuang (2000), Imre and Balazs (2005), Lipton and Regan (2014), Silva (2018), Johnston et al. (2019). A good introduction to the instructions of Open QASM is the famous article in Cross et al. (2017).

6.7 Exercises

6.1 Prove that the transformation of the Oracle is $O = I_{2^n,2^n} - 2 \times |x_0\rangle\langle x_0|$, where x_0 is one element in the domain of the Oracle and x_0 satisfies $O(x_0) = 1$.

6.2 Determine the matrix of the Oracle that is $O = I_{2^2,2^2} - 2 \times |x_0\rangle\langle x_0|$, where $x_0 = 2$ and x_0 satisfies $O(x_0) = 1$.

6.3 Show that the unitary operator U (inversion about the average) is equivalent to reflect its input state $|\phi_2\rangle$ over $|\phi_1\rangle$ to $|\phi_3\rangle$ that is a reflection about $|\phi_1\rangle$ in the two-dimensional geometrical interpretation of Fig. 6.7.

6.4 Compute the matrix of the Grover operator G in the basis of ($|\varphi\rangle$) and ($|\lambda\rangle$) in Fig. 6.7.

6.5 Calculate the eigenvalues and corresponding eigenvectors of the Grover operator G in the basis of ($|\varphi\rangle$) and ($|\lambda\rangle$) in Fig. 6.7.

References

Brassard, G., Hoyer, P., Tapp, A.: Quantum counting. In: The 25th International Colloquium on Automata, Languages, and Programming (ICALP), LNCS 1443, pp. 820–831 (1998)

Cleve, R., Ekert, A., Macciavello, C., Mosca, M.: Quantum algorithms revisited. Proc. Roy. Soc. Lond. Ser. A **454**, 339–354 (1998). E-print quant-ph/9708016

Coppersmith, D.: An approximate Fourier transform useful in quantum factoring. IBM Research Report, Technical Report RC 19642, December 1994. E-print quant-ph/0201067

Cormen, T.H., Leiserson, C.E., Rivest, R.L., Stein, C.: Introduction to Algorithms, 3rd edn. The MIT Press, Cambridge (2009). ISBN-13: 978-0262033848, ISBN-10: 9780262033848

Cross, A.W., Bishop, L.S., Smolin, J.A., Gambetta, J.M.: Open quantum assembly language. 2017. https://arxiv.org/abs/1707.03429

Griffiths, R.B., Niu, C.S.: Semiclassical Fourier transform for quantum computation. Phys. Rev. Lett. **76**(17), 3228–3231 (1996). E-print quant-ph/9511007

Horowitz, E., Sahni, S., Anderson-Freed, S.: Fundamentals of Data Structures in C, 12th edn. Computer Science Press, New York (2003). ISBN 0-7167-8250-2

Imre, S., Balazs, F.: Quantum Computation and Communications: An Engineering Approach. Wiley, UK (2007). ISBN-10: 047086902X, ISBN-13: 978-0470869024, 2005

Johnston, E.R., Harrigan, N., Gimeno-Segovia, M.: Programming Quantum Computers: Essential Algorithms and Code Samples. O'Reilly Media, Inc. (2019). ISBN-13: 978-1492039686, ISBN-10: 1492039683

Lipton, R.J., Regan, K.W.: Quantum Algorithms via Linear Algebra: A Primer. The MIT Press, Cambridge (2014). ISBN 978-0-262-02839-4

Nielsen, M.A., Chuang, I.L.: Quantum Computation and Quantum Information. Cambridge University Press, New York, NY (2000). ISBN-10: 9781107002173, ISBN-13: 978-1107002173

Shor, P.: Algorithms for quantum computation: discrete logarithms and factoring. In: Proceedings 35th Annual Symposium on Foundations of Computer Science, Santa Fe, pp. 124–134, 20–22 Nov 1994

Silva, V.: Practical Quantum Computing for Developers: Programming Quantum Rigs in the Cloud using Python, Quantum Assembly Language and IBM Q Experience. Apress, New York, 13 Dec 2018. ISBN-10: 1484242173, ISBN-13: 978-1484242179

Answer Key

Solutions to Chapter 1

1.1 The problem at hand is to find the parameters based on the following equality:

$$U3(\theta, \phi, \lambda) = \begin{bmatrix} \cos\frac{\theta}{2} & -e^{i\lambda}\sin\frac{\theta}{2} \\ e^{i\phi}\sin\frac{\theta}{2} & e^{i(\lambda+\phi)}\cos\frac{\theta}{2} \end{bmatrix} = \begin{bmatrix} 0 & 1 \\ 1 & 0 \end{bmatrix} = NOT$$

In the following we assume the interval $[0, \pi]$. Since $\cos\frac{\theta}{2} = 0$, $\frac{\theta}{2}$ must equal $\frac{\pi}{2}$, and thus $\theta = \pi$. Since $\sin\frac{\pi}{2} = 1$, $e^{i\phi} = 1$ as well, and hence $i\phi = 0$ and therefore also $\phi = 0$. Again, since $\sin\frac{\pi}{2} = 1$, $-e^{i\lambda} = 1$, and hence by squaring both sides we have $e^{i2\lambda} = 1$ and from this $\lambda = 0$.

1.2 Along the same lines as in Exercise 1.1

$$U3(\theta, \phi, \lambda) = \begin{bmatrix} \cos\frac{\theta}{2} & -e^{i\lambda}\sin\frac{\theta}{2} \\ e^{i\phi}\sin\frac{\theta}{2} & e^{i(\lambda+\phi)}\cos\frac{\theta}{2} \end{bmatrix} = \begin{bmatrix} \frac{1}{\sqrt{2}} & \frac{1}{\sqrt{2}} \\ \frac{1}{\sqrt{2}} & -\frac{1}{\sqrt{2}} \end{bmatrix} = H$$

In order for $\cos\frac{\theta}{2} = \frac{1}{\sqrt{2}}$, $\frac{\theta}{2} = \frac{\pi}{4}$ and therefore $\theta = \frac{\pi}{2}$. For this value of θ, $\sin\frac{\pi}{4} = \frac{1}{\sqrt{2}}$ as well, and hence $e^{i\phi} = 1$. From this it follows that $\phi = 0$. Since $e^{i(\lambda+\phi)} = -1$ and given that $\phi = 0$, we have that $e^{i\lambda} = -1$ and therefore $\lambda = \pi$.

1.3 As in both previous exercises, we set

$$U3(\theta, \phi, \lambda) = \begin{bmatrix} \cos\frac{\theta}{2} & -e^{i\lambda}\sin\frac{\theta}{2} \\ e^{i\phi}\sin\frac{\theta}{2} & e^{i(\lambda+\phi)}\cos\frac{\theta}{2} \end{bmatrix} = \begin{bmatrix} 1 & 0 \\ 0 & -1 \end{bmatrix} = Z$$

With $\cos\frac{\theta}{2} = 1$ it follows that $\theta = 0$. Since $\cos\frac{0}{2} = 1$, it must be the case that $e^{i(\lambda+\phi)} = -1$. It must therefore be the case that $\lambda + \phi = \pi$. With $\sin\frac{0}{2} = 0$

© The Author(s), under exclusive license to Springer Nature Switzerland AG 2021
W.-L. Chang and A. V. Vasilakos, *Fundamentals of Quantum Programming in IBM's Quantum Computers*, Studies in Big Data 81,
https://doi.org/10.1007/978-3-030-63583-1

the values of λ and ϕ cannot be established from the off-diagonal formulas. For that reason, the gate Z can have many parametrizations, including when $\phi = \lambda = \frac{\pi}{2}$.

1.4   Again, as above

$$U3(\theta, \phi, \lambda) = \begin{bmatrix} \cos\frac{\theta}{2} & -e^{i\lambda}\sin\frac{\theta}{2} \\ e^{i\phi}\sin\frac{\theta}{2} & e^{i(\lambda+\phi)}\cos\frac{\theta}{2} \end{bmatrix} = \begin{bmatrix} 0 & -i \\ i & 0 \end{bmatrix} = Y$$

With $\cos\frac{\theta}{2} = 0$, $\theta = \pi$. $\sin\frac{\pi}{2} = 1$. Hence, $e^{i\phi} = i$ and $-e^{i\lambda} = -i$. From these equations we obtain the values $\phi = \frac{\pi}{2}$ and $\lambda = \frac{\pi}{2}$.

1.5   As above,

$$U3(\theta, \phi, \lambda) = \begin{bmatrix} \cos\frac{\theta}{2} & -e^{i\lambda}\sin\frac{\theta}{2} \\ e^{i\phi}\sin\frac{\theta}{2} & e^{i(\lambda+\phi)}\cos\frac{\theta}{2} \end{bmatrix} = \begin{bmatrix} 1 & 0 \\ 0 & i \end{bmatrix} = S$$

With $\cos\frac{\theta}{2} = 1$, $\theta = 0$. Then $e^{i(\lambda+\phi)} = i$. This means that $\lambda + \phi = \frac{\pi}{2}$. Again, there are multiple parametrizations that achieve this, e.g., $\lambda = \phi = \frac{\pi}{4}$.

1.6   The S^+ gate differs from the S gate in the sign of the complex number i only. Therefore, the parameters are as follows: $\theta = 0$ and $\lambda + \phi = \frac{3\pi}{2}$.

1.7   The gate T differs from the S only by having the value $\frac{1+i}{\sqrt{2}}$ instead of i. Therefore, the parameter $\theta = 0$, while for the remaining two parameters there are multiple values for which holds that $e^{i(\lambda+\phi)} = \frac{1+i}{\sqrt{2}}$. To obtain the parameters we use trigonometric identities

$$e^{i(\lambda+\phi)} = e^{i\lambda}e^{i\phi} = (\cos\lambda + i\sin\lambda)(\cos\phi + i\sin\phi)$$

This corresponds to

$$\cos\lambda\cos\phi + i\sin\phi\cos\lambda + i\sin\lambda\cos\phi - \sin\phi\sin\lambda$$
$$= (\cos\lambda\cos\phi - \sin\lambda\sin\phi) + i(\sin\phi\cos\lambda + \sin\lambda\cos\phi)$$
$$= \cos(\lambda + \phi) + i\sin(\lambda + \phi) = \frac{1}{\sqrt{2}} + \frac{i}{\sqrt{2}}$$

Thus we have $\lambda + \phi = \frac{\pi}{4}$.

1.8   The reasoning regarding the T^+ gate is similar to that for the T gate. Instead of $\frac{1+i}{\sqrt{2}}$, the T^+ has its conjugate $\frac{1-i}{\sqrt{2}}$. The parameter $\theta = 0$. Using trigonometric identities

$$\cos(\lambda + \phi) + i\sin(\lambda + \phi) = \frac{1}{\sqrt{2}} - \frac{i}{\sqrt{2}}$$

From $\cos(\lambda + \phi) = \frac{1}{\sqrt{2}}$ and $\sin(\lambda + \phi) = -\frac{1}{\sqrt{2}}$ we have that $\lambda + \phi = 315° = \frac{7\pi}{4}$. Therefore, any combination of values for λ and ϕ that conforms to the condition $\lambda + \phi = \frac{7\pi}{4}$ is a valid parametrization.

1.9   Finding the parameters for the identity gate is straightforward. From $\cos\frac{\theta}{2} = 1$ it follows that $\theta = 0$. Then, from $e^{i(\lambda+\phi)} = 1$ it follows that $\lambda + \phi = 0$ and therefore $\lambda = \phi = 0$.

1.10 For the gate

$$U1(\lambda) = \begin{bmatrix} 1 & 0 \\ 0 & e^{i\lambda} \end{bmatrix}$$

$\theta = 0$, as above. Then, from $e^{i(\lambda+\phi)} = e^{i\lambda}$ it follows that $\phi = 0$.

1.11 For the gate $U2(\phi, \lambda)$ we have

$$U3(\theta, \phi, \lambda) = \begin{bmatrix} \cos\frac{\theta}{2} & -e^{i\lambda}\sin\frac{\theta}{2} \\ e^{i\phi}\sin\frac{\theta}{2} & e^{i(\lambda+\phi)}\cos\frac{\theta}{2} \end{bmatrix} = \begin{bmatrix} \frac{1}{\sqrt{2}} & -\frac{e^{i\lambda}}{\sqrt{2}} \\ \frac{e^{i\phi}}{\sqrt{2}} & \frac{e^{i(\lambda+\phi)}}{\sqrt{2}} \end{bmatrix} = U2(\phi, \lambda)$$

we have that $\cos\frac{\theta}{2} = \frac{1}{\sqrt{2}}$, and therefore $\theta = \frac{\pi}{2}$.

Solutions to Chapter 2

2.1   To obtain the required operation, 3 quantum bits and 3 classical bits are needed. In the following circuit, $q[0]$ represents the x operand, $q[1]$ represents the y operand, while $q[2]$ is the additional bit in the OR gate introduced in the material of this chapter. First, set the two operands into a superposition by applying to each a Hadamard gate H. As the OR gate requires that the auxiliary quantum bit be equal 1, apply the NOT gate to it. The OR gate would require a total of 4 NOT gates. It can however be simplified due to the fact that one of the operands, namely x, is negated. The equivalence of the two subcircuits can be easily checked by writing down truth tables for each of them. Note that the output is presented in the order $c[2]$, $c[1]$, $c[0]$.

The circuit:

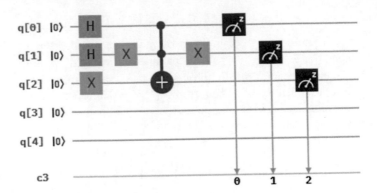

The code:

```
1.   OPENQASM 2.0;
2.   include "qelib1.inc";
3.   qreg q[5];
4.   creg c[3];
5.   h q[0];
6.   h q[1];
7.   x q[1];
8.   x q[2];
9.   ccx q[0], q[1], q[2];
10.  x q[1];
11.  measure q[0] → c[0];
12.  measure q[1] → c[1];
13.  measure q[2] → c[2];
```

The output:

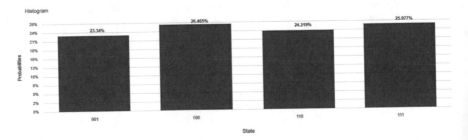

2.2   To obtain the required operation, 3 quantum bits and 3 classical bits are needed. $q[0]$ represents the x operand, $q[1]$ represents y, while $q[2]$ is the additional bit in the OR gate introduced in the material of this chapter. First, set the two operands into a superposition by applying to each a Hadamard gate H. As the

OR gate requires that the auxiliary quantum bit be equal 1, apply the Pauli X gate to it. The *OR* gate would require a total of 4 X gates. It can however be simplified due to the fact that one of the operands, namely y, is negated. Note that the output is presented in the order $c[2]$, $c[1]$, $c[0]$.

The circuit:

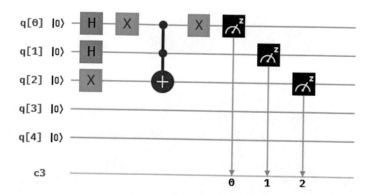

The code:

```
1.   OPENQASM 2.0;
2.   include "qelib1.inc";
3.   qreg q[5];
4.   creg c[3];
5.   h q[0];
6.   h q[1];
7.   x q[2];
8.   x q[0];
9.   ccx q[0], q[1], q[2];
10.  x q[0];
11.  measure q[0] → c[0];
12.  measure q[1] → c[1];
13.  measure q[2] → c[2];
14.  measure q[0] → c[0];
15.  measure q[1] → c[1];
16.  measure q[2] → c[2];
```

The output:

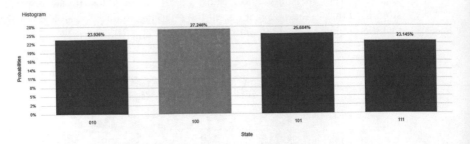

2.3 A possible implementation of this operation could be as below. To this end, 5 quantum bits and 3 classical bits are required. Observe that $\bar{y} \wedge (x \vee \bar{x}) = (\bar{y} \wedge x) \vee (\bar{y} \wedge \bar{x})$. Therefore, in the following, we separate the operations into three parts: $(\bar{y} \wedge x)$, $(\bar{y} \wedge \bar{x})$, and the *OR* operation on the two. The three operations are separated by barriers (barriers are only used for convenience and they do not contribute to the meaning of the circuit). Everything up to the first barrier, except the two Hadamard gates, encodes the clause $(\bar{y} \wedge x)$. Clause $(\bar{y} \wedge \bar{x})$ is encoded between the first and the second barrier. The *OR* operation on the two clauses is encoded between the second and the third barrier. In the following circuit, $q[0]$ represents x, $q[1]$ represents y, $q[2]$ is the auxialiary quantum bit for the *AND* operation in the term $(\bar{y} \wedge x)$, $q[3]$ is the auxiliary bit for the *AND* operation in the term $(\bar{y} \wedge \bar{x})$, while $q[4]$ is the auxiliary bit for the *OR* operation mentioned above. Note that the output is presented in the order $c[2]$, $c[1]$, $c[0]$. The circuit:

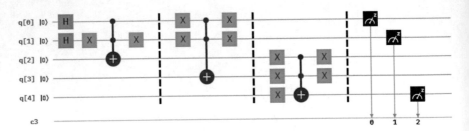

The code:

```
1.  OPENQASM 2.0;
2.  include "qelib1.inc";
3.  qreg q[5];
4.  creg c[3];
5.  h q[0];
6.  h q[1];
7.  x q[1];
```

8. ccx q[0], q[1], q[2];
9. x q[1];
10. barrier q[0], q[1], q[2], q[3], q[4];
11. x q[0];
12. x q[1];
13. ccx q[0], q[1], q[3];
14. x q[0];
15. x q[1];
16. barrier q[0], q[1], q[2], q[3], q[4];
17. x q[2];
18. x q[3];
19. x q[4];
20. ccx q[2], q[3], q[4];
21. x q[2];
22. x q[3];
23. barrier q[0], q[1], q[2], q[3], q[4];
24. measure q[0] → c[0];
25. measure q[1] → c[1];
26. measure q[4] → c[2];

The output:

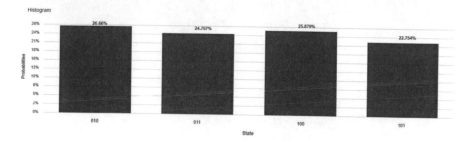

2.4 A possible implementation of this operation could be as below. To this end, 5 quantum bits and 3 classical bits are required. The separation of operations is along the same lines as in the previous exercise. Observe that $\bar{x} \wedge (y \vee \bar{y}) = (\bar{x} \wedge y) \vee (\bar{x} \wedge \bar{y})$. Therefore, in the following, we separate the operations into three parts: $(\bar{x} \wedge y)$, $(\bar{x} \wedge \bar{y})$, and the *OR* operation on the two. The three operations are separated by barriers (barriers are only used for convenience and they do not contribute to the meaning of the circuit). Everything up to the first barrier, except the two Hadamard gates, encodes the clause $(\bar{x} \wedge y)$. Clause $(\bar{x} \wedge \bar{y})$ is encoded between the first and the second barrier. The *OR* operation on the two clauses is encoded between the second and the third barrier. $q[0]$ represents x, $q[1]$ represents y, $q[2]$ is the auxiliary quantum bit for the *AND* operation in the term $(\bar{x} \wedge y)$, $q[3]$ is the auxiliary bit for the *AND* operation in the term $(\bar{x} \wedge \bar{y})$, while $q[4]$ is the auxiliary bit for the *OR* operation mentioned above. Note that the output is presented in the order $c[2]$, $c[1]$, $c[0]$.

The circuit:

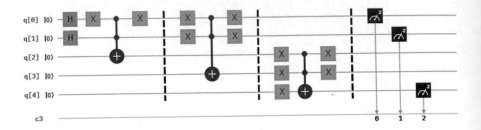

The code:

```
1.  OPENQASM 2.0;
2.  include "qelib1.inc";
3.  qreg q[5];
4.  creg c[3];
5.  h q[0];
6.  h q[1];
7.  x q[0];
8.  ccx q[0], q[1], q[2];
9.  x q[0];
10. barrier q[0], q[1], q[2], q[3], q[4];
11. x q[0];
12. x q[1];
13. ccx q[0], q[1], q[3];
14. x q[0];
15. x q[1];
16. barrier q[0], q[1], q[2], q[3], q[4];
17. x q[2];
18. x q[3];
19. x q[4];
20. ccx q[2], q[3], q[4];
21. x q[2];
22. x q[3];
23. barrier q[0], q[1], q[2], q[3], q[4];
24. measure q[0] → c[0];
25. measure q[1] → c[1];
26. measure q[4] → c[2];
```

The output:

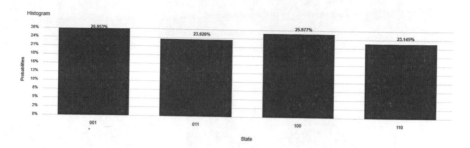

2.5 To obtain the required operation, 3 quantum bits and 3 classical bits are needed. $q[0]$ represents the x operand, $q[1]$ represents y, while $q[2]$ is the additional bit in the *AND* gate introduced in the material of this chapter. First, set the two operands into a superposition by applying to each a Hadamard gate H. The two X gates are to account for the fact that operand x is negated. Note that the output is presented in the order $c[2]$, $c[1]$, $c[0]$.

The circuit:

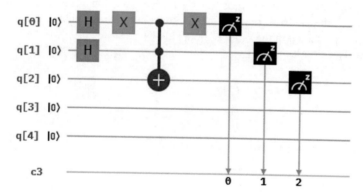

The code:

```
1.  OPENQASM 2.0;
2.  include "qelib1.inc";
3.  qreg q[5];
4.  creg c[3];
5.  h q[0];
6.  h q[1];
7.  x q[0];
8.  ccx q[0], q[1], q[2];
9.  x q[0];
10. measure q[0] → c[0];
```

11. measure q[1] → c[1];
12. measure q[2] → c[2];

The output:

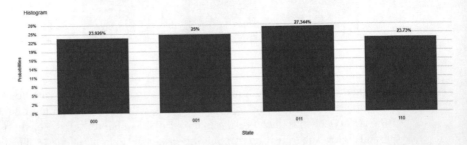

2.6 To obtain the required operation, 3 quantum bits and 3 classical bits are needed. $q[0]$ represents the x operand, $q[1]$ represents y, while $q[2]$ is the additional bit in the *AND* gate introduced in the material of this chapter. First, set the two operands into a superposition by applying to each a Hadamard gate H. The two X gates are to account for the fact that operand y is negated. Note that the output is presented in the order $c[2]$, $c[1]$, $c[0]$.
The circuit:

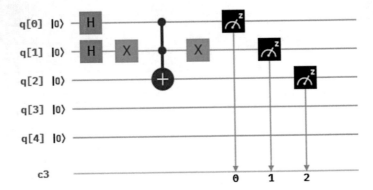

The code:

```
1.  OPENQASM 2.0;
2.  include "qelib1.inc";
3.  qreg q[5];
4.  creg c[3];
5.  h q[0];
6.  h q[1];
```

7. x q[1];
8. ccx q[0], q[1], q[2];
9. x q[1];
10. measure q[0] → c[0];
11. measure q[1] → c[1];
12. measure q[2] → c[2];

The output:

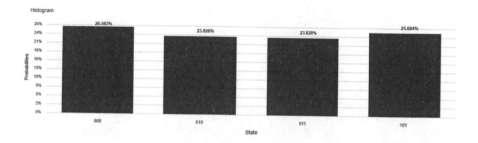

Solutions to Chapter 3

3.1 The circuit for the satisfiability problem specified by the oracle $F(x_1, x_2) = \overline{x_1} \wedge x_2$ is very similar to the one given for $F(x_1, x_2) = x_1 \wedge x_2$ in this chapter (see also the answer to Exercise 3.4). The difference is that the input x_1 is negated (that is $q[3]$). $q[0]$ is a workspace qubit, $q[1]$ is not used, $q[2]$ represents s_2, $q[3]$ is for x_1, while $q[4]$ is for x_2. We introduce barriers to separate different parts of the code. Everything up to the first barrier represents the initial encoding. $q[0]$ represents the state $|-\rangle = \frac{1}{\sqrt{2}}(|0\rangle - |1\rangle)$, as in the material, while $q[3]$ and $q[4]$ are set into a uniform superposition. The oracle that negates the amplitude of the qubit for which the condition $\overline{x_1} \wedge x_2 = 1$ is fulfilled is located between the first and the second barrier. To represent the fact that the value of x_1 is negated, the *NOT* gate is used on the qubit representing x_1. After the oracle has completed its task, another *NOT* gate is applied to x_1 to uncompute. The circuit part between the second and the third barriers has been introduced in the material of this chapter and encodes the Grover diffusion operator. The circuit:

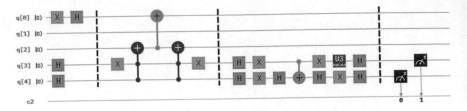

The code:

```
1.   OPENQASM 2.0;
2.   include "qelib1.inc";
3.   qreg q[5];
4.   creg c[2];
5.   x q[0];
6.   h q[3];
7.   h q[4];
8.   h q[0];
9.   barrier q[0], q[1], q[2], q[3], q[4];
10.  x q[3];
11.  ccx q[4], q[3], q[2];
12.  cx q[2], q[0];
13.  ccx q[4], q[3], q[2];
14.  x q[3];
15.  barrier q[0], q[1], q[2], q[3], q[4];
16.  h q[3];
17.  h q[4];
18.  x q[3];
19.  x q[4];
20.  h q[4];
21.  cx q[3], q[4];
22.  x q[3];
23.  h q[4];
24.  u3(2 * pi, 0, 0) q[3];
25.  x q[4];
26.  h q[3];
27.  h q[4];
28.  barrier q[0], q[1], q[2], q[3], q[4];
29.  measure q[4] → c[0];
30.  measure q[3] → c[1];
```

The output:

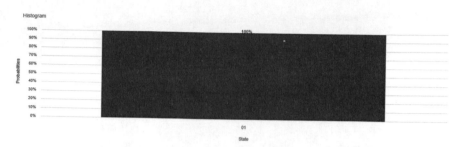

3.2 The circuit for the satisfiability problem specified by the oracle $F(x_1, x_2) = x_1 \wedge \overline{x_2}$ is very similar to the one given for $F(x_1, x_2) = \overline{x_1} \wedge x_2$ in Exercise 3.1. The difference is that the input x_2 is negated (that is $q[2]$) instead of x_1. $q[0]$ is a workspace qubit, $q[1]$ is not used, $q[2]$ represents s_2, $q[3]$ is for x_1, while $q[4]$ is for x_2. We introduce barriers to separate different parts of the code. Everything up to the first barrier represents the initial encoding. $q[0]$ represents the state $|-\rangle = \frac{1}{\sqrt{2}}(|0\rangle - |1\rangle)$, as in the material, while $q[3]$ and $q[4]$ are set into a uniform superposition. The oracle that negates the amplitude of the qubit for which the condition $x_1 \wedge \overline{x_2} = 1$ is fulfilled is located between the first and the second barrier. To represent the fact that the value of x_2 is negated, the NOT gate is used on the qubit representing x_2. After the oracle has completed its task, another NOT gate is applied to x_2 to uncompute. The circuit part between the second and the third barriers has been introduced in the material of this chapter and encodes the Grover diffusion operator.

The circuit:

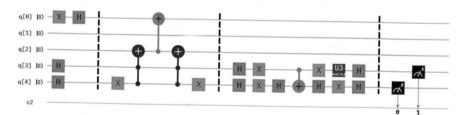

The code:

```
1.  OPENQASM 2.0;
2.  include "qelib1.inc";
3.  qreg q[5];
```

4. creg c[2];
5. x q[0];
6. h q[3];
7. h q[4];
8. h q[0];
9. barrier q[0], q[1], q[2], q[3], q[4];
10. x q[4];
11. ccx q[4], q[3], q[2];
12. cx q[2], q[0];
13. ccx q[4], q[3], q[2];
14. x q[4];
15. barrier q[0], q[1], q[2], q[3], q[4];
16. h q[3];
17. h q[4];
18. x q[3];
19. x q[4];
20. h q[4];
21. cx q[3], q[4];
22. x q[3];
23. h q[4];
24. u3(2 * pi, 0, 0) q[3];
25. x q[4];
26. h q[3];
27. h q[4];
28. barrier q[0], q[1], q[2], q[3], q[4];
29. measure q[4] → c[0];
30. measure q[3] → c[1];

The output:

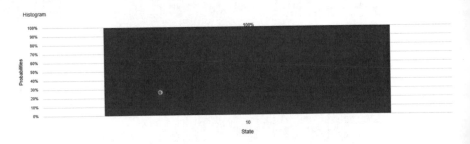

3.3 In theory, the circuit for the satisfiability problem specified by the oracle
    $F(x_1, x_2) = \overline{x_1 \wedge x_2}$ should be very similar to the one given for $F(x_1, x_2) =$

$x_1 \wedge x_2$ in this chapter (see also the answer to Exercise 3.4). The difference would be that the quantum bit $q[2]$ is negated in order to represent the *NAND* gate. $q[0]$ is a workspace qubit, $q[1]$ is not used, $q[2]$ represents s_2, $q[3]$ is for x_1, while $q[4]$ is for x_2. We introduce barriers to separate different parts of the code. Everything up to the first barrier represents the initial encoding. $q[0]$ represents the state $|-\rangle = \frac{1}{\sqrt{2}}(|0\rangle - |1\rangle)$, as in the material, while $q[3]$ and $q[4]$ are set into a uniform superposition. The oracle that negates the amplitude of the qubit for which the condition $\overline{x_1 \wedge x_2} = 1$ is fulfilled is located between the first and the second barrier. As mentioned above, the *NOT* gate is used on the qubit representing s_2. After the oracle has completed its task, another *NOT* gate is applied to s_2 to uncompute. The circuit part between the second and the third barriers has been introduced in the material of this chapter and encodes the Grover diffusion operator. After executing the code we notice that the obtained output is incorrect. In fact it shows the state $|11\rangle$ with probability 1. The correct answer should have included equal probabilities for three states: $|00\rangle$, $|01\rangle$, and $|10\rangle$. These three states are the correct outcomes for the *NAND* gate. The reason we obtained the wrong answer is that Grover's algorithm imposes a limitation on the ratio between the number of answers (A) and the number of all possible outcomes (O). This limitation is $\frac{A}{O} < \frac{1}{2}$. In this exercise, this ratio is $\frac{3}{4} > \frac{1}{2}$ however.

The circuit:

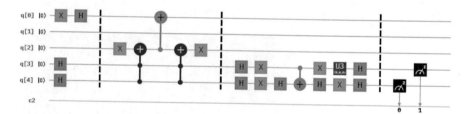

The code:

```
 1.  OPENQASM 2.0;
 2.  include "qelib1.inc";
 3.  qreg q[5];
 4.  creg c[2];
 5.  x q[0];
 6.  h q[3];
 7.  h q[4];
 8.  h q[0];
 9.  barrier q[0], q[1], q[2], q[3], q[4];
10.  x q[2];
11.  ccx q[4], q[3], q[2];
12.  cx q[2], q[0];
```

13. ccx q[4], q[3], q[2];
14. x q[2];
15. barrier q[0], q[1], q[2], q[3], q[4];
16. h q[3];
17. h q[4];
18. x q[3];
19. x q[4];
20. h q[4];
21. cx q[3], q[4];
22. x q[3];
23. h q[4];
24. u3(0, 0, 0) q[3];
25. x q[4];
26. h q[3];
27. h q[4];
28. barrier q[0], q[1], q[2], q[3], q[4];
29. measure q[4] → c[0];
30. measure q[3] → c[1];

The output:

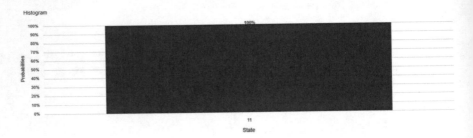

3.4 The circuit for the satisfiability problem specified by the oracle $F(x_1, x_2) = x_1 \wedge x_2$ which was given in the material of this chapter, can be simplified by replacing the twelve gates used to represent the *CCNOT* operation with the *CCNOT* gate itself (in code: *ccx*). As the procedure of deriving the circuit has been described in detail in the material, here we only present the corresponding circuit, the code and its output. As before, $q[0]$ is a workspace qubit, $q[1]$ is not used, $q[2]$ represents s_2, $q[3]$ is for x_1, while $q[4]$ is for x_2. We introduce barriers to separate different parts of the code. Everything up to the first barrier represents the initial encoding. $q[0]$ represents the state $|-\rangle = \frac{1}{\sqrt{2}}(|0\rangle - |1\rangle)$, as in the material, while $q[3]$ and $q[4]$ are set into a uniform superposition. The oracle that negates the amplitude of the qubit for which the condition $x_1 \wedge x_2 = 1$ is fulfilled is located between the first and the second barrier. The circuit part

between the second and the third barriers has been introduced in the material of this chapter and encodes the Grover diffusion operator.
The circuit:

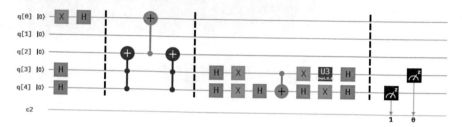

The code:

```
1.   OPENQASM 2.0;
2.   include "qelib1.inc";
3.   qreg q[5];
4.   creg c[2];
5.   x q[0];
6.   h q[3];
7.   h q[4];
8.   h q[0];
9.   barrier q[0], q[1], q[2], q[3], q[4];
10.  ccx q[4], q[3], q[2];
11.  cx q[2], q[0];
12.  ccx q[4], q[3], q[2];
13.  barrier q[0], q[1], q[2], q[3], q[4];
14.  h q[3];
15.  h q[4];
16.  x q[3];
17.  x q[4];
18.  h q[4];
19.  cx q[3], q[4];
20.  x q[3];
21.  h q[4];
22.  u3(2 * pi, 0, 0) q[3];
23.  x q[4];
24.  h q[3];
25.  h q[4];
26.  barrier q[0], q[1], q[2], q[3], q[4];
27.  measure q[4] → c[1];
28.  measure q[3] → c[0];
```

The output:

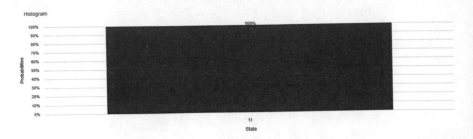

Solutions to Chapter 4

4.1 The mapping this oracle performs is $a_1 a_2 a_3 a_4 \rightarrow \frac{1}{4} e^{2\pi i \times 0.a_2 a_3 a_4} = \frac{1}{4} e^{2\pi i \left(\frac{a_2}{2} + \frac{a_3}{4} + \frac{a_4}{8} \right)}$. The following table shows the particular results. Input 0/1 indicates that the respective value is either 0 or 1.

a_1	a_2	a_3	a_4	$\frac{1}{4} e^{2\pi i \left(\frac{a_2}{2} + \frac{a_3}{4} + \frac{a_4}{8} \right)}$
0/1	0	0	0	$1/4$
0/1	0	0	1	$\sqrt{2}/8 + i\sqrt{2}/8$
0/1	0	1	0	$i/4$
0/1	0	1	1	$-\sqrt{2}/8 + i\sqrt{2}/8$
0/1	1	0	0	$-1/4$
0/1	1	0	1	$-\sqrt{2}/8 - i\sqrt{2}/8$
0/1	1	1	0	$-i/4$
0/1	1	1	1	$\sqrt{2}/8 - i\sqrt{2}/8$

This corresponds to the sequence of rotations $0°, 45°, 90°, 125°, 180°, 225°, 270°, 315°$. As these values (see the table above) appear twice, the expected frequency f is 2. With this the period is $r = 8$.

The solution follows closely that already introduced in the material of this chapter. Quantum bits $q[0]$, $q[1]$, $q[2]$, and $q[3]$ correspond to a_1, a_2, a_3, a_4, respectively. The first column indicates that all 4 quantum bits are set into a uniform superposition by applying to them the Hadamard gate. In the next column, the respective rotations are encoded. In accordance with the mapping formula, quantum bit $q[1]$ is rotated by $180°$, $q[2]$ by $90°$, and $q[3]$ by $45°$. These operations are delineated by the barrier. After the barrier, the inverse quantum Fourier transform is applied. This transformation is the same (starting in row 13 of the code) as introduced in the chapter material. What changes is the application of the $U1$ gates to the qubits $q[1]$, $q[2]$, $q[3]$ corresponding to a_2, a_3, a_4, respectively.

The circuit:

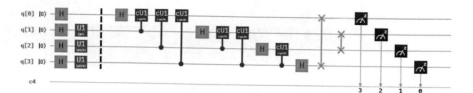

The code:

```
1.   OPENQASM 2.0;
2.   include "qelib1.inc";
3.   qreg q[4];
4.   creg c[4];
5.   h q[0];
6.   h q[1];
7.   h q[2];
8.   h q[3];
9.   u1(pi) q[1];
10.  u1(pi/2) q[2];
11.  u1(pi/4) q[3];
12.  barrier q[0], q[1], q[2], q[3];
13.  h q[0];
14.  cu1(−pi/2) q[1], q[0];
15.  cu1(−pi/4) q[2], q[0];
16.  cu1(−pi/8) q[3], q[0];
17.  h q[1];
18.  cu1(−pi/2) q[2], q[1];
19.  cu1(−pi/4) q[3], q[1];
20.  h q[2];
21.  cu1(−pi/2) q[3], q[2];
22.  h q[3];
23.  swap q[0], q[3];
24.  swap q[1], q[2];
25.  measure q[0] → c[3];
26.  measure q[1] → c[2];
27.  measure q[2] → c[1];
28.  measure q[3] → c[0];
```

The output:

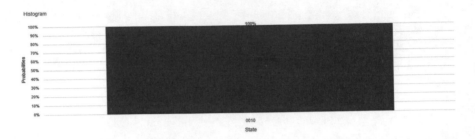

4.2 The mapping this oracle performs is $a_1a_2a_3 \rightarrow \frac{1}{2\sqrt{2}}e^{2\pi i \times 0.a_3} = \frac{1}{2\sqrt{2}}e^{2\pi i \left(\frac{a_3}{2}\right)} = \frac{1}{2\sqrt{2}}e^{i\pi a_3}$. The following table shows the particular results. Input 0/1 indicates that the respective value is either 0 or 1.

a_1a_2	a_3	$\frac{1}{2\sqrt{2}}e^{i\pi a_3}$
00/01/10/11	0	$1/2\sqrt{2}$
00/01/10/11	1	$-1/2\sqrt{2}$

The oracle oscillates between two values corresponding to rotations by $180°$. The frequency is 4, whereby the period equals $8/4 = 2$. Our solution follows closely that already introduced in the material of this chapter. Quantum bits $q[0]$, $q[1]$, and $q[2]$ correspond to a_1, a_2, a_3, respectively. 3 classical registers are needed to encode the output. The first column indicates that all 3 quantum bits are set into a uniform superposition by applying to them the Hadamard gate. In the next column, the respective rotations are encoded. In accordance with the mapping formula, quantum bit $q[2]$ is rotated by $180°$. These operations are delineated by the barrier. After the barrier, the inverse quantum Fourier transform is applied. This transformation is the same (starting in row 10 of the code) as in the chapter material, but applied to 3 quantum bits instead of 4. What changes is the application of the $U1$ gate to the qubit $q[2]$ corresponding to a_3. Moreover, the swapping operation takes place between the most significant digit and the least significant digit only.

The circuit:

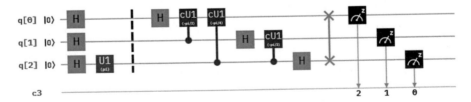

The code:

```
1.   OPENQASM 2.0;
2.   include "qelib1.inc";
3.   qreg q[3];
4.   creg c[3];
5.   h q[0];
6.   h q[1];
7.   h q[2];
8.   u1(pi) q[2];
9.   barrier q[0], q[1], q[2];
10.  h q[0];
11.  cu1(-pi/2) q[1], q[0];
12.  cu1(-pi/4) q[2], q[0];
13.  h q[1];
14.  cu1(-pi/2) q[2], q[1];
15.  h q[2];
16.  swap q[0], q[2];
17.  measure q[0] → c[2];
18.  measure q[1] → c[1];
19.  measure q[2] → c[0];
```

The output:

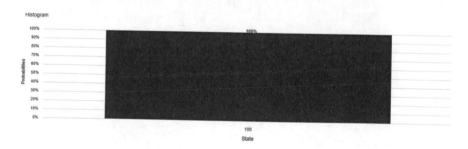

4.3 The mapping this oracle performs is $a_1 a_2 a_3 \rightarrow \frac{1}{2\sqrt{2}} e^{2\pi i \times 0.a_2 a_3} = \frac{1}{2\sqrt{2}} e^{2\pi i \left(\frac{a_2}{2} + \frac{a_3}{4}\right)}$. The following table shows the particular results. Input 0/1 indicates that the respective value is either 0 or 1.

a_1	a_2	a_3	$\frac{1}{2\sqrt{2}} e^{2\pi i \left(\frac{a_2}{2} + \frac{a_3}{4}\right)}$
0/1	0	0	$1/2\sqrt{2}$
0/1	0	1	$i/2\sqrt{2}$

(continued)

(continued)

a_1	a_2	a_3	$\frac{1}{2\sqrt{2}}e^{2\pi i\left(\frac{a_2}{2}+\frac{a_3}{4}\right)}$
0/1	1	0	$-1/2\sqrt{2}$
0/1	1	1	$-i/2\sqrt{2}$

The oracle oscillates between four values corresponding to rotations by 90°. The frequency is 2, whereby the period equals $8/2 = 4$. Our solution follows closely that already introduced in the material of this chapter. Quantum bits $q[0]$, $q[1]$, and $q[2]$ correspond to a_1, a_2, a_3, respectively. 3 classical registers are needed to encode the output. The first column indicates that all 3 quantum bits are set into a uniform superposition by applying to them the Hadamard gate. In the next column, the respective rotations are encoded. In accordance with the mapping formula, quantum bit $q[1]$ is rotated by 180°, while $q[2]$ is rotated by 90°. These operations are delineated by the barrier. After the barrier, the inverse quantum Fourier transform is applied. This transformation is the same (starting in row 11 of the code) as in the chapter material, but applied to 3 quantum bits instead of 4. What changes is the application of the $U1$ gate to the qubits $q[1]$ and $q[2]$ corresponding to a_2 and a_3, respectively. Moreover, the swapping operation takes place between the most significant digit and the least significant digit only.

The circuit:

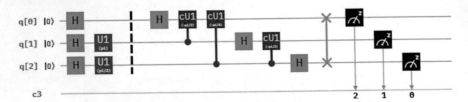

The code:

```
1.   OPENQASM 2.0;
2.   include "qelib1.inc";
3.   qreg q[3];
4.   creg c[3];
5.   h q[0];
6.   h q[1];
7.   h q[2];
8.   u1(pi) q[1];
9.   u1(pi/2) q[2];
10.  barrier q[0], q[1], q[2];
11.  h q[0];
12.  cu1(−pi/2) q[1], q[0];
```

13. cu1(−pi/4) q[2], q[0];
14. h q[1];
15. cu1(−pi/2) q[2], q[1];
16. h q[2];
18. swap q[0], q[2];
18. measure q[0] → c[2];
19. measure q[1] → c[1];
20. measure q[2] → c[0];

The output:

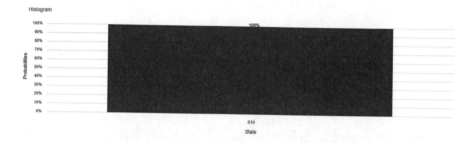

Solutions to Chapter 5

5.1 Since the remainder $r \neq 0$, $c = xd + r$, where xd is a multiple of d. From this it follows that $r = c - xd$. Let $b_1 = \gcd(c, d)$ and let $b_2 = \gcd(d, r)$. Then, on the one hand, $b_1|c$ and $b_1|d$, and also $b_1|xd$. With this, we have that $b_1|c - xd$ and thus $b_1|r$. Thus, $b_1|b_2$. On the other hand, $b_2|d$ and $b_2|r$. Hence, $b_2|xd$, from which it follows that $b_2|xd + r$ with $xd + r = c$. Thus, $b_2|b_1$. Therefore it must hold that $b_1 = b_2$.

5.2 Let

$$Y_i = \frac{Y}{y_i}$$

Since $\gcd(y_i, y_j) = 1$ for each $i \neq j$, it follows that $\gcd(Y_i, y_i) = 1$ as well. By Bézout's identity there exist two integers K_i and k_i such that $K_i Y_i + k_i y_i = \gcd(Y_i, y_i) = 1$. Then a solution can be constructed as follows:

$$z = \sum_{i=1}^{n} c_i K_i Y_i$$

Each congruence relation i is then represented as

$$z = c_i K_i Y_i = c_i (1 - k_i y_i) = c_i \pmod{y_i}$$

Moreover, any two solutions are equal modulo Y. To prove that let z_1 and z_2 be two solutions to the given system of equations. Since $y_1, y_2, \ldots, y_n$ are pairwise coprime, it follows that $y_1 | (z_1 - z_2)$, $y_2 | (z_1 - z_2)$, $\ldots y_n | (z_1 - z_2)$. Thus, $y_1 y_2 \ldots y_n | (z_1 - z_2)$, or, equivalently

$$z_1 = z_2 \pmod{y_1 y_2 \ldots y_n}$$

5.3 By the definition of binomial coefficients

$$\binom{p}{k} = \frac{p!}{k!(p-k)!} = \frac{p(p-1)\cdots(p-k+1)(p-k)!}{k!(p-k)!}$$
$$= \frac{p(p-1)\cdots(p-k+1)}{k!}$$

As binomial coefficients represent integers, $\frac{p(p-1)\cdots(p-k+1)}{k!}$ is an integer. From the above equation it can be verified that $\binom{p}{k}$ is a multiple of p. Hence, what must be shown is that $\frac{(p-1)\cdots(p-k+1)}{k!}$ is an integer. To that end we use Euclid's lemma. Euclid's lemma states that if a prime n divides the product ab, where both a and b are integers, then n must divide at least one of them. Since $\frac{p(p-1)\cdots(p-k+1)}{k!}$ is an integer, it follows that $k!$ divides $p(p-1)\cdots(p-k+1)$. Moreover, since $k!$ does not divide p because p is prime, then by Euclid's lemma, $k!$ must divide $(p-1)\cdots(p-k+1)$ and so $\frac{(p-1)\cdots(p-k+1)}{k!}$ is an integer. Therefore, p divides $\binom{p}{k}$.

5.4 This proof is due to James Ivory ("Demonstration of a theorem respecting prime numbers", *New Series of the Mathematical Depository*, 1(2): 6–8, 1806). Let a be an integer and p be prime that does not divide a. The list of the first $p-1$ positive multiples of a is

$$a, 2a, 3a, \ldots, (p-1)a$$

By reducing each element modulo p, a new list is obtained that consists of a permutation of the integers $1, 2, 3, \ldots, p-1$. Therefore, multiplying the elements

$$a \cdot 2a \cdot 3a \cdots (p-1)a = 1 \cdot 2 \cdot 3 \cdots (p-1) \pmod{p}$$

This corresponds to

$$(p-1)!a^{p-1} = (p-1)!(\bmod p)$$

and further to

$$a^{p-1} = 1(\bmod p)$$

Now let a be any integer and p be a prime. If $p|a$ then $a^p = a(\bmod p) = 0$. If p does not divide a then multiplying the above equation by a

$$a \cdot a^{p-1} = a(\bmod p)$$

gives the required result $a^p = a(\bmod p)$.

5.5 Consider the multiplicative group modulo n : $\mathbb{Z}/n\mathbb{Z}$. Every element in this group has a unique inverse. Let the elements of this group be denoted as $k_1, k_2, \ldots, k_{\phi(n)}$. Then for $a \in \mathbb{Z}/n\mathbb{Z}$ the elements $ak_1, ak_2, \ldots, ak_{\phi(n)}$ are also element of $\mathbb{Z}/n\mathbb{Z}$. In analogy to the proof of Fermat's little theorem, multiplying the elements k_i corresponds then to

$$k_1 \cdot k_2 \cdots k_{\phi(n)} = ak_1 \cdot ak_2 \cdots ak_{\phi(n)} = a^{\phi(n)}k_1 \cdot k_2 \cdots k_{\phi(n)}$$

Cancelling equal terms leads to $a^{\phi(n)} = 1(\bmod n)$.

5.6 The following proof has been adapted from the proof given in "Continued Fractions, Pell's equation, and other applications" by Jeremy Booher and holds for any rational fraction, not only $\frac{i}{2^n}$.

Assume $\frac{i}{2^n}$ is not a convergent of the continued fraction of $\frac{i}{2^n}$. Then r can be picked to lie between the denominators of two convergents $\frac{p_n}{q_n}$ and $\frac{p_{n+1}}{q_{n+1}}$ of $\frac{i}{2^n}$, that is, $q_n < r < q_{n+1}$. Suppose

$$\left| z - r\frac{i}{2^n} \right| \le \left| p_n - q_n\frac{i}{2^n} \right| \tag{1}$$

The determinant of the matrix in the following equation

$$\begin{pmatrix} p_n & p_{n+1} \\ q_n & q_{n+1} \end{pmatrix} \begin{pmatrix} u \\ v \end{pmatrix} = \begin{pmatrix} z \\ r \end{pmatrix}$$

is 1 or -1 based on the difference between two successive convergents, which is given by

$$\frac{p_{n+1}}{q_{n+1}} - \frac{p_n}{q_n} = \frac{(-1)^n}{q_n q_{n+1}}$$

Then for the system of equations

$$z = up_n + vp_{n+1} \tag{2}$$

$$r = uq_n + vq_{n+1}$$

holds that $uv \leq 0$. u and v cannot be both positive or both negative because this would imply that $|r| > |q_{n+1}|$ which contradicts our assumption that $q_n < r < q_{n+1}$.

With (2) we have

$$
\begin{aligned}
\left| z - r\frac{i}{2^n} \right| &= \left| (up_n + vp_{n+1}) - (uq_n + vq_{n+1})\frac{i}{2^n} \right| \\
&= \left| u\left(p_n - q_n\frac{i}{2^n} \right) + v\left(p_{n+1} - q_{n+1}\frac{i}{2^n} \right) \right|
\end{aligned}
$$

Since even convergents are increasing and odd convergents are decreasing with $\frac{i}{2^n}$ lying in between them, and given that $uv \leq 0$, it must be that either (1) $u\left(p_n - q_n\frac{i}{2^n} \right)$ and $v\left(p_{n+1} - q_{n+1}\frac{i}{2^n} \right)$ have the same sign, or (2) one of them is zero. Therefore,

$$\left| z - r\frac{i}{2^n} \right| = \left| u\left(p_n - q_n\frac{i}{2^n} \right) \right| + \left| v\left(p_{n+1} - q_{n+1}\frac{i}{2^n} \right) \right|$$

For

$$\left| z - r\frac{i}{2^n} \right| < \left| p_n - q_n\frac{i}{2^n} \right|$$

to be true, either (1) $|u| = 1$ and $v = 0$, or (2) $u = 0$. Assuming (1) is the case, then $\frac{z}{r} = \frac{p_n}{q_n}$ and is thus a convergent of $\frac{i}{2^n}$. Assuming (2) is the case, then $|r| = |vq_{n+1}|$, which contradicts our assumption that $r < q_{n+1}$.

As opposed to Eq. (1) suppose that

$$\left| z - r\frac{i}{2^n} \right| \geq \left| p_n - q_n\frac{i}{2^n} \right|$$

Then

$$\left| p_n - q_n\frac{i}{2^n} \right| < \frac{1}{2r}$$

From this it follows that

$$\left|\frac{z}{r} - \frac{p_n}{q_n}\right| \le \left|\frac{z}{r} - \frac{i}{2^n}\right| + \left|\frac{p_n}{q_n} - \frac{i}{2^n}\right| < \frac{1}{2r^2} + \frac{1}{q_n q_{n+1}} \le \frac{1}{2r q_n} + \frac{1}{2 q_n r} = \frac{1}{r q_n}$$
(3)

However

$$\left|\frac{z}{r} - \frac{p_n}{q_n}\right| = \left|\frac{z q_n - r p_n}{r q_n}\right|$$
(4)

Since due to (3)

$$|z q_n - r p_n| \le 1$$

and because we assumed that z and r are positive integers and therefore the nominator in (4) has to be an integer, we have

$$|z q_n - r p_n| = 1$$

which implies that $\frac{z}{r}$ is a convergent of $\frac{i}{2^n}$.

5.7 $\begin{aligned}|1 - e^{i\theta}|^2 &= (1 - e^{i\theta})(1 - e^{-i\theta}) = 1 - e^{-i\theta} - e^{i\theta} + 1 \\ &= 2 - (\cos\theta - i\sin\theta) - (\cos\theta + i\sin\theta) = 2 - 2\cos\theta\end{aligned}$

Let $\delta = \frac{\theta}{2}$. Then by trigonometric identity formula for double angles $\cos(2\theta) = 1 - 2\sin^2\frac{\theta}{2}$ we obtain

$$2 - 2\cos(2\delta) = 2 - 2(1 - 2\sin^2\delta) = 4\sin^2\delta = 4\sin^2\frac{\theta}{2}$$

Solutions to Chapter 6

6.1 Let $|x_0\rangle$ be the only solution of a given search problem. Then, recalling that the inner product $\langle x|x\rangle = 1$, we have

$$O|x_0\rangle = (I - 2|x_0\rangle\langle x_0|)|x_0\rangle = I|x_0\rangle - 2|x_0\rangle\langle x_0|x_0\rangle = |x_0\rangle - 2|x_0\rangle = -|x_0\rangle$$

6.2 For $|x_0\rangle = |2\rangle = |10\rangle$ the matrix has dimensions 4x4 and can be derived as follows:

$$\begin{bmatrix} 1 & 0 & 0 & 0 \\ 0 & 1 & 0 & 0 \\ 0 & 0 & 1 & 0 \\ 0 & 0 & 0 & 1 \end{bmatrix} - 2\begin{bmatrix} 0 \\ 0 \\ 1 \\ 0 \end{bmatrix}\begin{bmatrix} 0 & 0 & 1 & 0 \end{bmatrix} = \begin{bmatrix} 1 & 0 & 0 & 0 \\ 0 & 1 & 0 & 0 \\ 0 & 0 & -1 & 0 \\ 0 & 0 & 0 & 1 \end{bmatrix}$$

6.3 First, derive $U = 2|\phi_1\rangle\langle\phi_1| - I$

$$U = 2 \begin{bmatrix} \sqrt{\frac{N-S}{N}} \\ \sqrt{\frac{S}{N}} \end{bmatrix} \begin{bmatrix} \sqrt{\frac{N-S}{N}} & \sqrt{\frac{S}{N}} \end{bmatrix} - \begin{bmatrix} 1 & 0 \\ 0 & 1 \end{bmatrix} = \begin{bmatrix} \frac{N-2S}{N} & \frac{2\sqrt{S(N-S)}}{N} \\ \frac{2\sqrt{S(N-S)}}{N} & \frac{2S-N}{N} \end{bmatrix}$$

Then applying U to $|\phi_2\rangle$ we obtain $|\phi_3\rangle$

$$U \begin{bmatrix} \sqrt{\frac{N-S}{N}} \\ -\sqrt{\frac{S}{N}} \end{bmatrix} = \begin{bmatrix} \sqrt{\frac{N-S}{N}} \frac{N-4S}{N} \\ \sqrt{\frac{S}{N}} \frac{3N-4S}{N} \end{bmatrix}$$

Thus U has reflected vector $|\phi_2\rangle$ about $|\phi_1\rangle$ as shown in Fig. 6.7.

6.4 The matrix of G is

$$G = \begin{bmatrix} \cos\theta & -\sin\theta \\ \sin\theta & \cos\theta \end{bmatrix}$$

From the state $|\phi_1\rangle$ we have that $\sin\frac{\theta}{2} = \sqrt{\frac{S}{N}}$ and $\cos\frac{\theta}{2} = \sqrt{\frac{N-S}{N}}$. From the state $|\phi_3\rangle$ we have that $\sin\frac{3\theta}{2} = \frac{3N-4S}{N}\sqrt{\frac{S}{N}}$ while $\cos\frac{3\theta}{2} = \frac{N-4S}{N}\sqrt{\frac{N-S}{N}}$. The elements of Grovers operator can then be obtained from the following trigonometric identities

$$\sin\theta = \sin\left(\frac{3\theta}{2} - \frac{\theta}{2}\right) = \sin\frac{3\theta}{2}\cos\frac{\theta}{2} - \cos\frac{3\theta}{2}\sin\frac{\theta}{2} = \frac{2\sqrt{S(N-S)}}{N}$$

$$\cos\theta = \cos\left(\frac{3\theta}{2} - \frac{\theta}{2}\right) = \cos\frac{3\theta}{2}\cos\frac{\theta}{2} + \sin\frac{3\theta}{2}\sin\frac{\theta}{2} = \frac{N-2S}{N}$$

With the above the matrix has the form

$$G = \begin{bmatrix} \frac{N-2S}{N} & -\frac{2\sqrt{S(N-S)}}{N} \\ \frac{2\sqrt{S(N-S)}}{N} & \frac{N-2S}{N} \end{bmatrix}$$

6.5 We need to solve the equation

$$G|\psi\rangle = \lambda|\psi\rangle$$

where $|\psi\rangle$ is an eigenvector and λ is its corresponding eigenvalue. To that end we find the characteristic equation

$$|G - \lambda I| = 0$$

to be

$$\left|\begin{bmatrix} \frac{N-2S}{N} & -\frac{2\sqrt{S(N-S)}}{N} \\ \frac{2\sqrt{S(N-S)}}{N} & \frac{N-2S}{N} \end{bmatrix} - \begin{bmatrix} \lambda & 0 \\ 0 & \lambda \end{bmatrix}\right| = \left|\begin{bmatrix} \frac{N-2S}{N} - \lambda & -\frac{2\sqrt{S(N-S)}}{N} \\ \frac{2\sqrt{S(N-S)}}{N} & \frac{N-2S}{N} - \lambda \end{bmatrix}\right|$$

$$= \left(\frac{N-2S}{N} - \lambda\right)^2 + \left(\frac{2\sqrt{S(N-S)}}{N}\right)^2$$

$$= \lambda^2 - \frac{2N-4S}{N}\lambda + 1 = 0$$

The two eigenvalues are therefore (the complex value is due to the fact that $N \leq S$)

$$\lambda_1 = \frac{N - 2S + 2i\sqrt{S(N-S)}}{N}$$

and

$$\lambda_2 = \frac{N - 2S - 2i\sqrt{S(N-S)}}{N}$$

The corresponding eigenvectors are calculated as follows. For λ_1

$$\begin{bmatrix} \frac{N-2S}{N} - \lambda_1 & -\frac{2\sqrt{S(N-S)}}{N} \\ \frac{2\sqrt{S(N-S)}}{N} & \frac{N-2S}{N} - \lambda_1 \end{bmatrix}|\psi_1\rangle = \begin{bmatrix} -\frac{2i\sqrt{S(N-S)}}{N} & -\frac{2\sqrt{S(N-S)}}{N} \\ \frac{2\sqrt{S(N-S)}}{N} & -\frac{2i\sqrt{S(N-S)}}{N} \end{bmatrix}\begin{bmatrix} \psi_{1,1} \\ \psi_{1,2} \end{bmatrix} = 0$$

and hence $\psi_{1,1} = i\psi_{1,2}$. Therefore the eigenvector is $\frac{1}{\sqrt{2}}\begin{bmatrix} i \\ 1 \end{bmatrix}$. For λ_2

$$\begin{bmatrix} \frac{N-2S}{N} - \lambda_2 & -\frac{2\sqrt{S(N-S)}}{N} \\ \frac{2\sqrt{S(N-S)}}{N} & \frac{N-2S}{N} - \lambda_2 \end{bmatrix}|\psi_1\rangle = \begin{bmatrix} \frac{2i\sqrt{S(N-S)}}{N} & -\frac{2\sqrt{S(N-S)}}{N} \\ \frac{2\sqrt{S(N-S)}}{N} & \frac{2i\sqrt{S(N-S)}}{N} \end{bmatrix}\begin{bmatrix} \psi_{1,1} \\ \psi_{1,2} \end{bmatrix} = 0$$

and hence $\psi_{1,1} = -i\psi_{1,2}$. This corresponds to the eigenvector $\frac{1}{\sqrt{2}}\begin{bmatrix} -i \\ 1 \end{bmatrix}$.

Index

© The Author(s), under exclusive license to Springer Nature Switzerland AG 2021
W.-L. Chang and A. V. Vasilakos, *Fundamentals of Quantum Programming in IBM's Quantum Computers*, Studies in Big Data 81,
https://doi.org/10.1007/978-3-030-63583-1